Innovative Solutions for Sustainable Agriculture: From Waste to Biostimulants, Biofertilisers and Bioenergy

Innovative Solutions for Sustainable Agriculture: From Waste to Biostimulants, Biofertilisers and Bioenergy

Guest Editors

Daniele Del Buono
Giovanni Gigliotti
Alberto Maria Gambelli

MDPI

Basel • Beijing • Wuhan • Barcelona • Belgrade • Novi Sad • Cluj • Manchester

Guest Editors
Daniele Del Buono
Department of Agricultural, Food and Environmental Sciences
University of Perugia
Perugia
Italy

Giovanni Gigliotti
Department of Civil and Environmental Engineering
University of Perugia
Perugia
Italy

Alberto Maria Gambelli
Department of Civil and Environmental Engineering
University of Perugia
Perugia
Italy

Editorial Office
MDPI AG
Grosspeteranlage 5
4052 Basel, Switzerland

This is a reprint of the Special Issue, published open access by the journal *Agriculture* (ISSN 2077-0472), freely accessible at: https://www.mdpi.com/journal/agriculture/special_issues/Q1TY2FX81T.

For citation purposes, cite each article independently as indicated on the article page online and as indicated below:

Lastname, A.A.; Lastname, B.B. Article Title. *Journal Name* **Year**, *Volume Number*, Page Range.

ISBN 978-3-7258-6486-7 (Hbk)
ISBN 978-3-7258-6487-4 (PDF)
https://doi.org/10.3390/books978-3-7258-6487-4

Contents

About the Editors

Daniele Del Buono

Daniele Del Buono is an Associate Professor in the Department of Agricultural, Food, and Environmental Sciences (DSA3) at the University of Perugia, Italy. His research focuses on sustainable crop management, with particular attention to biostimulants and biofertilizers of biological origin derived from agro-industrial residues, circular bioeconomy approaches for waste valorization, and the biogenic synthesis of nanostructured materials for agricultural applications. He integrates physiological and biochemical analyses to study plant responses under normal and abiotic stress conditions. He has participated in several national and international research projects. He is the author of numerous publications in international journals in the disciplines of plant sciences, plant biochemistry, biomass recycling, the use of nanomaterials in agriculture, and environmental chemistry.

Giovanni Gigliotti

Giovanni Gigliotti is a retired Full Professor of Agricultural Chemistry at the University of Perugia's Department of Civil and Environmental Engineering. From the 1990/91 academic year to today, he has been teaching several disciplines related to Agricultural Chemistry.

He is the author of more than 300 publications: 110 indexed publications, about 4200 total citations, and an H-index of 36 (source SCOPUS and WOS). His current scientific activity is mainly focused on the following areas: Recovery of agricultural, zootechnical, industrial, and agro-industrial waste and by-products through biological treatments; degradation of xenobiotics during the biomass stabilization process; degradation of plastics and bioplastics in the soil and during biological waste treatment; third-generation biorefinery for the production of bioactive molecules, bioplastics, biofuels, and biofertilizers. He has been in charge of research projects financed by public institutions and private industries several times. He is a member of the Società Italiana di Chimica Agraria, the International Society of Soil Science, and the European Geosciences Union.

Alberto Maria Gambelli

Alberto Maria Gambelli covers the role of Assistant Professor in the Department of Civil and Environmental Engineering (DICA) at the University of Perugia. His research interests span the fields of Energy and Chemical Engineering, with particular attention to renewable and alternative energy sources. In detail, considering the themes deepened in the present Special Issue, his last research activities focused on the possibility of enhancing biogas production yield via co-digestion of different substrates in order to achieve the ideal C/N ratio and to promote microbial activity.

Preface

Agriculture faces significant challenges due to a complex context shaped by factors ranging from climate change to progressive soil degradation and growing demands for food quality and productivity. At the same time, farming systems need to reduce their environmental impact and dependence on synthetic chemicals. With this in mind, the valorization of agro-industrial and agricultural residues is emerging as a strategic approach aimed at strengthening the sustainability and resilience of agriculture, including the development of innovative materials and bioenergy production. This Reprint brings together the contributions published in the Special Issue "Innovative Solutions for Sustainable Agriculture: From Waste to Biostimulants, Biofertilisers and Bioenergy". It was conceived to collect and present scientific contributions on innovative, bioeconomy-based approaches aimed at transforming waste biomass into valuable products, ranging from biostimulants and biofertilizers to organic soil improvers and mineral–organic materials, and even bioenergy.

Therefore, this Reprint focuses on the importance of considering "waste" as a valuable starting point from which different classes of bio-inputs can be derived to support crop performance, product quality, and stress tolerance, while helping reduce conventional agrochemical inputs. In addition, this Reprint also focuses on soil-centered strategies to improve its management, and the importance of applying integrated "cascade" approaches to waste materials, in which biomass, after being processed to extract bio-inputs, is valorized through conversion into bioenergy.

Daniele Del Buono, Giovanni Gigliotti, and Alberto Maria Gambelli
Guest Editors

agriculture

MDPI

Editorial

Innovative Solutions for Sustainable Agriculture: From Waste to Biostimulants, Biofertilisers and Bioenergy

Daniele Del Buono [1,*], Alberto Maria Gambelli [2] and Giovanni Gigliotti [2]

1 Department of Agricultural, Food and Environmental Sciences, University of Perugia, Borgo XX Giugno 74, 06121 Perugia, Italy
2 Civil and Environmental Engineering Department, University of Perugia, Via G. Duranti 93, 06125 Perugia, Italy; albertomaria.gambelli@unipg.it (A.M.G.); giovanni.gigliotti@unipg.it (G.G.)
* Correspondence: daniele.delbuono@unipg.it; Tel.: +39-0755856229

Due to its intensive nature, modern agriculture has a significant environmental impact; therefore, in the coming years, the main challenge will be to address numerous problems related to its lack of environmental, social, and economic sustainability. For instance, about one-third of anthropogenic greenhouse gas (GHG) emissions derive from agri-food systems [1]. This has been estimated by considering the production and subsequent release of GHG from farms and livestock-related activities, the pre- and post-production phases of food, and the associated land-use changes due to agricultural expansion [1]. In addition, it has been documented that agricultural activities are increasingly energy-intensive, with consumption accounting for almost 30% of global energy demand [2]. A significant portion of this energy demand is met by non-renewable sources, such as fossil fuels [2]. Cultivation systems also place significant pressure on natural resources; approximately 92% of humanity's water footprint is linked to agriculture [3]. The biodiversity of agricultural environments is also affected by the use of land and ecosystems for intensive agriculture, with significant declines in species richness [4]. Furthermore, the ongoing climate change is characterized by sudden and extreme events and is also contributing to rapid soil degradation and progressive salinization [5]. These phenomena are significantly hampering the productivity of cropping systems by compromising the health of agricultural land [6,7]. It has been documented that since 1980, the climate crisis has progressively reduced yields of major crops; corn and wheat have shown reductions in production ranging from 4 to over 5.5% [8]. Without adequate countermeasures, climate change is projected to reduce the yields of some crops worldwide by 3–12% by mid-century and by 11–25% by the end of the century [9]. Another critical factor is the use of synthetic fertilisers. In fact, in addition to being a source of GHG, they can have a substantial environmental impact and cause many problems for natural ecosystems due to their mobility. Consumption of these products has increased significantly over the last fifty years, enabling significant growth in agricultural production, but at the same time causing significant environmental pollution [10]. All the situations depicted above should be framed in terms of a compelling demand for a sustainable increase in primary production to match the growth of the world population [11].

Among the various approaches to reduce the environmental impact of agriculture, the valorisation of agro-industrial waste through a circular model offers innovative and effective solutions. Many wastes and by-products have a chemical composition that indicates a wealth of substances that, once recovered, can be used in various applications and also provide valuable materials for agriculture itself [12]. Among the substances of particular interest in waste are biostimulants [13]. These substances, when applied to

https://doi.org/10.3390/agriculture15232499

plants, soil, or the rhizosphere, improve crop performance by stimulating biochemical, physiological, and adaptive processes that enhance nutrient use efficiency and increase tolerance to environmental stresses [6,13–15]. In addition, agro-industrial residues from various sources can be used to produce compost, which can serve as valuable organic fertilisers, helping improve soil organic matter, optimise plant nutrition, and enhance the nutrient cycle [16]. To complement the above approach, the anaerobic digestion (AD) of agro-industrial residues or residual biomass represents another important aspect of circularity [17]. By means of AD, it is possible to further valorise residues by converting them into clean bioenergy, enabling the further exploitation of the bio-derived resource and facilitating a significant transition from fossil fuels to alternative and renewable forms of energy [18]. Moreover, once AD is completed, recent studies have shown that digestate remains a source of added value, as it can be further used as a biofertiliser for plant nutrition [17].

Considering the aspects listed above and the critical problems presented, this Special Issue aimed to gather scientific contributions that explore multiple avenues for applying and even transforming waste from various sources into valuable bio-inputs to make agriculture more sustainable, productive, and resilient to environmental pressures. On the whole, some main lines have been identified that highlight how the circular approach, based on the valorisation of waste biomass, aims to mitigate agricultural impact without losing sight of the quality of primary production. This circular-based approach represents an alternative to the linear model of resource utilisation and, as shown in this Special Issue by the contributions, some of the most suitable and promising solutions to improve the environmental sustainability of agriculture are the valorisation of residues to obtain biostimulants (microbial and non-microbial), compost, fertilizers and the conversion of waste biomass into bioenergy.

A series of scientific contributions have generally focused on the use of biostimulants and biofertilisers of non-microbial and microbial origin. These studies aimed to improve various aspects of cropping systems, including yield, crop quality, and tolerance to stressors. At the same time, these contributions emphasised the need to find solutions that reduce the use of synthetic fertilisers and plant protection products. Afonso et al. [19] studied the use of biostimulants on sweet cherries before harvesting as a strategy to improve fruit yield and quality and reduce the use of conventional agrochemicals. To this end, the authors studied the effect of biostimulants based on glycine betaine and Ecklonia maxima extract on two different sweet cherry cultivars. In general, biostimulant treatments improved the parameters studied, although responses varied by cultivar. Therefore, the authors propose applying the substances mentioned above to achieve significant improvements in the key quality characteristics of sweet cherries, offering clear advantages in terms of economic and environmental sustainability. Valente et al. [20] investigated the effect of inoculating common wheat with *Methylobacterium symbioticum*, aiming to reduce nitrogen input to the crop. The study showed that, in some cases, the bacterium promoted root length density, delayed leaf senescence, and improved photosynthetic activity, stomatal conductance, and photosystem II efficiency, even at reduced nitrogen levels. Therefore, the results indicate that *Methylobacterium symbioticum* improved nitrogen metabolism, offering a viable approach to reducing chemical fertilisation. A further contribution investigated the effectiveness of the nitrogen-fixing bacterium *Gluconacetobacter diazotrophicus* [21]. The trials evaluated the application of GIBI029 with or without fertilisation, and the results were compared with control plants unfertilised or fertilised with nitrogen and phosphorus. Yields and fruit numbers were higher with GIBI029 alone, with results similar to those of plants subjected to complete nitrogen and phosphorus fertilisation. This study highlights how this strategy can reduce or replace synthetic N and P use, offering a green alternative with no environmental impact. Nowak et al. [22] addressed the problems caused by the pathogenic fungus *Rhizoctonia solani*, which is typically controlled with pesticides. The authors proposed the bacterium *Priestia megaterium* (KW16), obtained from *Poa pratensis* L., for oilseed rape. Preliminary in vitro tests showed that KW16 inhibited the growth of *R. solani*. In healthy plants, KW16 improved plant growth, while in infected plants, it contrasted the pathogen effects. These results position KW16 as a powerful biological alternative to synthetic fungicides. This first group of works can be associated with a review that detailed the problems associated with the prolonged and intensive use of various chemical inputs in agriculture, as well as the environmental and agronomic challenges they have generated [23]. This contribution, therefore, highlighted the need for viable, sustainable alternatives. Among the proposed solutions, particular emphasis was placed on the potential of microalgae and cyanobacteria, which offer several specific advantages. These include the ability to capture CO_2, assimilate essential micro- and macroelements, and, thanks to their metabolic versatility, produce bioactive substances with biostimulant and biocontrol properties. Finally, this review also highlighted gaps in the use of these microorganisms and offers perspectives for the development of sustainable agriculture.

Another group of scientific articles in this Special Issue focused on the use of organic residues and mineral-organic matrices to improve fertilisation and soil organic matter, to reduce the use of synthetic fertilisers and offer solutions to mitigate the environmental impact of contaminants. Ramos-Romero et al. [24] addressed the problem of valorising liquid waste from whey (CW) derived from cheese production by co-composting it with solid organic waste, such as crop residues and cow manure, using a turned-pile composting system. The results showed that CW had a high organic load and macronutrient concentrations, and low concentrations of heavy metals. It was observed that co-composting CW with agro-zootechnical waste was a sustainable strategy for producing compost with stabilised and humified organic matter and considerable agricultural value. Li et al. [25] tested Chinese astragalus and buried rice straw, using three nitrogen levels (0%, 60%, and 100% of the usual dose) to improve rice cultivation. The results indicated that the best combination for rice was obtained when astragalus was grown with straw and 60% nitrogen, yielding levels equal to those of the crop grown with 100% conventional nitrogen. In addition, the authors found that this combination stimulated microbial biomass and increased soil ammonium content. This strategy is therefore viable for maintaining crop yield and reducing the risk of nitrogen loss to the environment. Another study published in this Special Issue aimed at valorising Cupuaçu, a fruit native to the Amazon, widely used in the food industry, and which generates large quantities of shell and seed residues [26]. After characterising these residues, the authors tested their potential as substrates for seedling production. Chemical analyses showed that these residues contained significant amounts of macro- and micronutrients, exceeding those in the control soil. In addition, the substrates produced from the residues had lower densities and higher porosity than the soil. Jarosz et al. [27] studied the influence of mineral-organic mixtures containing zeolite compounds obtained from fly ash and lignite or leonardite on the organic matter of a sandy loam soil in pot experiments on maize plants. SOM and the evolution of compounds were analysed, as well as the effect of these mixtures on soil carbon reserves. The addition of these materials improved SOM stability and significantly increased total organic carbon and nitrogen content, indicating greater soil fertility. Aguilar et al. [28] investigated the effect of Ni on foliar fertilisation with urea, as this element is a cofactor of the urease enzyme. The results showed that the addition of Ni increased urease activity, improving urea assimilation, photosynthetic activity, and pigment content, and stimulating the biosynthesis of nitrogen-containing compounds. Loffredo et al. [29] studied the use of compost derived from digestate in two agricultural soils in southern Italy. In particular, the authors of this study found that compost increased the adsorption of the fungicide penconazole, the herbicide S-metolachlor and BPA, with very limited desorption. The results confirm the key role of organic matter (native and anthropogenic) in retaining xenobiotics and limiting their transfer to water and crops.

The last two contributions of this Special Issue paid attention to waste valorisation through its conversion into biochar and biogas. Chun et al. [30] obtained biochar from poultry manure and then tested it on *Salicornia herbacea*. The biochar was produced by pyrolysis at different temperatures. The biochar obtained at 500 °C was then applied to the soil in different doses. At the end of the trial, the authors found that biochar derived from poultry manure improved soil health and Salicornia performance. The last contribution by Di Mario et al. [31] focused on the valorisation of various biomasses to produce biogas. The authors evaluated biogas production from olive pomace, the pulp obtained from it via an ionic liquid-based procedure, and olive mill wastewater processed via freeze-drying (FDOW). The FDOW showed the highest biogas production, while the pulp produced the lowest. To address this issue, the author co-digested the pulp with the brewery's spent grain (BSG), thereby enhancing biogas production. The same happened in the case of FDOW and BSG co-digestion.

Author Contributions: Conceptualization, D.D.B., A.M.G. and G.G.; writing—original draft preparation, D.D.B., A.M.G. and G.G.; writing—review and editing, D.D.B., A.M.G. and G.G. All authors have read and agreed to the published version of the manuscript.

Conflicts of Interest: The authors declare no conflicts of interest.

References

1. Crippa, M.; Solazzo, E.; Guizzardi, D.; Monforti-Ferrario, F.; Tubiello, F.N.; Leip, A. Food Systems Are Responsible for a Third of Global Anthropogenic GHG Emissions. *Nat. Food* **2021**, *2*, 198–209. [CrossRef]
2. Chandio, A.A.; Gokmenoglu, K.K.; Dash, D.P.; Khan, I.; Ahmad, F.; Jiang, Y. Exploring the Energy-Climate-Agriculture (ECA) Nexus: A Roadmap toward Agricultural Sustainability in Asian Countries. *Environ. Dev. Sustain.* **2025**, *27*, 12769–12795. [CrossRef]
3. Hoekstra, A.Y.; Mekonnen, M.M. The Water Footprint of Humanity. *Proc. Natl. Acad. Sci. USA* **2012**, *109*, 3232–3237. [CrossRef] [PubMed]
4. Newbold, T.; Hudson, L.N.; Hill, S.L.L.; Contu, S.; Lysenko, I.; Senior, R.A.; Börger, L.; Bennett, D.J.; Choimes, A.; Collen, B.; et al. Global Effects of Land Use on Local Terrestrial Biodiversity. *Nature* **2015**, *520*, 45–50. [CrossRef] [PubMed]
5. Hopmans, J.W.; Qureshi, A.S.; Kisekka, I.; Munns, R.; Grattan, S.R.; Rengasamy, P.; Ben-Gal, A.; Assouline, S.; Javaux, M.; Minhas, P.S.; et al. Chapter One-Critical Knowledge Gaps and Research Priorities in Global Soil Salinity. In *Advances in Agronomy*; Sparks, D.L., Ed.; Academic Press: Cambridge, MA, USA, 2021; Volume 169, pp. 1–191, ISBN 0065-2113.
6. Del Buono, D. Can Biostimulants Be Used to Mitigate the Effect of Anthropogenic Climate Change on Agriculture? It Is Time to Respond. *Sci. Total Environ.* **2021**, *751*, 141763. [CrossRef]
7. Del Buono, D.; Regni, L.; Del Pino, A.M.; Bartucca, M.L.; Palmerini, C.A.; Proietti, P. Effects of Megafol on the Olive Cultivar 'Arbequina' Grown Under Severe Saline Stress in Terms of Physiological Traits, Oxidative Stress, Antioxidant Defenses, and Cytosolic Ca^{2+}. *Front. Plant Sci.* **2021**, *11*, 603576. [CrossRef]
8. Lobell, D.B.; Schlenker, W.; Costa-Roberts, J. Climate Trends and Global Crop Production since 1980. *Science* **2011**, *333*, 616–620. [CrossRef]
9. Wing, I.S.; De Cian, E.; Mistry, M.N. Global Vulnerability of Crop Yields to Climate Change. *J. Environ. Econ. Manag.* **2021**, *109*, 102462. [CrossRef]
10. Lubkowski, K. Environmental Impact of Fertilizer Use and Slow Release of Mineral Nutrients as a Response to This Challenge. *Pol. J. Chem. Technol.* **2016**, *18*, 72–79. [CrossRef]
11. Gerten, D.; Kummu, M. Feeding the World in a Narrowing Safe Operating Space. *One Earth* **2021**, *4*, 1193–1196. [CrossRef]
12. Raftery, A.E.; Li, N.; Ševčíková, H.; Gerland, P.; Heilig, G.K. Bayesian Probabilistic Population Projections for All Countries. *Proc. Natl. Acad. Sci. USA* **2012**, *109*, 13915–13921. [CrossRef]
13. Rouphael, Y.; Colla, G. Editorial: Biostimulants in Agriculture. *Front. Plant Sci.* **2020**, *11*, 40. [CrossRef] [PubMed]
14. Abd El-Mageed, T.A.; Semida, W.M.; Rady, M.M. Moringa Leaf Extract as Biostimulant Improves Water Use Efficiency, Physio-Biochemical Attributes of Squash Plants under Deficit Irrigation. *Agric. Water Manag.* **2017**, *193*, 46–54. [CrossRef]

15. Ahmad, A.; Blasco, B.; Martos, V. Combating Salinity Through Natural Plant Extracts Based Biostimulants: A Review. *Front. Plant Sci.* **2022**, *13*, 862034. [CrossRef] [PubMed]
16. Azim, K.; Soudi, B.; Boukhari, S.; Perissol, C.; Roussos, S.; Thami Alami, I. Composting Parameters and Compost Quality: A Literature Review. *Org. Agric.* **2018**, *8*, 141–158. [CrossRef]
17. Zhang, C.; Su, H.; Baeyens, J.; Tan, T. Reviewing the Anaerobic Digestion of Food Waste for Biogas Production. *Renew. Sustain. Energy Rev.* **2014**, *38*, 383–392. [CrossRef]
18. Kim, D.-H.; Oh, S.-E. Continuous High-Solids Anaerobic Co-Digestion of Organic Solid Wastes under Mesophilic Conditions. *Waste Manag.* **2011**, *31*, 1943–1948. [CrossRef]
19. Afonso, S.; Oliveira, I.; Ribeiro, C.; Vilela, A.; Meyer, A.S.; Gonçalves, B. Exploring the Role of Biostimulants in Sweet Cherry (*Prunus avium* L.) Fruit Quality Traits. *Agriculture* **2024**, *14*, 1521. [CrossRef]
20. Valente, F.; Panozzo, A.; Bozzolin, F.; Barion, G.; Bolla, P.K.; Bertin, V.; Potestio, S.; Visioli, G.; Wang, Y.; Vamerali, T. Growth, Photosynthesis and Yield Responses of Common Wheat to Foliar Application of *Methylobacterium symbioticum* under Decreasing Chemical Nitrogen Fertilization. *Agriculture* **2024**, *14*, 1670. [CrossRef]
21. Ceballos-Aguirre, N.; Restrepo, G.M.; Patiño, S.; Cuéllar, J.A.; Sánchez, Ó.J. Utilization of *Gluconacetobacter diazotrophicus* in Tomato Crop: Interaction with Nitrogen and Phosphorus Fertilization. *Agriculture* **2025**, *15*, 1191. [CrossRef]
22. Nowak, B.; Chlebek, D.; Hupert-Kocurek, K. *Priestia megaterium* KW16: A Novel Plant Growth-Promoting and Biocontrol Agent Against *Rhizoctonia solani* in Oilseed Rape (*Brassica napus* L.)—Functional and Genomic Insights. *Agriculture* **2025**, *15*, 1435. [CrossRef]
23. Jurado-Flores, A.; Heredia-Martínez, L.G.; Torres-Cortes, G.; Díaz-Santos, E. Harnessing Microalgae and Cyanobacteria for Sustainable Agriculture: Mechanistic Insights and Applications as Biostimulants, Biofertilizers and Biocontrol Agents. *Agriculture* **2025**, *15*, 1842. [CrossRef]
24. Ramos-Romero, S.; Gavilanes-Terán, I.; Idrovo-Novillo, J.; Idrovo-Gavilanes, A.; Valverde-Orozco, V.; Paredes, C. Cheese Whey Characterization for Co-Composting with Solid Organic Wastes and the Agronomic Value of the Compost Obtained. *Agriculture* **2025**, *15*, 513. [CrossRef]
25. Li, P.; Zhao, L.; Li, D.; Leng, Q.; Geng, M.; Zhu, Q. Replacing Nitrogen Fertilizers with Incorporation of Rice Straw and Chinese Milk Vetch Maintained Rice Productivity. *Agriculture* **2025**, *15*, 623. [CrossRef]
26. Sousa Filho, I.M.R.D.; Santos, C.A.A.D.; Paulino, G.D.S.; Araújo, A.J.C.; Sousa, W.L.D.; Alves, H.D.S.; Vieira, T.A.; Lustosa, D.C. Characterization of Cupuaçu (*Theobroma grandiflorum*) Waste for Substrate in Seedling Production. *Agriculture* **2025**, *15*, 870. [CrossRef]
27. Jarosz, R.; Kowalska, J.B.; Gondek, K.; Bejger, R.; Mielnik, L.; Lahori, A.H.; Mierzwa-Hersztek, M. The Effect of New Zeolite Composites from Fly Ashes Mixed with Leonardite and Lignite in Enhancing Soil Organic Matter. *Agriculture* **2025**, *15*, 786. [CrossRef]
28. Aguilar, J.V.; Lapaz, A.D.M.; Bomfim, N.C.P.; Mendes, T.F.S.; Souza, L.A.; Furlani Júnior, E.; Camargos, L.S. Photosynthetic Performance and Urea Metabolism After Foliar Fertilization with Nickel and Urea in Cotton Plants. *Agriculture* **2025**, *15*, 699. [CrossRef]
29. Loffredo, E.; Campanale, E.; Cocozza, C.; Denora, N. Digestate-Derived Compost Modulates the Retention/Release Process of Organic Xenobiotics in Amended Soil. *Agriculture* **2025**, *15*, 1925. [CrossRef]
30. Chun, D.; Cho, H.; Hahm, V.J.; Kim, M.; Im, S.W.; Kim, H.G.; Kim, Y.S. Effects of Poultry Manure Biochar on *Salicornia herbacea* L. Growth and Carbon Sequestration. *Agriculture* **2024**, *14*, 1590. [CrossRef]
31. Di Mario, J.; Ranucci, A.; Gambelli, A.M.; Rallini, M.; Priolo, D.; Brienza, M.; Puglia, D.; Del Buono, D.; Gigliotti, G. Biogas Production from Olive Oil Mill Byproducts: A Comparative Study of Two Treatments for Pursuing a Biorefinery Approach. *Agriculture* **2025**, *15*, 2204. [CrossRef]

agriculture

MDPI

Article

Biogas Production from Olive Oil Mill Byproducts: A Comparative Study of Two Treatments for Pursuing a Biorefinery Approach

Jessica Di Mario [1], Antonella Ranucci [1], Alberto Maria Gambelli [1,*], Marco Rallini [2], Dario Priolo [3], Monica Brienza [4], Debora Puglia [2], Daniele Del Buono [3] and Giovanni Gigliotti [1]

1 Department of Civil and Environmental Engineering, University of Perugia, Via G. Duranti 93, 06125 Perugia, Italy; jessica.dimario@dottorandi.unipg.it (J.D.M.); antonella.ranucci@unipg.it (A.R.); giovanni.gigliotti@unipg.it (G.G.)

2 Consorzio Interuniversitario Nazionale per la Scienza e Tecnologia dei Materiali (INSTM)—University of Perugia Research Unit (UdR Perugia), Department of Civil and Environmental Engineering, University of Perugia, Strada di Pentima 4, 05100 Terni, Italy; marco.rallini@unipg.it (M.R.); debora.puglia@unipg.it (D.P.)

3 Department of Agricultural, Food and Environmental Sciences, University of Perugia, Borgo XX Giugno 74, 06121 Perugia, Italy; dario.priolo@unipg.it (D.P.); daniele.delbuono@unipg.it (D.D.B.)

4 Department of Sciences, University of Basilicata, Via dell'Ateneo Lucano 10, 85100 Potenza, Italy; monica.brienza@unibas.it

* Correspondence: albertomaria.gambelli@unipg.it

Abstract: Olive cultivation is one of the most widespread agro-industrial activities in the Mediterranean area. However, required pretreatments often affect the anaerobic digestion process, promoting or inhibiting the overall yield. Therefore, the efficiency of Anaerobic Digestion (AD) processes cannot be established in advance but needs to be experimentally validated for each biomass-pretreatment combination. Following the present purpose, these biomasses were firstly treated: the olive pomace (OP) with a procedure based on the use of an ionic liquid (IL) composed of triethylamine and sulfuric acid [Et_3N][HSO_4] to remove hemicellulose and lignin and recover the insolubilized OP, while olive mill wastewater (OW) was processed via freeze-drying. The resulting materials, the pulp from olive pomace (POP) and freeze-dried OW (FDOW), were then digested using lab-scale anaerobic reactors. The biogas production was then compared with the quantity obtained by digesting the same untreated biomasses (OW and OP). The FDOW showed the highest biogas production due to the freeze-drying treatment that led to some morphological and structural surface modifications of OW (respectively, 658 mL vs. 79 mL/g for the two matrices), prompting microorganism activity. Conversely, the method based on the use of IL significantly reduced the nitrogen content of POP, thus resulting in the lowest biogas production, which ceased by the second day. To address this issue, we co-digested POP with the brewery's spent grain, a biomass rich in nitrogen. This step enhanced the biogas yield of POP, resulting in an extended anaerobic digestion period and the production of 466 mL/g. Additionally, we tested FDOW in co-digestion with BSG to evaluate improvements in production. The codigestion of the two matrices increased the biogas yield of FDOW from 944 to 1131 mL/g.

Keywords: bioeconomy; biorefinery; brewery's spent grain; olive mill wastewater; olive pomace

1. Introduction

Today, the overall area for olive crop cultivation reaches about 11.5 Mha globally, covering 58 nations [1]. According to the International Oil Council (IOC), 97% of worldwide olive oil production occurs around the Mediterranean lands. As evidence of this, Spain, Italy, Greece, and Portugal account for 69% of the olives produced on a planetary scale [1].

Regarding the manufacturing processes, olive oil is extracted from olives mechanically. As the first step, olives undergo air blowing and washing with water to remove impurities, then a grinding process to crush the olives and a malaxation process for recovering oil droplets from the aqueous and solid parts [2]. After that, olive oil extraction is carried out by a three-phase centrifugation system that produces olive mill wastewater (OW) (38–48%) and olive pomace (OP) (35–45%) as waste. Another extractive technique is based on a two-phase centrifugation system that produces a wetter olive pomace as a waste [3].

In the OW, the main component is water, and it generally shows an acidic pH (4.2–5.9), a high content of phenolic compounds (0.5 to 25 g L^{-1}), and the presence of other organic compounds, such as protein, fats, and the fibrous fraction (cellulose, hemicellulose, and lignin). OW can substantially negatively affect the environment due to the high content of some phenolic compounds, which, in some cases, can be highly toxic. As such, inappropriate and irresponsible disposal of OW can lead to environmental issues like soil deterioration in terms of physical and chemical attributes, microbiome alterations, and plant phytotoxicity. This has more than one effect of concern; for instance, the impairment of the ecosystem services of the soil, poses, in addition, more than one risk to biodiversity [4,5]. The conventional disposal of this kind of waste provides treatment with calcium oxide for neutralization and a coagulation process, followed by storing it in waterproof lagoons. However, this method has some drawbacks, including creating unpleasant odors and high transfer costs [6].

As far as the OP is concerned, this waste usually shows a pH in the range of 4.8–5.2 and a high organic matter content, mainly due to the remarkable content of phenolic compounds (200–300 mg/100 g), which, as already mentioned above, can create more than one environmental hazard. Furthermore, it should be pointed out that this waste also contains significant amounts of copper, zinc, and manganese [7].

Conversely, if adequately treated, wastes from the olive oil supply chain can be a promising source of valuable and interesting compounds. Indeed, biomass can undergo some treatment methods with the scope of achieving high market-value compounds and materials under a biorefinery and green chemistry approach. As an example, the OP has an oily component (8–12% *w*/*w*), which can be exploited to recover the oil by chemical solvent extraction (e.g., hexane), producing olive pomace oil [8]. Furthermore, these biomasses are rich in bioactive compounds; for instance, it is possible to recover and purify the phenolic fraction for applications as antioxidants in food, cosmetics, and medicine [9,10]. For instance, the OP treatment by pressurized liquid extraction allows for obtaining and recovering molecules not found in the maceration process [11]. In addition, El-Abassi et al. [12] claim that a good phenolic compound extraction yield (66.5 ± 3.2%) was achieved from OW by ultrafiltration, which permitted the removal of a remarkable amount of total suspended solids, and then proceeded to cloud point extraction.

Applying a biorefinery approach to the wastes of the olive oil production chain can allow us to convert them into high-value-added products and biogas. One recent and promising option consists of using these residuals (especially OP) for the production of protein hydrolysates, which can then be used as soil improvers and bio-stimulants [13]. Given the abundance of organic matter shown by this biomass, the anaerobic digestion (AD) process is exploited [14]. However, OP and OW are rich in substances that can slow down AD. In particular, lignin, among the most abundant components of agro-industrial waste, can be an obstacle to AD due to its recalcitrant nature. From a biological point of view, lignin is one of the components of the plant cell walls, thus conferring mechanical support, strength, and protection against pathogens, among others [15]. In addition, this biopolymer is also involved in controlling plant nutrients and water transport [16]. From a structural and chemical point of view, lignin is an amorphous biopolymer composed of some phenolic units: hydroxycinnamoyl, coniferyl, sinapyl, and coumaryl alcohols. These molecules are linked by covalent carbon-carbon and carbon-oxygen bonds, which determine a very complex and articulated three-dimensional structure [17]. As already pointed out, during the AD process, lignin is nearly non-degradable because of its structural complexity and plenty of strong covalent bonds that cover hemicellulose and cellulose. This requires hydrolytic enzymes, which need oxygen to depolymerize lignin [18,19]. As a result, lignin limits the hydrolysis step, which is one of the four main stages of AD (hydrolysis, acidogenesis, acetogenesis, and methanogenesis), thereby restricting enzymatic activity during this phase and reducing the efficiency of the entire AD.

In OP, lignin constitutes approximately 25–30% of the dry weight [20]. The removal of lignin from OP can be considered a crucial pretreatment, as it helps maximize biogas production by mitigating its inhibitory effects. In a study evaluating the impact of ultrasound treatment on OP for biogas production, pretreated OP showed a significant increase in biogas yield compared to the untreated sample, rising from 300 mL to approximately 450 mL of biogas per 5 g of OP [21]. This highlights the effectiveness of pretreatments in enhancing OP's biogas potential. Another study [22] investigated the co-digestion of OP with manure, demonstrating that the inclusion of OP led to a higher biogas yield in co-digested samples. These findings confirm that OP is a valuable biomass for biogas production. Therefore, in this study, we evaluated the potential methane production of OP and OP after lignin extraction (OP-Pulp, referred to as POP in the text). This is essential to understanding the effect of lignin removal and fully utilizing the pomace.

In addition, OW was considered for our AD experiments [23], taking into account its capacity to limit bacterial growth and inhibit their activity [24], ascribable to its high content of phenolic compounds [25]. There is a threshold concentration below which inhibition caused by polyphenols begins: polyphenol concentrations above 2 g L^{-1} have been shown to completely inhibit microbial activity in AD, with impairments already starting below 1 g L^{-1} [26]. In previous studies [27], the biogas production of olive mill wastewater (OW) was assessed after ultrasound pretreatment, which positively increased the biogas yield. The increase was even more significant when OW was also diluted, confirming that polyphenol concentration plays a crucial role. For this reason, in this work, biogas production was tested for both unmodified and freeze-dried OW. The freeze-dried OW was studied to evaluate the effects of this pretreatment, as freeze-drying can stabilize the substrate, enhance the physical characteristics of the matrix by creating a more porous structure, and mitigate the negative impacts associated with traditional thermal drying. In addition, freeze-drying can be considered a more environmentally friendly technique compared to alkaline pre-treatment, as it does not require the use of any chemicals.

Summarizing the above, this work evaluated and compared biomethane production from different treated waste materials—POP and freeze-dried OW—with untreated biomasses, OP and OW. The main aim was to enhance the biogas yield of these olive mill by-products. In one case, lignin was extracted, a component that can be reused for various purposes. For olive mill wastewater (OW), a pretreatment was tested to improve the morphological characteristics of the substrate and assess its impact on biogas production. In addition, co-digestion was tested by using the brewery's spent grain (BSG), the main waste material from beer production, to increase biogas production in treatments with low yields. In addition, co-digestion was tested using brewery spent grain (BSG), the main by-product of beer production, to enhance biogas yields in treatments with low methane production. Organic wastes, such as food waste and food-industry residues, are commonly used for anaerobic digestion. Llanos-Lizcano et al. (2024) [28] investigated the chemical biomethane potential of various organic wastes, including canteen leftovers, dairy industry residues, brewery by-products, and cardboard from eggs. Their results showed that waste from the beer industry achieved the highest biomethane yield as well as the greatest degree of biodegradability.

2. Materials and Methods

Experiments described in this study were carried out in order to accurately determine the production of biogas and the corresponding biomethane content, to ensure the reliability of results and their comparability with the current literature. For the scope, the experimental procedure was also defined, considering the methodology widely recognized and documented in the literature [29–31].

2.1. Materials, Procedures, and Analysis

This study was carried out with the same digestate previously used and described in Montegiove et al. (2024) [32]. The corresponding biogas/biomethane production was therefore already evaluated and detracted from the results here described.

The main characteristics of the digestate employed in this study are shown in Table 1.

Table 1. Main characteristics of the inoculum used to prepare the various samples.

Parameter	Digestate
Moisture [%]	88.08 ± 0.05
VS [%]	72.34 ± 0.05
pH	8.08 ± 0.05
TOC [% of DM]	53.05 ± 0.01
TKN [% of DM]	5.47 ± 0.01
Total P [g kg^{-1} DM]	3.11 ± 0.01
Total K [g kg^{-1} DM]	75.62 ± 0.01
WEOC [g kg^{-1} DM]	110.51 ± 0.01
WEN [g kg^{-1} DM]	67.86 ± 0.01

DM = dry matter; TOC = Total Organic Carbon; TKN = Total Kjeldahl Nitrogen; WEOC = Water-Extractable Organic Carbon; WEN = Water-Extractable Nitrogen.

2.1.1. Analytic Methods for Biomass Characterization

Sample humidity and volatile solids were determined according to the official protocol with some modifications [31]. To this end, two grams of each biomass were dried in an oven (TCN 50 Plus, Argolab, Modena, Italy) at 105 °C. When the weight was constant, the moisture content was calculated as the difference between the initial wet weight and the dried weight. The final value is expressed as a percentage. The humidity content for each biomass was measured in triplicate. For vs. content, two grams of each previously dried sample were weighed in ceramic crucibles in triplicate. The samples were then placed in a muffle (FM13, Falc, Bergamo, Italy) where the temperature was gradually increased to 550 °C and maintained for 24 h. That time period refers to the whole process, while the samples were kept at the target temperature only for 5 h. Subsequently, the crucibles were placed in a desiccator with silica gel to cool the sample before weighing. The volatile solids content was estimated as the difference between the initial and incinerated weight.

The pH was measured following the protocol [33] after water extraction of samples (1:2.5 *w*/*v*) and using a glass electrode (60 VioLab, XS, Modena, Italy). WEOC and WEN have been analyzed using an elemental analyzer (multi N/C 2100, Analytik Jena GmbH, Überligen, Germany). Total organic C was analyzed by the Walkley-Black method using an HT1300 (Analytik Jena GmbH, Überligen, Germany), while total N was measured after Kjeldahl digestion followed by ammonia distillation (UDK 129, Velp, Monza Brianza, Italy) and titration with 0.0357 N sulfuric acid [34]. WEOC and WEN were extracted from samples with deionized water (1:2 w/v) for 24 h × 200 rpm at room temperature. After filtration, water extracts were analyzed using multi N/C 2100S® (Analytik Jena GmbH, Überligen, Germany) [35]. Total P was analyzed following the Olsen method [34]; total K was quantified by means of ICP-OES (Optima 2100 DV, PerkinElmer, Springfield, IL, USA).

All measurements were carried out in triplicate, and average values are reported.

2.1.2. Production of Pomace Pulp

An ionic liquid (IL) composed of triethylamine and sulfuric acid $[Et_3N][HSO_4]$ was synthesized according to Cequier et al. [36] and then used to treat PO to solubilize and then remove hemicellulose and lignin, leaving as much cellulose as possible in the pulp (solid fraction after the treatment). To this end, 2 g of pomace were treated with a solution containing 90% of $[Et_3N][HSO_4]$ and 10% water by weight and an additional 0.009 mol of H_2SO_4 to remove as much lignin and hemicellulose as possible [37]. All the reagents were provided by Merck Life Science S.r.l. (Milan, Italy) and used as received without additional purification. The mixture obtained was left to react at 120 °C for 4 h, then cooled to room temperature, and EtOH was added to solubilize lignin and hemicellulose [37]. Finally, the solid fractions (the cellulose-enriched portion) were recovered by vacuum filtration and washed repeatedly with ethanol, followed by water.

2.1.3. Characterization of Produced Pulp

The resulting pulps were characterized by Fourier transform infrared (FT-IR) analyses carried out using a Jasco FT-IR 615 spectrometer (Jasco Corporation, Tokyo, Japan) in the 4000–600 cm^{-1} range in ATR mode. Powdered raw pomace and the pulp obtained as described above were tested using KBr powder to prepare disks.

The materials' morphological characteristics were investigated using a Field Emission Scanning Electron Microscope (FESEM, Supra 25-Zeiss, Oberkochen, Germany), gold coating the powders with an ion sputter coater, and observing the samples with the gun operating at 5 kV. Thermal stability was assessed by heating the samples from 30 to 900 °C at a rate of 10 °C per minute under a nitrogen atmosphere (250 mL/min), using a TGA Seiko Exstar 6300 (Seiko Instruments Inc., Chiba, Japan).

2.1.4. Production of Freeze-Dried Olive Mill Wastewater

The freeze-drying process is divided into three different steps. The first is the so-called "pre-freezing" phase, where the biomass is frozen at low-temperature conditions. The freezing step is carried out at temperatures within −20/−80 °C; it is crucial to immobilize the various components and prevent foaming during the vacuum step. The second step consists of the "primary drying" and mainly depends on the sublimation of ice [32]. Finally, "secondary drying" occurs after the temperature moves below the critical temperature of solid-to-gas transition; liquid water is obtained via condensation and is transferred to a cold trap. Freeze-dried OW (FDOW in the text) was previously tested after polyphenol extraction [32]; for completeness, the achieved results, which motivated the testing criteria adopted in this study, were briefly discussed in Section 3.4.

Prior to the extraction of polyphenols, the samples of OW were defatted with n-hexane (1:2).

A mixture of $MeOH/H_2O/MeOH\text{-}HCOOH$ 0.1%/MeCN (1:1:8:5) was used at the ratio sample/solvent 1:1 for 1 h in an ultrasonic bath at room temperature. The chemicals used were provided by Merck KGaA (Darmstadt, Germany). The procedure was repeated three times. After centrifugation, the supernatants were collected, and the solvent was evaporated under vacuum.

2.2. Anaerobic Bioreactors

Biogas production was carried out in small-scale batch and unstirred bioreactors, where 50 mL of internal volume was filled with the mixture consisting of inoculum + substrate. The temperature was kept at 37 °C throughout the production period. The volume of biogas and biomethane produced was measured using the volumetric method, and for the determination of biomethane production, an alkaline trap (0.5 M NaOH with thymolphthalein as a pH indicator) was employed [29,38]. As shown in Figure 1, the volumetric method involves a vessel where the biomass is placed, a bottle with water, and a beaker for water collection for biogas measurement. For biomethane determination, an additional bottle containing the NaOH solution is required to separate CO_2 from the biogas mixture. In the first vessel, the inoculum and the biomass to be tested are mixed, and the atmosphere is saturated by introducing nitrogen. This vessel is connected via a tube to a second container, which contains either water (for biogas measurement) or the alkaline trap (for biomethane determination). In the biogas measurement setup, the produced gas displaces water into the collecting beaker, with the displaced volume corresponding to the volume of biogas generated. In the biomethane determination setup, CO_2 in the biogas reacts with NaOH (as illustrated in Figure 1), forming sodium carbonate and leaving methane as the remaining gas. The methane, and also the other components included in the gas mixture, for instance, nitrogen [39], then pass through the water bottle, and the displaced water volume corresponds to the methane volume produced. It should be noted that the alkaline trap and the following vessel filled with water also allow the capture of the other liquid and water-soluble species normally contained (at low concentrations or in traces) in biogas mixtures. Therefore, the role of these species in the determination of the final volume of gas produced can be considered negligible for the scope of this study.

Throughout the entire period of biogas production, the pH of the digestate was periodically measured to ensure that it was within the optimal pH range for AD (6.5–7.5) [40].

Each sample consisted of 75% of the dry weight from the inoculum and the remaining 25% from the dry biomass tested. The control consisted solely of the inoculum, and the corresponding results are not reported in the dedicated section, as they were consistent with those obtained in previous experiments [32]. The following are the treatments investigated in terms of the composition of the biomass introduced into the reactors:

- Control consisting of sole inoculum;
- Sample OW: composed of $^3/_4$ (dw) inoculum and $^1/_4$ (dw) OW;
- Sample FDOW: composed of $^3/_4$ (dw) inoculum and $^1/_4$ (dw) freeze-dried OW (referred as FDOW);
- Sample OP: composed of $^3/_4$ (dw) inoculum and $^1/_4$ (dw) OP;
- Sample POP: composed of $^3/_4$ (dw) inoculum and $^1/_4$ (dw) POP;
- Sample BSG: composed of $^3/_4$ (dw) inoculum and $^1/_4$ (dw) BSG;
- Sample POP+BSG: composed of $^3/_4$ (dw) inoculum and $^1/_4$ (dw) POP+BSG.
- Sample FDOW+BSG: composed of $^3/_4$ (dw) inoculum and $^1/_4$ (dw) FDOW+BSG.

The relative weight of the inoculum and the added biomass corresponds to 1.8 g for the $^3/_4$ in dry weight and 0.6 g for the $^1/_4$ in dry weight for the other biomass. In the case of codigestion, such as POP+BSG and FDOW+BSG, the dry weight was 0.3 g for each biomass for a total of 0.6 g.

All treatments were run in triplicate.

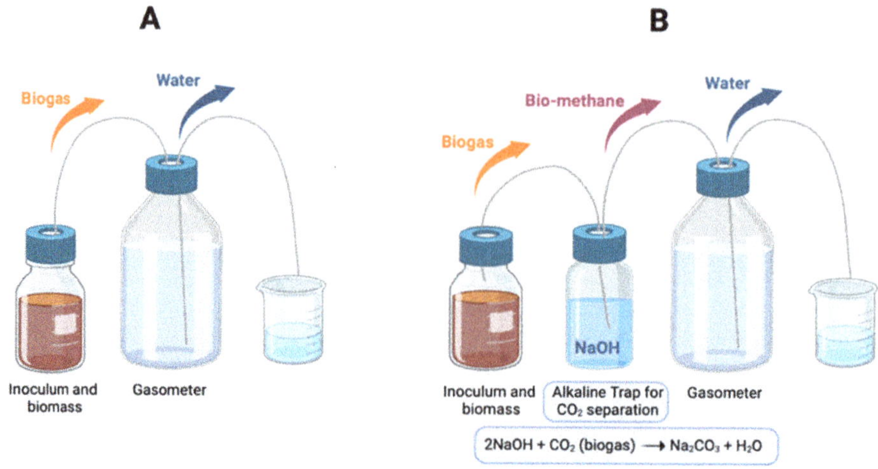

Figure 1. A batch bioreactor used in the experiment. (**A**) A bioreactor used in this study for biogas production. (**B**) A bioreactor used for the bio-methane determination. The reaction that allows the separation of CO_2 from the biogas is shown.

2.3. Statistical Analysis

In order to ensure the reliability and repeatability of results, each experiment was carried out in triplicate. In the experimental section, for each sample, the cumulative production of biogas was defined as the mean of the quantities measured in the three corresponding tests. The error bars allow us to identify the region where the three related tests fall within.

Differences among treatments were evaluated using one-way analysis of variance (ANOVA), considering biomass type as a fixed factor and biogas production as the dependent variable. When the ANOVA indicated significant effects, mean comparisons were performed using Tukey's post hoc test at a significance level of $p < 0.05$. Statistical analyses were carried out in R (R version 4.4.3 (28 February 2025 ucrt)). Results of Tukey's test are reported in Supplementary Materials (Table S1 and Figure S1).

3. Results and Discussion

3.1. Characterization of Biomasses

Freeze-drying and IL treatment were selected as pretreatment procedures for raw OW and OP, respectively. The related biomasses were characterized, and the main results are summarized in Table 2.

Table 2. Main characteristics of the biomasses considered in the present study (TS = Total Solids; VS = Volatile Solids).

Biomass	Moisture (%)	TS (%)	VS (%)	pH	TKN (% on DM)	TOC (% on DM)
OW	88.08 ± 0.05	11.92 ± 0.05	90.80 ± 0.05	5.10 ± 0.05	0.53 ± 0.01	61.20 ± 0.01
FDOW	8.08 ± 0.05	91.92 ± 0.05	92.60 ± 0.05	5.40 ± 0.05	1.22 ± 0.01	30.86 ± 0.01
OP	53.05 ± 0.05	46.95 ± 0.05	97.00 ± 0.05	5.50 ± 0.05	0.70 ± 0.01	47.70 ± 0.01
POP	5.47 ± 0.05	94.53 ± 0.05	95.39 ± 0.05	3.01 ± 0.05	0.31 ± 0.01	31.28 ± 0.01

The SEM images (Figure 2a,b) show that FDOW presents a heterogeneous surface with small embedded particles [40,41] and a spongy appearance with numerous hollows. OW contains high concentrations of organic substances, including carbohydrates, pectins, mucilage, lignin, tannins, lipids, and inorganic substances. Several factors affect this composition, including olive cultivar, degree of ripeness, harvest time, climate, farming practices, and extraction technique [42,43]. The FTIR spectrum analysis (Figure 2c) of FDOW revealed an absorption band in the range of 3700–3000 cm^{-1}, corresponding to the O–H stretching vibrations of alcohols, phenolic compounds, and carboxylic groups. The peaks observed at 2900 and 2800 cm^{-1} correspond to long-chain aliphatic methylene groups (–CH, –CH_2, and –CH_3), indicative of the presence of long-chain lipids in the OW. The main peak assignments in the FDOW in the fingerprint region show the presence of a sharp peak at 1745 cm^{-1}, related to the stretching vibration C=O of the ketones, aldehydes, and carboxylic acid groups and ester functional groups (acidic nature of the OMSW). The signal at 1646 cm^{-1} is assigned to C=C (ν) (aromatics) and C=O stretching [44]. The signal at 1458 cm^{-1} is due to CH_2 asymmetric bending (scissoring), which is strong in cellulose I, while the CH_2 bending mode in cellulose can be found at 1374 cm^{-1}. At 1231 cm^{-1}, we can find C-O syringyl nuclei in lignin. The signal at 1161 cm^{-1} is related to the C–O–C asymmetric stretch vibration in cellulose. The band at 1117 cm^{-1} is indicative of CH and CO deformation or stretching vibrations in different groups of lignin and carbohydrates. The signals at 1095 cm^{-1} and 725 cm^{-1} are due, respectively, to C–O–C skeletal vibration of the polysaccharide ring and rocking vibration of the –CH_2 group in cellulose [45]. In Figure 2d, a minor mass reduction due to moisture evaporation was observed up to around 150 °C, while in stage II, the volatilization of solid residues took place [46]. The second and third zones were ranged between 150 °C and 300 °C and 300 °C and 600 °C, respectively. The degradation of hemicellulose and cellulose is responsible for the first two peaks, while the third peak may stem from the rapid decomposition of lignin or soluble organic compounds like alcohols and aromatic substances.

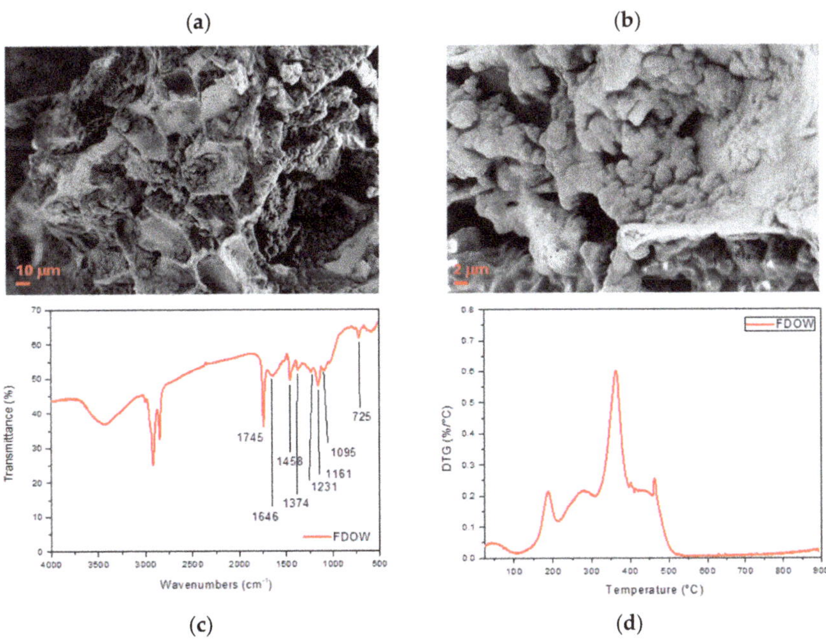

Figure 2. FESEM images at different magnifications (**a**,**b**), FT-IR spectrum (**c**), and DTG curve (**d**) for the FDOW sample.

The results of the morphological, chemical, and thermal characterization of POP are included in Figure 3.

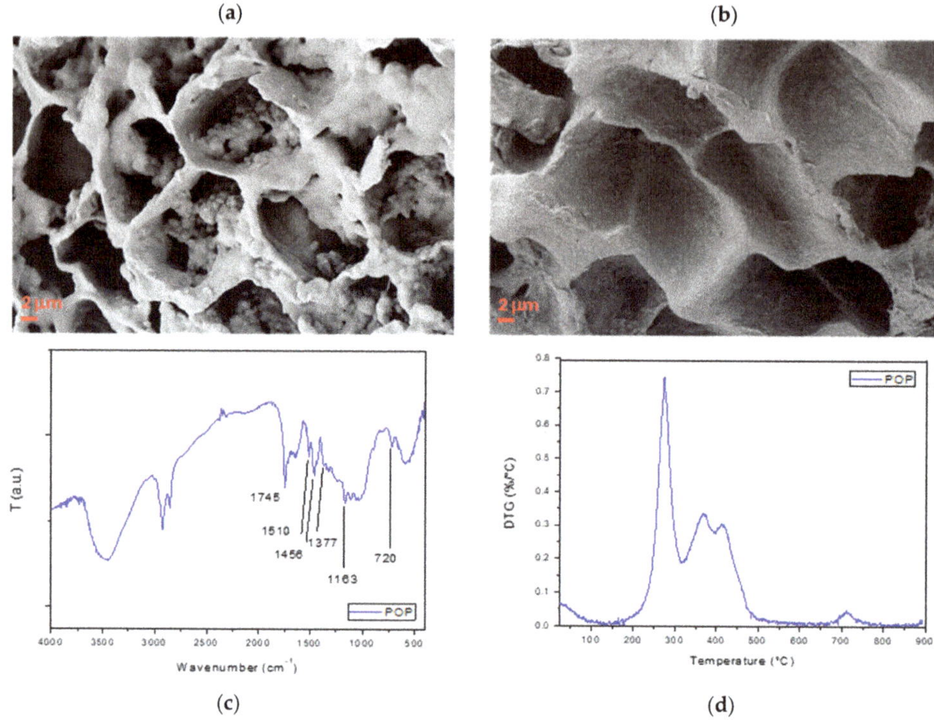

Figure 3. FESEM images of untreated olive pomace (OP) (**a**), treated olive pomace (POP) (**b**), FTIR spectrum (**c**), and DTG curve (**d**) for the POP sample.

Untreated OP (Figure 3a) showed a rough and heterogeneous surface similar to most biomass structures. After the IL treatment (Figure 3b), the compactness of the surface of POP diminished, compared to OP, making the cell structure more evident due to the removal of hemicellulose and lignin. Both microstructures (FDOW and POP) evidenced the typical biomass morphology, composed of cells and walls. However, FDOW showed less clear pores, slightly thicker and denser cell walls, and a less smooth surface than POP, which, on the contrary, presents a sharper and more structured porous morphology.

As shown in the FTIR spectrum (Figure 3c), a broad absorption band between 3600 and 3000 cm^{-1} is linked to O–H stretching in alcohols and phenols. Peaks between 3000 and 2700 cm^{-1} are related to C–H stretching [47]. In the fingerprint region of POP, the distinct peak at 1745 cm^{-1} indicates the presence of carbonyl groups from cellulose and lignin. C=C vibrations of aromatic rings in lignin were recorded at 1509 cm^{-1}, confirming the possible presence of residual lignin in the IL-treated OP [48]. The signals at 1456 cm^{-1} and 1377 cm^{-1} are due to CH_2 bending in cellulose I. The signals at 1163 cm^{-1} and 720 cm^{-1} can be assigned to C–O–C asymmetric stretching and rocking vibration of the –CH_2 group in cellulose, respectively. While the cellulosic component is present in both fractions, the main differences in the FTIR spectra of the two biomasses can be found in the fingerprint region and, specifically, in the presence of more visible peaks related to pectin, heteromannans, and heteroxylans at 1740 cm^{-1} and 1015 cm^{-1}, proteins at 1231 cm^{-1} in FDOW, while the bands at 1456 and 1520 cm^{-1}, attributed to lignin, are more intense in the POP spectrum.

The DTG curve (Figure 3d) for POP revealed three typical degradation stages, attributed to moisture evaporation, the main pyrolysis stage, and char formation (>500 °C). In the second step, three shoulders appear, likely corresponding to the distinct decomposition of the main cellulose components (first two peaks) and a considerably smaller fraction of lignin, which begins to degrade above 400 °C [49]. Finally, as supported by numerous studies, the IL-treated biomass can show decreased thermal stability compared to raw biomass.

3.2. Biogas Yield for the Different Biomasses

Figures 4 and 5 describe the daily biogas production for the four samples in Table 2.

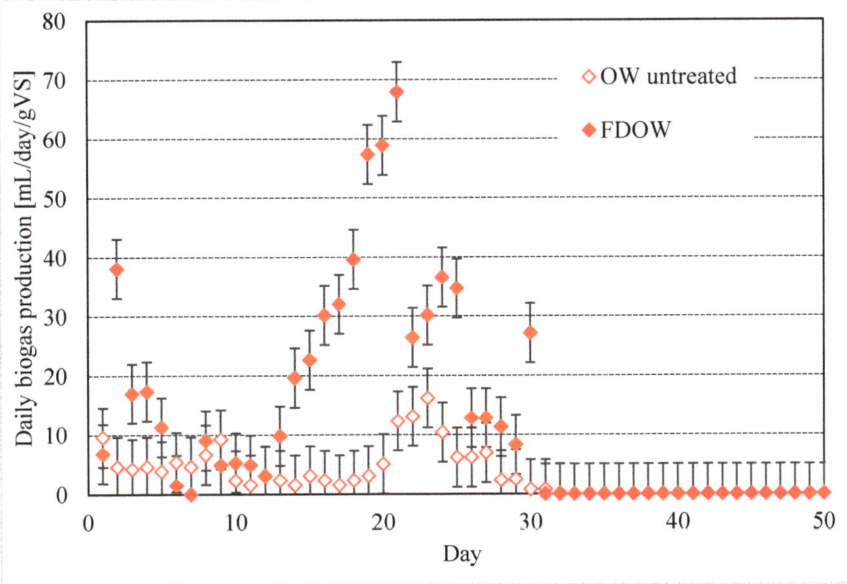

Figure 4. Daily biogas production (expressed in mL/day/gVS) measured for OW and FDOW.

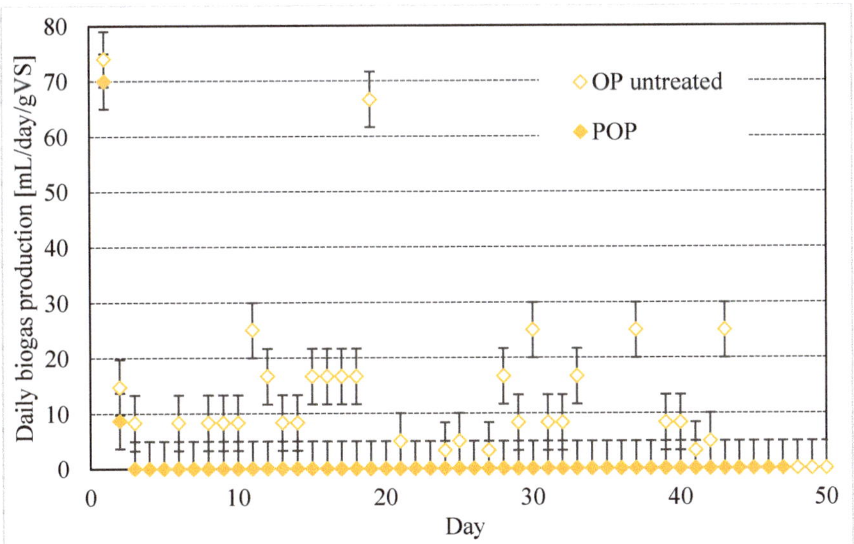

Figure 5. Daily biogas production (expressed in mL/day/gVS) measured for OP and POP.

The daily production curves are referred to a single test for each sample and allow for easy determination of the process evolution for each different matrix investigated. For each value, the maximum error possible is equal to the accuracy of the graduated scale of vessels containing water and is equal to ±5 mL.

Biogas production from FDOW and OW, occurred in 30 and 32 days, respectively. No differences in production time were observed in the treated and untreated biomass. Conversely, untreated OP required 43 days to complete the process, while POP produced biogas only for 2 days. Nonetheless, the production of this matrix cannot be effectively compared with the other samples since the process ended prematurely, and the reasons behind this behavior will be discussed later in the text.

OP and OW showed low biogas production, 451 mL and 79 mL, respectively. In contrast, the production period differed: 43 days for OP and 32 days for OW. The pulp production process based on the use of IL drastically hindered the anaerobic bio-digestion, significantly reducing the biogas yield. Conversely, the freeze-drying process significantly enhanced the biogas yield, and the total production increased from 79 mL (untreated OW) to 658 mL. Even for this kind of treatment, a specific section is dedicated to deepening the reasons behind such a promoting action.

The production trend widely differed between the two different biomasses. For both OW and FDOW, an initial production peak was observed, which immediately dropped. Finally, a second peak was observed, which was responsible for almost the whole quantity of biogas produced. Then, production decreased again, and the process ended. The consistent difference in biogas production between OW and FDOW may be attributed to the effects of the freeze-drying process. This process reduces the moisture content of the biomass and enhances its preservation [50]. Freeze-drying is an effective procedure for dehydrating labile products via vacuum desiccation [50]. With this procedure, water is completely removed, and the remaining dry substance maintains most of the viable microbes and active substances present in the untreated biomass [51]. The high level of preservation of the initial microbial content and the ensured long shelf life make freeze-drying a widely applied procedure [52,53]. Additionally, the C/N ratio of FDOW (25.29) falls within the optimal range for anaerobic digestion, which is 20–30:1 [54]. This could further explain its higher production compared to other biomasses. In contrast, the C/N ratio of POP is 101, which could account for its lack of biogas production.

Also, the bioreactors containing OP and POP showed an initial peak, which immediately decreased. Then, OP assumed a constant production trend, which continued until day 43 and then definitively stopped. Based on the results of analyses carried out in this study, this latter trend can be associated with the low content of nitrogen in these two matrices (respectively, 0.70% and 0.31% on DM), which immediately acted as a limiting factor.

Table 3 shows the methane content of the biogas mixtures produced by the different samples. The results were calculated by considering the total biogas and biomethane production.

Table 3 indicates the total biogas production, the associated methane content, and the production period for four samples. Results are normalized as a function of VS.

Table 3. Total production of biogas, corresponding methane content, and days of production for untreated and treated OP and OW.

Biomass Tested	Total Biogas [mL]	Total CH_4 [mL]	Production Time [Days]
OP	451 ± 18	299 ± 12	43
POP	79 ± 5	39 ± 2	2
OW	79 ± 6	45 ± 4	32
FDOW	658 ± 33	458 ± 23	30

The following average methane concentrations were measured:

(1) OP: 66.30 vol%;
(2) POP: 49.75 vol%;
(3) OW: 56.99 vol%;
(4) FDOW: 69.31 vol%.

3.3. Co-Digestion of Biomasses to Improve Nutrient Composition

To better understand the reason behind such a difference, biogas was produced again from a matrix containing the same biomasses previously studied but mixed with a third typology of biomass, consisting of a brewery's spent grain resulting from beer production waste (in this text referred to as BGS). BGS biomass was selected since it is capable of abundant biogas production and shows acceptable moisture and high nitrogen content. The main characteristics of this latter biomass are indicated in Table 4, together with the corresponding biogas and biomethane yields and the production time. Similar to the matrices previously described, the structural characterization of BSG was performed and is provided in Figure 6. The daily biogas production, obtained with the sole BGS and with the same procedure previously adopted for the other matrices, is described in Figure 7. The quantities of biogas shown in these diagrams and, more in general, in the whole experimental section of this study, together with the corresponding amounts of biomethane, are referred to standard conditions for gases.

Table 4. Main characteristics measured for BGS, together with the corresponding total production of biogas, the corresponding methane content, and the days of production.

	Brewery's Spent Grain
Moisture [%]	78.25 ± 0.05
pH	5.06 ± 0.05
TOC [% on dry matter]	26.66 ± 0.01
TKN [% on dry matter]	3.89 ± 0.01
Total biogas [mL/gVS]	944 ± 156.4
Total CH_4 [mL/gVS]	519 ± 57
Production time [days]	61

TOC = Total Organic Carbon; TKN = Total Kjeldahl Nitrogen.

Figure 6. *Cont.*

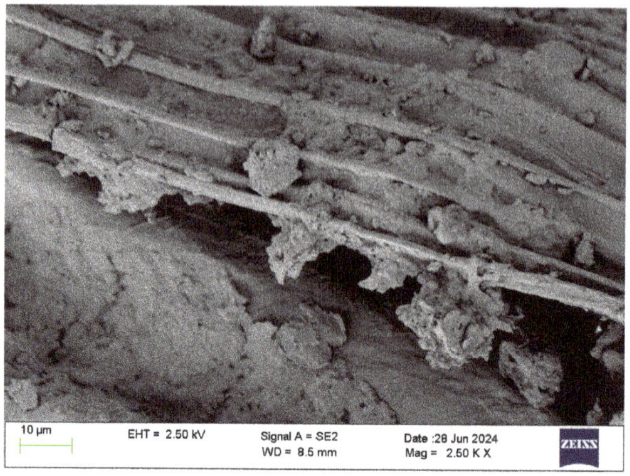

Figure 6. FESEM images of the brewery's spent grain (BSG) [13].

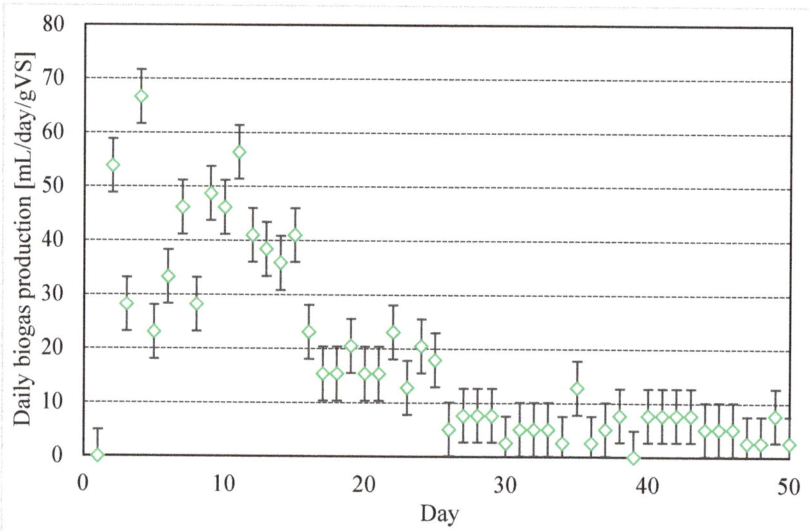

Figure 7. Daily biogas production (expressed in mL/day/gVS) was measured for BSG.

In addition to the data reported in Tables 2 and 4, fibers were measured for OP and BSG. Expressed as a percentage of the total composition, OP showed 30.9% lignin, 21.3% cellulose, and 15.6% hemicellulose. Similarly, the following percentages were measured for BSG: 18.2% lignin, 22.5% cellulose, and 59.9% hemicellulose.

By comparing the data shown in Tables 2 and 4, there is a clear link between the ability of biomass to produce biogas and nitrogen content. Compared to OP and OW, those with a higher nitrogen load performed better, as in the case of BSG. The same can be asserted when comparing FDOW, OW, OP, and POP. This allowed us to postulate a clear need to provide nitrogen to the deficient biomass as an essential nutrient to support microbial activity. BSG was considered a suitable biomass to ascertain the need for nitrogen in biomass and to increase its biogas yield.

The following diagram (Figure 8) describes the cumulative biogas production of all the matrices tested in this work, reporting those already discussed for some biomasses in the previous sections. The diagram shows the cumulative trends for the following:

(i) Untreated olive pomace (OP);
(ii) Untreated olive mill wastewater (OW);
(iii) Untreated brewery's spent grain (BSG);
(iv) Pulp from olive pomace (POP);
(v) Freeze-dried olive mill wastewater (FDOW);
(vi) POP+BSG;
(vii) FDOW+BSG.

As specified in Section 2, one curve for each sample is here plotted and discussed. AD experiments conducted in triplicate yielded mean values for each sample, which were used to plot the subsequent curves. The error bars denote the region where the three curves, obtained for each typology of matrix, fall within.

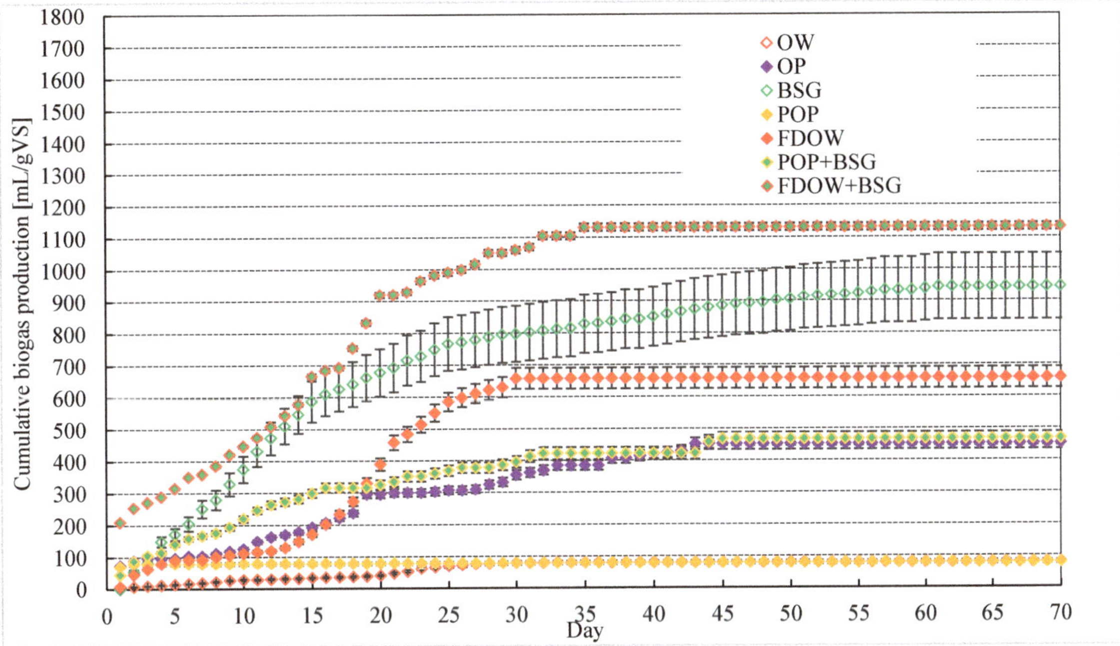

Figure 8. Cumulative biogas production (expressed in mL/gVS) for the following samples: untreated olive pomace (OP), untreated olive mill wastewater (OW), untreated brewery's spent grain (BSG), pulp from olive pomace (POP), freeze-dried olive mill wastewater (FDOW), POP+BSG, FDOW+BSG.

The results indicate that biogas production due to the FDOW+BSG combination was significantly more performant than the singles of OW and FDOW (In Section 3.4, these results are compared with those of all the other OW-based matrices). The production period of 34 days was intermediate between that of FDOW (30 days) and that of BSG (61 days). The reason can be found in the TKN value: the freeze-drying process increased such a value from 0.53 (untreated OW) to 1.22 in FDOW, with the consequent more massive biogas production. From the data, it emerges that, when producing biogas from BSG, the addition of smaller amounts of FDOW significantly shortens production time, resulting in reduced operational costs.

A similar trend was observed for POP+BSG, which showed a biogas production of 46 mL in 44 days. The production fell between those related to BSG (920 mL) and POP (75 mL). Also, the production period was intermediate: 44 vs. 62 days for BSG and 2 days for POP. The addition of BSG significantly improved the process due to its high nitrogen content. The following two sections provide more details about the biogas production from OW and OW-based samples (Section 3.4) and OP and OP-based samples (Section 3.5).

Finally, it should be noted that raw materials having a C/N ratio within the 20–30 range reached the highest production observed in this study.

3.4. Biogas Production from OW and OW-Based Matrices

The production of biogas from untreated olive mill wastewater was not comparable with that of freeze-dried OW. By themselves, OWs are not used for biogas production; however, it is widely established that their addition to other matrices is often capable of enhancing the anaerobic digestion of biomass [55]. Several issues must be overcome before using exclusively untreated OW, such as the low pH and nitrogen content and the excessive presence of inhibiting compounds [56]. The main reason behind such low production was often attributed to polyphenols. However, recent studies revealed that these compounds, together with other substances such as flavonoids, are helpful for the process and assume the role of inhibitors only when their concentration is too elevated [57,58]. That explains why OW is an effective additive but cannot be used for high-efficiency biogas production [32,59]. Moreover, the humidity content of OW must often be reduced to achieve acceptable biogas production per mass unit of used biomass.

The freeze-drying process solved most of the previously mentioned problems. In addition to the complete removal of moisture, it increased the nitrogen content and acted on the morphology of the whole matrix, providing it with the ideal porosity for bacterial activity [51–53]. Moreover, the process also damages the microbial entities present in OW. Ice crystals can disrupt the permeability, integrity, and fluidity of cell membranes. Also, part of metabolism-related enzymes is inevitably eliminated, with the consequent reduction in the number of viable microbes [60]. These effects could help keep the concentration of these substances below the threshold required to achieve high-efficiency biogas production.

The following diagram (see Figure 9) shows the cumulative quantity of biogas produced from FDOW after the extraction of polyphenols and flavonoids (FDOW-PF) previously obtained with the same apparatus by Montegiove and colleagues [32].

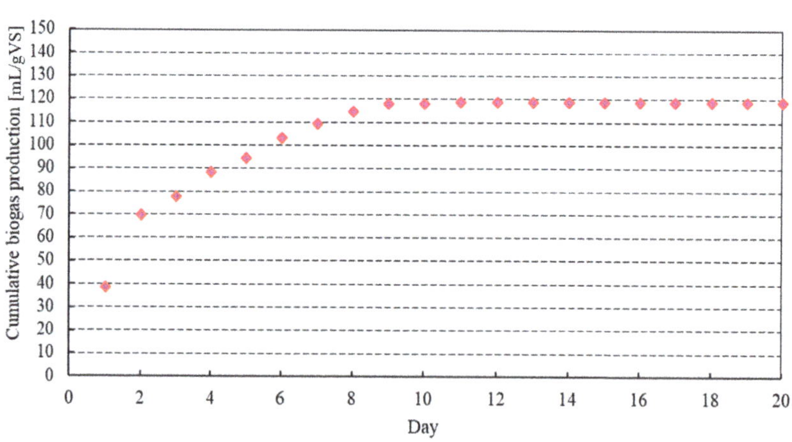

Figure 9. Cumulative biogas production (expressed in mL/day) for FDOW after the extraction of polyphenols and flavonoids (FDOW-PF).

Comparing the results of this study with those obtained in [32], it is evident that freeze-drying the biomass before anaerobic digestion enhances biogas production. In contrast, the extraction of polyphenols and flavonoids can significantly hinder the process. The total quantity of biogas produced was equal to 119 mL/gVS, while the same quantity of

FDOW, but without the extraction of high-added-value products, led to the production of 658 mL/gVS of biogas.

Typical concentrations of polyphenols in OW range from 0.5 to 24 g L^{-1} [61,62], and concentrations varying from 0.5 to 2 g/L are enough to reduce the biogas production [63] significantly and consequently, to provide a meaningful contribution, OW should be co-digested with other substrates.

Based on the results discussed in this section, the freeze-drying process could lower the concentration of those species without removing them completely, thus enhancing the biogas yield.

3.5. Biogas Production from OP and OP-Based Matrices

The production of biogas from olive pomace (OP) was markedly higher than what was obtained with the untreated OW: 451 mL of biogas were produced in 43 days. The main difference with OW was the methane content. In addition, the methane content of biogas produced with OP was higher than that measured for OW: 66.3 vol% for OP and 56.99 vol% for OW. Differences in biomethane concentration are closely related to the chemical composition of the biomass, which influences all stages of anaerobic digestion (AD), resulting in different biogas composition [64]. The process ended after two days, suggesting the exhaustion of a required ingredient. Based on the TKN analyses, the lack of nitrogen (TKN equal to 0.31% on DM) caused the premature interruption of the process. The addition of BSG (having TKN equal to 3.9% on dry matter) significantly increased the production: 442 mL of biogas were produced in 44 days. It should be noted that once a source of nitrogen was provided to the OP-based substrate, the production period became equal to the one observed for OP, thus further confirming the lack of nitrogen as the main drawback for biogas production from POP. This reduction of nitrogen ascertained for POP can be considered the consequence of a typical feature of many ionic liquids, which can solubilize the protein from the treated biomass. In particular, ILs can increase the solubility of these biomolecules by directly interacting through their ions with the functional groups of the proteins [59]. The solubilizing effect may alternatively result from the interfacial interactions of strongly hydrated IL anions close to protein surfaces, whereas the intensely hydrated cations tend to bind with protein amide groups [65].

Finally, as previously indicated, the POP+BSG samples produced 466 mL of biogas and contained 0.3 g (dry matter) of BSG.

The following Table 5 summarizes the overall quantity of biogas produced for each substrate, associated with the corresponding volatile solids (VS) quantity. The results were calculated according to Section 2.1.1. The quantities of biogas produced per unit mass of VS were reported with the corresponding standard deviations.

Table 5. For each substrate tested in this work, the table shows (from left to right) the total dry matter inserted in the bioreactor, the volatile solids, both as a percentage and in grams, and the quantity of biogas produced per unit of mass of VS.

Substrate	Dry Matter [g]	VS [%]	VS [g]	Biogas/VS [mL/g]
FDOW+BSG	1.000 ± 0.005	0.950 ± 0.005	0.950 ± 0.005	1131 ± 14.1
BSG	1.005 ± 0.005	0.975 ± 0.005	0.980 ± 0.005	944 ± 156.4
FDOW	2.860 ± 0.005	0.926 ± 0.005	2.648 ± 0.005	658 ± 15.8
POP+BSG	1.000 ± 0.005	0.948 ± 0.005	0.948 ± 0.005	466 ± 23.9
OP	1.000 ± 0.005	0.970 ± 0.005	0.970 ± 0.005	451 ± 8.3
OW	5.720 ± 0.005	0.908 ± 0.005	5.194 ± 0.005	79 ± 2.1
POP	1.000 ± 0.005	0.954 ± 0.005	0.954 ± 0.005	79 ± 1.5
FDOW-PF	5.720 ± 0.005	0.920 ± 0.005	5.262 ± 0.005	32 ± 8.8

Regarding OW and OW-based substrates, Table 5 and Figure 8 results confirmed the promoting effect due to the freeze-drying process: FDOW produced 658 mL/g of biogas, while untreated OW produced 79 mL/g. Similarly, the low production of FDOW-PF (32 mL/g) proved that the complete extraction of polyphenols and flavonoids can reduce biogas production [66]. Also, the results obtained for OP and OP-based substrates confirmed what was previously asserted. OP produced 451 mL/g of biogas, while only 79 mL/g were obtained with POP. Adding BSG to POP improved biogas production.

The normalization of biogas produced as a function of vs. revealed the synergistic effect obtained by coupling FDOW with BGS. This substrate led to the highest production reached in this work, equal to 1131 mL/g, and can consequently be considered, among the samples tested, the most viable pretreated mixture for biogas production.

4. Conclusions

The present research dealt with the reuse of agro-industrial residuals, derived from the oil and beer production sectors, aimed at favoring the environmental and economic suitability of the whole supply chains. By itself, crop production could not reach the standard fixed to carry out the biorefinery approach; the reuse of residuals and, mostly, the research aimed at optimizing their exploitation are therefore mandatory.

Based on their properties and chemical composition, OW and OP were first treated by applying two different procedures (IL treatment for OP and freeze-drying for OW, respectively) and then mixed with the same typology of digestate in lab-scale batch anaerobic reactors. FESEM images revealed that IL treatment reduced the compactness of the OP surface due to the removal of hemicellulose and lignin fraction; partial removal of lignin was also detected by FTIR, while typical peaks for the cellulose component were found unaltered after IL treatment. The DTG curve showed a reduction in thermal stability after the IL treatment in comparison with the raw biomass.

The IL treatment significantly reduced the biogas production of OP, which decreased from 451 mL of biogas per gram of VS to 79 mL in the treated OP. Based on TKN analyses, the main reason for this drop was ascribed to a very low nitrogen content shown by POP. Further experiments were conducted by mixing POP with a different kind of biomass, the BSG, with a high TKN, to ascertain the above hypothesis. This combination decisively improved the biogas yield of POP, demonstrating the importance of nitrogen in biogas production. In addition, it is important to consider that mixing different biomasses can alter microbial communities, including species involved in the methanogenesis process. Therefore, further studies should focus on investigating these microbial communities to understand their changes and their relationship with biogas and biomethane yields.

Conversely, the freeze-drying process strongly enhanced the OW biogas production. Indeed, FDOW yields 658 mL of biogas per gram of VS, compared to the untreated OW, which produced only 79 mL of biogas per gram of VS. Additionally, in the co-digestion of FDOW with BSG, the biogas production of BSG was enhanced, increasing from 944 mL/g of vs. to 1131 mL in the co-digested BSG. Based on the results achieved in this work and the state of the art, the freeze-drying process made the raw biomass more attachable by microorganisms. In addition, the freeze-drying also inevitably reduced the content of active biomolecules of OW, making their concentrations closer to the optimal ranges defined in the literature.

Supplementary Materials: The following supporting information can be downloaded at: https://www.mdpi.com/article/10.3390/agriculture15212204/s1.

Author Contributions: All authors contributed to the study conceptualization and design. Experimental data were produced by J.D.M., A.R. and D.P. (Dario Priolo). Analyses were performed by

M.R., D.P. (Dario Priolo), D.D.B. and D.P. (Debora Puglia). G.G. validated and supervised the whole study. The first draft was written by J.D.M., A.M.G. and D.D.B. All authors have read and agreed to the published version of the manuscript.

Funding: This work was funded by the European Union—NextGenerationEU as part of the National Innovation Ecosystem grant ECS00000041—VITALITY promoted by the Ministero dell'Università e della Ricerca (MUR). We thank the University of Perugia and the MUR for their support within the VITALITY project.

Institutional Review Board Statement: Not applicable.

Data Availability Statement: The datasets generated and analyzed during the current study are available from the corresponding author on reasonable request.

Conflicts of Interest: The authors declare no conflicts of interest.

References

1. IOC [International Olive Council]. Olive Oil Production by Country. 2019. Available online: https://www.internationaloliveoil.org/olive-oil-estimates-2019-20-crop-year/ (accessed on 20 October 2025).
2. Di Giacomo, G.; Romano, P. Evolution of the Olive Oil Industry along the Entire Production Chain and Related Waste Management. *Energies* **2022**, *15*, 465. [CrossRef]
3. Rapa, M.; Ciano, S. A Review on Life Cycle Assessment of the Olive Oil Production. *Sustainability* **2022**, *14*, 654. [CrossRef]
4. Abu-Lafi, S.; Al-Natsheh, M.S.; Yaghmoor, R.; Al-Rimawi, F. Enrichment of Phenolic Compounds from Olive Mill Wastewater and In Vitro Evaluation of Their Antimicrobial Activities. *J. Evid. Based Complement. Altern. Med.* **2017**, *2017*, 3706915. [CrossRef] [PubMed]
5. Albalasmeh, A.A.; Alajlouni, M.A.; Ghariabeh, M.A.; Rusan, M.J. Short-Term Effects of Olive Mill Wastewater Land Spreading on Soil Physical and Hydraulic Properties. *Water Air Soil Pollut.* **2019**, *230*, 208. [CrossRef]
6. Galloni, M.G.; Ferrara, E.; Falletta, E.; Bianchi, C.L. Olive Mill Wastewater Remediation: From Conventional Approaches to Photocatalytic Processes by Easily Recoverable Materials. *Catalysts* **2022**, *12*, 923. [CrossRef]
7. Antónia Nunes, M.; Pawlowski, S.; Costa, A.S.G.; Alves, R.C.; Oliveira, M.B.P.P.; Velizarov, S. Valorization of Olive Pomace by a Green Integrated Approach Applying Sustainable Extraction and Membrane-Assisted Concentration. *Sci. Total Environ.* **2019**, *652*, 40–47. [CrossRef] [PubMed]
8. Chanioti, S.; Tzia, C. Processing Parameters on the Extraction of Olive Pomace Oil and Its Bioactive Compounds: A Kinetic and Thermodynamic Study. *J. Am. Oil Chem. Soc.* **2018**, *95*, 371–382. [CrossRef]
9. Gullón, P.; Gullón, B.; Astray, G.; Carpena, M.; Fraga-Corral, M.; Prieto, M.A.; Simal-Gandara, J. Valorization of By-Products from Olive Oil Industry and Added-Value Applications for Innovative Functional Foods. *Food Res. Int.* **2020**, *137*, 109683. [CrossRef]
10. Caporaso, N.; Formisano, D.; Genovese, A. Use of Phenolic Compounds from Olive Mill Wastewater as Valuable Ingredients for Functional Foods. *Crit. Rev. Food Sci. Nutr.* **2018**, *58*, 2829–2841. [CrossRef]
11. Cea Pavez, I.; Lozano-Sánchez, J.; Borrás-Linares, I.; Nuñez, H.; Robert, P.; Segura-Carretero, A. Obtaining an Extract Rich in Phenolic Compounds from Olive Pomace by Pressurized Liquid Extraction. *Molecules* **2019**, *24*, 3108. [CrossRef]
12. El-Abbassi, A.; Kiai, H.; Raiti, J.; Hafidi, A. Cloud Point Extraction of Phenolic Compounds from Pretreated Olive Mill Wastewater. *J. Environ. Chem. Eng.* **2014**, *2*, 1480–1486. [CrossRef]
13. Di Mario, J.; Bertoldi, A.; Priolo, D.; Calzoni, E.; Gambelli, A.M.; Dominici, F.; Rallini, M.; Del Buono, D.; Puglia, D.; Emiliani, C.; et al. Characterization of processes aimed at maximizing the reuse of Brewery's Spent Grain: Novel biocomposite materials, high-added-value molecule extraction, codigestion and composting. *Recycling* **2025**, *10*, 124. [CrossRef]
14. Del Mar Contreras, M.; Romero, I.; Moya, M.; Castro, E. Olive-Derived Biomass as a Renewable Source of Value-Added Products. *Process Biochem.* **2020**, *97*, 43–56. [CrossRef]
15. Poletto, M. *Lignin: Trends and Applications*; BoD—Books on Demand: Norderstedt, Germany, 2018.
16. Laurichesse, S.; Avérous, L. Chemical Modification of Lignins: Towards Biobased Polymers. *Prog. Polym. Sci.* **2014**, *39*, 1266–1290. [CrossRef]
17. Munk, L.; Sitarz, A.K.; Kalyani, D.C.; Mikkelsen, J.D.; Meyer, A.S. Can Laccases Catalyze Bond Cleavage in Lignin? *Biotechnol. Adv.* **2015**, *33*, 13–24. [CrossRef]
18. Ruiz-Dueñas, F.J.; Martínez, Á.T. Microbial Degradation of Lignin: How a Bulky Recalcitrant Polymer Is Efficiently Recycled in Nature and How We Can Take Advantage of This. *Microb. Biotechnol.* **2009**, *2*, 164–177. [CrossRef]
19. Fan, Y.; Van Klemeš, J.J.; Perry, S.; Lee, C.T. Anaerobic Digestion of Lignocellulosic Waste: Environmental Impact and Economic Assessment. *J. Environ. Manag.* **2019**, *231*, 352–363. [CrossRef]

20. Di Mario, J.; Gambelli, A.M.; Gigliotti, G. Biomethane production from untreated and treated Brewery's Spent Grain: Feasibility of anaerobic digestion after pretreatments according to biogas yield and energy efficiency. *Agronomy* **2024**, *14*, 2980. [CrossRef]
21. Amirante, R.; Demastro, G.; Distaso, E.; Hassaan, M.A.; Mormando, A.; Pantaleo, A.M.; Tamburrano, P.; Tedone, L.; Clodoveo, M.L. Effects of Ultrasound and Green Synthesis ZnO Nanoparticles on Biogas Production from Olive Pomace. *Energy Procedia* **2018**, *148*, 940–947. [CrossRef]
22. Sözer, S. A research on biogas production from a mixture of olive pomace and cattle manure. *Biomass Convers. Biorefin.* **2024**, *14*, 10651–10659. [CrossRef]
23. Turano, E.; Curcio, S.; De Paola, M.G.; Calabrò, V.; Iorio, G. An Integrated Centrifugation–Ultrafiltration System in the Treatment of Olive Mill Wastewater. *J. Membr. Sci.* **2002**, *209*, 519–531. [CrossRef]
24. Pereira, J.A.; Pereira, A.P.G.; Ferreira, I.C.F.R.; Valentão, P.; Andrade, P.B.; Seabra, R.; Estevinho, L.; Bento, A. Table Olives from Portugal: Phenolic Compounds, Antioxidant Potential, and Antimicrobial Activity. *J. Agric. Food Chem.* **2006**, *54*, 8425–8431. [CrossRef] [PubMed]
25. Didry, N.; Seidel, V.; Dubreuil, L.; Tillequin, F.; Bailleul, F. Isolation and Antibacterial Activity of Phenylpropanoid Derivatives from Ballota Nigra. *J. Ethnopharmacol.* **1999**, *67*, 197–202. [CrossRef]
26. Calabrò, P.S.; Fòlino, A.; Tamburino, V.; Zappia, G.; Zema, D.A. Increasing the Tolerance to Polyphenols of the Anaerobic Digestion of Olive Wastewater through Microbial Adaptation. *Biosyst. Eng.* **2018**, *172*, 19–28. [CrossRef]
27. Oz, N.A.; Uzun, A.C. Ultrasound pretreatment for enhanced biogas production from olive mill wastewater. *Ultrason. Sonochem.* **2015**, *22*, 565–572. [CrossRef]
28. Llanos-Lizcano, R.; Senila, L.; Modoi, O.C. Evaluation of Biochemical Methane Potential and Kinetics of Organic Waste Streams for Enhanced Biogas Production. *Agronomy* **2024**, *14*, 2546. [CrossRef]
29. Raposo, F.; Fernández-Cegrì, V.; De la Rubia, M.A.; Borja, R.; Béline, F.; Cavinato, C.; Demirer, G.; Fernández, B.; Fernández-Polanco, M.; Frigon, J.C.; et al. Biochemical methane potential (BMP) of solid organic substrates: Evaluation of anaerobic biodegradability using data from an international interlaboratory study. *J. Chem. Technol. Biotechnol.* **2011**, *86*, 1088–1098. [CrossRef]
30. Porqueddu, I.; Ficara, E.; Alibardi, L.; Bona, D.; Brina, A.; Calabrò, P.S.; Casaletta, E.; Cavinato, C.; Daffonchio, D.; De Gioannis, G.; et al. Results of an Italian inter-laboratory study on biochemical methane potential. In Proceedings of the 13th World Congress on Anaerobic Digestion-Recovering (bio) Resources for the World, Santiago de Compostela, Spain, 25–28 June 2013.
31. Hafner, S.D.; de Laclos, H.F.; Koch, K.; Hollinger, C. Improving inter-laboratory reproducibility in measurement of Biochemical Methane Potential (BMP). *Water* **2020**, *12*, 1752. [CrossRef]
32. Montegiove, N.; Gambelli, A.M.; Calzoni, E.; Bertoldi, A.; Puglia, D.; Zadra, C.; Emiliani, C.; Gigliotti, G. Biogas production with residual deriving from olive mill wastewater and olive pomace wastes: Quantification of produced energy, spent energy, and process efficiency. *Agronomy* **2024**, *14*, 531. [CrossRef]
33. Carter Martin, R.; Gregorich, E.G. *Soil Sampling and Methods of Analysis*; CRC Press: Boca Raton, FL, USA, 2007.
34. Barili, S.; Bernetti, A.; Sannino, C.; Montegiove, N.; Calzoni, E.; Cesaretti, A.; Pinchuk, I.; Pezzolla, D.; Turchetti, B.; Buzzini, P.; et al. Impact of PVC microplastics on soil chemical and microbiological parameters. *Environ. Res.* **2023**, *229*, 115891. [CrossRef] [PubMed]
35. Olsen, S.R. *Estimation of Available Phosphorus in Soils by Extraction with Sodium Bicarbonate*; No. 939; US Department of Agriculture: Washington, DC, USA, 1954.
36. Cequier, E.; Aguilera, J.; Balcells, M.; Canela-Garayoa, R. Extraction and characterization of lignin from olive pomace: A comparison study among ionic liquid, sulfuric acid, and alkaline treatments. *Biomass Convers. Biorefinery* **2019**, *9*, 241–252. [CrossRef]
37. Tolisano, C.; Luzi, F.; Regni, L.; Proietti, P.; Puglia, D.; Gigliotti, G.; Di Michele, A.; Priolo, D.; Del Buono, D. A way to valorize pomace from olive oil production: Lignin nanoparticles to biostimulate maize plants. *Environ. Technol. Innov.* **2023**, *31*, 103216. [CrossRef]
38. Holder, N.; Mota-Meira, M.; Born, J.; Sutrina, S.L. New Small Scale Bioreactor System for the Determination of the Biochemical Methane Potential. *Waste Biomass Valoriz.* **2019**, *10*, 1083–1090. [CrossRef]
39. Paz, L.; Gentil, S.; Fierro, V.; Celzard, A. Activated carbons outperform other sorbents for biogas desulfurization. *Chem. Eng. J.* **2025**, *506*, 160304. [CrossRef]
40. Liu, C.; Yuan, X.; Zeng, G.; Li, W.; Li, J. Prediction of methane yield at optimum pH for anaerobic digestion of organic fraction of municipal solid waste. *Bioresour. Technol.* **2008**, *99*, 882–888. [CrossRef]
41. Ge, S.; Han, J.; Sun, Q.; Zhou, Q.; Ye, Z.; Li, P.; Gu, Q. Research progress on improving the freeze-drying resistance of probiotics. *Trends Food Sci.* **2024**, *147*, 104425. [CrossRef]
42. Azzaz, A.A.; Jeguirim, M.; Kinigopoulou, V.; Doulgeris, C.; Goddard, M.L.; Jellali, S.; Ghimbeu, C.M. Olive mill wastewater: From a pollutant to green fuels, agricultural and water source and bio-fertilizer—Hydrotermal carbonization. *Sci. Total Environ.* **2020**, *733*, 139314. [CrossRef] [PubMed]

43. Solomakou, N.; Goula, A.M. Treatment of olive mill wastewater by adsorption of phenolic compounds. *Rev. Environ. Sci. BioTechnol.* **2021**, *20*, 839–863. [CrossRef]
44. Dali, I.; Aydi, A.; Stamenic, M.; Kolsi, L.; Grachem, K.; Zizovic, I.; Manef, A.; Delgado, D.R. Extraction of lyophilized olive mill wastewater using supercritical CO_2 processes. *Alex. Eng. J.* **2022**, *61*, 237–246. [CrossRef]
45. Mekersi, N.; Addad, D.; Kadi, K.; Casini, S.; Hackenberger, D.K.; Boumazaa, A.; Lekmine, S. Effects of olive mill wastewater and olive mill pomace on soil physicochemical properties and soil polyphenols. *J. Mater. Cycles Waste Manag.* **2023**, *25*, 1404–1416. [CrossRef]
46. El Hajjouji, H.; Bailly, J.R.; Winterton, P.; Merlina, G.; Revel, J.C.; Hafidi, M. Chemical and spectroscopic analysis of olive mill waste water during a biological treatment. *Bioresour. Technol.* **2008**, *99*, 4958–4965. [CrossRef]
47. Wzorek, M.; Junga, R.; Yilmaz, E.; Bozhenko, B. Thermal decomposition of olive-mill byproducts: A TG-FTIR approach. *Energies* **2021**, *14*, 4123. [CrossRef]
48. Dorado, F.; Sanchez, P.; Alcazar-Ruiz, A.; Sanchez-Silva, L. Fast pyrolysis as an alternative to the valorization of olive mill wastes. *J. Sci. Food Agric.* **2021**, *101*, 2650–2658. [CrossRef] [PubMed]
49. Khan, A.S.; Man, Z.; Bustam, M.A.; Kait, C.F.; Nasrullah, A.; Ullah, Z.; Muhammad, N. Dicationic ionic liquids as sustainable approach for direct conversion of cellulose to levulinic acid. *J. Clean. Prod.* **2018**, *170*, 591–600. [CrossRef]
50. Ward, K.R.; Matejtschuk, P. The principles of freeze-drying and application of analytical technologies. In *Cryopreservation and Freeze-Drying Protocols*; Wolkers, W.F., Oldenhof, H., Eds.; Springer: New York, NY, USA, 2015; pp. 99–127.
51. Liu, Y.W.; Pu, H.B.; Sun, D.W. Hyperspectral imaging technique for evaluating food quality and safety during various processes: A review of recent applications. *Trends Food Sci.* **2017**, *69*, 25–35. [CrossRef]
52. Buahom, J.; Siripornadulsil, S.; Sukon, P.; Sooksawat, T.; Siripornadulsil, W. Survivability of freeze- and spray-dried probiotics and their effect on the growth and health performance of broilers. *Vet. World* **2023**, *16*, 1849–1865. [CrossRef]
53. Tian, Y.; He, Z.; He, L.; Li, C.; Qiao, S.; Tao, H.; Wang, X.; Zeng, X.; Tian, Y. Effect of freeze-dried protectants on the survival rate and fermentation performance of fermented milk's directed vat set starters. *Cryobiology* **2024**, *114*, 104811. [CrossRef]
54. Khanal, S.K.; Nindhia, T.G.T.; Nitayavardhana, S. Chapter 11—Biogas from Wastes: Processes and Applications. In *Sustainable Resource Recovery and Zero Waste Approaches*; Elsevier: Amsterdam, The Netherlands, 2016; pp. 165–174.
55. Siciliano, A.; Stillitano, M.A.; De Rosa, S. Biogas production from wet olive mill wastes pretreated with hydrogen peroxide in alkaline conditions. *Renew. Energy* **2016**, *85*, 903–916. [CrossRef]
56. Orive, M.; Cebrian, M.; Zufia, J. Techno-economic anaerobic co-digestion feasibility study for two-phase olive oil mill pomace and pig slurry. *Renew. Energy* **2020**, *97*, 532–540. [CrossRef]
57. Chapleur, O.; Madigou, C.; Civade, R.; Rodolphe, Y.; Mazeas, L.; Bouchez, T. Increasing concentrations of phenol progressively affect anaerobic digestion of cellulose microbial communities. *Biodegradation* **2016**, *27*, 15–27. [CrossRef]
58. Wang, Y.T.; Gabbard, H.D.; Pai, P.G. Inhibition of acetate methanogenesis by phenols. *J. Environ. Eng.* **1991**, *117*, 487–500. [CrossRef]
59. Field, J.A.; Lettinga, G. The methanogenic toxicity and anaerobic degradability of a hydrolysable tannin. *Water Res.* **1987**, *20*, 367–374. [CrossRef]
60. Sang, Y.; Wang, J.; Zhang, Y.X.; Gao, H.N.; Ge, S.Y.; Feng, H.; Zhang, Y.; Ren, F.; Wen, P.; Wang, R. Influence of temperature during freeze-drying process on the viability of Bifidobacterium Longum BB68S. *Microorganisms* **2023**, *11*, 181. [CrossRef]
61. Borja, R.; Pelillo, M.; Rincon, B.; Raposo, F.; Martin, A. Mathematical modelling of the aerobic degradation of two-phase olive mill effluents in a batch reactor. *Biochem. Eng. J.* **2006**, *30*, 308–315. [CrossRef]
62. Gonzalez-Gonzalez, A.; Cuadros, F. Effect of aerobic pretreatment on anaerobic digestion of olive mill wastewater (OMWW): An ecoefficient treatment. *J. Environ. Sci. Health* **1994**, *29*, 851–865. [CrossRef]
63. Borja, R.; Banks, J.C.; Albe, J.; Maestro, R. The effect of the most important phenolic constituents of OMW on batch anaerobic methanogenis. *Environ. Technol.* **1996**, *17*, 167–174. [CrossRef]
64. Rasi, S. Biogas Composition and Upgrading to Biomethane. Ph.D. Thesis, University of Jyväskylä, Jyväskylä, Finland, 2009. No. 202.
65. Tarannum, A.; Rao, J.R.; Fathima, N.N. Insights into protein-ionic liquid interaction: A comprehensive overview on theoretical and experimental approaches. *Int. J. Biol. Macromol.* **2022**, *209*, 498–505. [CrossRef] [PubMed]
66. Di Mario, J.; Montegiove, N.; Gambelli, A.M.; Brienza, M.; Zadra, C.; Gigliotti, G. Waste biomass pretreatments for biogas yield optimization and for the extraction of valuable high-added-value products: Possible combinations of the two processes toward a biorefinery purpose. *Biomass* **2024**, *4*, 865–885. [CrossRef]

agriculture

Article

Digestate-Derived Compost Modulates the Retention/Release Process of Organic Xenobiotics in Amended Soil

Elisabetta Loffredo *, Emanuela Campanale, Claudio Cocozza and Nicola Denora

Dipartimento di Scienze del Suolo, della Pianta e degli Alimenti, Università degli Studi di Bari Aldo Moro, Via Amendola 165/A, 70126 Bari, Italy; e.campanale3@studenti.uniba.it (E.C.); claudio.cocozza@uniba.it (C.C.); nicola.denora@uniba.it (N.D.)

* Correspondence: elisabetta.loffredo@uniba.it

Abstract: This study examined the effects of 2, 4 and 8% digestate-derived compost (DCP) on the retention/release of the fungicide penconazole (PEN), the herbicide S-metolachlor (S-MET) and the endocrine disruptor bisphenol A (BPA) in two agricultural soils sampled in Valenzano (SOV) and Trani (SOT), in Sothern Italy. DCP alone showed a conspicuous adsorption of the three xenobiotics, followed by their slow and scarce release. Sorption isotherm data of the compounds on unamended and DCP-amended soils were well described by the Freundlich model. Compared to unamended soil, the addition of the highest dose (8%) DCP to SOV increased the distribution coefficient, Kd, values of PEN, S-MET and BPA by 281%, 192% and 176%, respectively, while for SOT, the increases were 972%, 786% and 563%, respectively. Desorption of PEN and S-MET from all treatments was slow and partial (hysteresis), and only slightly reduced or unaffected by the addition of DCP, whereas BPA was almost entirely undesorbed in all treatments. Highly significant correlations between the adsorption coefficients of the three compounds in all soil treatments and the corresponding organic C contents confirm the prominent role of native and anthropogenic OM in the adsorption of contaminants and, consequently, in the control of their transfer into natural waters and/or entry in crop plants.

Keywords: soil amendment; organic contaminant; adsorption; desorption; sorption model; organic matter; biobased adsorbent

1. Introduction

From the perspective of the current concept of circular economy, expressed as the "reduce, reuse, recycle" paradigm, researchers are focusing their scientific efforts on the valorisation of organic wastes, trying to convert them into a valuable resource for the energy sector and modern agriculture [1]. Some current technologies allow operators to respond to multiple community needs through a single sustainable process [2]. This is the case of anaerobic digestion (AD), which has gained increasing interest since it combines multiple benefits, such as the disposal of agrozootechnical waste, the production of bioenergy and the contrast to climate change through carbon (C) sequestration in the by-product digestate (DG).

The effectiveness of DG as a soil improver is widely recognized both when used raw and after its conversion into safer and more profitable products such as compost (CP), vermicompost and biochar [3]. On the other hand, it is well known that intensive agricultural use of soil has led to a progressive depletion of the native organic fraction, thus exposing the soil to degradation (erosion and desertification) and resulting in a reduction in the quantity and quality of crops [4]. Soil organic matter (OM) is undoubtedly the

https://doi.org/10.3390/agriculture15181925

protagonist of the overall soil fertility, and, in addition to ensuring good edaphic and trophic conditions for plants, it regulates the level of inorganic and organic contaminants present in pore water, thus controlling their availability to all soil-dwelling organisms. The control action is also exerted on the dynamics of pollutants in the entire soil–plant-water system [5].

In addition to biogas, AD releases a raw semisolid product from which, after a solid–liquid separation, a liquid DG and a solid DG are obtained. The latter can be converted into high-quality biofertilizer, namely DG-derived compost (DCP) and vermicompost (VC), through aerobic treatment lasting several weeks [6,7]. During this time, the lignocellulosic components of DG undergo biodegradation, and a humified fraction is gradually formed until the product is fully mature. Soil application of DCP or VC has multiple beneficial effects in agricultural systems as they provide fairly stable OM that improves soil fertility by stimulating plant growth and microbial activity [8,9]. Therefore, AD and subsequent aerobic treatment constitute a virtuous recycling of biowaste, generating both renewable energy and zero-waste soil improvers. Thanks to DCP properties such as its humic component, loose structure, large specific surface area, and nutrient content, this material is very suitable as a soil amendment. A recent study reports interesting applications of DCP in bioremediation and soil detoxification from industrial xenobiotics [10].

Although the European Green Deal and the Sustainable Development Goals of 2030 Agenda are aimed at significantly reducing pesticide use, the quantities of these products currently employed are still very high and only slowly declining, which causes soil and water pollution in areas where intensive and super-intensive agriculture is practised. In 2022, the global use of such products in agriculture amounted to 3.70 Mt of active substances, with 4% increase compared to the previous year [11]. While the more water-soluble compounds are particularly dangerous to the quality of surface and groundwater, the more hydrophobic ones are the most dangerous to human and animal health, as they can accumulate in vital organs and cause disease and death. Furthermore, hydrophobic compounds are by far the most persistent in the environment as they can bind tightly to OM in soil and sediments and are resistant to biodegradation [12].

Penconazole [1-[2-(2,4-dichlorophenyl)pentyl]-1,2,4-triazole, PEN] is a systemic fungicide belonging to the triazole class. It is worldwide used to protect crops, such as grapes, fruit trees, vegetables and ornamental plants, from a broad spectrum of pathogenic fungi and is particularly effective against powdery mildew [13]. This fungicide inhibits the biosynthesis of ergosterol, an essential component of fungal cell membranes [14]. It is toxic for soil microorganisms and animals and can exert endocrine-disrupting activity [15]. S-Metolachlor [2-chloro-N-(2-ethyl-6-methylphenyl)-N-[(2S)-1-methoxy-2-propanyl]acetamide, S-MET] is a chloroacetamide pre-emergent herbicide widely adopted to control broadleaf weeds in a variety of crops [16]. In Italy, in 2022, S-MET was detected in 240 water samples with a contamination frequency of more than 27% [17]. Its mode of action consists of the inhibition of long-chain fatty acids biosynthesis which disrupts the formation of new cell membrane and consequently cell division during seed germination [18]. S-MET has been reported as a suspected human carcinogen [19] and endocrine disrupting chemical [20].

Intensive human activities result in a continuous discharge of industrial byproducts into aquatic and terrestrial ecosystems. Many of these chemicals, of which bisphenol A [4,4′-(2,2-propandiyl)diphenol, BPA] is a very representative molecule, are endocrine disrupting chemicals with estrogen-like and anti-androgen activity. BPA has been frequently reported in soil and natural waters, especially in highly urbanized and industrialized areas [21]. The global production of BPA is approximately 8 Mt per year and is constantly growing; it is predicted that the volume of BPA produced worldwide will exceed 10 Mt in the next years [22]. As a component of materials such as polycarbonate and epoxy resins, it is present in many commonly used products, including electrical and electronic components, medical devices, food and beverage packaging, and so on [23]. In soil, BPA binds to OM, but, due to its moderate water solubility, it can leach and contaminate groundwater, especially in saturated soil. Various organic amendments are employed in the process of remediating soil and water from BPA contamination [24].

Despite extensive evidence of the potential of soil amendments to retain organic compounds from aqueous media [25,26], very few studies have focused on the role of CP and its humic fraction as a bioadsorbent of soil inorganic and organic pollutants [27,28]. C-rich materials resulting from waste biomass recycling can make a significant contribution to the containment of pollution, especially where soil OM is scarce, and fertility is reduced. Differently from soil amendments obtained through the thermochemical conversion of waste biomass, such as biochar and hydrochar, DCP, which is biologically produced, has the added value of a humic (or humic-like) component of great importance for the overall soil fertility. Similar to native soil OM, this humic fraction interacts with organic xenobiotics and immobilizes them, thereby reducing their concentration in soil pore water and their transfer to natural water [29]. Furthermore, since the ingestate of AD is generally composed of animal waste alone or plant residues alone or their mixture, DCP is less likely to contain hazardous organic compounds or heavy metals compared to CPs prepared from the organic fraction of municipal solid waste or sewage sludge. This is also true when DCP is compared to soil amendments originated from thermochemical processes, which, despite being more efficient as adsorbents, can contain alarming levels of toxic compounds and elements.

The employment of DCP in agriculture as a bioadsorbent represents a valuable way to recover and valorise a by-product (DG) which in turn derives from waste (mainly agro-waste). In this case, the generation of renewable energy (biogas) is virtuously combined with the production of a low-cost multifunctional material (DCP), beneficial for agriculture and the environment, fully respecting the circular economy. Anyway, the addition of DCP to the soil requires a thorough understanding of the modifications it induces in the entire soil–plant system. Most studies on the agricultural use of DCP are limited to their effects on soil physical, chemical, and biological fertility [30]. Furthermore, characterization of DCPs focuses primarily on their content of plant nutrients, the absence of potentially toxic elements, and biological activity [31,32], neglecting their adsorbent properties. Several published studies have focused on adsorption/desorption of organic pollutants on/from amendments alone or soil alone, while investigation on changes following soil amendment has been scarcely explored. Finally, although there is evidence in the literature that soil amendment can attenuate pesticide leaching, systematic data on the adsorption/desorption behaviour of DCP at different application rates and in the coexistence of multiple classes of organic pollutants (fungicides, herbicides, and endocrine disruptors) are still lacking, particularly in soils with low organic matter levels.

In light of all this, the present study aimed to evaluate the role of different DCP dosages on the adsorption and release of three organic xenobiotics, PEN, S-MET and BPA, in two agricultural soils.

2. Materials and Methods

2.1. Chemicals, Compost and Soil

The compounds considered in this study, namely the fungicide PEN with purity $\geq$ 98%, the herbicide S-MET with purity $\geq$ 98%, and the endocrine disruptor BPA with purity $\geq$ 99% were purchased from Merck KGaA, Darmstadt, Germany. Molecular mass, water solubility at 25 °C, and log Kow were, respectively, 284.2 g mol^{-1}, 1.6 and 4.67 for PEN, 283.8, 50.9 and 3.24 for S-MET, and 228.3, 300 and 3.32 for BPA [33]. Single methanolic solutions (HPLC grade) of the compounds were prepared at a concentration of 2000 μg mL^{-1} and subsequently diluted with double-distilled H_2O to obtain the aqueous mixtures used in the experiments.

The DCP was collected from a biogas-producing plant located in Capaccio, South Italy belonging to C&F Energy, Società Agricola S.r.l., Salerno, Italy. The ingestate was a mixture of buffalo manure (80%, *w*/*w*), olive mill wastewater (15%, *w*/*w*), agrifood industry residues (3%, *w*/*w*) and poultry manure (2%, *w*/*w*), and was processed under thermophilic conditions (55–60 °C). After the release, the solid DG was subjected to heat treatment at a temperature of 65 °C for 5 days and then exposed to the air under controlled conditions for approximately 2 months. Before use, the DCP was air-dried and thoroughly mixed. The complete characterization of DCP is reported in the recently published article [34]. Some DCP properties are summarized as follows: 17.2% moisture, pH 7.9, EC 1.17 dS m^{-1} at 25 °C, 10.3% ash, 452 g kg^{-1} total organic C, 19.7 g kg^{-1} total N, C/N 23 [34]. Major elements of DCP were Ca (10,802 mg kg^{-1}), K (7765 mg kg^{-1}), P (7180 mg kg^{-1}), S (1940 mg kg^{-1}), Fe (1172 mg kg^{-1}); potentially toxic metals, Cu, Zn, Ni and Pb, were all below the limits fixed by the EU Regulation for organic fertilizers [34].

Two loamy Apulian soils with different OM contents were used. One soil (SOV) was sampled at 0–20 cm depth in an olive grove located in the Martucci experimental station of the University of Bari, in Valenzano. The other soil (SOT) is cultivated with vegetables and was sampled at 0–20 cm depth in a farm located in Trani. After sampling, each soil was air-dried at room temperature (approximately 20 °C), sieved with a 2 mm sieve to remove the skeleton and homogenized.

2.2. Soil Characterization

The properties of SOV were determined using conventional methods described in the work of Colatorti et al. [35] and shown in Table 1. SOT properties were determined according to Italian official methods of soil analysis [36] that are briefly reported as follows. Soil moisture was measured after heating the soil at 105 °C overnight; the pH was potentiometrically measured in a 1:2.5 (*w*/*v*) soil-to-H_2O suspension; electrical conductivity (EC) was measured at 25 °C in a 1:2 (*w*/*v*) soil-to-H_2O slurry; organic C and total N were determined using the Walkley-Black method and the Kjeldahl method, respectively. SOT properties are shown in Table 1.

Table 1. Main properties of the two soils.

Parameter	SOV [a]	SOT
Sand (%)	36	42
Silt (%)	43	36
Clay (%)	21	22
pH [b]	7.45 ± 0.03	8.23 ± 0.03
EC (dS m^{-1}) [c]	0.20 ± 0.003	0.11± 0.001
Moisture (%)	4.4 ± 0.500	2.8 ± 0.003
Organic C (g kg^{-1})	37.9 ± 0.08	9.4 ± 0.28
Total N (g kg^{-1})	2.98 ± 0.080	0.78 ± 0.003
C/N	12.7	12.1

[a] From Colatorti et al. [35]; [b] soil/H_2O: 1:2.5 (*w*/*v*); [c] soil/H_2O: 1:2 (*w*/*v*).

2.3. Adsorption and Desorption Experiments

The adsorption study of the three molecules on DCP was conducted at different solution/DCP ratios in batch tests. Quantities of 2, 5, 10 and 20 mg of DCP were interacted with 20 mL of an aqueous mixture of PEN, S-MET and BPA, each at a concentration of 2 μg mL^{-1}, thus obtaining solution/adsorbent ratios of 10,000, 4000, 2000 and 1000. The pH of the aqueous mixture was 6.92. To assess possible matrix effects, blank samples were also prepared by adding 20 mL of double-distilled H_2O to DCP. To reach the adsorption equilibrium, the samples were kept under mechanical shaking at 20 °C in the dark for 24 h. Preliminary experiments in which DCP was interacted with the aqueous mixture of the compounds for 4, 8, and 16 h showed that, compared to the 8 h sampling, after 16 h, the variations in adsorbed PEN, S-MET and BPA were 3.9, −0.8 and 3.8%, respectively. Therefore, 24 h was considered a sufficiently long time for all compounds to reach equilibrium. Subsequently, the suspensions were centrifuged at 10,000× *g* at 10 °C for 10 min. A 10 mL volume of supernatant solution of each sample was filtered through 0.45 µm cellulose acetate filters (Ø 25 mm, Incofar s.r.l., Modena, Italy), and the equilibrium concentration of each molecule was measured by high-performance liquid chromatography (HPLC). All experiments were triplicated. The amount of compound adsorbed after a contact of 24 h per unit mass of DCP (q_t, µg g^{-1}) was calculated using the following equation:

$$q_t = (C_0 - C_t)\,V/m \quad (1)$$

where C_0 (µg mL^{-1}) is the initial concentration of the molecule in solution, C_t (µg mL^{-1}) is the concentration of the molecule after 24 h, V (mL) is the volume of the solution, and m (g) is the mass of the adsorbent.

Adsorption isotherms were studied to quantify the adsorption of PEN, S-MET, and BPA on unamended soil (SOV and SOT) and DCP-amended soil at doses of 2% (SOV-DCP2 and SOT-DCP2), 4% (SOV-DCP4 and SOT-DCP4) and 8% (*w*/*w*) (SOV-DCP8 and SOT-DCP8). After the addition of DCP, the organic C content in SOV-DCP2, SOV-DCP4, SOV-DCP8, SOT-DCP2, SOT-DCP4, SOT-DCP8 were 4.62, 5.45, 7.11, 1.83, 2.71 and 4.48, respectively. An aliquot of 2 g of unamended or amended soil was interacted with 20 mL aqueous mixtures of the compounds at individual concentrations of 0 (blank samples), 0.1, 0.2, 0.4, 0.5, 1, and 2 µg mL^{-1}. For the two pesticides, the concentration range tested was chosen based on the doses commonly applied to soil and in agreement with previous studies concerning PEN [37] and S-MET [38]; identical concentrations were used for BPA for comparison purposes. To reach equilibrium, the suspensions were stirred at 20 ± 1 °C, in the dark, for 48 h. In previous experiments, SOV and SOT were interacted with the aqueous mixture of the compounds for 12, 24 and 48 h; compared to the 24 h sampling, after 48 h, the variations in adsorbed PEN, S-MET and BPA on SOV were −4.2, −0.1 and 1.5%, respectively, while on SOT they were 2.9, −0.6 and 3.2%, respectively. On this basis, an equilibration time of 48 h was considered appropriate for all compounds. Before equilibration, pH values of triplicated suspensions of SOV and SOT (2 g) with 2 µg mL^{-1} solution (20 mL, pH 6.92) were 7.23 ± 0.09 and 8.16 ± 0.15, respectively; after equilibration (48 h), the pH values, in the same order, were 6.96 ± 0.05 and 7.02 ± 0.01. Subsequently, the samples were processed and analyzed as already described in session 2.3. All experiments were triplicated.

The desorption tests of the three molecules from the unamended and the soil amended with 4 and 8% DCP were started immediately after adsorption using the samples added with the three compounds at concentration of 2 µg mL^{-1}. After each desorption cycle, 10 mL of the supernatant solution was replaced with the same volume of double-distilled H_2O; the sample was shaken again for 48 h at 20 ± 1 °C and then centrifuged, filtered and analyzed under the conditions described in the next session. Three desorption cycles were performed in triplicate for each treatment. All experiments conducted in this study were performed from mid-2024 to early 2025.

2.4. Analytical Protocol

Samples were analyzed using a Nexera HPLC system (Shimadzu Corporation, Kyoto, Japan). All instrumental components, including the detectors, were from Shimadzu company. Samples were loaded into a SIL-40C XR autosampler connected to an LC-40D XR pump, a DGU-40S degasser, and a Shim-pack GIST-HP C18 column (150 mm × 3.0 mm × 3 μm) inserted into a CTO-40C thermoregulation system. The mobile phase was water (A) and acetonitrile (B). The gradient elution was set as follows: 0–7 min, 70% B, flow rate 0.5 mL min^{-1}; 7–12 min, 75% B, flow rate 0.7 mL min^{-1}; 12–15 min, 70% B, flow rate 0.5 mL min^{-1}. Under these conditions, the retention times of PEN, S-MET, and BPA were 11.1 min, 10.7 min, and 6.6 min, respectively. PEN and S-MET were detected with an SPD-M40 diode array detector operating at a wavelength of 211 nm, while BPA was detected with an RF-20AXS fluorescence detector operating at wavelengths of 200 and 290 nm for excitation and emission, respectively.

The external standard method was adopted to quantify the molecules. Blank samples were prepared for DCP and soil treatments in duplicate. No interfering peaks from the matrix were observed and no significant changes in the retention times of the compounds were recorded. To quantify the concentration of each compound in solution, calibration curves at five concentration levels (0.1, 0.2, 0.4, 0.5, 1 and 2 μg mL^{-1}) were performed; linearity was expressed by the determination coefficients, R^2, which were: 0.99973, 0.99997 and 0.99996 for PEN, S-MET and BPA, respectively.

2.5. Theoretical Models

To find the best suitable model to describe the adsorption process of the compounds, adsorption isotherms data were fitted into the two-parameter nonlinear Freundlich, Langmuir and Temkin equations, and the linear Henry model. The Freundlich isotherm [39] is given by

$$q_e = K_F\, C_e^{1/n} \tag{2}$$

where q_e (μg g^{-1}) is the amount of solute adsorbed at equilibrium per unit of adsorbent, C_e (μg mL^{-1}) is the equilibrium concentration of the compound in solution, $1/n$ indicates the degree of nonlinearity between solution concentration and amount adsorbed, while the reciprocal, n, expresses the adsorption intensity, and K_F (Freundlich adsorption constant, mL g^{-1}) expresses the adsorption capacity of the adsorbent. The Langmuir equation [40] is

$$q_e = (K_L C_e b)/(1 + K_L C_e) \tag{3}$$

where q_e and C_e are defined as in the Freundlich equation, b (μg g^{-1}) is the maximum adsorption capacity, i.e., the amount of solute that forms a monolayer on the adsorbent, and K_L (L mg^{-1}, Langmuir constant) expresses the energy of adsorption and the affinity of the solute for the adsorbent. The Temkin model [41] is expressed as

$$q_e = B \ln (A_T C_e) \tag{4}$$

where q_e and C_e have been already defined for the previous equations, A_T (mL g^{-1}) is the Temkin isotherm constant, $B = RT/b_T$, where b_T (J mol^{-1}) is the Temkin constant, which expresses the process enthalpy, T is the absolute temperature (K) and R is the universal gas constant (8.314 J mol^{-1} K^{-1}).

The parameters of the nonlinear models were calculated by the nonlinear regression method using the solver add-in component of Microsoft Excel. The correlation coefficient, r, was calculated as

$$r = \sqrt{\frac{\Sigma(q_t m - \overline{q_t})^2}{\Sigma(q_t m - \overline{q_t})^2 + \Sigma(q_t m - q_t)^2}} \quad (5)$$

where $q_t m$ is the theoretical adsorbed concentration of the compound (μg g^{-1}) at time t, q_t is the experimental concentration (μg g^{-1}), and $\overline{q_t}$ is the mean q_t.

The linear Henry model:

$$q_e = K_d C_e \quad (6)$$

was used to calculate the distribution coefficient, K_d, from the slope, and the partition coefficient, K_{OC}, as: $K_{OC} = (K_d/(\% \text{ OC})) \times 100$. The latter parameter expresses the amount of compound adsorbed per organic C unit of the adsorbent.

Finally, to estimate the deviation between theoretical and experimental data, two error functions were considered: the correlation coefficient, r, and the sum of squared residuals, SSR.

3. Results and Discussion

3.1. DCP Sample

The concentrations of each molecule adsorbed at equilibrium on DCP at various solution/adsorbent ratios are shown in Figure 1. The ratios adopted are comparable with those used in similar studies on the adsorption of xenobiotics on C-rich matrices. Increasing the solution/DCP ratio, the adsorbed amount of each compound significantly increased at each successive ratio tested (Figure 1). A 10-fold increase in the ratio (from 1000 to 10,000) increased the concentration of absorbed PEN, S-MET and BPA by 2.2, 3.4 and 4.1 times. It can be assumed that at higher ratios, i.e., at smaller amounts of compound for the same solution volume, a greater number of interaction sites are available, including the more internal sites that are generally more difficult to reach for more hydrophobic molecules or more shielded by water molecules for less hydrophobic compounds. At the highest ratio (10,000), the concentration of adsorbed PEN, S-MET and BPA were 2067, 809, and 2356 μg g^{-1}, respectively. Based on these results, the highest ratio was chosen for subsequent adsorption experiments.

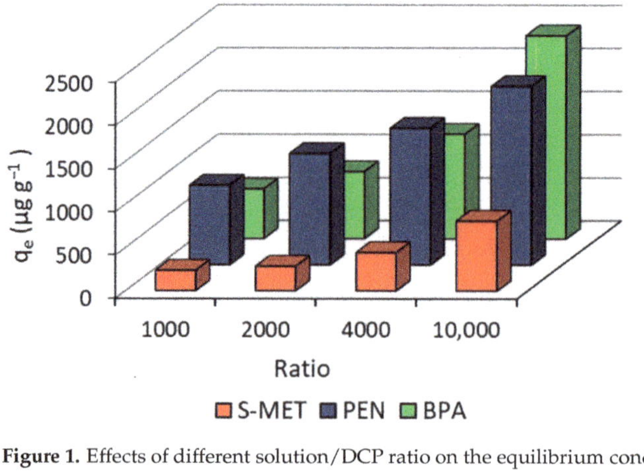

Figure 1. Effects of different solution/DCP ratio on the equilibrium concentration of the adsorbed compound.

The relatively high adsorption capacity shown by this DCP was quite predictable based on the results of its previous extensive characterization [34]. Structural properties of DCP are briefly summarized here: (i) the BET specific surface area, an essential parameter for adsorbent materials, was approximately 1 m^2 g^{-1}; (ii) SEM analysis showed a heterogeneous surface, diffuse irregular ridges, channels, cavities, and low porosity typical of organic matrices originated from biological treatments of biomass; (iii) the FTIR spectrum showed numerous functional groups—such as alcoholic and phenolic OH, C=O of amide groups and conjugated ketones, COO^- groups of carboxylic acids, aromatic C=C from quinones of lignin-like bulks, and other absorption peaks typical of polysaccharides or polysaccharide-like moieties possibly associated with silicate impurities, clay minerals and humic acids; (iv) Raman spectroscopy highlighted the presence of CH, CC, CO and CN bonds which confirmed the dual nature of this material, i.e., aliphatic and aromatic. All these features are a clear indication of a remarkable potential of DCP to stably interact with polar and hydrophobic compounds [34].

The desorption study demonstrated a considerable retention of PEN, S-MET and BPA on DCP even after three dilution steps of the liquid phase, when approximately 84, 33 and 55% of PEN, S-MET and BPA, respectively, were still adsorbed (Figure 2). These results demonstrate that DCP, in addition to effectively adsorb the compounds, has a good capacity to retain them when the liquid phase of the soil diluted.

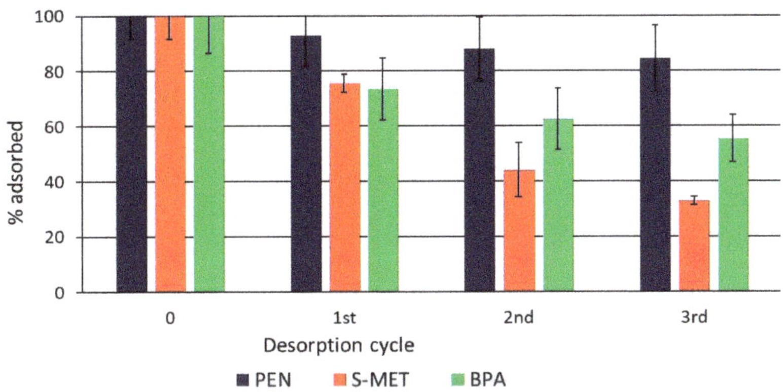

Figure 2. Percentage of compound that remains adsorbed on DCP after each desorption cycle.

3.2. Soil

Although CP amendment is a widespread agricultural practise, the behaviour of this material as biosorbent of pollutants has certainly been overlooked. The adsorption isotherm describes the relationship between the concentration of a solute on an adsorbent and its concentration in solution at constant temperature. To gain insight into the adsorption mechanisms, experimental isotherm data were described by various theoretical models, namely the two-parameter Freundlich, Langmuir and Temkin models and the linear Henry equation. The Freundlich model (Equation (2)) describes multilayer adsorption on heterogeneous surfaces. Conversely, the Langmuir model (Equation (3)) is more appropriate for homogeneous surfaces with single-layer adsorption. The Temkin isotherm (Equation (4)) is appropriate for intermediate solute concentrations and heterogeneous adsorbents. It is based on the assumption that there is a linear reduction in the heat of adsorption of all solute molecules in a layer with increasing adsorbent surface coverage, and that the process occurs with a uniform distribution of the binding energy [41]. The Henry model (Equation (6)) is appropriate at low solute concentrations and for hydrophobic substances in aqueous solution. The model assumes a linear partitioning of the solute between the adsorbent and the solution over the entire concentration range explored; it is used to calculate the distribution coefficient, Kd, which is a reliable parameter to express the sorption efficiency of a substrate, and the organic C partition coefficient, K_{OC}. The accordance between experimental adsorption data of PEN, S-MET and BPA on unamended and amended soil and the theoretical models was assessed based on the values of the correlation coefficient, r, and the sum of squared residues (SSR).

The adsorption isotherms obtained for the three compounds on unamended and amended SOV and SOT are depicted in Figures 3 and 4, respectively, while Tables 2 and 3 show the values of the adsorption parameters obtained by fitting the experimental data into the theoretical models. Based on the r and SSR values, with very few exceptions, the experimental sorption data of the three molecules on all treatments were best described by the Freundlich model, although in several cases there was also a good match with one or more of the other models tested. On both soils, the Freundlich parameter 1/n was generally less than or approximately equal to unity. This parameter indicates the type of Giles isotherm, i.e., L-type (1/n < 1), C-type (1/n~1) and S-type (1/n > 1), and is related to the intensity of adsorption and the type of process. When 1/n < 1, the interaction is predominantly physical, while it is predominantly chemical when it is greater than 1, and both processes contribute when it approximates to unity [42].

SOV

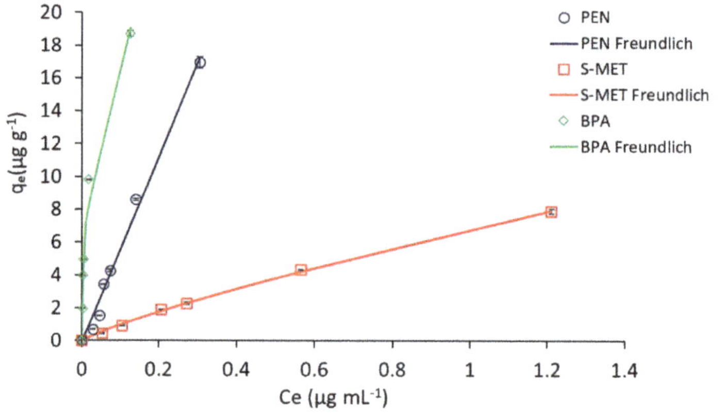

SOV-DCP2

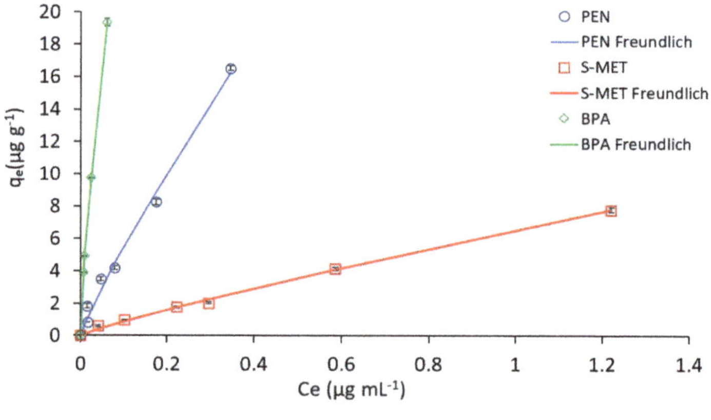

Figure 3. *Cont.*

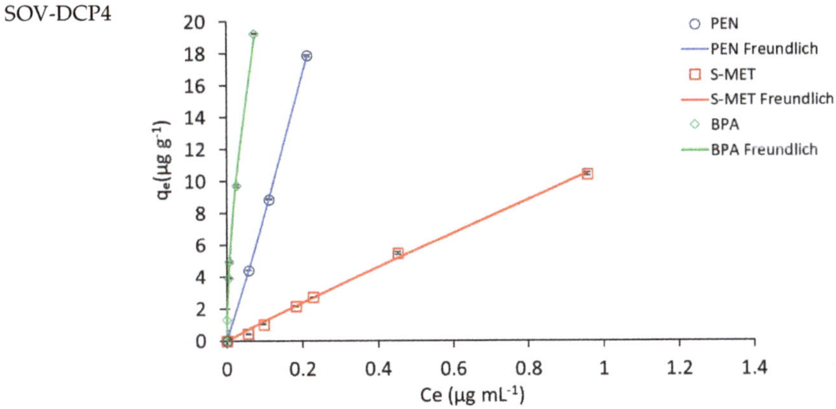

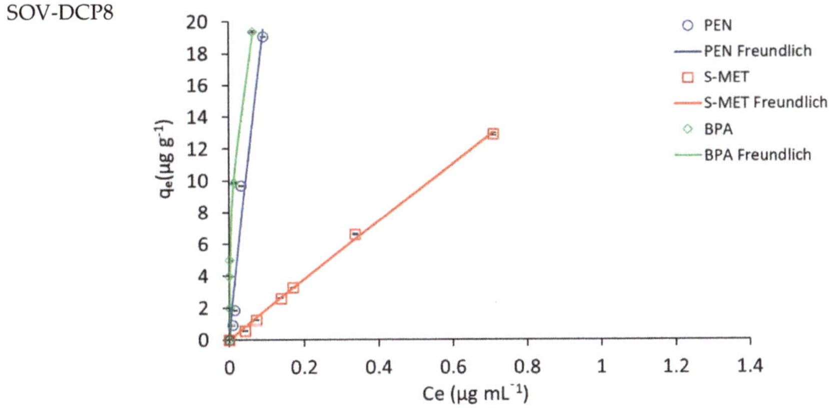

Figure 3. Adsorption isotherms of the compounds on unamended Valenzano soil (SOV) and SOV amended with 2, 4 and 8% DCP (SOV-DPC2, SOV-DPC4, SOV-DPC8). Standard error is reported as vertical bar on each point (n = 3).

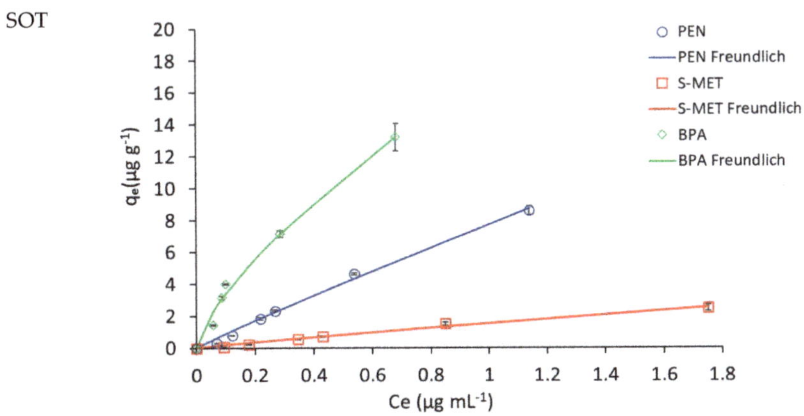

Figure 4. *Cont.*

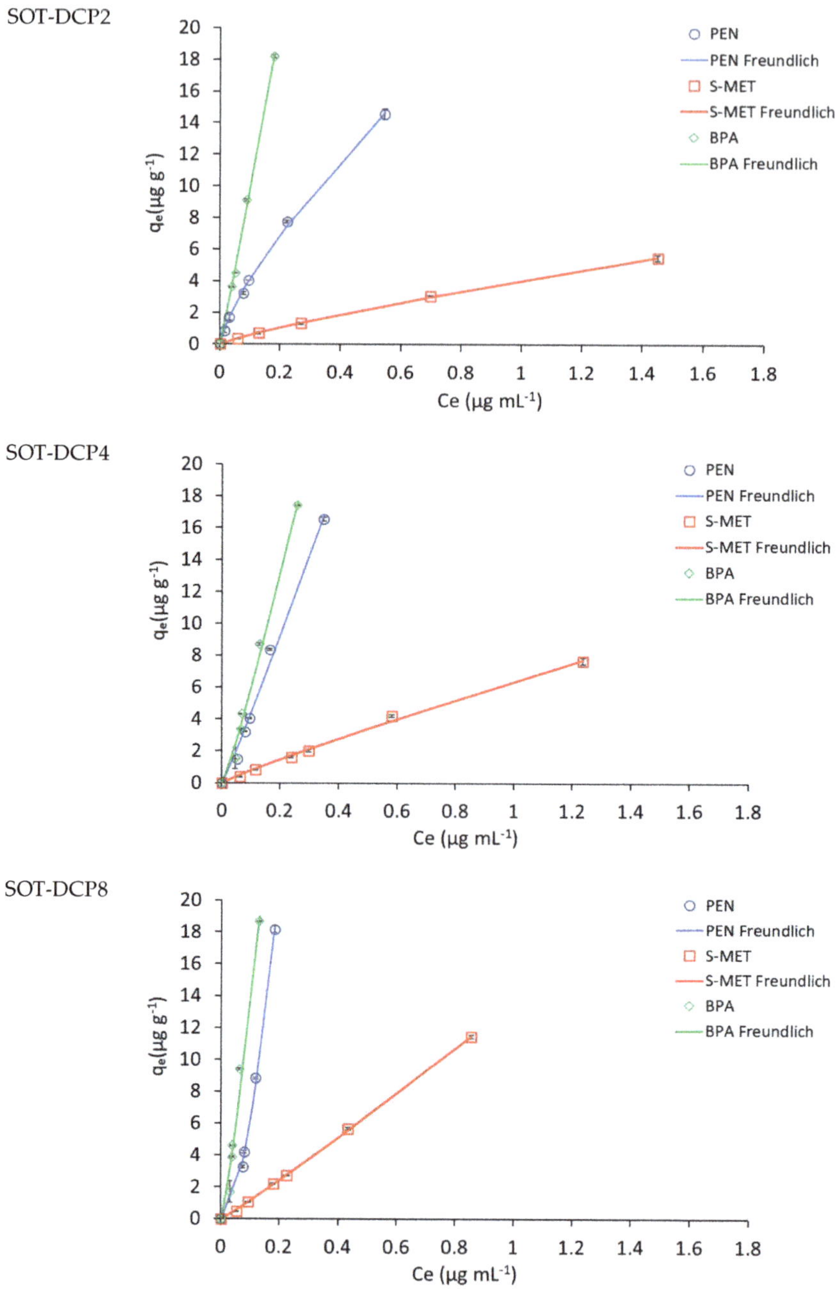

Figure 4. Adsorption isotherms of the compounds on unamended Trani soil (SOT) and SOT amended with 2, 4 and 8% DCP (SOT-DPC2, SOT-DPC4, SOT-DPC8). Standard error is reported as vertical bar on each point ($n = 3$).

Table 2. Adsorption parameters of the compounds on SOV obtained by applying theoretical models via nonlinear regression.

	Henry				Freundlich				Langmuir				*Temkin*				
Substrate	r	SSR	Kd $mL\ g^{-1}$	K_{OC} $mL\ g^{-1}$	r	SSR	K_F $mL\ g^{-1}$	1/n	r	SSR	b $\mu g\ g^{-1}$	K_L $L\ mg^{-1}$	r	SSR	A_T $mL\ g^{-1}$	B	b_T $J\ mol^{-1}$
PEN																	
SOV	0.997	2.59	56.1	1476	0.993	2.54	57.8	1.02	0.992	2.65	332	0.176	0.973	9.82	29.6	6.98	348
SOV-DCP2	0.997	2.88	68.3	1045	0.996	1.28	56.2	0.882	0.994	1.76	117	0.522	0.946	28.1	64.9	4.24	587
SOV-DCP4	1.00	0.021	87.6	1564	1.00	0.001	95.5	1.08	0.999	0.038	4845	0.016	1.00	3.49	32.8	6.87	236
SOV-DCP8	0.975	10.4	214	3010	0.976	9.27	168	0.902	0.979	7.80	76.1	3.70	0.984	6.76	116	7.76	316
S-MET																	
SOV	0.996	0.309	6.34	167	1.00	0.054	6.78	0.831	0.998	0.143	33.2	0.272	0.943	4.26	14.7	2.38	1050
SOV-DCP2	0.997	0.450	6.59	142	0.999	0.099	6.54	0.876	0.998	0.150	39.4	0.202	0.913	7.19	17.6	2.00	1227
SOV-DCP4	0.998	0.314	11.0	196	0.998	0.204	11.0	0.944	0.999	0.128	79.0	0.159	0.947	7.19	13.7	3.45	711
SOV-DCP8	0.999	0.265	18.5	260	0.999	0.198	18.1	0.967	0.999	0.140	148	0.135	0.941	12.1	18.1	4.25	577
BPA																	
SOV	0.945	19.3	119	3112	0.985	5.28	45.9	0.425	0.995	1.76	21.5	50.7	0.994	2.11	673	4.14	588
SOV-DCP2	0.975	6.91	139	3015	0.999	1.96	62.0	0.530	0.998	0.181	26.2	24.0	0.993	5.19	203	7.24	388
SOV-DCP4	0.990	3.76	230	4111	0.998	0.643	91.0	0.600	0.990	4.72	31.9	19.7	0.894	40.1	1273	3.35	732
SOV-DCP8	0.931	68.5	328	4613	0.991	3.56	99.2	0.406	0.965	18.2	22.0	86.4	0.955	17.1	3218	3.28	749

Table 3. Adsorption parameters of the compounds on SOT obtained by applying theoretical models via nonlinear regression.

	Henry				Freundlich				Langmuir				*Temkin*				
Substrate	r	SSR	Kd $mL\ g^{-1}$	K_{OC} $mL\ g^{-1}$	r	SSR	K_F $mL\ g^{-1}$	1/n	r	SSR	b $\mu g\ g^{-1}$	K_L $L\ mg^{-1}$	r	SSR	A_T $mL\ g^{-1}$	B	b_T $J\ mol^{-1}$
PEN																	
SOT	0.998	0.426	7.79	865	0.997	0.313	7.73	0.939	0.997	0.223	56.5	0.158	0.955	4.21	11.0	2.95	826
SOT-DCP2	0.989	6.35	28.5	1603	0.999	0.161	23.1	0.750	0.999	0.133	54.3	1.35	0.938	16.8	48.8	3.69	664
SOT-DCP4	0.997	1.863	47.5	1778	0.995	1.40	52.2	1.07	0.992	2.00	683	0.070	0.983	4.90	19.5	8.12	299
SOT-DCP8	0.971	24.7	83.5	1881	0.999	0.296	387	1.79	0.826	25.0	3369	0.024	0.987	3.59	16.0	16.3	149
S-MET																	
SOT	0.995	0.084	1.49	164	0.992	0.058	1.53	0.903	0.995	0.039	11.5	0.160	0.952	0.39	7.84	0.82	2963
SOT-DCP2	0.997	0.238	3.93	221	1.00	0.006	4.03	0.843	1.00	0.005	20.4	0.254	0.938	1.91	13.7	1.57	1559
SOT-DCP4	0.998	0.345	6.41	240	0.998	0.144	6.38	0.908	0.998	0.077	41.7	0.182	0.935	4.63	12.2	2.36	1031
SOT-DCP8	0.999	0.190	13.2	297	1.00	0.026	13.6	1.06	0.998	0.246	587	0.022	0.923	12.4	13.9	3.76	648
BPA																	
SOT	0.984	7.95	20.8	2311	0.994	1.066	17.4	0.711	0.994	1.12	27.0	1.38	0.985	2.54	23.0	4.50	541
SOT-DCP2	0.999	0.824	99.9	5613	0.998	0.469	113	1.06	0.996	1.01	656	0.155	0.990	4.33	32.3	9.87	248
SOT-DCP4	0.996	3.16	96.2	2477	0.995	1.53	82.1	1.13	0.987	3.34	1249	0.053	0.988	4.00	23.5	9.16	266
SOT-DCP8	0.990	9.07	138	3117	0.985	5.10	216	1.19	0.966	9.25	2201	0.063	0.998	0.39	38.2	11.5	211

The values of the adsorption constants Kd, K_F and K_{OC} indicated a general higher adsorption of PEN and BPA, compared to S-MET, on both soils. In particular, in SOV treatments, Kd, K_F and K_{OC} values for BPA (119, 46 and 3112 mL g^{-1}, respectively) were all higher than those obtained for PEN (56, 58 and 1476 mL g^{-1}, respectively) and S-MET (6, 7 and 167 mL g^{-1}, respectively) (Table 2). The same trend was observed in SOT treatments, where Kd, K_F and K_{OC} values for BPA (21, 17 and 2311 mL g^{-1}, respectively) were all higher than those observed for PEN (8, 8 and 865 mL g^{-1}, respectively) and S-MET (1, 1 and 164 mL g^{-1}, respectively) (Table 3).

As expected, a significant increase in the adsorption constants was observed for both DCP-amended soils, compared to unamended soils (Figures 3 and 4 and Tables 2 and 3). Based on the Kd values, the adsorption efficiency followed the trend: unamended soil < soil-DCP2 < soil-DCP4 < soil-DCP8. Compared to SOV, the addition of 2, 4 and 8% DCP increased the Kd values of PEN by 22, 56 and 281%, of S-MET by 4, 74 and 192%, and of BPA by 17, 93 and 176%, respectively; while the same additions to SOT increased Kd values of PEN by 266, 510 and 972%, of S-MET by 164, 330 and 786%, and of BPA by 380, 363 and 563%, respectively. Comparing the two soils, it is evident that all SOV treatments present higher values of the adsorption constants, which can be reasonably attributed to the significantly higher organic C content of SOV (37.9%) compared to SOT (9.4%). The K_F values obtained in this study for unamended and amended SOV were very similar to those reported by Jiang et al. [37] who added a sugarcane bagasse CP and a chicken manure CP at dosages of 2.5 and 5% in a Chinese soil.

The incorporation of DCP into the soil produced an increase in organic C content, which improved the sorption of each compound. It is known that OM is the main soil component controlling the retention of organic pollutants, especially hydrophobic compounds, although clay minerals can also play a significant role in the process, especially for polar compounds. However, in addition to the quantity of OM present in the soil, its origin and nature are certainly important since they determine its compositional, structural and functional characteristics. Soluble OM could even reduce or have no effects on the sorption of high-water-soluble compounds because of the competition between the two types of sorbates [43]. In soil, stable humic compounds interact with readily decomposable OM of fresh plant and animal residues and/or various organic amendments. The two OM pools have different chemical structure, composition, and accessibility. Their combination, as occurs in the case of anthropogenic OM inputs to the soil, could modify the dynamics of xenobiotics even in unexpected ways. Mitchell and Simpson [44] examined the adsorption affinity of BPA to five soils of varying OM composition and structure, and concluded that polar components of soil OM, such as O-alkyl components, can impair adsorption by blocking high-affinity sorption sites. It is a matter of fact that the adsorption capacity of soil humic substances is far superior to that of many organic amendments, including CP [25].

The possible relationships between Kd values of the compounds and the corresponding organic C percentages of all soil treatments (unamended and DCP-amended SOV and SOT) were explored using linear regressions (Figure 5). The results indicated that for each compound, the correlation was positive and highly significant (Figure 5), confirming the prominent role of soil OM in the retention of the compounds. This is consistent with the findings of other researchers who demonstrated that the adsorption of xenobiotics in soil is mainly controlled by the level of OM [45]. Ibrahim and Shalaby [46] monitored pesticide residues, including PEN, in many agricultural soils and reported positive correlation between OM content and pesticide retention in soil.

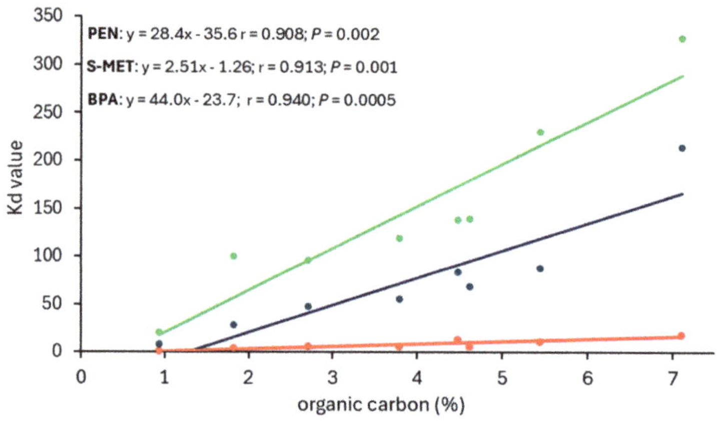

Figure 5. Relationship between distribution coefficients (Kd) of the compounds and organic carbon percentages of all soil treatments (unamended and amended SOV and SOT).

Desorption experiments started immediately after adsorption. After the three desorption cycles, slow and limited desorption was observed, indicating the occurrence of a hysteresis process (Figures 6 and 7). This phenomenon is related to the binding modes of the sorbate onto the adsorbent that, in turn, depend on the chemical groups of the adsorbent involved in the interaction. Complex macromolecules present in soil, such as humic and fulvic acids, can form both weak and strong bonds with xenobiotics. Weak binding will favour rapid release of the compounds, whereas strong binding will determine hysteretic conditions. The significant hysteresis observed in this study for all soil treatments and all molecules, especially BPA, can be attributed to the interaction of the compounds with the large number of hydrophobic and hydrophilic sites and chemically reactive functional groups—carboxylic and phenolic OH, alcoholic OH, carbonyl (quinonoid and ketonic) C=O, amino groups and so on—of native soil OM, especially the humic fraction, and of CP.

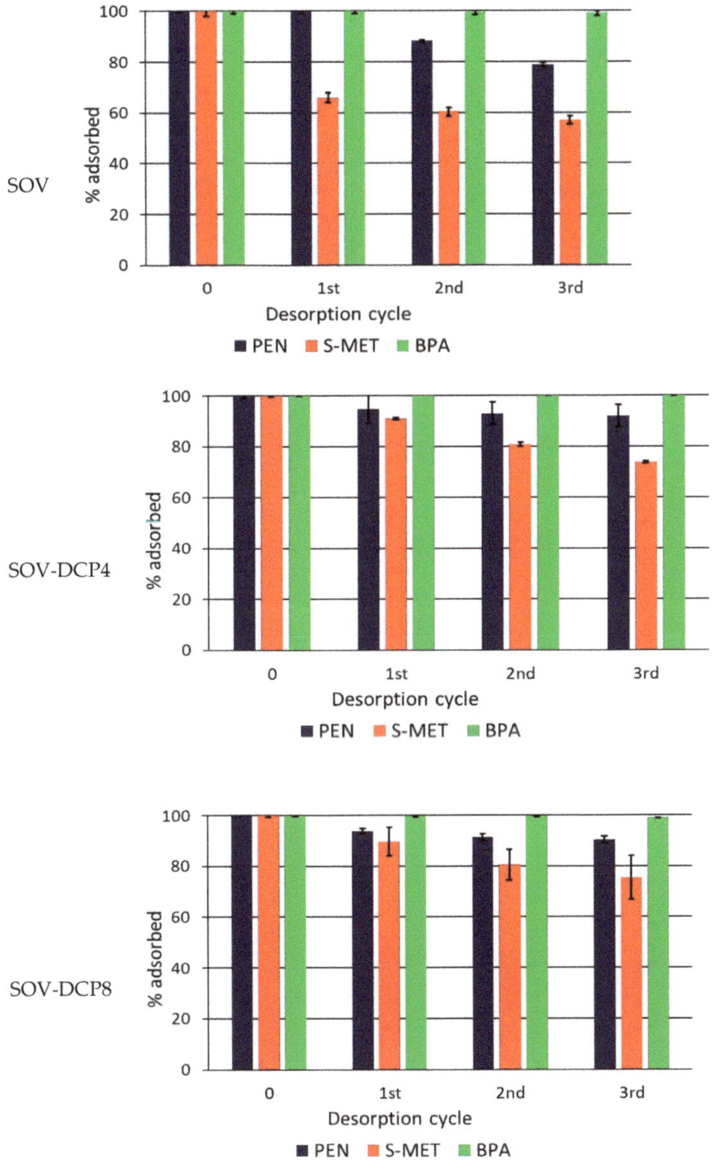

Figure 6. Percentage of compound that remains adsorbed on Valenzano soil (SOV) and SOV amended with 4 and 8% DCP after each desorption cycle. Standard error is reported as a vertical line on each bar ($n = 3$).

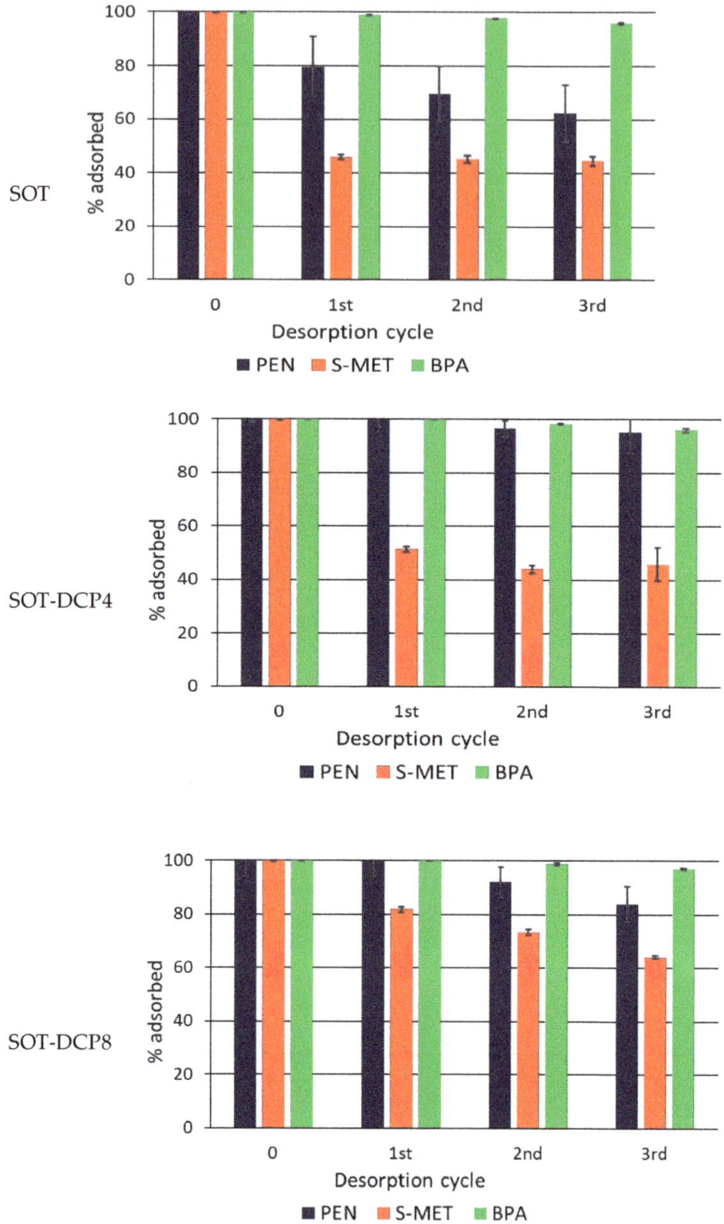

Figure 7. Percentage of compound that remains adsorbed on Trani soil (SOT) and SOT amended with 4 and 8% DCP after each desorption cycle. Standard error is reported as a vertical line on each bar (n = 3).

While the addition of DCP caused a noticeable increase in each adsorbed compound, the effects were less evident in the desorption patterns, where less evident variations were observed between amended and unamended soil. It is plausible that the strong bonds (chemisorption) formed between the compounds and native soil OM, particularly humic acids, were primarily responsible for the desorption trend, while DCP was less conditioning.

The release of PEN and, especially, of S-MET from CP-amended soil appeared generally reduced compared to the unamended soil, whereas only a negligible desorption was measured for BPA in all soil treatments (Figures 6 and 7). At the end of experiments (third dilution) the percentages of PEN, S-MET and BPA still adsorbed on SOV were 79, 57 and 99%, respectively, while they were 92, 74 and 100% on SOV-DCP4 and 90, 75 and 99% on SOV-DCP8 (Figure 6). In the case of SOT, after the third desorption cycle, the percentages of adsorbed PEN, S-MET and BPA on SOT were 62, 45 and 96%, respectively, while they were 95, 46 and 96% on SOT-DCP4 and 84, 64 and 97% on SOT-DCP8 (Figure 7).

In a previous study, Lei et al. [37] observed that the adsorption capacity of a loam soil for PEN greatly increased after the addition of two different CP, whereas the desorption of the compound was drastically reduced. The authors commented that these results were more evident as CP dosage increased. In a recent study, Bushra et al. [47] reported that the poor release of BPA from soil may be due to the entrapment of the compound in the micropores of OM or to the formation of strong bonds with the hydrophobic domain of native OM. The behaviour of BPA, characterized by significant adsorption and negligible release, would suggest a low risk for the environment from the molecule. However, as soil conditions vary, it is possible that at least part of the pollutant load will be leached into deeper soil layers, with potential risks of groundwater contamination. This scenario is particularly real and critical in agricultural areas characterized by intense rainfall or frequent irrigation which could facilitate the transfer of the contaminant into natural waters. Previous studies on the occurrence and bioaccumulation of BPA in the environment highlighted the need to carefully monitor the presence of this contaminant in cultivated soils and suggested the adoption of preventive measures, such as soil organic enrichment, to limit BPA mobility and impact on groundwater resources [48].

Considering the results obtained, it appears clear that CP amendment can significantly influence the retention/release of xenobiotics in soil, whether they are agrochemicals or organic contaminants accidentally present in the soil. The reduced release of xenobiotics in soil amended with CP can be attributed to the increased OM content, which, establishing bonds of varying strength with these compounds, hinders their desorption. This phenomenon is frequently observed in the adsorption of hydrophobic contaminants and is closely related to the affinity between the solute and soil components, especially OM. Hysteresis slows the removal of pesticides, favouring the immobilization of residues in soil even for extended periods. Although this condition can sometimes cause unexpected phytotoxic effects on sensitive plant species included in the crop rotation, it generally ensures prolonged protection for tolerant species. Furthermore, high and long-term pesticide retention in the soil helps limit xenobiotic leaching into groundwater, transport into surface water bodies, and uptake by plants and microorganisms, thus significantly reducing environmental risk. In a very recent work, Bushra et al. [47] reported that the factors influencing BPA adsorption are primarily organic C content, but also soil mineralogy, including clay type and mineral surface area. These factors can influence hydrophobic partitioning and other interactions between BPA and soil. Similar conclusions are reported in the work of Lei et al. [37] regarding the effects of a sugarcane bagasse CP and a chicken manure CP on PEN adsorption/desorption in an amended soil.

4. Conclusions

A CP obtained from aerobic treatment of a mixed biomass digestate, DCP, was evaluated for its contribution to the retention capacity of two different soils for three organic xenobiotics, namely the pesticides PEN and S-MET and the industrial contaminant BPA. DCP alone showed significant capacity to adsorb xenobiotics. The addition of 2, 4, and 8% DCP significantly increased the capacity of each soil to adsorb the compounds. The adsorption constants, calculated from equilibrium adsorption isotherm data and according to various theoretical models, were significantly higher for DCP-amended soils than for soil alone. Overall, the adsorption of each molecule on soil treatments preferentially followed the nonlinear Freundlich model, indicating multilayer adsorption on heterogeneous surfaces. A significant positive correlation was observed between the extent of adsorption of each compound and the organic carbon content of the treatments. The desorption of the three molecules from all soil samples was rather slow and incomplete, indicating a hysteresis phenomenon. Furthermore, the addition of DCP to the soil did not substantially alter the desorption pattern nor the quantitative release of the compounds. The overall results obtained support the use of DCP, which, in addition to its known effects on soil fertility, can be an effective adsorbent for soil contaminants. This role is of fundamental importance in preventing the leaching and transfer of contaminants into natural waters, as well as in limiting their entry into plants, and consequently into the human and animal food chain. Finally, although this study was the first to highlight a generally overlooked aspect of DCP, namely its ability to immobilize organic soil contaminants, it certainly suffers from the limitations of having been conducted solely on a laboratory scale, considering a limited number of contaminants and only two soils, albeit with quite different properties. At the field scale, further evaluation is required to assess the long-term stability of DCP's adsorption performance under alternating wet-dry cycles and rhizosphere compartment pulses.

Author Contributions: E.L.: conceptualization, methodology, supervision, writing, funding acquisition. E.C.: material preparation, investigation, formal analysis. C.C.: material preparation, investigation, data analysis. N.D.: material preparation, investigation, data analysis. All authors have read and agreed to the published version of the manuscript.

Funding: The present work contributes to the project PRIN 2022 PNRR entitled 'Microbially-enriched biosorbents from waste recycling and soil-resident fungi as novel and sustainable tools to mitigate soil pollution by chemicals of emerging concern and prevent their entry and accumulation in vegetables', Call 1409, dated 14 September 2022, financially supported by the European Union—NextGenerationEU—PNRR—Mission 4, Component 2, Investment 1.1. CUP: H53D23010520001. This manuscript reflects only the authors' views and opinions. Neither the European Union nor the European Commission can be considered responsible for them.

Data Availability Statement: The data presented in this study are available on request from the corresponding author.

Acknowledgments: The authors thank C&F Energy Società Agricola s.r.l., Altavilla Silentina, Italy, for providing the compost sample used in this study. Sincere thanks to the reviewers of the article for their valuable comments and suggestions.

Conflicts of Interest: The authors declare no competing interests.

References

1. Global Bioenergy Statistics (GBS). 2024. Available online: https://www.worldbioenergy.org/uploads/241023%20GBS%20Report%20Short%20Version.pdf (accessed on 15 April 2025).
2. Singh, L.; Kalia, V.C. *Waste Biomass Management—A Holistic Approach*; Springer: Cham, Switzerland, 2017; 392p. [CrossRef]
3. Adnane, I.; Taoumi, H.; Lahrech, K.; Fertahi, S.; Ghodbane, M. From waste to resource: Biogas and digestate valorization strategies for sustainable energy and agriculture. *Biomass Bioenergy* **2025**, *200*, 108006. [CrossRef]
4. Diacono, M.; Montemurro, F. Long-term effects of organic amendments on soil fertility. A review. *Agron. Sustain. Dev.* **2010**, *30*, 401–422. [CrossRef]
5. Senesi, N.; Loffredo, E. The chemistry of soil organic matter. In *Soil Physical Chemistry*, 2nd ed.; Sparks, D.L., Ed.; CRC Press: Boca Raton, FL, USA, 2018; pp. 239–370. [CrossRef]
6. Cesaro, A. The valorization of the anaerobic digestate from the organic fractions of municipal solid waste: Challenges and perspectives. *J. Environ. Manag.* **2021**, *280*, 111742. [CrossRef]

7. Wang, W.; Lee, D.-J. Valorization of anaerobic digestion digestate: A prospect review. *Bioresour. Technol.* **2021**, *323*, 124626. [CrossRef]
8. Traversa, A.; Loffredo, E.; Gattullo, C.E.; Palazzo, A.; Bashore, T.L.; Senesi, N. Comparative evaluation of compost humic acids and their effects on the germination of switchgrass (*Panicum vigatum* L.). *J. Soils Sediments* **2014**, *14*, 432–440. [CrossRef]
9. Chen, Y.; Camps-Arbestain, M.; Shen, Q.; Singh, B.; Cayuela, M.L. The long-term role of organic amendments in building soil nutrient fertility: A meta analysis and review. *Nutr. Cycl. Agroecosyst.* **2018**, *111*, 103–125. [CrossRef]
10. Palansooriya, K.N.; Dissanayake, P.D.; Igalavithana, A.D.; Tang, R.; Cai, Y.; Chang, S.X. Converting food waste into soil amendments for improving soil sustainability and crop productivity: A review. *Sci. Total Environ.* **2023**, *881*, 163311. [CrossRef]
11. Food and Agriculture Organization (FAO). Pesticides Use and Trade 1990–2022. *Food and Agriculture Organization of the United Nations.* 2022. Available online: https://www.fao.org/statistics/highlights-archive/highlights-detail/pesticides-use-and-trade-1990-2022/en (accessed on 20 April 2025).
12. Khan, S.; Naushad, M.; Govarthanan, M.; Iqbal, J.; Alfadul, S.M. Emerging contaminants of high concern for the environment: Current trends and future research. *Environ. Res.* **2022**, *207*, 112609. [CrossRef] [PubMed]
13. Song, C.; Zeng, C.; Qin, T.; Lv, T.; Xu, Z.; Xun, Z.; Wang, L.; Chen, X.; Liu, B.; Peng, X. A dual-state-emission chalcone-based supramolecular probe for ratiometric detection of penconazole in environmental samples. *Chem. Eng. J.* **2023**, *468*, 143610. [CrossRef]
14. Mercadante, R.; Polledri, E.; Scurati, S.; Moretto, A.; Fustinoni, S. Identification of Metabolites of the Fungicide Penconazole in HumanUrine. *Chem. Res. Toxicol.* **2016**, *29*, 1179–1186. [CrossRef]
15. Perdichizzi, S.; Mascolo, M.S.; Silingardi, P.; Morandi, E.; Rotondo, F.; Guerrini, A.; Prete, L.; Vaccari, M.; Colacci, A. Cancer-related genes transcriptionally induced by the fungicide penconazole. *Toxicol. Vitr.* **2014**, *28*, 125–130. [CrossRef] [PubMed]
16. Kouame, K.B.J.; Savin, M.C.; Willett, C.D.; Bertucci, M.B.; Butts, T.R.; Grantz, E.; Roma-Burgos, N. S-metolachlor persistence in soil as influenced by within-season and inter-annual herbicide use. *Environ. Adv.* **2022**, *9*, 100318. [CrossRef]
17. ISPRA. Rapporto Nazionale Pesticidi Nelle Acque. Istituto Superiore per la Protezione e la Ricerca Ambientale. 2023. Available online: https://www.isprambiente.gov.it/files2022/pubblicazioni/rapporti/rapporto_371_2022.pdf (accessed on 10 April 2025).
18. Rangani, G.; Noguera, M.; Salas-Perez, R.; Benedetti, L.; Roma-Burgos, N. Mechanism of Resistance to S-metolachlor in Palmer amaranth. *Front. Plant Sci.* **2021**, *12*, 652581. [CrossRef]
19. European Chemical Agency (ECHA). 2024. Available online: https://echa.europa.eu/search?p_p_id=com_liferay_portal_search_web_portlet_SearchPortlet&p_p_lifecycle=0&p_p_state=maximized&p_p_mode=view&_com_liferay_portal_search_web_portlet_SearchPortlet_mvcPath=%2Fsearch.jsp&_com_liferay_portal_search_web_portlet_SearchPortlet_redirect=%2Fweb%2Fguest%2Fsearch%3Fp_p_id%3Dcom_liferay_portal_search_web_portlet_SearchPortlet%26p_p_lifecycle%3D0%26p_p_state%3Dnormal%26p_p_mode%3Dview&_com_liferay_portal_search_web_portlet_SearchPortlet_scope=this-site&p_auth= (accessed on 20 August 2025).
20. Ou-Yang, K.; Feng, T.; Han, Y.; Li, G.; Li, J.; Ma, H. Bioaccumulation, metabolism and endocrine-reproductive effects of metolachlor and its S-enantiomer in adult zebrafish (Danio rerio). *Sci. Total Environ.* **2022**, *802*, 149826. [CrossRef]
21. Pelch, K.E.; Li, Y.; Perera, L.; Thayer, K.A.; Korach, K.S. Characterization of Estrogenic and Androgenic Activities for Bisphenol A-like Chemicals (BPs): In Vitro Estrogen and Androgen Receptors Transcriptional Activation, Gene Regulation, and Binding Profiles. *Toxicol. Sci.* **2019**, *172*, 23–37. [CrossRef]
22. Manzoor, M.F.; Tariq, T.; Fatima, B.; Sahar, A.; Tariq, F.; Munir, S.; Khan, S.; Ranjha, M.M.A.N.; Sameen, A.; Zeng, X.-A.; et al. An insight into bisphenol A, food exposure and its adverse effects on health: A review. *Front. Nutr.* **2022**, *9*, 1047827. [CrossRef]
23. Metcalfe, C.; Bayen, S.; Desrosiers, M.; Muñoz, G.; Sauvé, S.; Yargeau, V. An introduction to the sources, fate, occurrence and effects of endocrine disrupting chemicals released into the environment. *Environ Res.* **2022**, *207*, 112658. [CrossRef]
24. Loffredo, E. Recent advances on innovative materials from biowaste recycling for the removal of environmental estrogens from water and soil. *Materials* **2022**, *15*, 1894. [CrossRef]
25. Senesi, N.; Loffredo, E.; D'Orazio, V.; Brunetti, G.; Miano, T.M.; La Cava, P. Adsorption of pesticides by humic acids from organic amendments and soils. In *Humic Substances and Chemical Contaminants*; Clapp, C.E., Hayes, M.H.B., Senesi, N., Bloom, P.R., Jardine, P.M., Eds.; ASA, CSSA, SSSA Books: Chichester, UK, 2015; pp. 129–153. [CrossRef]
26. Fouad, M.R.; El-Aswad, A.F.; Badawy, M.E.I.; Aly, M.I. Effect of soil organic amendments on sorption behavior of two insecticides and two herbicides. *Curr. Chem. Lett.* **2024**, *13*, 377–390. [CrossRef]
27. Gamiz, B.; Pignatello, J.J.; Cox, L.; Hermosín, M.C.; Celis, R. Environmental fate of the fungicide metalaxyl in soil amended with composted olive-mill waste and its biochar: An enantioselective study. *Sci. Total Environ.* **2016**, *54*, 776–783. [CrossRef] [PubMed]
28. Zanin Lima, J.; Monici Raimondi Nauerth, I.; Ferreira da Silva, E.; José Pejon, O.; Guimarães Silvestre Rodrigues, V. Competitive sorption and desorption of cadmium, lead, and zinc onto peat, compost, and biochar. *J. Environ. Manag.* **2023**, *344*, 118515. [CrossRef] [PubMed]
29. Loffredo, E.; Picca, G.; Parlavecchia, M. Single and combined use of *Cannabis sativa* L. and carbon-rich materials for the removal of pesticides and endocrine-disrupting chemicals from water and soil. *Environ. Sci. Pollut. Res.* **2021**, *28*, 3601–3616. [CrossRef]

30. Caracciolo, A.B.; Bustamante, M.A.; Nogues, I.; Di Lenola, M.; Luprano, M.L.; Grenni, P. Changes in microbial community structure and functioning of a semiarid soil due to the use of anaerobic digestate derived composts and rosemary plants. *Geoderma* **2015**, *245–246*, 89–97. [CrossRef]
31. Vitti, A.; Elshafie, H.S.; Logozzo, G.; Marzario, S.; Scopa, A.; Camele, I.; Nuzzaci, M. Physico-chemical characterization and biological activities of a digestate and a more stabilized digestate-derived compost from agro-waste. *Plants* **2021**, *10*, 389. [CrossRef] [PubMed]
32. Nogués, I.; Rumpel, C.; Sebilo, M.; Vaury, V.; Moral, R.; Bustamante, M.A. Stable C and N isotope variation during anaerobic digestate composting and in the compost-amended soil-plant system. *J. Environ. Manag.* **2023**, *329*, 117063. [CrossRef]
33. ChemSpider. Available online: https://www.chemspider.com (accessed on 5 April 2025).
34. Loffredo, E.; Vona, D.; Porfido, C.; Giangregorio, M.M.; Gelsomino, A. Compositional and structural characterization of bioenergy digestate and its aerobic derivatives compost and vermicompost. *J. Sustain. Agric. Environ.* **2024**, *3*, e70002. [CrossRef]
35. Colatorti, N.; Digregorio, N.V.; Camposeo, S.; Loffredo, E. Solid fraction of digestate from olive pomace modulates abiotic and biotic processes in soil: Retention of agrochemicals and inhibition of fungal pathogens. *Sci. Hortic.* **2024**, *337*, 113545. [CrossRef]
36. Gazzetta Ufficiale della Repubblica Italiana—GU Serie Generale n.248 del 21-10-1999—Suppl. Ordinario n. 185. Available online: https://www.gazzettaufficiale.it/eli/id/1999/10/21/099A8497/sg (accessed on 20 April 2025).
37. Jiang, L.; Lin, J.L.; Jia, L.X.; Liu, Y.; Pan, B.; Yang, Y.; Lin, Y. Effects of two different organic amendments addition to soil on sorption-desorption, leaching, bioavailability of penconazole and the growth of wheat (*Triticum aestivum* L.). *J. Environ. Manag.* **2016**, *167*, 130–138. [CrossRef]
38. Douibi, M.; Rodríguez-Cruz, M.S.; Sánchez-Martín, M.J.; Marín-Benito, J.M. Sustainable agricultural practices influence s-metolachlor, foramsulfuron and thiencarbazone-methyl degradation and their metabolites formation. *Sci. Total. Environ.* **2024**, *945*, 174039. [CrossRef]
39. Freundlich, H.M.F. Uber die adsorption in losungen. *J. Phys. Chem.* **1906**, *57*, 385–470. [CrossRef]
40. Langmuir, I. The adsorption of gases on plane surface of glass, mica and platinum. *J. Am. Chem. Soc.* **1918**, *40*, 1361–1403. [CrossRef]
41. Hamdaoui, O.; Naffrechoux, E. Modeling of adsorption isotherms of phenol and chlorophenols onto granular activated carbon: Part I. Two-parameter models and equations allowing determination of thermodynamic parameters. *J. Hazard. Mater.* **2007**, *147*, 381–394. [CrossRef] [PubMed]
42. Prasannamedha, G.; Senthil Kumar, P.; Mehala, R.; Sharumitha, T.J.; Surendhar, D. Enhanced adsorptive removal of sulfamethoxazole from water using biochar derived from hydrothermal carbonization of sugarcane bagasse. *J. Hazard. Mater.* **2021**, *407*, 124825. [CrossRef] [PubMed]
43. Fernandes, M.C.; Cox, L.; Hermosín, M.C.; Cornejo, J. Organic amendments affecting sorption, leaching and dissipation of fungicides in soils. *Pest Manag. Sci.* **2006**, *62*, 1207–1215. [CrossRef]
44. Mitchell, P.J.; Simpson, M.J. High affinity sorption domains in soil are blocked by polar soil organic matter components. *Environ. Sci. Technol.* **2013**, *47*, 412–419. [CrossRef]
45. Parlavecchia, M.; D'orazio, V.; Loffredo, E. Wood biochars and vermicomposts from digestate modulate the extent of adsorption-desorption of the fungicide metalaxyl-m in a silty soil. *Environ. Sci. Pollut. Res.* **2019**, *26*, 35924–35934. [CrossRef]
46. Ibrahim, E.A.; Shalaby, S.E.M. Screening and assessing of pesticide residues and their health risks in vegetable field soils from the Eastern Nile Delta, Egypt. *Toxicol. Rep.* **2022**, *9*, 1281–1290. [CrossRef]
47. Bushra, K.; Javaid, I.; Shazia, M.; Muhammad, N.A.; Aitezaz, A.K.; Farwa, J.; Tariq, A.; Saleh, A.A.; Abdulhakeem, S.A.; Majid, A. Sorption and desorption of bisphenol A on agricultural soils and its implications for surface and groundwater contamination. *Desalin. Water Treat.* **2025**, *322*, 101180. [CrossRef]
48. Corrales, J.; Kristofco, L.A.; Steele, W.B.; Yates, B.S.; Breed, C.S.; Williams, E.S.; Brooks, B.W. Global Assessment of Bisphenol A in the Environment: Review and Analysis of Its Occurrence and Bioaccumulation. *Dose-Response* **2015**, *13*, 1559325815598308. [CrossRef]

agriculture

MDPI

Review

Harnessing Microalgae and Cyanobacteria for Sustainable Agriculture: Mechanistic Insights and Applications as Biostimulants, Biofertilizers and Biocontrol Agents

Ana Jurado-Flores [1,†], Luis G. Heredia-Martínez [2,†], Gloria Torres-Cortes [3,*] and Encarnación Díaz-Santos [4,*]

1 Instituto de Bioquímica Vegetal y Fotosíntesis (IBVF/CSIC), cicCartuja, Calle Américo Vespucio, 49, 41092 Sevilla, Spain; ana.jurado@ibvf.csic.es
2 Institut de Biologie Physico-Chimique (IBPC), CNRS, 75005 Paris, France; lheredia@us.es
3 Innoplant S.L., Calle Reina Sofía 66, Viznar, 18179 Granada, Spain
4 Laboratory of Biochemistry, Faculty of Experimental Sciences, Marine International Campus of Excellence and REMSMA, University of Huelva, 21071 Huelva, Spain
* Correspondence: gloriatorresco@gmail.com (G.T.-C.); ediazsantos25@gmail.com or ediaz6@us.es (E.D.-S.)
† These authors contributed equally to this work.

Abstract: The prolonged and intensive use of chemical inputs in agriculture, particularly synthetic fertilizers, has generated a variety of environmental and agronomic challenges. This has intensified the need for alternative, viable, and sustainable solutions. Plant-associated microbes have emerged as promising candidates in this regard. While research has largely focused on bacteria and fungi, comparatively less attention has been paid to other microbial groups such as microalgae and cyanobacteria. These photosynthetic microorganisms offer multiple agronomic benefits, including the ability to capture CO_2, assimilate essential micro- and macroelements, and synthesize a wide range of high-value metabolites. Their metabolic versatility enables the production of bioactive molecules with biostimulant and biocontrol properties, as well as biofertilizer potential through their intrinsic nutrient content. Additionally, several cyanobacterial species can fix atmospheric nitrogen, further enhancing their agricultural relevance. This review aims to summarize the potential of these microorganisms and their application in the agriculture sector, focusing primarily on their biofertilization, biostimulation, and biocontrol capabilities and presents a compilation of the products currently available on the market that are derived from these microorganisms. The present work also identifies the gaps in the use of these microorganisms and provides prospects for developing a suitable solution for today's agriculture.

Keywords: microalgae; cyanobacteria; biofertilizer; biostimulant; biocontrol agents; sustainable agriculture; circular bioeconomy; plant growth-promoting microorganisms

1. Introduction

The rapid growth of the global population underscores an urgent need for sustainable solutions to ensure future food security, with FAO projections indicating that global food production must increase by 60% by 2050 [1]. Achieving this target requires the expansion of agriculture within a circular economy framework, where resource efficiency and environmental stewardship are prioritized [2]. However, the overexploitation of land, inefficient irrigation practices, and excessive use of chemical fertilizers and pesticides have degraded soil health, reduced biodiversity, and increased greenhouse gas emissions. Ultimately, these factors undermine the very crop productivity they were meant to support [3–5]. Soil health (defined by its physical, chemical, and biological properties) is fundamental to sustainable agriculture. It supports plant growth, drives nutrient cycling, and strengthens resilience against pests, diseases, and climate change [6]. Traditional practices such as crop rotation, reduced tillage, organic fertilization, and cover cropping have long been used to maintain soil quality. However, these measures alone are not always sufficient. Innovative, nature-based microbial solutions are now emerging. These microorganisms can help restore soil structure, improve nutrient availability, and stimulate beneficial microbial activity, offering new tools for sustainable soil management. These approaches not only boost yields but also align with regenerative and low-impact farming methods essential for long-term food security and environmental sustainability [7,8]. Prioritizing soil health is essential for ensuring food security, conserving biodiversity, and promoting environmentally sustainable farming practices [6].

Plant and soil-associated microbial communities are fundamental to sustainable agriculture, as they support plant growth, drive nutrient cycling, and enhance resilience against pests, diseases, and climate change [9,10]. Although bacterial and fungal communities have been extensively studied over the past decades for these purposes, much less is known about other important microorganisms such as microalgae and cyanobacteria. These microscopic photosynthetic organisms inhabit soil, freshwater, and marine ecosystems and represent one of the most ancient and metabolically diverse groups of microorganisms on Earth. They possess an extraordinary capacity to convert carbon dioxide, sunlight, and inorganic nutrients into a wide array of valuable biomolecules, making them highly relevant candidates for agricultural innovation [11,12]. Thus, microalgae and cyanobacteria have been extensively studied for applications ranging from biofuels to pharmaceuticals [13–15]. Their high photosynthetic efficiency, fast growth rates, and ability to thrive in non-arable land and wastewater environments distinguish them from terrestrial crops and make them ideal candidates for circular bioeconomy models [16,17].

In recent years, microalgae and cyanobacteria have attracted growing interest also in the agricultural sector due to their multifunctional roles in enhancing plant productivity, improving soil health, and contributing to climate change mitigation through CO_2 sequestration [15,18]. Reflecting this potential, the global agricultural biologicals market is projected to reach $14.6 billion by 2025, with algal and microbial biostimulants among the fastest-growing research areas [19]. In agriculture, a biostimulant is defined as a substance or microorganism that, when applied in small amounts, stimulates natural processes to enhance nutrient uptake efficiency, tolerance to abiotic stress, and/or crop quality, regardless of its nutrient content [20]. Biostimulants primarily act by improving plant physiological responses, enabling better growth and resilience under stress. The biochemical versatility of microalgae and cyanobacteria makes them particularly effective in this role [21] as they produce phytohormones, amino acids, polysaccharides, and antioxidants that boost photosynthetic performance, enhance antioxidant defenses, and promote the accumulation of osmoprotectants, especially under stress conditions such as drought and salinity [22]. Recent studies, including Brito-Lopez et al., 2025 [5] have confirmed their capacity to enhance plant growth and resilience across diverse environments, reinforcing their potential as climate-smart agricultural inputs. Compared to other commonly used biostimulants such as plant extracts, humic substances, or microbial inoculants based on bacteria and fungi, microalgae and cyanobacteria offer unique advantages. Their capacity to synthesize a wide range of bioactive metabolites, their high metabolic plasticity, and, in the case of some cyanobacteria, their ability to fix atmospheric nitrogen, enable them to function simultaneously as biofertilizers, biostimulants, and biocontrol agents [11].

The multifunctionality is particularly relevant given one of the central challenges in modern agriculture: the overreliance on synthetic nitrogen-based fertilizers and chemical

NPK inputs. While these compounds were key to the Green Revolution and have contributed significantly to global food security, their intensive and prolonged use has led to serious environmental and health consequences, including nitrate leaching, eutrophication of aquatic ecosystems, nitrous oxide emissions, and the decline of soil biodiversity [23]. These drawbacks highlight the urgent need for sustainable and environmentally compatible alternatives. Microalgae and cyanobacteria address this need by acting as biofertilizers that colonize the plant rhizosphere or tissues and enhance nutrient uptake [24–26]. Biofertilizers are natural preparations containing living microorganisms or other nat-ural substances that, when applied to seeds, plant surfaces, or soil, promote plant growth by increasing the supply or availability of essential nutrients [27]. Their benefits include nutrient enrichment (e.g., nitrogen, phosphorus, potassium, and trace elements), production of plant growth–promoting substances (auxins, gibberellins, cytokinins), stimulation of beneficial soil microbial communities, and improved tolerance to both biotic and abiotic stresses [22]. Nitrogen-fixing cyanobacteria provide a compelling alternative to synthetic nitrogen fertilizers by mitigating the negative environmental impacts of conventional practices while maintaining or even enhancing soil fertility and crop productivity [28]. As the world faces the dual pressures of climate change and food security, harnessing the potential of microalgae and cyanobacteria represents a strategic pathway toward more sustainable and resilient food systems.

Building on this promise, microalgae and cyanobacteria offer specific practical solutions that further underscore their value in sustainable agriculture. Not only can they fix atmospheric carbon and accumulate nitrogen and phosphorus from various sources (including wastewater), but their application to soil can also release nutrients in bioavailable forms, reducing the risk of leaching and volatilization [29]. Furthermore, microalgal and cyanobacteria biomass can be integrated into regenerative agricultural systems, supporting nutrient recycling, organic matter enrichment, and improved soil structure [30–32]. Certain species such as *Chlorella sorokiniana*, *Scenedesmus obliquus*, and *Spirulina maxima* have already shown considerable promise as biofertilizers in field trials, with positive impacts on yield, root morphology, and chlorophyll content in a variety of crops including wheat, maize, tomatoes, and rice [5,26,30]. These benefits are further supported by recent findings showing significant reductions in synthetic fertilizer input without compromising crop yield [33].

Beyond their roles as biofertilizers and biostimulants, microalgae and cyanobacteria also exhibit significant potential as biopesticides. Many species produce a wide range of secondary metabolites (such as alkaloids, phenolics, peptides, and fatty acids) with antimicrobial, antifungal, and antiviral properties. These compounds can inhibit the germination of fungal spores, suppress the growth of phytopathogenic bacteria, and reduce the spread of viral diseases in crops [34]. Certain cyanobacteria additionally produce allelopathic compounds that interfere with pest development and reproduction, providing natural crop protection [35]. Overall, integrating microalgae- and cyanobacteria-based biopesticides into pest management programs offers a sustainable alternative for reducing chemical pesticide dependence and enhancing agroecosystem resilience.

Moreover, from a sustainability perspective, cultivating microalgae for agricultural applications supports circular economy and zero-waste approaches strategies [36]. Algal biomass can be produced using agricultural runoff, animal slurry, or even urban wastewater, thereby reducing waste streams and recovering nutrients that might otherwise cause pollution [31,32]. Additionally, in comparison to synthetic fertilizers, microalgal products can be produced with lower energy inputs and have significantly reduced carbon footprints when incorporated into biorefinery systems [37]. Despite these advantages, significant challenges remain for the large-scale implementation of microorganisms in agriculture, such as economic feasibility, the scalability of production systems, and the need for standardized protocols and regulatory frameworks. However, ongoing advances in biotechnology, strain selection, photobioreactor design, and downstream processing are steadily improving the viability of microalgae-based agricultural products [38].

In summary, microalgae and cyanobacteria are emerging as powerful, nature-based solutions for advancing sustainable agriculture. These photosynthetic microorganisms possess the ability to produce a wide range of bioactive compounds that promote plant growth, enhance nutrient use efficiency, and contribute to soil health through nutrient cycling [15,18]. Their multifunctional roles as biofertilizers, biostimulants, and biocontrol agents make them attractive candidates for reducing dependency on synthetic agrochemicals, offering a path toward reduced environmental impact and improved agricultural resilience. Nevertheless, several challenges remain, particularly regarding cost-effective large-scale production and formulation. In this review, we address these gaps through a comprehensive and in-depth bibliographic search, synthesizing current scientific knowledge, compiling existing commercial products, and outlining future research priorities. By doing so, we provide a mechanistic and application-oriented framework intended to accelerate the integration of microalgae- and cyanobacteria-based solutions into sustainable agricultural practices. An overview of the present revision article is illustrated in Figure 1.

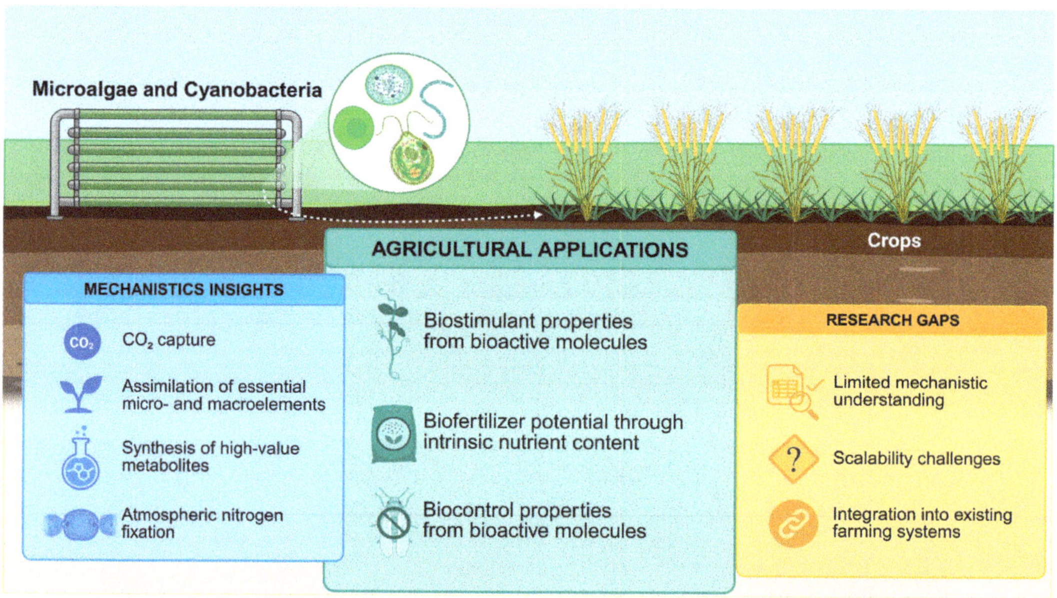

Figure 1. Overview of the present review. The figure illustrates the central agricultural applications of microalgae and cyanobacteria (as biofertilizers, biostimulants, and biocontrol agents), together with mechanistic insights (CO_2 capture, nutrient assimilation, metabolite synthesis, and nitrogen fixation) and the main research gaps (limited mechanistic understanding, scalability challenges, and integration into farming systems). Created in BioRender. Diaz Santos, E. (2025) https://BioRender.com/3xt9h1a, accessed on 19 March 2025.

2. The Use of Microalgae and Cyanobacteria as Biostimulants

Plant biostimulants are defined as substances or microorganisms that, when applied in small amounts, enhance plant nutrition processes, stress tolerance, growth, or quality, independently of direct nutrient supply [20]. Algae are recognized as some of the earliest plant growth promotion compounds used in agriculture, with historical records dating back to the Roman era, where they were applied as manure to enrich soil fertility and promote plant growth, an application that has continued through the centuries. Thanks to their remarkable diversity and metabolic flexibility, microalgae and cyanobacteria serve as a rich source of high-value metabolites, including proteins, amino acids, enzymes, pigments, polyunsaturated fatty acids, polysaccharides, vitamins, antioxidants, and phytohormones [17,39]. This biochemical richness explains why these photosynthetic organisms are gaining increasing attention as sustainable inputs in agriculture as biostimulants [40]. Their demonstrated growth-promoting properties and environmentally friendly profile have contributed to their rising use particularly among growers seeking alternatives to synthetic agrochemicals. In the following section key factors in the development of microalgae- and cyanobacteria-based biostimulants products will be explored.

2.1. Cyanobacteria and Microalgae Strain Selection

Although microalgae and cyanobacteria hold great promise for agricultural applications, the selection of the most effective and context-appropriate strains remains a challenging task. This process requires access to well-characterized strains with a demonstrated high capacity for producing target bioactive compounds of agricultural relevance, while also being amenable to cost-effective, large-scale cultivation. Additionally, strains should exhibit adaptability to variable environmental conditions, stable productivity over time, and compatibility with existing agricultural practices to ensure practical implementation and long-term sustainability. Thus, among the key criteria for strain selection, two are particularly fundamental in determining whether a strain is viable for further development. The first is the strain's ability to grow rapidly and uniformly while maintaining high productivity (preferably in nutrient rich media or, in the best case, using waste-derived resources) to ensure the production of sufficient biomass. In this way, the strains must be easily cultivated to generate sufficient biomass [41,42]. Species such as *Chlorella vulgaris* and *Scenedesmus* spp. are well known for their fast growth rates and short doubling time [43]. The second essential criterion is the strain's physiological and biochemical potential, particularly its capacity to produce bioactive compounds such as exopolysaccharides, amino acids, proteins, vitamins, and phytohormone-like substances (e.g., auxins and cytokinins). In practice, an effective decision-making matrix for strain selection should integrate both production performance metrics (such as growth rate, biomass yield, and cultivation efficiency) and detailed biochemical profiles, including the concentration and diversity of target bioactive compounds. This integrated approach enables the systematic comparison of candidate strains, ensuring that the final selection balances productivity, functional quality, and suitability for large-scale, cost-effective biostimulant development (Figure 2). Based on these parameters, cyanobacteria and green algae currently represent the widely used groups in biostimulant applications and will be further discussed. In contrast, the application of diatoms in this context is still in its early stages of development.

2.2. Modes of Action and Mechanisms of Microalgae- and Cyanobacteria-Derived Biostimulants

Cyanobacterial and microalgal cellular extracts and hydrolysates are widely recognized as effective biostimulants capable of enhancing plant growth and increasing crop yields [44]. Their activity results from a combination of biochemical, physiological, and ecological mechanisms that collectively improve plant performance. Rather than acting through a single pathway, these microorganisms influence a network of processes that boost growth, optimize nutrient use efficiency, and strengthen resilience against environmental stresses. Application methods vary according to crop requirements and cultivation practices. Common approaches include incorporating live or dried biomass into soil, priming seeds with microalgal extracts, and applying root drenches. The choice of method is often determined by whether the crop is directly sown or first cultivated in a nursery before transplantation to the field [45].

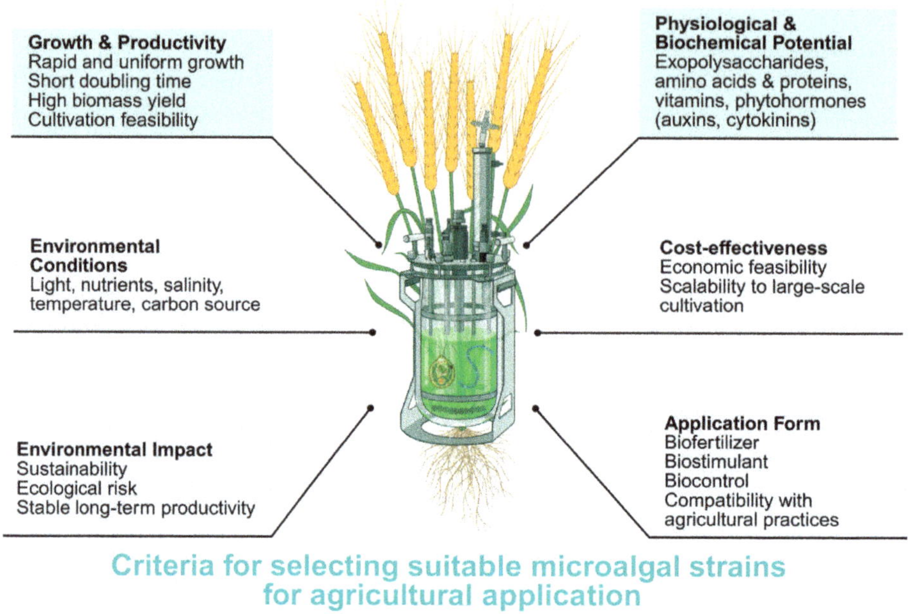

Figure 2. Selection criteria for the most suitable microalgal strain for agricultural application. The choice of an appropriate strain requires an integrated evaluation of multiple factors. Fundamental criteria include growth and productivity (growth rate, biomass yield, and cultivation feasibility) and physiological/biochemical potential (capacity to produce exopolysaccharides, amino acids, proteins, vitamins, and phytohormone-like substances). Additional factors such as adaptability to environmental conditions (light, nutrients, salinity, temperature), cost-effectiveness, application form (biofertilizer, biostimulant, biocontrol), and environmental impact (sustainability, ecological risks) further refine the decision-making process. Together, these criteria ensure the selection of strains that are not only productive and functionally relevant but also scalable, cost-effective, and sustainable for agricultural applications. Created in BioRender. Diaz Santos, E. (2025) https://BioRender.com/kt10r85, accessed on 19 March 2025.

The mechanisms (here defined as the specific biochemical, molecular, or physiological processes responsible for the observed effects) underlying the action of biostimulants in plants are still not fully elucidated. Their complexity makes them challenging to study, especially in the case of microalgae and cyanobacteria, whose extracts contain a wide variety of bioactive compounds. This diversity complicates the identification of the individual constituents responsible for specific effects. Nonetheless, recent advances in analytical and omics technologies, including next-generation sequencing (e.g., transcriptomics), deep phenotyping, and metabolomics, provide valuable insights into these intricate interactions [46].

The biostimulant effects of microalgae and cyanobacteria emerge from multi-layered interactions between their bioactive components and plant physiological and biochemical pathways. Thus, their modes of action (the overall functional outcomes through which these organisms exert their beneficial effects) are diverse, but can be broadly categorized as follows:

(1) Enhancement of Nutrient Uptake and Assimilation

Although biostimulants are not primarily nutrient suppliers (a role fulfilled by biofertilizers, discussed in the next section), microalgae and cyanobacteria contain compounds that facilitate nutrient acquisition. Organic acids, chelating agents, and certain polysaccharides can mobilize micronutrients such as iron, zinc, and manganese in the rhizosphere [47]. Additionally, phytohormone-like substances (e.g., auxin- and cytokinin-like compounds) stimulate root architecture, increasing the absorptive surface area and indirectly improving nutrient capture.

(2) Modulation of Plant Hormonal Balance

Bioactive molecules (including indole-3-acetic acid, gibberellin analogs, and brassinosteroid-like compounds) can modify the plant's endogenous hormone profile [48–50]. This modulation influences key developmental processes such as cell elongation, differentiation, and flowering, thereby improving plant vigor and productivity [51]. Certain hormones (particularly auxins, cytokinins, and ethylene) modulate root system architecture, increasing root surface area, branching, and exudation. These changes enhance the plant's capacity to explore the rhizosphere. In parallel, hormone-induced root exudates can stimulate beneficial soil microbiota, including nitrogen-fixing and phosphate-solubilizing microorganisms, creating a feedback loop that improves soil fertility. For instance, inoculation with *Calothrix elenkinii* stimulated the plant microbiome [52]. Additionally, phytohormones can upregulate the expression of nutrient transporter genes in roots, further facilitating the uptake of macro- and micronutrients [50]. Recent studies have shed light on brassinosteroid role in modulating agronomic traits that directly contribute to grain yield in rice (*Oryza sativa*) [53].

(3) Activation of Stress-Response Pathways

Polysaccharides, proteins, and secondary metabolites in microalgal and cyanobacterial extracts can activate systemic tolerance mechanisms [5,54]. These include the upregulation of antioxidant enzymes (e.g., superoxide dismutase, catalase, and peroxidases) and the accumulation of osmoprotectants (e.g., proline, glycine betaine), which reduce oxidative damage and help maintain cellular homeostasis under drought, salinity, or extreme temperatures [54].

(4) Improvement of Soil Microbial Communities

When applied to soil, whole biomass or extracts can serve as carbon and energy sources for beneficial microorganisms, fostering a more diverse and balanced microbiome for example, *Chlorella fusca* strawberry growth promotion correlates with changes in the plant microbiota, particularly the abundance of beneficial bacteria in the rhizosphere [55]. These bacteria can enhance nutrient cycling, particularly phosphate solubility, contributing to improved plant health, soil structure and stability.

2.3. *Cyanobacteria and Microalgae as Biostimulants in Agriculture*

Cyanobacteria are gaining attention in agriculture thanks to their ability to fix atmospheric nitrogen, solubilize phosphorus, and produce a wide variety of beneficial compounds (like phytohormones) that support plant health. Their unique physiological traits and metabolic flexibility have positioned them as promising biostimulants in crop production systems [56]. Certain genera, including *Anabaenopsis*, *Calothrix*, and *Anabaena*, have demonstrated the capacity to enhance seed germination and promote plant development through the synthesis of phytohormones like cytokinins, gibberellins, and auxins [57–59].

Beyond promoting plant growth, cyanobacteria also help plants cope with stress. For example, they produce extracellular polysaccharides, antioxidants, and signaling molecules that help mitigate the damaging effects of salinity [25]. Under drought conditions, their benefits extend further: cyanobacteria can regulate ion export, modulate the surrounding microbial community, and even improve germination rates under limited water availability. These stress-alleviating traits have been documented across a wide range of crops, as reviewed by Sánchez-Quintero et al., 2023 [60]. Alongside their direct effects on plants, cyanobacteria also play a crucial role in soil health. In addition to acting as biofertilizers (enhancing nitrogen fixation and phosphorus availability, as discussed in the next section), they foster symbiotic relationships with other soil microorganisms, thereby reinforcing their value in sustainable and regenerative agriculture [2,61].

Moreover, unlike seaweed, which are typically collected from the wild, cyanobacteria can be cultivated in controlled environments like open ponds or photobioreactors. This allows for more consistent quality and optimized conditions for producing bioactive compounds [56]. Taken together, the contributions of cyanobacteria to sustainable agriculture are multifaceted. They enhance soil structure by secreting polysaccharides, improve fertility through nitrogen fixation, increasing porosity and water retention, and release growth-promoting substances. Moreover, they aid in reducing salinity, making phosphorus more bioavailable, and facilitating the recycling of agricultural residues. Whether applied as live cell suspensions or processed extracts, cyanobacteria consistently demonstrate significant potential to enhance plant vitality, productivity, and resilience.

Green microalgae have demonstrated a wide range of beneficial effects on crop growth and are increasingly recognized as promising inputs for sustainable agriculture. Microalgae-based products can enhance plant nutrition, improve overall crop performance, support key physiological processes, and increase tolerance to abiotic stresses such as drought and salinity [62].

Among the various genera studied, *Chlorella* stands out as the most used in agricultural applications [63,64]. Numerous trials using both fresh biomass and extracts from *Chlorella* strains have reported notable improvements in plant development and yield [65,66]. For instance, treatments with *Chlorella fusca* have been shown to enhance the growth of Chinese chives under field conditions [67]. In another example, soil-drenching with live *Chlorella* cells significantly increased biomass production in *Medicago truncatula* [66]. Complementary findings have been reported for other *Chlorella* species as well. *C. vulgaris* and *C. pyrenoidosa*, for example, have proven effective as biostimulants in salt-affected soils, supporting the growth of crops such as lettuce, rice, eggplant, and cucumber under saline stress [68]. Likewise, applications of *Spirulina platensis* have demonstrated growth-promoting effects on leafy vegetables like rocket, red bayam, and pak choi [69], further reinforcing the versatility of microalgae across diverse crop types. A key mechanism by which microalgae enhance plant performance is through the production and secretion of phytohormones, including auxins and cytokinins. In addition, they release exopolysaccharides that contribute to drought and salinity tolerance by improving soil water retention and nutrient availability. These compounds also supply organic carbon to beneficial soil microbes, fostering microbial activity and nutrient cycling. Expanding the range of beneficial species, inoculation with filamentous cyanobacteria such as *Calothrix elenkinii* has been found to stimulate both the phyllosphere and rhizosphere microbiomes, further supporting plant health and productivity [70]. While research continues to identify and characterize the most effective microalgae strains, current evidence clearly underscores the potential of green microalgae as valuable biostimulants. Their capacity to enhance plant growth, improve resilience to environmental stresses, and promote beneficial soil microbial communities makes them a sustainable and impactful tool for advancing modern agriculture.

2.4. *Diatoms as Biostimulants in Agriculture*

Diatoms are an incredibly diverse group of microorganisms, with around 100,000 known species. They are easy and cost-effective to cultivate, which makes them an attractive source of bioactive compounds, many of which are already used in the pharmaceutical industry [71]. Although their application as biostimulants in agriculture is still limited, their unique structural and biochemical characteristics offer great potential. Studies have shown that diatoms, particularly *Navicula* species, can positively influence plant growth in species such as *Salix viminalis*, *Helianthus tuberosus*, and *Sida hermaphrodita* [72]. One of the most distinctive features of diatoms is their silica-based cell wall and the secretion of mucilage rich in sugars like mannose, fucose, and galactose. These properties can be utilized in the development of diatom-based fertilizers, which may enhance plant biomass and crop yields. Diatoms also possess phytoremediation abilities, as they can absorb heavy metals from the environment (including lead, zinc, nickel, cadmium, and titanium) [73]. As part of the next generation of sustainable fertilizers, diatoms could help improve plant growth and resilience, especially under stress conditions such as extreme climate events [72,74].

2.5. *Microalgae-Bacteria Consortia*

Finally, the use of microalgae–microorganism consortia has emerged as a promising alternative to the traditional application of single microalgal or microbial strains for promoting plant growth. In such systems, microalgae supply oxygen and organic compounds that stimulate beneficial bacteria, while bacteria recycle nutrients that enhance algal growth [75]. These synergistic partnerships offer significant advantages by providing plants with a broader spectrum of essential nutrients and molecules (such as nitrogen, phosphorus, and potassium) through complementary metabolic interactions [75,76]. Thus, when microalgae are combined with nitrogen-fixing bacteria, the consortium can perform complex biological functions that individual strains cannot achieve alone [77]. For example, co-inoculation of *Anabaena cylindrica* with diazotrophic *Azospirillum* increased maize productivity [78]. This cooperative behavior not only enhances nutrient availability but also supports processes with valuable biotechnological applications [79]. Furthermore, the combined activity of microalgae and beneficial microbes can activate plant defense mechanisms, leading to the production of fungal enzymes and antibiotics that protect plants from pests and diseases [80]. Beyond agricultural benefits, microalgae–bacteria consortia, especially those involving nitrogen-fixing species, also hold great promise in the fields of biotechnology [79].

3. Biofertilizers

A biofertilizer is a substance containing specific microorganisms or other natural preparations that, when applied to seeds, plant surfaces, or soil, promote plant growth by increasing the supply or availability of essential nutrients [27]. These microorganisms or preparations improve plant growth by colonizing the rhizosphere or interior of plants. This process enhances the supply or availability of nutrients to the host plant. The product is administered to seeds, plants, or soil [11,81,82]. In addition, these eco-friendly approaches have been shown to improve crop productivity, nutrient profile, and plant tolerance to abiotic and biotic stress. The use of biofertilizers reduces the problems associated with chemical fertilizers and leads to more sustainable agriculture [27].

The utilization and marketing of biofertilizers commenced in the 18th century with the patenting of 'Nitragin', the first *Rhizobium* formulation, by Hiltner and Nobbe [83]. The classification of biofertilizers is based on their function and mechanism of action. The most widely employed biofertilizers comprise nitrogen-fixers (N-fixers), potassium solubilizers (K solubilizers), phosphorus solubilizers (P solubilizers), zinc solubilizers (Zn solubilizers), and iron solubilizers (Fe solubilizers) [84–86].

While traditionally most biofertilizers have been based on heterotrophic plant growth–promoting bacteria (PGPB), such as *Rhizobium* or *Azotobacter*, in recent years photosynthetic microorganisms (microalgae and cyanobacteria) have emerged as highly promising candidates for next-generation biofertilizers. These photosynthetic microorganisms are rapidly emerging as valuable alternatives or supplements to traditional biofertilizers, owing to their multiple functions, including the improvement of soil properties through biofilm formation and the enhancement of organic matter. Their multifunctional nature, combined with their ability to thrive in a wide range of environmental conditions, makes them a promising tool for promoting sustainable crop production [87]. Moreover, these biofertilizers can be categorized into two broad groups: (i) photosynthetic microorganism-based biofertilizers (microalgae and cyanobacteria), and (ii) non-photosynthetic microorganism-based biofertilizers (classical PGPR and fungi). In the following subsections, the discussion will focus on the roles of photosynthetic microorganism-based biofertilizers.

3.1. *Cyanobacteria as Biofertilizers Beyond Nitrogen Fixation*

Nitrogen deficiency is one of the most critical constraints to agricultural productivity worldwide, since nitrogen is an essential macronutrient for plant growth and development. To overcome this limitation, sustainable alternatives to synthetic fertilizers have been widely investigated. Among them, nitrogen-fixing biofertilizers have traditionally attracted major research interest in leguminous crops, where rhizobial inoculants establish highly efficient symbiotic associations within root nodules to convert atmospheric nitrogen (N_2) into plant-available forms [79,82]. However, the use of rhizobia is restricted to legumes, limiting their applicability across broader agricultural systems. In addition, several free-living or associative diazotrophic bacteria (such as *Azospirillum*, *Azoarcus*, *Burkholderia*, *Gluconacetobacter diazotrophicus*, *Herbaspirillum*, *Azotobacter* and *Paenibacillus polymyxa*) have been demonstrated to fix nitrogen, and some have even been successfully formulated into commercial biofertilizers [27,83]. Yet, their nitrogen fixation efficiency is considerably lower compared to symbiotic microbes, and their performance is often inconsistent under field conditions due to competition with native soil microbiota, sensitivity to environmental stresses, and limited persistence in the rhizosphere [88].

By contrast, photosynthetic diazotrophic microorganisms such as cyanobacteria represent a highly promising and more versatile solution [89,90]. Members of the order *Nostocales* are photoautotrophic prokaryotes capable of fixing atmospheric nitrogen into ammonia (NH_3^+), which can be directly assimilated by plants [91]. This process not only reduces dependence on chemical fertilizers but also contributes to soil fertility and ecological balance. Unlike rhizobia, cyanobacteria are not restricted to specific host plants, and unlike most free-living diazotrophs, they combine nitrogen fixation with photosynthesis, making them largely self-sufficient. Under nitrogen starvation conditions, *Nostoc* and related genera develop heterocysts—specialized thick-walled cells that create a microoxic environment enabling nitrogenase activity. These heterocysts are strategically distributed along the filamentous chains of vegetative cells, ensuring an efficient supply of fixed nitrogen to the colony and, ultimately, to the surrounding rhizosphere [92]. This dual capacity to fix nitrogen and support broader soil fertility highlights cyanobacteria as a powerful alternative for developing next generation biofertilizers.

The development of cyanobacteria-based biofertilizer (CBF) has been enabled by the potential characteristics of these prokaryotic photoautotrophic microbes [93,94]. It is worth noting that using cyanobacteria as biofertilizers is not a completely new concept in agriculture (Table 1). These organisms have already been studied extensively in various crop systems. For instance, notable advancements have been made in rice cultivation over the past few years, such as in Andalusian paddy fields [94]. Re-inoculation of cyanobacterial isolates from this area as CBF has had a positive, significant effect on plant growth, with a significant increase in plant length of 127% recorded, as well as significant increases in grain weight and number per panicle. Similarly, a recent study investigated the potential of two cyanobacteria, *Anabaena vaginicola* ISB42 and *Nostoc spongiaeforme* var. *tenue* ISB65, as promising candidates for producing environmentally friendly biofertilizers for sustainable peppermint cultivation because they improve the quantity and quality of essential oils (EOs) by upregulating the key genes involved in the menthol biosynthetic pathway in *Mentha piperita* [95]. Furthermore, their potential as biofertilizers for cotton crops has recently been evaluated, demonstrating that reintegrating these beneficial species into agricultural ecosystems can enhance crop growth and maintain a balanced microbial environment [96]. Taken together, these results suggest that cyanobacterial biofertilizers could be a promising way to sustain rice production [94]. This contributes to the broader goal of achieving sustainable agriculture on a global scale [96].

Among eukaryotic microalgae, the use of *Chlorella vulgaris* and *Scenedesmus obliquus* suspensions, cultivated in maize drainage water, has been demonstrated to be a cost-effective, slow-release organic fertilizer on farmland, when applied to lettuce. The replacement of 50% of the nitrogen mineral fertilizer applied to lettuce by microalgae suspensions resulted in a significant increase in biomass production, reaching up to 2-fold. It has been demonstrated that this approach has the capacity to enhance lettuce fresh biomass and improve soil health [97]. Five species of algae were also evaluated in greenhouse conditions with pea plants (*Pisum sativum*) and in open field plots with spring wheat: the cyanobacterium *Arthrospira platensis* (*Spirulina*), the unicellular green microalga *Chlorella* sp., the red alga *Palmaria palmata*, and the brown algae *Laminaria digitata* and *Ascophyllum nodosum*. The results showed that *Chlorella* sp. and *Spirulina* increased total nitrogen and available phosphorus in the soil, with *Spirulina* also significantly enhancing nitrate levels. *Palmaria palmata* and *Laminaria digitata* significantly increased the concentrations of inorganic nitrogen compounds (NH_4^+ and NO_3^-). Moreover, *Chlorella* sp. was demonstrated to improve total phosphorus, nitrogen, and carbon contents in the soil, as well as available phosphorus, ammonium (NH_4^+), nitrate (NO_3^-), and pea crop yield [98].

Beyond nitrogen fixation, cyanobacteria provide additional plant growth-promoting benefits. They improve soil structure through biofilm formation [99]. In addition, they can increase soil porosity and decrease soil salinity [7]. *Nostoc* species possess the ability to produce phytohormones (auxins, cytochromes, gibberellins and ethylene), siderophores (iron binders) and mineral solubilizers (e.g., phosphorus, potassium and zinc) [100,101]. These symbiotic cyanobacteria have a wide diversity of associations with plants distributed throughout the plant kingdom such as spore-forming bryophytes, ferns, cycads or rice [102–104]. Special mention should be given to *Arthrospira platensis*, an edible cyanobacterium known worldwide for its high nutritional value, as well as for the interest in its biological activity and bioactive compounds [105]. Supplementing *Arthrospira platensis* by drenching the soil resulted in increased growth and productivity of chia plants cultivated under alkaline soil conditions, as well as increased antioxidant levels in the chia seeds. Following microalgae application, the oil content increased, as did the proportion of omega-3 [106]. More recently, according to [107], applying a biofertilizer containing *Spirulina maxima*, marine *Lactobacillus plantarum*, molasses, and industrial organic waste (IOW) at a concentration of 0.1% can enhance the growth, development, and nutrient uptake of rosemary plants by generating bioactive compounds, including vitamins, carbohydrates, and phytohormones (auxins, gibberellins, and cytokinins).

These findings highlight the potential of cyanobacterial biofertilizers as a multifunctional and sustainable tool for agriculture [94]. Their role extends far beyond nitrogen fixation, encompassing improvements in soil health, nutrient availability, plant physiology, and crop productivity. Their application in staple crops such as rice suggests that cyanobacterial biofertilizers could become a cornerstone of sustainable food production in the years ahead.

3.2. Microalgae and Cyanoboacteria as Biofertilizers: Nutrient Solubilization

Phosphorus (P) and potassium (K) are often limiting factors in agricultural soils, as both elements are commonly present in insoluble mineral complexes unavailable for plant uptake. This has led to growing interest in the development of phosphorus-solubilizing biofertilizers (PSB) [108]. Traditionally, various bacteria and fungi have been applied as PSB, including *Pseudomonas aeruginosa* in rice, *Pantoea agglomerans* in maize, *Pseudomonas* sp. in chili pepper, *Enterobacter* in soybeans, *Aspergillus niger* in beans, *Burkholderia cepacia* in peanuts, and *Azospirillum* in wheat [109–115]. However, beyond these classical microbial inoculants, photosynthetic microorganisms such as cyanobacteria and microalgae are increasingly recognized for their remarkable capacity to solubilize nutrients, representing a promising trend in sustainable agriculture. Species like *Nostoc* and *Anabaena* secrete organic acids and phosphatases that mobilize insoluble phosphates, while their extracellular polymeric substances (EPS) can chelate potassium and enhance its availability within soil aggregates. In addition, many microalgae not only solubilize P and K but also release bioactive metabolites and phytohormones, amplifying their plant growth-promoting effects [25,116]. This dual ability to improve nutrient bioavailability while simultaneously stimulating plant physiology gives cyanobacteria and microalgae a significant advantage over conventional PSB, positioning them as multifunctional biofertilizers with broad applications in sustainable agriculture.

Eukaryotic microalgae (*Chlorella*, *Scenedesmus*, *Chlamydomonas*) have also shown strong potential as P- and K-solubilizers. Although *Chlamydomonas* is not a strong nutrient-mobilizing species on its own, it can participate in associative biofertilization when co-applied with nutrient-solubilizing consortia; in such consortia, algal-derived metabolites support microbial activity that increases nitrogen and phosphorus availability to plants [75]. In addition, both microalgae can stimulate root growth through auxin-like activity while contributing to the mobilization or improved uptake of phosphorus and potassium, leading to better growth under nutrient-limited or stress environments [45,82].

3.3. Microalgae and Cyanobacteria as Biofertilizers: Siderophore-Mediated Growth Promotion

Micronutrient deficiencies, particularly zinc (Zn) and iron (Fe), severely limit crop productivity. In this context, Zinc-solubilizing biofertilizers have become increasingly important in the context of crop production. For instance, the process of zinc solubilization by certain bacterial species, including *Azospirillum*, *Azotobacter*, *Pseudomonas*, and *Rhizobium*, has been demonstrated to enhance zinc assimilation in wheat [117]. Furthermore, three selected different bacteria, *Acinetobacter calcoaceticus*, *Bacillus proteolyticus and Stenotrophomonas pavanii*, formulated in both free and encapsulated forms, showed improved plant growth parameters and enhanced zinc content in *Zea mays* and can be applied as biofertilizers to enhance soil fertility [118]. Zinc solubilization has also been observed in several *Chlorella* strains, where algal secretions mobilize insoluble Zn compounds, thereby enhancing plant Zn uptake [119–121]. The application of these algal inoculants could help address widespread micronutrient deficiencies in human diets by biofortifying staple crops such as rice and wheat.

Iron is involved in a variety of metabolic pathways within the cell, including photosynthesis, thus rendering it an essential element for plant life [122]. It is noteworthy, however, that siderophores represent a distinct group of low-molecular-weight compounds (less than 1.5 kDa) that exhibit a high affinity for Fe in environments with low Fe concentrations [61]. The synthesis and secretion of these compounds by different microbial strains occurs under specific conditions, thereby increasing and regulating the bioavailability of Fe. Consequently, siderophore-producing bacteria have attracted considerable scientific interest because of their potential application as biofertilizers. These bacteria have been demonstrated to enhance soil fertility and increase plant biomass, a finding that is of significant relevance for the development of sustainable agricultural practices [123,124]. For instance, the promotional effect of the AS19 strain, a bacterium capable of producing high levels of siderophores and facilitating the absorption of Fe^{3+} by seeds and plants, on the germination of pepper and maize seeds and the development of shoots and leaves of *Gynura divaricata* (Linn.) has been demonstrated [61]. In addition, a recent study has demonstrated the significant potential of Fe-solubilizing rhizobacteria (*Bacillus* spp.) isolated from the maize rhizosphere in calcareous soils as effective biofertilizers. These bacterial strains, namely *Bacillus pyramidoids, Bacillus firmicutes,* and *Bacillus cereus,* have the potential to mitigate Fe deficiency in crops, thereby promoting sustainable agriculture practices [125].

In addition to these examples, cyanobacteria such as *Synechococcus* and *Nostoc*, and microalgae like *Dunaliella* and *Chlorella*, can synthesize siderophores, which bind Fe^{3+} and facilitate its uptake by plants. For example, *Synechococcus mundulus*–derived siderophores improved Fe bioavailability in maize, significantly enhancing chlorophyll content and photosynthetic efficiency [126]. Moreover, in relation to the *Cyanobacteria* spp., Brick et al., in 2025 [126] highlighted the significant potential of *Synechococcus mundulus*-derived siderophore in stimulating *Zea mays* physicochemical growth parameters and iron uptake. The results of this study indicate the capacity of cyanobacteria to synthesize siderophores as a sustainable substitute for synthetic iron chelators, and their role in the management of plant stress [126].

Notably, microalgae have been demonstrated to play a pivotal role in maintaining ion homeostasis. *Dunaliella*, a genus of algae, has demonstrated a remarkable capacity for adaptation to environments characterized by low iron levels. Several species of *Dunaliella*, namely *Dunaliella tertiolecta, Dunaliella salina* and *Dunaliella bardawil*, have been identified as originating from radically divergent environments. These species have been found to possess a unique family of siderophore-iron-uptake proteins [127]. Moreover, it has been demonstrated that *Dunaliella salina* extracts, particularly exopolysaccharides, have the capacity to promote the germination and growth of *Triticum aestivum L.* seedlings under conditions of salt stress, thus offering a potentially viable solution for enhancing the resilience of crops in salt-affected environments [128].

To conclude this section, Table 1 provides a summary of the main microalgae and cyanobacteria species with biofertilizer potential.

Table 1. Main microalgae and cyanobacteria species involved in biofertilization.

Species	Group	Biofertilization Mechanism	Plant/System	References
Nostoc sp.	Cyanobacteria	Nitrogen fixation, phytohormone production, solubilization of P/K/Zn	Wheat/In vitro Rize/Soil	[94,100]
Anabaena vaginicola ISB42	Cyanobacteria	Phytohormone-linked nutrient uptake, peppermint oil enhancement	Mentha/Greenhouse conditions	[95]
Nostoc spongiaeforme var. *tenue* ISB65	Cyanobacteria	Phytohormone-linked nutrient uptake, peppermint oil enhancement	Mentha/Greenhouse conditions	[95]
Synechococcus mundulus	Cyanobacteria	Siderophore production, enhanced Fe uptake in maize	Maize/In vitro	[126]
Arthrospira platensis	Cyanobacteria	Phytohormone production, improved nutrient acquisition in chia	Chia/Soil	[106]
Spirulina maxima	Cyanobacteria (marketed as microalgae)	Bioactive compounds enhancing growth and nutrient uptake in rosemary	Rosemary/Soil	[107]

Table 1. *Cont.*

Species	Group	Biofertilization Mechanism	Plant/System	References
Chlorella vulgaris	Microalgae	Nitrogen fixation Phosphorus and potassium solubilization, auxin-like activity	Lettuce/Soil	[45]
Scenedesmus obliquus	Microalgae	Nitrogen fixation P/K mobilization, root stimulation under stress	Lettuce/Soil	[45,129]
Dunaliella salina	Microalgae	Exopolysaccharides Siderophore-mediated Fe uptake under deficiency	Wheat/In vitro	[128]

4. Biocontrol Agents

Plants are constantly exposed to a wide range of biotic stressors, including fungi, bacteria, nematodes, insects, and viruses. In response, they have evolved complex and sophisticated defense mechanisms. The production of secondary metabolites plays a crucial role due to their effectiveness in pathogen and pest resistance [130]. Compounds such as saponin have been widely recognized for their antifungal activity and are considered key components of plant defense systems. These molecules can be harnessed to develop novel, eco-friendly strategies for disease control, reducing the environmental impact of chemical pesticides [131,132]. While many of these compounds are traditionally derived from plants, recent studies have identified the potential of microalgae and cyanobacteria as a promising alternative source of secondary metabolites, including alkaloids, flavonoids, terpenes, and essential oils [133,134]. Microalgae and cyanobacteria form beneficial associations with plants, which can enhance the production of secondary metabolites, especially under abiotic stress conditions. Among these metabolites, allelochemicals produced by these microorganisms have garnered particular interest due to their potential applications in sustainable crop protection and biocontrol strategies [134]. Microalgae and specially cyanobacteria, represent a prolific source of biologically active compounds involved in allelopathic interactions, many of which could be utilized for pest control and crop protection [96] (Table 2).

4.1. Phytopathogen Resistance

Allelochemicals from microalgae have demonstrated strong antimicrobial activity against a wide spectrum of phytopathogens. For instance, hapalindole T (an antibacterial alkaloid from *Fischerella* sp.), nostofungicin (a fungicidal lipopeptide from *Nostoc commune*), and eicosapentaenoic acid (antimicrobial fatty acid from *Phaeodactylum tricornutum*) are notable examples [135–138]. Moreover, cell extracts from *Chlorella vulgaris* and *Tetradesmus obliquus* have been successfully applied to protect spinach crops against *Fusarium oxysporum* infections [139]. Gene editing in *Chlamydomonas reinhardtii* has also been used to enhance bacterial resistance in tobacco plants [140].

Several studies have evaluated the antifungal activity of microalgal extracts across diverse strains. *Anabaena* HSSASE11 and *Oscillatoria nigroviridis* HSSASE15 showed inhibitory effects against *Botryodiplodia theobromae* and *Pythium ultimum*, respectively, while *Dunaliella* HSSASE13 was effective against *Fusarium solani*. Similar antifungal results were reported for *Scenedesmus obliquus* extracts against *Sclerotium rolfsi* [141]. Interestingly, some microalgal extracts also show nematicidal activity. For example, *Scenedesmus obliquus*, *Chlorella vulgaris*, and *Anabaena oryzae* were able to inhibit the banana pathogen *Meloidogyne incognita* [142]. The antimicrobial activity of these extracts is largely attributed to phenolics, alkaloids, and peptides. However, in many cases, the specific active compounds have not been fully identified or characterized. While the mechanisms remain under investigation, the antifungal effects of phenolic compounds may involve interference with fungal cell wall biosynthesis [143].

4.2. Microalgae and Cyanobacteria as Herbicides

The herbicidal potential of microalgae and cyanobacteria metabolites is an emerging field with encouraging preliminary findings. Several allelochemicals—mainly from cyanobacteria—have demonstrated phytotoxic activity. For instance, cyanobacterin, a phenolic compound produced by *Scytonema hofmanni*, inhibits photosynthetic electron transport. Similarly, nostocyclamide (a peptide from *Nostoc* sp.) and fischerellins (alkaloids from *Fischerella* sp.) interfere with photosystem II [144,145].

Other compounds like microcystins (peptides that inhibit protein phosphatases) and cryptophycins (polyketides that block microtubule polymerization) also show potential as herbicides [144,146]. Despite these findings, research on herbicidal compounds from eukaryotic microalgae remains limited, presenting an underexplored area with high potential for sustainable weed management.

Table 2. Main secondary metabolites from microalgae and cyanobacteria and their potential applications as biocontrol agents in agriculture.

Class of Compound	Characteristics	Microorganism	Underlying Mechanism of Biocontrol	Potential Target Pathogens/Pests	Potential Use in Agriculture	Experimental Evidence	References
Alkaloids	Nitrogen-containing heterocyclic compounds	*Fischerella* sp., *Calothrix* sp., *Desertifilum dzianense*	Interfere with DNA replication and protein synthesis in pathogens; disruption of cell walls.	Insects (neurotoxin), fungi (*Agroathelia rolfsii*), and broad microbial pathogens	Natural bioinsecticides or antimicrobial agents for biocontrol	in vitro bioassays	[147,148]
Polyketides	Structurally diverse metabolites derived from carboxylic acid precursors	*Gambierdiscus toxicus*, *Karenia brevis*	Inhibition of ion channels, disruption of cell signaling and membrane integrity	Plant-pathogenic fungi, bacteria; brevetoxins also toxic to invertebrates	Broad-spectrum fungicides or bactericides for crops	both in vivo (animal models) and in vitro; in vitro assays (channel agonist)	[149,150]
Fatty acids	Extracellular free fatty acids with allelopathic activity	*Chlorella vulgaris*, *Botryococcus braunii*	Membrane destabilization; inhibition of seed germination through allelopathy	Competing algae (*Pseudokirchneriella subcapitata*), weeds	Natural weed growth inhibitors (bioherbicides)	in vitro inhibition assays; in vitro allelopathic tests	[151,152]
Peptides	Non-ribosomal peptides biosynthesized by multifunctional enzyme complexes	*Anabaena* sp. PCC7120, *Microcystis* sp., *Planktothrix* sp., *Oscillatoria limosa*, *Synechococcus lividus*	Pore formation in membranes; inhibition of protein phosphatases; induction of oxidative stress in pathogens	Other cyanobacteria, aquatic weeds, possible cross-toxicity to pathogenic fungi/bacteria	Plant defense promoters or biostimulants	review (mostly in vitro); in vivo detection in hot springs; in vitro isolation/assays; in situ environmental surveys	[153–156]
Terpenoids	Organic compounds derived from C5 precursors with toxicity to invertebrates	*Nostoc commune*, *Calothrix* sp. PCC7507	Neurotoxic and deterrent effects on insects; oxidative stress induction	Bacteria (*Bacillus cerus*, *S. epidermidis*, *E. coli*), Insect pests (e.g., Lepidoptera larvae), nematodes; general herbivory deterrenc	Natural insecticides or pest repellents	in vitro bioassays; review (includes in vitro and some in vivo reports)	[157,158]

4.3. Microalgae and Cyanobacteria as Insecticides

Among all properties shown from microalgae and cyanobacteria extracts throughout this review, insecticide activity is another one proposed. There are several studies in which microalgae have demonstrated insecticide properties. In the case of diatoms or chlorophytes, the potential insecticide activity has been investigated. For example, the exploitation of *Chlamydomonas reinhardtii* extracts in the development of preparations combined with microparticles of zinc oxide was able to improve the larvicidal potential of *Tenebrio molitor* compared to zinc oxide alone treatment [159]. Another study has demonstrated that the extract from *Amphora coffeaeformis* and *Scenedesmus oliquus* presented larvicidal activity against *Culex pipiens* [160].

In another approach, biofilm-forming cyanobacteria were found to enhance plant defense mechanisms against insects. Some studies have demonstrated that biofilm from *Fischerella* ATCC 43239 increased the mortality of larvae from *Chirinimus riparius*, demonstrating that the biofilm improves the production of allelochemicals with insecticide activity [161–164]. Moreover, biomass extracts from *Microcystis aeruginosa* 205 and *Anabaena circinalis* 86 showed high toxicity against larvae of *Aedes aegypti* [165]. Alternatively, some researchers have described that both unsaturated and saturated fatty acids are responsible for the insecticide activity of microalgae extracts against larvae from different species. Thus, the possible mechanism of larvicidal activity of fatty acids was investigated. The authors tested the inhibition properties of various fatty acids, demonstrating that linolenic and linoleic acids might have a dual mode of action against octopamine signaling pathways [166,167].

As a summary, Table 3 provides an overview of the main biocontrol-related agents identified with phytopathogenic resistance and herbicide and insecticide potential, together with their producing species, mechanisms of action, and target organisms.

Table 3. Overview of the main biocontrol-related agents identified with phytopathogenic resistance, herbicide and insecticide potential from microalgae and cyanobacteria. The table summarizes their mechanisms of action, producing species, and target pathogens or pests, highlighting their potential applications as natural alternatives to synthetic pesticides.

Biocontrol Mechanisms	Microalgae/Cyanobacteria	Produced Molecules	Mode of Action	Target Pathogens/Organisms	References
Phytopathogen resistance	*Fischerella* sp.	Hapalindole T (alkaloid)	Antibacterial activity	Phytopathogenic bacteria	[135–138]
	Nostoc commune	Nostofungicin (lipopeptide)	Fungicidal	Phytopathogenic fungi	[135–138]
	Phaeodactylum tricornutum	Eicosapentaenoic acid (EPA, fatty acid)	Antimicrobial	Various pathogens	[135–138]
	Chlorella vulgaris, *Tetradesmus obliquus*	Cell extracts (phenolics, peptides, not fully identified)	Inhibition of fungal growth	*Fusarium oxysporum* (spinach)	[139]
	Chlamydomonas reinhardtii	Genetic modification	Enhanced bacterial resistance via gene editing	Tobacco plants	[140]
	Anabaena HSSASE11	Phenolic/peptide extracts	Antifungal	*Botryodiplodia theobromae*	[141]
	Oscillatoria nigroviridis HSSASE15	Phenolic/peptide extracts	Antifungal	*Pythium ultimum*	[141]
	Dunaliella HSSASE13	Phenolic/peptide extracts	Antifungal	*Fusarium solani*	[141]
	Scenedesmus obliquus	Phenolic extracts	Antifungal	*Sclerotium rolfsii*	[141]
	Scenedesmus obliquus, *Chlorella vulgaris*, *Anabaena oryzae*	Phenolics, alkaloids, peptides	Nematicidal; possible inhibition of fungal cell wall biosynthesis	*Meloidogyne incognita* (banana pathogen)	[142]
Herbicides	*Scytonema hofmanni*	Cyanobacterin (phenolic)	Inhibition of photosynthetic electron transport	Weeds (phytotoxic effect)	[144,145]
	Nostoc sp.	Nostocyclamide (peptide)	Inhibition of photosystem II	Weeds	[144,145]
	Fischerella sp.	Fischerellins (alkaloids)	Inhibition of photosystem II	Weeds	[144,145]
	Microcystis sp.	Microcystins (peptides)	Inhibition of protein phosphatases	Weeds	[144,146]
	Various cyanobacteria	Cryptophycins (polyketides)	Blockage of microtubule polymerization	Weeds	[144,146]

Table 3. *Cont.*

Biocontrol Mechanisms	Microalgae/Cyanobacteria	Produced Molecules	Mode of Action	Target Pathogens/Organisms	References
Insecticides	*Chlamydomonas reinhardtii* + ZnO	Cell extracts	Enhanced larvicidal effect when combined with ZnO	*Tenebrio molitor*	[159]
	Amphora coffeaeformis, *Scenedesmus obliquus*	Cell extracts (fatty acids)	Larvicidal activity	*Culex pipiens*	[160]
	Fischerella ATCC 43239 (biofilm)	Biofilm-induced allelochemicals	Increased larval mortality	*Chironomus riparius*	[161–164]
	Microcystis aeruginosa 205, *Anabaena circinalis* 86	Biomass extracts	Toxic	*Aedes aegypti*	[165]
	Various microalgae	Unsaturated fatty acids (linolenic, linoleic acids)	Disruption of octopamine signaling (dual mechanism)	Insect larvae (various species)	[166,167]

5. Commercialized Microalgae- and Cyanobacteria-Based Products

An increasing amount of scientific research demonstrating the effectiveness of microalgae in agriculture has driven the development of an emerging and fast-growing market for microalgae-derived bioproducts. Commercial formulations such as Algafert, PhycoTerra®, and Spiralgrow have already entered the global market, targeting both conventional and organic farming sectors (Table 4). These products are offered as liquid concentrates, dry powders, or granules, and are often marketed for their ability to improve root development, nutrient uptake, and crop resilience under stress. Start-ups and aggrotech companies are also investing in vertically integrated production systems that combine algae cultivation with carbon capture and waste valorization, aligning with global sustainability goals [60]. According to recent market projections, the global microalgae-based biofertilizer segment is expected to grow at a compound annual growth rate (CAGR) exceeding 8% over the next five years, driven by regulatory shifts toward low-input agriculture and increasing demand for eco-certified inputs. This momentum underscores the transition of microalgae from an experimental innovation to a commercially viable and scalable component of sustainable agriculture [45].

Table 4. Examples of some commercialized microalgae- and cyanobacteria-based products.

Product Name	Manufacturer	Microalgae Used	Formulation	Main Effects	URL or Reference
Algafert	Biorizon Biotech	*Spirulina* spp.	Dry powder	Provides macro and micronutrients, promotes chlorophyll synthesis.	https://www.biorizon.es/en/products/biostimulants-y-bioenhancers/algafert/ (accessed on 11 August 2025)
AgriAlgae	AlgaEnergy	*Nannochloropsis* spp.	Liquid biostimulant	Enhances photosynthesis, nutrient uptake, and crop vigor.	https://ag.algaenergy.com/es/product-category/agrialgae-premium/?lang=it (accessed on 11 August 2025)
Allfertis	Allmicroalgae	*Chlorella* spp.	Powder	Promote resistance to biotic and abiotic agents; Increases the size and fruit weight.	https://www.allmicroalgae.com/en/agro/#Allfertis (accessed on 11 August 2025)
Biofertilizer by MicroAlgaex	Microalgaex	Not specified—described as "microalgae formulations"	Liquid	Enhanced plant growth and yield, improved defense against abiotic stress, and increased nutrient absorption.	https://microalgaex.com/biofertilizer/ (accessed on 11 August 2025)
Ecotop	Herogra	*Ascophyllum nodosum* and blend of other microalgae	Liquid biostimulant	Enhances plant vigor, growth, and resilience to abiotic/biotic stress.	https://herograespeciales.com/en/productos/bioestimulantes/ecotop/ (accessed on 11 August 2025)
Kelpak	Kelpak	*Ecklonia maxima* (macroalga, used in synergy with microalgae)	Liquid biostimulant	Promotes root and shoot development, stress tolerance.	https://www.kelpak.com/kelpakintro.html (accessed on 11 August 2025)
AlgaGrow	Plagron	Proprietary blend including cyanobacteria	Liquid biostimulant	Increases nutrient uptake and crop yield.	https://plagron.com/en/hobby/products/alga-grow (accessed on 11 August 2025)

Table 4. *Cont.*

Product Name	Manufacturer	Microalgae Used	Formulation	Main Effects	URL or Reference
Seasol	Seasol International (Australia)	Blend of seaweed and microalgae extracts	Liquid concentrate	Broad-spectrum plant tonic.	https://www.seasol.com.au/products/seasol/ (accessed on 11 August 2025)
Weed-Max	Trade S.A.E. Company (Egypt)	Cyanobacteria extract in powder phase	Dry powder	Suppress soil-borne fungi and enhance the antagonistic abilities of other bioagents.	[168]
Oligo-X algal	Arabian Group for Agricultural Service	Blue-green algal extracts in liquid phase	Liquid concentrate	Suppress soil-borne fungi.	[168]

6. Research Bottlenecks and Future Perspectives

Microalgae and cyanobacteria have emerged as highly attractive candidates for sustainable agriculture and bio-based industries due to their distinctive biological and ecological traits. These photosynthetic microorganisms exhibit rapid and continuous growth, short generation times, and remarkable metabolic flexibility. Importantly, their cultivation does not compete with traditional agriculture for critical resources. They can be grown on non-arable land using saline water or even wastewater and rely primarily on sunlight as an energy source. These characteristics underscore their potential as complementary components to existing agri-food systems and as contributors to the circular bioeconomy, particularly in the production of food, feed, and high-value fertilizer products.

However, translating this potential into large-scale, economically viable operations presents significant challenges (Figure 3). Among the foremost constraints are the costs and efficiencies associated with cultivation, harvesting, and biomass processing. On the technical front, optimization of both upstream (e.g., growth conditions, nutrient supply, and biomass accumulation) and downstream (e.g., harvesting, extraction, and formulation) processes is essential to ensure yield consistency, biomass quality, and cost-effectiveness [38]. Infrastructure costs (especially for photobioreactors and open pond systems) are substantial, and operational expenditures such as lighting, temperature regulation, fluid circulation, cleaning, and biomass recovery remain high [169]. Closed systems, while offering advantages in terms of contamination control and process stability, are particularly cost intensive. Inoculum preparation constitutes another critical bottleneck; the reliability of large-scale operations depends on the production of high-quality starter cultures. Any microbial contamination or physiological variability during this stage can lead to serious disruptions, compromising the productivity of the entire system.

Operational stability is further challenged by routine maintenance, system downtime, and unexpected failures. Incorporating automation technologies offers a partial solution, reducing labor costs and human error while enabling real-time monitoring, improved safety, and enhanced process control. Nevertheless, successful deployment of these systems requires a multidisciplinary workforce with expertise in microbiology, bioprocess engineering, chemistry, and systems maintenance. Experienced personnel are indispensable for troubleshooting, process optimization, and quality control.

In parallel, a deeper mechanistic understanding of how microalgae and cyanobacteria interact with plant physiology and soil processes is critical to optimizing their use in agricultural contexts. Elucidating their modes of action (such as modulation of plant hormone levels, nutrient uptake, stress resilience, and microbiome interactions) is key to tailoring applications for specific crops, soil types, and climatic conditions. Such knowledge will improve consistency and efficacy under real-world field conditions. To overcome the current limitations, several strategies are under investigation. Continuous cultivation systems provide a stable operational platform and reduce the need for frequent re-inoculation. Increasing biomass density is another focus area, as it minimizes the volume that must be handled during harvesting. Additionally, phycoprospecting efforts aim to identify naturally robust, high-performing strains, while genetic engineering approaches target enhanced productivity, stress resistance, and metabolite synthesis. Integrated biorefinery models are also gaining prominence, wherein multiple valuable compounds (such as proteins, pigments, fatty acids, and polysaccharides) are co-extracted from a single biomass stream, thereby increasing the overall economic viability of production systems.

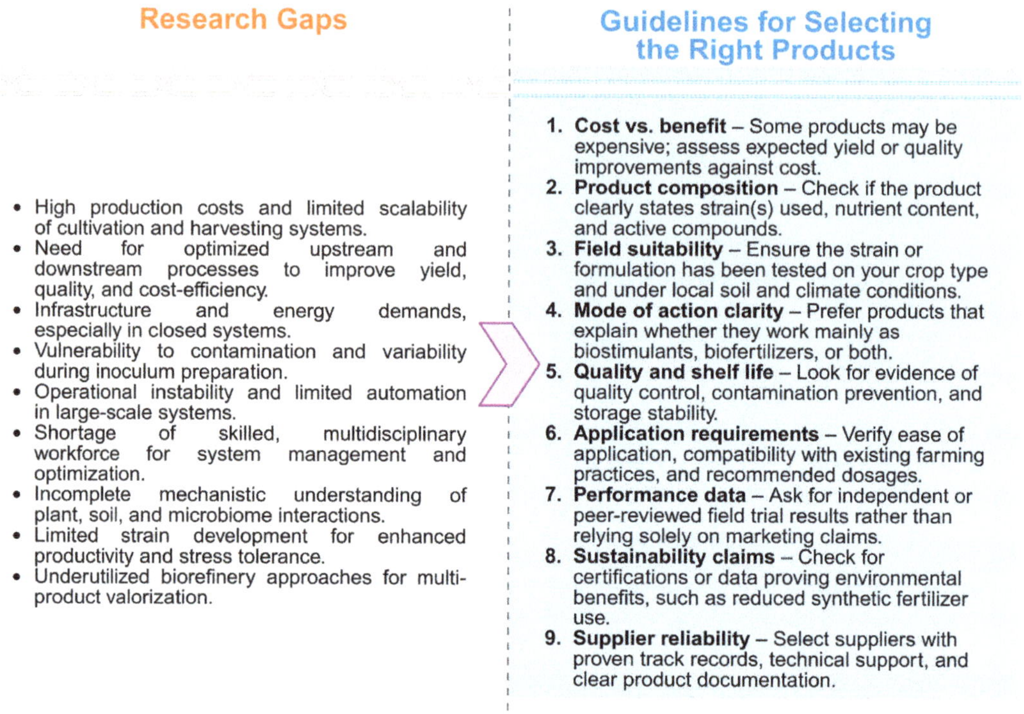

Figure 3. Research gaps and practical guidelines for the selection of microalgae- and cyanobacteria-based products in agriculture. The (**left panel**) highlights current limitations in large-scale development, including high production costs, limited scalability, contamination risks, operational instability, and insufficient mechanistic understanding of plant–microbe interactions. The (**right panel**) provides guidelines for selecting appropriate products, emphasizing cost–benefit assessment, product composition, field suitability, mode of action clarity, quality and shelf life, application requirements, performance validation, sustainability claims, and supplier reliability. Together, these perspectives illustrate the challenges and practical considerations necessary for effective strain selection and product implementation. Created in BioRender. Diaz Santos, E. (2025) https://BioRender.com/8wya7ti, accessed on 19 March 2025.

In summary, while the upscaling of microalgae and cyanobacteria cultivation is constrained by technical and economic barriers, it also presents a promising frontier for sustainable agriculture and industrial biotechnology. With continued innovation in system design, strain selection, process automation, and bioproduct valorization, these microorganisms have the potential to become foundational components of resilient, resource-efficient, and climate-smart production systems.

7. Conclusions

The present review highlights the significant potential of microalgae and cyanobacteria as multifunctional biological resources in agriculture. Their diverse roles as biofertilizers, biostimulants, and biocontrol agents provide multiple agronomic benefits, including the improvement of nutrient availability, enhancement of plant growth and yield, and reinforcement of tolerance to biotic and abiotic stresses. At the same time, these microorganisms contribute to broader sustainability goals through nutrient recycling, carbon sequestration, and a measurable reduction in the dependence on synthetic fertilizers and pesticides. Their capacity to generate a wide variety of bioactive compounds, such as phytohormones, exopolysaccharides, and allelochemicals, underscores their versatility and makes them promising candidates for integration into environmentally friendly farming systems. Together, these findings confirm their relevance as key drivers in the transition toward sustainable and resilient agriculture.

Nevertheless, the translation of laboratory findings into large-scale applications remains limited. Several critical challenges must be addressed before their potential can be fully realized. Future research should prioritize:

1. Scalable cultivation systems—the development of cost-effective and energy-efficient production platforms, ideally integrating waste streams and renewable resources to minimize costs and environmental impact.
2. Formulation and delivery strategies—optimizing stable, field-ready formulations (e.g., encapsulation, consortia-based inoculants, or liquid suspensions) that ensure consistent performance under variable agronomic conditions.
3. Mechanistic understanding—advancing molecular and physiological studies to clarify the interactions between plants and microalgae/cyanobacteria, thereby identifying the key pathways responsible for plant growth promotion and stress alleviation.
4. Regulatory frameworks and standardization—establishing clear guidelines for safety, efficacy testing, and product approval to accelerate the transition from experimental studies to market-ready solutions.
5. Integration into circular economy models—exploring their role in waste valorization, nutrient recovery, and soil regeneration as part of holistic and climate-smart farming practices.

Overall, microalgae and cyanobacteria should not be regarded merely as alternative agricultural inputs but as integral components of next-generation, nature-based agricultural strategies. By coupling mechanistic research with technological innovation and policy development, their promise can be translated into scalable, commercially viable applications that contribute to global food security while reducing environmental impact.

Author Contributions: Conceptualization, E.D.-S.; Manuscript writing—original draft preparation, A.J.-F., L.G.H.-M., G.T.-C. and E.D.-S.; writing—review and editing, G.T.-C. and E.D.-S.; manuscript supervision, G.T.-C. and E.D.-S.; A.J.-F. and L.G.H.-M. contributed equally to this work. The manuscript was corrected, revised, and approved by all authors. All authors have read and agreed to the published version of the manuscript.

Funding: This research received no external funding.

Institutional Review Board Statement: Not applicable.

Acknowledgments: Figures 1–3 were created using BioRender (URL: https://www.biorender.com/; accessed on 19 March 2025). During the preparation of this manuscript, the authors used the GenAI tools: Google Translator (URL: https://translate.google.com; accessed on 27 March 2025), ChatGPT (GPT-4.5 Orion) and DeepL Write (URL: https://www.deepl.com/es/write; accessed on 16 May 2025) for the purposes of improving some English expressions. The authors have reviewed and edited the output and take full responsibility for the content of this publication.

Conflicts of Interest: Author Gloria Torres-Cortes was employed by the company Innoplant S.L. The remaining authors declare that the research was conducted in the absence of any commercial or financial relationships that could be construed as a potential conflict of interest.

References

1. FAO. *The Future of Food and Agriculture—Drivers and Triggers for Transformation*; FAO: Rome, Italy, 2022.
2. Ren, C.G.; Kong, C.C.; Liu, Z.Y.; Zhong, Z.H.; Yang, J.C.; Wang, X.L.; Qin, S. A Perspective on Developing a Plant 'Holobiont' for Future Saline Agriculture. *Front. Microbiol.* **2022**, *13*, 763014. [CrossRef]
3. Singh, A. Soil Salinization Management for Sustainable Development: A Review. *J. Environ. Manag.* **2021**, *277*, 111383. [CrossRef]
4. Cuevas, J.; Daliakopoulos, I.N.; Del Moral, F.; Hueso, J.J.; Tsanis, I.K. A Review of Soil-Improving Cropping Systems for Soil Salinization. *Agronomy* **2019**, *9*, 295. [CrossRef]
5. Brito-Lopez, C.; Van Der Wielen, N.; Barbosa, M.; Karlova, R. Plant Growth-Promoting Microbes and Microalgae-Based Biostimulants: Sustainable Strategy for Agriculture and Abiotic Stress Resilience. *Philos. Trans. R. Soc. B Biol. Sci.* **2025**, *380*, 20240251. [CrossRef]
6. Banerjee, S.; van der Heijden, M.G.A. Soil Microbiomes and One Health. *Nat. Rev. Microbiol.* **2023**, *21*, 6–20. [CrossRef]
7. Crystal-Ornelas, R.; Thapa, R.; Tully, K.L. Soil Organic Carbon Is Affected by Organic Amendments, Conservation Tillage, and Cover Cropping in Organic Farming Systems: A Meta-Analysis. *Agric. Ecosyst. Environ.* **2021**, *312*, 107356. [CrossRef]
8. Raihan, A. A Review of Climate Change Mitigation and Agriculture Sustainability through Soil Carbon Sequestration. *J. Agric. Sustain. Environ.* **2023**, *2*, 23–56. [CrossRef]
9. Gonzalez-Gonzalez, L.M.; de-Bashan, L.E. The Potential of Microalgae–Bacteria Consortia to Restore Degraded Soils. *Biology* **2023**, *12*, 693. [CrossRef] [PubMed]
10. Ramakrishnan, B.; Maddela, N.R.; Venkateswarlu, K.; Megharaj, M. Potential of Microalgae and Cyanobacteria to Improve Soil Health and Agricultural Productivity: A Critical View. *Environ. Sci. Adv.* **2023**, *2*, 586–611. [CrossRef]
11. Alvarez, A.L.; Weyers, S.L.; Goemann, H.M.; Peyton, B.M.; Gardner, R.D. Microalgae, Soil and Plants: A Critical Review of Microalgae as Renewable Resources for Agriculture. *Algal Res.* **2021**, *54*, 102200. [CrossRef]
12. Chauton, M.S.; Reitan, K.I.; Norsker, N.H.; Tveterås, R.; Kleivdal, H.T. A Techno-Economic Analysis of Industrial Production of Marine Microalgae as a Source of EPA and DHA-Rich Raw Material for Aquafeed: Research Challenges and Possibilities. *Aquaculture* **2015**, *436*, 95–103. [CrossRef]
13. Spinola, M.V.; Díaz-Santos, E. Microalgae Nutraceuticals: The Role of Lutein in Human Health. In *Microalgae Biotechnology for Food, Health and High Value Products*; Springer: Singapore, 2020.
14. León-Bañares, R.; González-Ballester, D.; Galván, A.; Fernández, E. Transgenic Microalgae as Green Cell-Factories. *Trends Biotechnol.* **2004**, *22*, 45–52. [CrossRef]
15. Osathanunkul, M.; Thanaporn, S.; Karapetsi, L.; Nteve, G.M.; Pratsinakis, E.; Stefanidou, E.; Lagiotis, G.; Avramidou, E.; Zorxzobokou, L.; Tsintzou, G.; et al. Diversity of Bioactive Compounds in Microalgae: Key Classes and Functional Applications. *Mar. Drugs* **2025**, *23*, 222. [CrossRef]
16. Maurya, N.; Sharma, A.; Sundaram, S. The Role of PGPB-Microalgae Interaction in Alleviating Salt Stress in Plants. *Curr. Microbiol.* **2024**, *81*, 1–14. [CrossRef] [PubMed]
17. Chiaiese, P.; Corrado, G.; Colla, G.; Kyriacou, M.C.; Rouphael, Y. Renewable Sources of Plant Biostimulation: Microalgae as a Sustainable Means to Improve Crop Performance. *Front. Plant Sci.* **2018**, *871*, 1782. [CrossRef] [PubMed]
18. Farda, B.; Djebaili, R.; Sabbi, E.; Pagnani, G.; Cacchio, P.; Pellegrini, M. Isolation and Characterization of Cyanobacteria and Microalgae from a Sulfuric Pond: Plant Growth-Promoting and Soil Bioconsolidation Activities. *AIMS Microbiol.* **2024**, *10*, 944–972. [CrossRef]
19. Farmonaut® Agriculture Biologicals Market: 2025 Growth & Research Guide. Available online: https://farmonaut.com/news/agriculture-biologicals-market-2025-growth-research (accessed on 14 July 2025).
20. du Jardin, P. Plant Biostimulants: Definition, Concept, Main Categories and Regulation. *Sci. Hortic.* **2015**, *196*, 3–14. [CrossRef]
21. Barkia, I.; Saari, N.; Manning, S.R. Microalgae for High-Value Products towards Human Health and Nutrition. *Mar. Drugs* **2019**, *17*, 304. [CrossRef]
22. Vangenechten, B.; De Coninck, B.; Ceusters, J. How to Improve the Potential of Microalgal Biostimulants for Abiotic Stress Mitigation in Plants? *Front. Plant Sci.* **2025**, *16*, 1568423. [CrossRef]
23. Tyagi, J.; Ahmad, S.; Malik, M. Nitrogenous Fertilizers: Impact on Environment Sustainability, Mitigation Strategies, and Challenges. *Int. J. Environ. Sci. Technol.* **2022**, *19*, 11649–11672. [CrossRef]
24. Lorenzi, A.S.; Chia, M.A. Cyanobacteria's Power Trio: Auxin, Siderophores, and Nitrogen Fixation to Foster Thriving Agriculture. *World J. Microbiol. Biotechnol.* **2024**, *40*, 1–18. [CrossRef]

25. Nawaz, T.; Saud, S.; Gu, L.; Khan, I.; Fahad, S.; Zhou, R. Cyanobacteria: Harnessing the Power of Microorganisms for Plant Growth Promotion, Stress Alleviation, and Phytoremediation in the Era of Sustainable Agriculture. *Plant Stress* **2024**, *11*, 100399. [CrossRef]
26. Mignoni, D.S.B.; Nonato, J.D.S.; Alves, J.S.; Michelon, W.; de Oliveira Nunes, E.; Pandolfi, J.R.; Luchessi, A.D. Agriculture application, comparison, and functional association between macrophytes and microalgae: A review. *Discov. Agric.* **2025**, *3*, 1–17. [CrossRef]
27. Bhardwaj, D.; Ansari, M.W.; Sahoo, R.K.; Tuteja, N. Biofertilizers Function as Key Player in Sustainable Agriculture by Improving Soil Fertility, Plant Tolerance and Crop Productivity. *Microb. Cell Fact.* **2014**, *13*, 1–10. [CrossRef]
28. Prasanna, R.; Sood, A.; Ratha, S.K.; Singh, P.K. Cyanobacteria as a "Green" Option for Sustainable Agriculture. In *Cyanobacteria: An Economic Perspective*; John Wiley & Sons Inc.: Hoboken, NJ, USA, 2013.
29. Ferreira, A.; Bastos, C.R.V.; Marques-dos-Santos, C.; Acién-Fernandez, F.G.; Gouveia, L. Algaeculture for Agriculture: From Past to Future. *Front. Agron.* **2023**, *5*, 1064041. [CrossRef]
30. Miranda, A.M.; Hernandez-Tenorio, F.; Villalta, F.; Vargas, G.J.; Sáez, A.A. Advances in the Development of Biofertilizers and Biostimulants from Microalgae. *Biology* **2024**, *13*, 199. [CrossRef] [PubMed]
31. Rifna, E.J.; Rajauria, G.; Dwivedi, M.; Tiwari, B.K. Circular Economy Approaches for the Production of High-Value Polysaccharides from Microalgal Biomass Grown on Industrial Fish Processing Wastewater: A Review. *Int. J. Biol. Macromol.* **2024**, *254*, 126887. [CrossRef]
32. Álvarez-González, A.; Uggetti, E.; Serrano, L.; Gorchs, G.; Ferrer, I.; Díez-Montero, R. Can Microalgae Grown in Wastewater Reduce the Use of Inorganic Fertilizers? *J. Environ. Manag.* **2022**, *323*, 116224. [CrossRef] [PubMed]
33. Yuan, H.; Huang, P.; Yu, J.; Wang, P.; Jiang, H.B.; Jiang, Y.; Deng, S.; Huang, Z.; Yu, J.; Zhu, W. Efficient Wastewater Treatment and Biomass Co-Production Using Energy Microalgae to Fix C, N, and P. *RSC Adv.* **2025**, *15*, 14030–14041. [CrossRef] [PubMed]
34. Asimakis, E.; Shehata, A.A.; Eisenreich, W.; Acheuk, F.; Lasram, S.; Basiouni, S.; Emekci, M.; Ntougias, S.; Taner, G.; May-Simera, H.; et al. Algae and Their Metabolites as Potential Bio-Pesticides. *Microorganisms* **2022**, *10*, 307. [CrossRef]
35. Chaïb, S.; Pistevos, J.C.A.; Bertrand, C.; Bonnard, I. Allelopathy and Allelochemicals from Microalgae: An Innovative Source for Bio-Herbicidal Compounds and Biocontrol Research. *Algal Res.* **2021**, *54*, 1–15. [CrossRef]
36. Nur, M.M.A.; Mahreni; Murni, S.W.; Setyoningrum, T.M.; Hadi, F.; Widayati, T.W.; Jaya, D.; Sulistyawati, R.R.E.; Puspitaningrum, D.A.; Dewi, R.N.; et al. Innovative Strategies for Utilizing Microalgae as Dual-Purpose Biofertilizers and Phycoremediators in Agroecosystems. *Biotechnol. Rep.* **2025**, *45*, e00870.
37. Onyeaka, H.; Miri, T.; Obileke, K.C.; Hart, A.; Anumudu, C.; Al-Sharify, Z.T. Minimizing Carbon Footprint via Microalgae as a Biological Capture. *Carbon. Capture Sci. Technol.* **2021**, *1*, 1–14. [CrossRef]
38. Novoveská, L.; Nielsen, S.L.; Eroldoğan, O.T.; Haznedaroglu, B.Z.; Rinkevich, B.; Fazi, S.; Robbens, J.; Vasquez, M.; Einarsson, H. Overview and Challenges of Large-Scale Cultivation of Photosynthetic Microalgae and Cyanobacteria. *Mar. Drugs* **2023**, *21*, 445. [CrossRef]
39. Ronga, D.; Biazzi, E.; Parati, K.; Carminati, D.; Carminati, E.; Tava, A. Microalgal Biostimulants and Biofertilisers in Crop Productions. *Agronomy* **2019**, *9*, 192. [CrossRef]
40. Chabili, A.; Minaoui, F.; Hakkoum, Z.; Douma, M.; Meddich, A.; Loudiki, M. A Comprehensive Review of Microalgae and Cyanobacteria-Based Biostimulants for Agriculture Uses. *Plants* **2024**, *13*, 159. [CrossRef]
41. León-Vaz, A.; León, R.; Díaz-Santos, E.; Vigara, J.; Raposo, S. Using Agro-Industrial Wastes for Mixotrophic Growth and Lipids Production by the Green Microalga *Chlorella sorokiniana*. *N. Biotechnol.* **2019**, *51*, 31–38. [CrossRef]
42. Rozenberg, J.M.; Sorokin, B.A.; Mukhambetova, A.N.; Emelianova, A.A.; Kuzmin, V.V.; Belogurova-Ovchinnikova, O.Y.; Kuzmin, D.V. Recent Advances and Fundamentals of Microalgae Cultivation Technology. *Biotechnol. J.* **2024**, *19*, e2300725. [CrossRef]
43. Janssen, M.; Kuijpers, T.C.; Veldhoen, B.; Ternbach, M.B.; Tramper, J.; Mur, L.R.; Wijffels, R.H. Specific Growth Rate of *Chlamydomonas reinhardtii* and *Chlorella sorokiniana* Under Medium Duration Light/Dark Cycles: 13–87 s. *Prog. Ind. Microbiol.* **1999**, *35*, 323–333. [CrossRef]
44. Parmar, P.; Kumar, R.; Neha, Y.; Srivatsan, V. Microalgae as next Generation Plant Growth Additives: Functions, Applications, Challenges and Circular Bioeconomy Based Solutions. *Front. Plant Sci.* **2023**, *14*, 1073546. [CrossRef]
45. Renuka, N.; Guldhe, A.; Prasanna, R.; Singh, P.; Bux, F. Microalgae as Multi-Functional Options in Modern Agriculture: Current Trends, Prospects and Challenges. *Biotechnol. Adv.* **2018**, *36*, 1255–1273. [CrossRef]
46. González-Morales, S.; Solís-Gaona, S.; Valdés-Caballero, M.V.; Juárez-Maldonado, A.; Loredo-Treviño, A.; Benavides-Mendoza, A. Transcriptomics of Biostimulation of Plants Under Abiotic Stress. *Front. Genet.* **2021**, *12*, 583888. [CrossRef]
47. Chanda, M.J.; Merghoub, N.; El Arroussi, H. Microalgae Polysaccharides: The New Sustainable Bioactive Products for the Development of Plant Bio-Stimulants? *World J. Microbiol. Biotechnol.* **2019**, *35*, 1–10. [CrossRef]
48. Stirk, W.A.; Ordog, V.; Novák, O.; Rolčík, J.; Strnad, M.; Balint, P.; van Staden, J. Auxin and Cytokinin Relationships in Twenty-Four Microalgae Strains. *J. Phycol.* **2013**, *49*, 459–467. [CrossRef] [PubMed]

49. Stirk, W.A.; Bálint, P.; Tarkowská, D.; Novák, O.; Strnad, M.; Ördög, V.; van Staden, J. Hormone Profiles in Microalgae: Gibberellins and Brassinosteroids. *Plant Physiol. Biochem.* **2013**, *70*, 348–353. [CrossRef]
50. Nakano, M.; Omae, N.; Tsuda, K. Inter-Organismal Phytohormone Networks in Plant-Microbe Interactions. *Curr. Opin. Plant Biol.* **2022**, *68*, 1–12. [CrossRef] [PubMed]
51. Mutale-joan, C.; Redouane, B.; Najib, E.; Yassine, K.; Lyamlouli, K.; Laila, S.; Zeroual, Y.; Hicham, E.A. Screening of Microalgae Liquid Extracts for Their Bio Stimulant Properties on Plant Growth, Nutrient Uptake and Metabolite Profile of *Solanum lycopersicum* L. *Sci. Rep.* **2020**, *10*, 1–12. [CrossRef]
52. Manjunath, M.; Kanchan, A.; Ranjan, K.; Venkatachalam, S.; Prasanna, R.; Ramakrishnan, B.; Hossain, F.; Nain, L.; Shivay, Y.S.; Rai, A.B.; et al. Beneficial Cyanobacteria and Eubacteria Synergistically Enhance Bioavailability of Soil Nutrients and Yield of Okra. *Heliyon* **2016**, *2*, 1–28. [CrossRef]
53. Yin, W.; Dong, N.; Li, X.; Yang, Y.; Lu, Z.; Zhou, W.; Qian, Q.; Chu, C.; Tong, H. Understanding Brassinosteroid-Centric Phytohormone Interactions for Crop Improvement. *J. Integr. Plant Biol.* **2025**, *67*, 563–581. [CrossRef] [PubMed]
54. Kumar, S.; Korra, T.; Singh, U.B.; Singh, S.; Bisen, K. Microalgal Based Biostimulants as Alleviator of Biotic and Abiotic Stresses in Crop Plants. In *New and Future Developments in Microbial Biotechnology and Bioengineering*; Elsevier: Amsterdam, The Netherlands, 2022; pp. 195–216. [CrossRef]
55. Lee, H.J.; Cho, G.; Kwak, Y.S. Role of *Chlorella fusca* in Promoting Plant Health through Microbiota Regulation in Strawberry. *J. Appl. Microbiol.* **2025**, *136*, lxaf060. [CrossRef]
56. Santini, G.; Biondi, N.; Rodolfi, L.; Tredici, M.R. Plant Biostimulants from Cyanobacteria: An Emerging Strategy to Improve Yields and Sustainability in Agriculture. *Plants* **2021**, *10*, 643. [CrossRef] [PubMed]
57. Jithesh, M.N.; Shukla, P.S.; Kant, P.; Joshi, J.; Critchley, A.T.; Prithiviraj, B. Physiological and Transcriptomics Analyses Reveal That *Ascophyllum nodosum* Extracts Induce Salinity Tolerance in *Arabidopsis* by Regulating the Expression of Stress Responsive Genes. *J. Plant Growth Regul.* **2019**, *38*, 463–478. [CrossRef]
58. Singh, J.S.; Kumar, A.; Rai, A.N.; Singh, D.P. Cyanobacteria: A Precious Bio-Resource in Agriculture, Ecosystem, and Environmental Sustainability. *Front. Microbiol.* **2016**, *7*, 529. [CrossRef]
59. Nishanth, S.; Kokila, V.; Prasanna, R. Metabolite Profiling of Plant Growth Promoting Cyanobacteria—*Anabaena laxa* and *Calothrix elenkinii*, Using Untargeted Metabolomics. *3 Biotech* **2024**, *14*, 1–13. [CrossRef]
60. Sánchez-Quintero, Á.; Fernandes, S.C.M.; Beigbeder, J.B. Overview of Microalgae and Cyanobacteria-Based Biostimulants Produced from Wastewater and CO2 Streams towards Sustainable Agriculture: A Review. *Microbiol. Res.* **2023**, *277*, 1–14. [CrossRef]
61. Wang, Y.; Zhang, G.; Huang, Y.; Guo, M.; Song, J.; Zhang, T.; Long, Y.; Wang, B.; Liu, H. A Potential Biofertilizer—Siderophilic Bacteria Isolated From the Rhizosphere of *Paris. polyphylla* Var. yunnanensis. *Front. Microbiol.* **2022**, *13*, 870413. [CrossRef]
62. Fiorentino, S.; Bellani, L.; Santin, M.; Castagna, A.; Echeverria, M.C.; Giorgetti, L. Effects of Microalgae as Biostimulants on Plant Growth, Content of Antioxidant Molecules and Total Antioxidant Capacity in *Chenopodium quinoa* Exposed to Salt Stress. *Plants* **2025**, *14*, 781. [CrossRef] [PubMed]
63. Moon, J.; Park, Y.J.; Choi, Y.B.; Truong, T.Q.; Huynh, P.K.; Kim, Y.B.; Kim, S.M. Physiological Effects and Mechanisms of *Chlorella vulgaris* as a Biostimulant on the Growth and Drought Tolerance of *Arabidopsis thaliana*. *Plants* **2024**, *13*, 3012. [CrossRef]
64. Minaoui, F.; Hakkoum, Z.; Chabili, A.; Douma, M.; Mouhri, K.; Loudiki, M. Biostimulant Effect of Green Soil Microalgae *Chlorella vulgaris* Suspensions on Germination and Growth of Wheat (*Triticum aestivum* Var. achtar) and Soil Fertility. *Algal Res.* **2024**, *82*, 103655. [CrossRef]
65. La Bella, E.; Baglieri, A.; Rovetto, E.I.; Stevanato, P.; Puglisi, I. Foliar Spray Application of *Chlorella vulgaris* Extract: Effect on the Growth of Lettuce Seedlings. *Agronomy* **2021**, *11*, 308. [CrossRef]
66. Gitau, M.M.; Farkas, A.; Balla, B.; Ördög, V.; Futó, Z.; Maróti, G. Strain-Specific Biostimulant Effects of *Chlorella* and *Chlamydomonas* Green Microalgae on *Medicago truncatula*. *Plants* **2021**, *10*, 1060. [CrossRef] [PubMed]
67. Kim, M.J.; Shim, C.K.; Kim, Y.K.; Ko, B.G.; Park, J.H.; Hwang, S.G.; Kim, B.H. Effect of Biostimulator *Chlorella fusca* on Improving Growth and Qualities of Chinese Chives and Spinach in Organic Farm. *Plant Pathol. J.* **2018**, *34*, 567–574. [CrossRef]
68. Abdel-Tawwab, M.; Mousa, M.A.A.; Mamoon, A.; Abdelghany, M.F.; Abdel-Hamid, E.A.A.; Abdel-Razek, N.; Ali, F.S.; Shady, S.H.H.; Gewida, A.G.A. Dietary *Chlorella vulgaris* Modulates the Performance, Antioxidant Capacity, Innate Immunity, and Disease Resistance Capability of Nile Tilapia Fingerlings Fed on Plant-Based Diets. *Anim. Feed. Sci. Technol.* **2022**, *283*, 115181. [CrossRef]
69. Wuang, S.C.; Khin, M.C.; Chua, P.Q.D.; Luo, Y.D. Use of *Spirulina* Biomass Produced from Treatment of Aquaculture Wastewater as Agricultural Fertilizers. *Algal Res.* **2016**, *15*, 59–64. [CrossRef]
70. Priya, H.; Prasanna, R.; Ramakrishnan, B.; Bidyarani, N.; Babu, S.; Thapa, S.; Renuka, N. Influence of Cyanobacterial Inoculation on the Culturable Microbiome and Growth of Rice. *Microbiol. Res.* **2015**, *171*, 78–89. [CrossRef]

71. Fimbres-Olivarria, D.; Carvajal-Millan, E.; Lopez-Elias, J.A.; Martinez-Robinson, K.G.; Miranda-Baeza, A.; Martinez-Cordova, L.R.; Enriquez-Ocaña, F.; Valdez-Holguin, J.E. Chemical Characterization and Antioxidant Activity of Sulfated Polysaccharides from *Navicula* sp. *Food Hydrocoll.* **2018**, *75*, 229–236. [CrossRef]
72. Piotrowski, K.; Romanowska-Duda, Z.; Messyasz, B. Cultivation of Energy Crops by Ecological Methods Under the Conditions of Global Climate and Environmental Changes with the Use of Diatom Extract as a Natural Source of Chemical Compounds. *Acta Physiol. Plant.* **2020**, *42*, 1–13. [CrossRef]
73. Kiran Marella, T.; Saxena, A.; Tiwari, A. Diatom Mediated Heavy Metal Remediation: A Review. *Bioresour. Technol.* **2020**, *305*, 1–11. [CrossRef]
74. Elhamji, S.; Haydari, I.; Sbihi, K.; Aziz, K.; Elleuch, J.; Kurniawan, T.A.; Chen, Z.; Yap, P.S.; Aziz, F. Uncovering Applicability of *Navicula permitis* Algae in Removing Phenolic Compounds: A Promising Solution for Olive Mill Wastewater Treatment. *J. Water Process Eng.* **2023**, *56*, 1–9. [CrossRef]
75. Udaypal, U.; Goswami, R.K.; Verma, P. Strategies for Improvement of Bioactive Compounds Production Using Microalgal Consortia: An Emerging Concept for Current and Future Perspective. *Algal Res.* **2024**, *82*, 103664. [CrossRef]
76. Calatrava, V.; Hom, E.F.Y.; Guan, Q.; Llamas, A.; Fernández, E.; Galván, A. Genetic Evidence for Algal Auxin Production in *Chlamydomonas* and Its Role in Algal-Bacterial Mutualism. *iScience* **2024**, *27*, 1–16. [CrossRef] [PubMed]
77. Torres-Tiji, Y.; Fields, F.J.; Yang, Y.; Heredia, V.; Horn, S.J.; Keremane, S.R.; Jin, M.M.; Mayfield, S.P. Optimized Production of a Bioactive Human Recombinant Protein from the Microalgae *Chlamydomonas reinhardtii* Grown at High Density in a Fed-Batch Bioreactor. *Algal Res.* **2022**, *66*, 102786. [CrossRef]
78. Gavilanes, F.Z.; Souza Andrade, D.; Zucareli, C.; Horácio, E.H.; Sarkis Yunes, J.; Barbosa, A.P.; Alves, L.A.R.; Cruzatty, L.G.; Maddela, N.R.; Guimarães, M. de F. Co-Inoculation of *Anabaena cylindrica* with *Azospirillum brasilense* Increases Grain Yield of Maize Hybrids. *Rhizosphere* **2020**, *15*, 1–8. [CrossRef]
79. Llamas, A.; Leon-Miranda, E.; Tejada-Jimenez, M. Microalgal and Nitrogen-Fixing Bacterial Consortia: From Interaction to Biotechnological Potential. *Plants* **2023**, *12*, 2476. [CrossRef]
80. Michalak, I.; Chojnacka, K. Algae as Production Systems of Bioactive Compounds. *Eng. Life Sci.* **2015**, *15*, 160–176. [CrossRef]
81. Vessey, J.K. Plant Growth Promoting Rhizobacteria as Biofertilizers. *Plant Soil* **2003**, *255*, 571–586. [CrossRef]
82. Mahanty, T.; Bhattacharjee, S.; Goswami, M.; Bhattacharyya, P.; Das, B.; Ghosh, A.; Tribedi, P. Biofertilizers: A Potential Approach for Sustainable Agriculture Development. *Environ. Sci. Pollut. Res.* **2017**, *24*, 3315–3335. [CrossRef] [PubMed]
83. Nobbe, F.; Hiltner, L. Inoculation of the Soil for Cultivating Leguminous Plants. U.S. Patent US570813A, 3 November 1896.
84. Nosheen, S.; Ajmal, I.; Song, Y. Microbes as Biofertilizers, a Potential Approach for Sustainable Crop Production. *Sustainability* **2021**, *13*, 1868. [CrossRef]
85. Aloo, B.N.; Tripathi, V.; Makumba, B.A.; Mbega, E.R. Plant Growth-Promoting Rhizobacterial Biofertilizers for Crop Production: The Past, Present, and Future. *Front. Plant Sci.* **2022**, *13*, 1002448. [CrossRef]
86. Daniel, A.I.; Fadaka, A.O.; Gokul, A.; Bakare, O.O.; Aina, O.; Fisher, S.; Burt, A.F.; Mavumengwana, V.; Keyster, M.; Klein, A. Biofertilizer: The Future of Food Security and Food Safety. *Microorganisms* **2022**, *10*, 1220. [CrossRef] [PubMed]
87. Sadvakasova, A.K.; Bauenova, M.O.; Kossalbayev, B.D.; Zayadan, B.K.; Huang, Z.; Wang, J.; Balouch, H.; Alharby, H.F.; Chang, J.S.; Allakhverdiev, S.I. Synthetic Algocyanobacterial Consortium as an Alternative to Chemical Fertilizers. *Environ. Res.* **2023**, *233*, 1–15. [CrossRef] [PubMed]
88. Ladha, J.K.; Peoples, M.B.; Reddy, P.M.; Biswas, J.C.; Bennett, A.; Jat, M.L.; Krupnik, T.J. Biological Nitrogen Fixation and Prospects for Ecological Intensification in Cereal-Based Cropping Systems. *Field Crops Res.* **2022**, *283*, 1–34. [CrossRef]
89. Álvarez, C.; Jimenez-Ríos, L.; Iniesta-Pallares, M.; Jurado-Flores, A.; Molina-Heredia, F.P.; Ng, C.K.Y.; Mariscal, V. Symbiosis between Cyanobacteria and Plants: From Molecular Studies to Agronomic Applications. *J. Exp. Bot.* **2023**, *74*, 6145–6157. [CrossRef] [PubMed]
90. Ma, F.; Li, Y.; Han, X.; Li, K.; Zhao, M.; Guo, L.; Li, S.; Wang, K.; Qin, K.; Duan, J.; et al. Microalgae-Based Biofertilizer Improves Fruit Yield and Controls Greenhouse Gas Emissions in a Hawthorn Orchard. *PLoS ONE* **2024**, *19*, e0307774. [CrossRef]
91. Fujita, Y.; Uesaka, K. Nitrogen Fixation in Cyanobacteria. In *Cyanobacterial Physiology: From Fundamentals to Biotechnology*; Elsevier: Amsterdam, The Netherlands, 2022; pp. 29–45, ISBN 9780323961066.
92. Neilson, A.; Rippka, I.; Kunisawa, R. *Heterocyst Formation and Nitrogenase Synthesis in* Anabaena *sp. A Kinetic Study*; Springer-Verlag: Berlin/Heidelberg, Germany, 1971; Volume 76.
93. Bibi, S.; Saadaoui, I.; Bibi, A.; Al-Ghouti, M.; Abu-Dieyeh, M.H. Applications, Advancements, and Challenges of Cyanobacteria-Based Biofertilizers for Sustainable Agro and Ecosystems in Arid Climates. *Bioresour. Technol. Rep.* **2024**, *25*, 1–18. [CrossRef]
94. Iniesta-Pallarés, M.; Álvarez, C.; Gordillo-Cantón, F.M.; Ramírez-Moncayo, C.; Alves-Martínez, P.; Molina-Heredia, F.P.; Mariscal, V. Sustaining Rice Production through Biofertilization with N2-Fixing Cyanobacteria. *Appl. Sci.* **2021**, *11*, 4628. [CrossRef]
95. Ghotbi-Ravandi, A.A.; Shariatmadari, Z.; Riahi, H.; Hassani, S.B.; Heidari, F.; Nohooji, M.G. Enhancement of Essential Oil Production and Expression of Some Menthol Biosynthesis-Related Genes in *Mentha piperita* Using Cyanobacteria. *Iran. J. Biotechnol.* **2023**, *21*, 96–107. [CrossRef]

96. Jiménez-Ríos, L.; Torrado, A.; González-Pimentel, J.L.; Iniesta-Pallarés, M.; Molina-Heredia, F.P.; Mariscal, V.; Álvarez, C. Emerging Nitrogen-Fixing Cyanobacteria for Sustainable Cotton Cultivation. *Sci. Total Environ.* **2024**, *924*, 1–13. [CrossRef]
97. Alvarenga, P.; Martins, M.; Ribeiro, H.; Mota, M.; Guerra, I.; Cardoso, H.; Silva, J.L. Evaluation of the Fertilizer Potential of *Chlorella vulgaris* and *Scenedesmus obliquus* Grown in Agricultural Drainage Water from Maize Fields. *Sci. Total Environ.* **2023**, *861*, 1–10. [CrossRef]
98. Alobwede, E.; Leake, J.R.; Pandhal, J. Circular Economy Fertilization: Testing Micro and Macro Algal Species as Soil Improvers and Nutrient Sources for Crop Production in Greenhouse and Field Conditions. *Geoderma* **2019**, *334*, 113–123. [CrossRef]
99. Kollmen, J.; Strieth, D. The Beneficial Effects of Cyanobacterial Co-Culture on Plant Growth. *Life* **2022**, *12*, 223. [CrossRef]
100. Hussain, A.; Shah, S.T.; Rahman, H.; Irshad, M.; Iqbal, A. Effect of IAA on in Vitro Growth and Colonization of *Nostoc* in Plant Roots. *Front. Plant Sci.* **2015**, *6*, 46. [CrossRef]
101. Toribio, A.J.; Suárez-Estrella, F.; Jurado, M.M.; López, M.J.; López-González, J.A.; Moreno, J. Prospection of Cyanobacteria Producing Bioactive Substances and Their Application as Potential Phytostimulating Agents. *Biotechnol. Rep.* **2020**, *26*, e00449. [CrossRef] [PubMed]
102. Kimura, J.; Nakano, T. Reconstitution of a *Blasia-Nostoc* Symbiotic Association Under Axenic Conditions. *Nova Hedwig.* **1990**, *50*, 191–200. [CrossRef]
103. Eily, A.N.; Pryer, K.M.; Li, F.W. A First Glimpse at Genes Important to the *Azolla–Nostoc* Symbiosis. *Symbiosis* **2019**, *78*, 149–162. [CrossRef]
104. Alvarez, C.; Navarro, J.A.; Molina-Heredia, F.P.; Mariscal, V. Endophytic Colonization of Rice (*Oryza sativa* L.) by the Symbiotic Strain *Nostoc punctiforme* PCC 73102. *Mol. Plant Microbe Interact.* **2020**, *33*, 1040–1045. [CrossRef]
105. García-Encinas, J.P.; Ruiz-Cruz, S.; Juárez, J.; Ornelas-Paz, J.D.J.; Del Toro-Sánchez, C.L.; Márquez-Ríos, E. Proteins from Microalgae: Nutritional, Functional and Bioactive Properties. *Foods* **2025**, *14*, 921. [CrossRef]
106. Youssef, S.M.; El-Serafy, R.S.; Ghanem, K.Z.; Elhakem, A.; Abdel Aal, A.A. Foliar Spray or Soil Drench: Microalgae Application Impacts on Soil Microbiology, Morpho-Physiological and Biochemical Responses, Oil and Fatty Acid Profiles of Chia Plants Under Alkaline Stress. *Biology* **2022**, *11*, 1844. [CrossRef]
107. El-Sayed, H.S.; Hassan, A.; Barakat, K.M.; Ghonam, H.E.B. Improvement of Growth and Biochemical Constituents of *Rosmarinus officinalis* by Fermented *Spirulina maxima* Biofertilizer. *Plant Physiol. Biochem.* **2024**, *208*, 1–7. [CrossRef] [PubMed]
108. Cordell, D.; Drangert, J.O.; White, S. The Story of Phosphorus: Global Food Security and Food for Thought. *Glob. Environ. Change* **2009**, *19*, 292–305. [CrossRef]
109. Gureev, A.P.; Kryukova, V.A.; Eremina, A.A.; Alimova, A.A.; Kirillova, M.S.; Filatova, O.A.; Moskvitina, M.I.; Kozin, S.V.; Lyasota, O.M.; Gureeva, M.V. Plant-Growth Promoting Rhizobacteria *Azospirillum* Partially Alleviate Pesticide-Induced Growth Retardation and Oxidative Stress in Wheat (*Triticum aestivum* L.). *Plant Growth Regul.* **2024**, *104*, 503–521. [CrossRef]
110. Gupta, R.; Kumari, A.; Sharma, S.; Alzahrani, O.M.; Noureldeen, A.; Darwish, H. Identification, Characterization and Optimization of Phosphate Solubilizing Rhizobacteria (PSRB) from Rice Rhizosphere. *Saudi J. Biol. Sci.* **2022**, *29*, 35–42. [CrossRef]
111. Sarikhani, M.R.; Khoshru, B.; Greiner, R. Isolation and Identification of Temperature Tolerant Phosphate Solubilizing Bacteria as a Potential Microbial Fertilizer. *World J. Microbiol. Biotechnol.* **2019**, *35*, 1–10. [CrossRef] [PubMed]
112. Patel, S.K.; Singh, S.; Benjamin, J.C.; Singh, V.R.; Bisht, D.; Lal, R.K. Plant Growth-Promoting Activities of *Serratia marcescens* and *Pseudomonas fluorescens* on *Capsicum annuum* L. Plants. *Ecol. Front.* **2024**, *44*, 654–663. [CrossRef]
113. Tariq, M.; Tahreem, N.; Zafar, M.; Raza, G.; Shahid, M.; Zunair, M.; Iram, W.; Zahra, S.T. Occurrence of Diverse Plant Growth Promoting Bacteria in Soybean [*Glycine max* (L.) Merrill] Root Nodules and Their Prospective Role in Enhancing Crop Yield. *Biocatal. Agric. Biotechnol.* **2024**, *57*, 1–13. [CrossRef]
114. Ibrahim, E.; Ahmad, A.A.; Abdo, E.S.; Bakr, M.A.; Khalil, M.A.; Abdallah, Y.; Ogunyemi, S.O.; Mohany, M.; Al-Rejaie, S.S.; Shou, L.; et al. Suppression of Root Rot Fungal Diseases in Common Beans (*Phaseolus vulgaris* L.) through the Application of Biologically Synthesized Silver Nanoparticles. *Nanomaterials* **2024**, *14*, 710. [CrossRef] [PubMed]
115. Han, L.; Zhang, H.; Bai, X.; Jiang, B. The Peanut Root Exudate Increases the Transport and Metabolism of Nutrients and Enhances the Plant Growth-Promoting Effects of *Burkholderia pyrrocinia* Strain P10. *BMC Microbiol.* **2023**, *23*, 85. [CrossRef]
116. Chowaniec, K.; Zubek, S.; Skubała, K. Exopolysaccharides in Biological Soil Crusts Are Important Contributors to Carbon and Nutrient Storage After the Restoration of Inland Sand Dunes. *Plant Soil* **2025**, 1–14. [CrossRef]
117. Hussain Shah, A.; Naz, I.; Ahmad, H.; Nasreen Khokhar, S.; Khan, K. Impact of Zinc Solubilizing Bacteria on Zinc Contents of Wheat. *J. Agric. Environ. Sci.* **2016**, *16*, 449–454.
118. Sultan, A.A.Y.A.; Gebreel, H.M.; Youssef, H.A.I.A.E. Biofertilizer Effect of Some Zinc Dissolving Bacteria Free and Encapsulated on *Zea mays* Growth. *Arch. Microbiol.* **2023**, *205*, 1–13. [CrossRef]
119. Hassler, C.S.; Behra, R.; Wilkinson, K.J. Impact of Zinc Acclimation on Bioaccumulation and Homeostasis in *Chlorella kesslerii*. *Aquat. Toxicol.* **2005**, *74*, 139–149. [CrossRef]
120. Li, Y.; Liu, S.; Ji, Z.; Sun, J.; Liu, X. Distinct Responses of *Chlorella vulgaris* upon Combined Exposure to Microplastics and Bivalent Zinc. *J. Hazard. Mater.* **2023**, *442*, 1–13. [CrossRef]

121. Wan Maznah, W.O.; Al-Fawwaz, A.T.; Surif, M. Biosorption of Copper and Zinc by Immobilised and Free Algal Biomass, and the Effects of Metal Biosorption on the Growth and Cellular Structure of *Chlorella* sp. and *Chlamydomonas* sp. Isolated from Rivers in Penang, Malaysia. *J. Environ. Sci.* **2012**, *24*, 1386–1393. [CrossRef] [PubMed]
122. Briat, J.F.; Dubos, C.; Gaymard, F. Iron Nutrition, Biomass Production, and Plant Product Quality. *Trends Plant Sci.* **2015**, *20*, 33–40. [CrossRef] [PubMed]
123. Chaudhary, S.; Sindhu, S.S. Iron Sensing, Signalling and Acquisition by Microbes and Plants Under Environmental Stress: Use of Iron-Solubilizing Bacteria in Crop Biofortification for Sustainable Agriculture. *Plant Sci.* **2025**, *356*, 1–26. [CrossRef]
124. Timofeeva, A.M.; Galyamova, M.R.; Sedykh, S.E. Bacterial Siderophores: Classification, Biosynthesis, Perspectives of Use in Agriculture. *Plants* **2022**, *11*, 3065. [CrossRef]
125. Ghazanfar, S.; Hussain, A.; Dar, A.; Ahmad, M.; Anwar, H.; Al Farraj, D.A.; Rizwan, M.; Iqbal, R. Prospects of Iron Solubilizing Bacillus Species for Improving Growth and Iron in Maize (*Zea mays* L.) Under Axenic Conditions. *Sci. Rep.* **2024**, *14*, 1–11. [CrossRef] [PubMed]
126. Brick, M.B.; Hussein, M.H.; Mowafy, A.M.; Hamouda, R.A.; Ayyad, A.M.; Refaay, D.A. Significance of Siderophore-Producing Cyanobacteria on Enhancing Iron Uptake Potentiality of Maize Plants Grown Under Iron-Deficiency. *Microb. Cell Fact.* **2025**, *24*, 1–18. [CrossRef]
127. Davidi, L.; Gallaher, S.D.; Ben-David, E.; Purvine, S.O.; Fillmore, T.L.; Nicora, C.D.; Craig, R.J.; Schmollinger, S.; Roje, S.; Blaby-Haas, C.E.; et al. Pumping Iron: A Multi-Omics Analysis of Two Extremophilic Algae Reveals Iron Economy Management. *Proc. Natl. Acad. Sci. USA* **2023**, *120*, e2305495120. [CrossRef]
128. El Arroussi, H.; Elbaouchi, A.; Benhima, R.; Bendaou, N.; Smouni, A.; Wahby, I. Halophilic Microalgae *Dunaliella salina* Extracts Improve Seed Germination and Seedling Growth of *Triticum aestivum* L. Under Salt Stress. In Proceedings of the Acta Horticulturae, International Society for Horticultural Science, Florence, Italy, 18 November 2016; Volume 1148, pp. 13–26.
129. Renuka, N.; Guldhe, A.; Singh, P.; Ansari, F.A.; Rawat, I.; Bux, F. Evaluating the Potential of Cytokinins for Biomass and Lipid Enhancement in Microalga *Acutodesmus obliquus* Under Nitrogen Stress. *Energy Convers. Manag.* **2017**, *140*, 14–23. [CrossRef]
130. Marchiosi, R.; dos Santos, W.D.; Constantin, R.P.; de Lima, R.B.; Soares, A.R.; Finger-Teixeira, A.; Mota, T.R.; de Oliveira, D.M.; Foletto-Felipe, M.P.; Abrahão, J.; et al. Biosynthesis and Metabolic Actions of Simple Phenolic Acids in Plants. *Phytochem. Rev.* **2020**, *19*, 865–906. [CrossRef]
131. Xu, H.; Tang, Z.; Yang, D.; Dai, X.; Chen, H. Enhanced Growth and Auto-Flocculation of *Scenedesmus quadricauda* in Anaerobic Digestate Using High Light Intensity and Nanosilica: A Biomineralization-Inspired Strategy. *Water Res.* **2023**, *235*, 1–11. [CrossRef] [PubMed]
132. Su, Y.; Ren, Y.; Wang, G.; Li, J.; Zhang, H.; Yang, Y.; Pang, X.; Han, J. Microalgae and Microbial Inoculant as Partial Substitutes for Chemical Fertilizer Enhance *Polygala tenuifolia* Yield and Quality by Improving Soil Microorganisms. *Front. Plant Sci.* **2024**, *15*, 1499966. [CrossRef]
133. Iovinella, M.; Palmieri, M.; Papa, S.; Auciello, C.; Ventura, R.; Lombardo, F.; Race, M.; Lubritto, C.; di Cicco, M.R.; Davis, S.J.; et al. Biosorption of Rare Earth Elements from Luminophores by *G. sulphuraria* (Cyanidiophytina, Rhodophyta). *Environ. Res.* **2023**, *239*, 117281. [CrossRef]
134. Shahbaz, A.; Hussain, N.; Saba, S.; Bilal, M. Actinomycetes, Cyanobacteria, and Fungi: A Rich Source of Bioactive Molecules. In *Microbial Biomolecules: Emerging Approach in Agriculture, Pharmaceuticals and Environment Management*; Elsevier: Amsterdam, The Netherlands, 2022; pp. 113–133, ISBN 9780323994767.
135. Asthana, R.K.; Srivastava, A.; Singh, A.P.; Deepali; Singh, S.P.; Nath, G.; Srivastava, R.; Srivastava, B.S. Identification of an Antimicrobial Entity from the Cyanobacterium *Fischerella* sp. Isolated from Bark of *Azadirachta indica* (Neem) Tree. *J. Appl. Phycol.* **2006**, *18*, 33–39. [CrossRef]
136. Kajiyama, S.; Kanzakt, H.; Kawazu, K.; Kobayashi, A. *Nostofungicidine, an Antifungal Lipopeptide from the Field-Grown Terrestrial Blue-Green Alga Nostoc commune*; Elsevier: Amsterdam, The Netherlands, 1998; Volume 39.
137. Desbois, A.P.; Mearns-Spragg, A.; Smith, V.J. A Fatty Acid from the Diatom *Phaeodactylum tricornutum* Is Antibacterial against Diverse Bacteria Including Multi-Resistant *Staphylococcus aureus* (MRSA). *Mar. Biotechnol.* **2009**, *11*, 45–52. [CrossRef]
138. Ouaddi, O.; Oukarroum, A.; Bouharroud, R.; Alouani, M.; EL blidi, A.; Qessaoui, R.; Radouane, N.; Errafii, K.; Hijri, M.; Hamadi, F.; et al. A Study of Antibacterial Efficacy of *Tetraselmis rubens* Extracts against Tomato Phytopathogenic *Pseudomonas corrugata*. *Algal Res.* **2025**, *86*, 1–11. [CrossRef]
139. Viana, C.; Genevace, M.; Gama, F.; Coelho, L.; Pereira, H.; Varela, J.; Reis, M. *Chlorella vulgaris* and *Tetradesmus obliquus* Protect Spinach (*Spinacia oleracea* L.) against *Fusarium oxysporum*. *Plants* **2024**, *13*, 1697. [CrossRef]
140. Sandor, R.; Wagh, S.G.; Kelterborn, S.; Großkinsky, D.K.; Novak, O.; Olsen, N.; Paul, B.; Petřík, I.; Wu, S.; Hegemann, P.; et al. Cytokinin-Deficient *Chlamydomonas reinhardtii* CRISPR-Cas9 Mutants Show Reduced Ability to Prime Resistance of Tobacco against Bacterial Infection. *Physiol. Plant* **2024**, *176*, 1–16. [CrossRef]
141. Schmid, B.; Coelho, L.; Schulze, P.S.C.; Pereira, H.; Santos, T.; Maia, I.B.; Reis, M.; Varela, J. Antifungal Properties of Aqueous Microalgal Extracts. *Bioresour. Technol. Rep.* **2022**, *18*, 1–8. [CrossRef]

142. Hamouda, R.; Hamouda, R.A.; El-Ansary, M.S.M. Potential of Plant-Parasitic Nematode Control in Banana Plants by Microalgae as a New Approach Towards Resistance. *Egypt. J. Biol. Pest Contro.* **2017**, *27*, 165–172.
143. Bang, K.H.; Lee, D.W.; Park, H.M.; Rhee, Y.H. Inhibition of Fungal Cell Wall Synthesizing Enzymes by Trans-Cinnamaldehyde. *Biosci. Biotechnol. Biochem.* **2000**, *64*, 1061–1063. [CrossRef]
144. Berry, J. Marine and Freshwater Microalgae as a Potential Source of Novel Herbicides. In *Herbicides and Environment*; InTech: Toulon, France, 2011.
145. Berry, J.P. Cyanobacterial Toxins as Allelochemicals with Potential Applications as Algaecides, Herbicides and Insecticides. *Mar. Drugs* **2008**, *6*, 117–146. [CrossRef]
146. Koodkaew, I.; Sunohara, Y.; Matsuyama, S.; Matsumoto, H. Isolation of Ambiguine D Isonitrile from *Hapalosiphon* sp. and Characterization of Its Phytotoxic Activity. *Plant Growth Regul.* **2012**, *68*, 141–150. [CrossRef]
147. Doan, N.T.; Rickards, R.W.; Rothschild, J.M.; Smith, G.D. Allelopathic Actions of the Alkaloid 12-Epi-Hapalindole E Isonitrile and Calothrixin A from Cyanobacteria of the Genera *Fischerella* and *Calothrix*. *J. Appl. Phycol.* **2000**, *12*, 409–416. [CrossRef]
148. Singh, L.; Nayak, A.; Behera, M.; Sethi, G.; Alheswairini, S.S.; Barasarathi, J.; Sayyed, R.; Behera, D.K.; Adarsh, V. Antifungal Potential of Cyanobacterium *Desertifilum dzianense* against Collar Rot Pathogen in Chickpea. *Sci. Rep.* **2025**, *15*, 1–12. [CrossRef] [PubMed]
149. Baden, D.G.; Bourdelais, A.J.; Jacocks, H.; Michelliza, S.; Naar, J. Natural and Derivative Brevetoxins: Historical Background, Multiplicity, and Effects. *Environ. Health Perspect.* **2005**, *113*, 621–625. [CrossRef]
150. Yokoyama, A.; Murata, M.; Oshima, Y.; Iwashita, T.; Yasumoto, T. Some Chemical Properties of Maitotoxin, a Putative Calcium Channel Agonist Isolated from a Marine Dinoflagellate. *J. Biochem.* **1988**, *104*, 184–187. [CrossRef]
151. DellaGreca, M.; Zarrelli, A.; Fergola, P.; Cerasuolo, M.; Pollio, A.; Pinto, G. Fatty Acids Released by *Chlorella vulgaris* and Their Role in Interference with *Pseudokirchneriella subcapitata*: Experiments and Modelling. *J. Chem. Ecol.* **2010**, *36*, 339–349. [CrossRef]
152. Chiang, I.Z.; Huang, W.Y.; Wu, J.T. Allelochemicals of *Botryococcus braunii* (Chlorophyceae). *J. Phycol.* **2004**, *40*, 474–480. [CrossRef]
153. Rastogi, R.P.; Sinha, R.P.; Incharoensakdi, A. The Cyanotoxin-Microcystins: Current Overview. *Rev. Environ. Sci. Biotechnol.* **2014**, *13*, 215–249. [CrossRef]
154. Mohamed, Z.A. Toxic Cyanobacteria and Cyanotoxins in Public Hot Springs in Saudi Arabia. *Toxicon* **2008**, *51*, 17–27. [CrossRef]
155. Christiansen, G.; Yoshida, W.Y.; Blom, J.F.; Portmann, C.; Gademann, K.; Hemscheidt, T.; Kurmayer, R. Isolation and Structure Determination of Two Microcystins and Sequence Comparison of the McyABC Adenylation Domains in *Planktothrix* Species. *J. Nat. Prod.* **2008**, *71*, 1881–1886. [CrossRef] [PubMed]
156. Hodoki, Y.; Ohbayashi, K.; Kobayashi, Y.; Okuda, N.; Nakano, S. ichi Detection and Identification of Potentially Toxic Cyanobacteria: Ubiquitous Distribution of *Microcystis aeruginosa* and *Cuspidothrix issatschenkoi* in Japanese Lakes. *Harmful Algae* **2012**, *16*, 49–57. [CrossRef]
157. Jaki, B.; Orjala, J.; Heilmann, J.; Linden, A.; Vogler, B.; Sticher, O. Novel Extracellular Diterpenoids with Biological Activity from the Cyanobacterium *Nostoc commune*. *J. Nat. Prod.* **2000**, *63*, 339–343. [CrossRef]
158. Casanova, L.M.; Macrae, A.; de Souza, J.E.; Neves Junior, A.; Vermelho, A.B. The Potential of Allelochemicals from Microalgae for Biopesticides. *Plants* **2023**, *12*, 1896. [CrossRef]
159. Rankic, I.; Zelinka, R.; Ridoskova, A.; Gagic, M.; Pelcova, P.; Huska, D. Nano/Microparticles in Conjunction with Microalgae Extract as Novel Insecticides against Mealworm Beetles, *Tenebrio molitor*. *Sci. Rep.* **2021**, *11*, 1–9. [CrossRef]
160. Hassan, M.E.; Mohafrash, S.M.M.; Fallatah, S.A.; El-Sayed, A.E.K.B.; Mossa, A.T.H. Eco-Friendly Larvicide of *Amphora coffeaeformis* and *Scenedesmus obliquus* Microalgae Extracts against *Culex pipiens*. *J. Appl. Phycol.* **2021**, *33*, 2683–2693. [CrossRef]
161. Becher, P.G.; Jüttner, F. Insecticidal Compounds of the Biofilm-Forming Cyanobacterium *Fischerella* sp. (ATCC 43239). *Environ. Toxicol.* **2005**, *20*, 363–372. [CrossRef] [PubMed]
162. Becher, P.G.; Keller, S.; Jung, G.; Süssmuth, R.D.; Jüttner, F. Insecticidal Activity of 12-Epi-Hapalindole J Isonitrile. *Phytochemistry* **2007**, *68*, 2493–2497. [CrossRef]
163. Becher, P.G.; Jüttner, F. Insecticidal Activity—A New Bioactive Property of the Cyanobacterium *Fischerella*. *Pol. J. Ecol.* **2006**, *54*, 653–662.
164. Höckelmann, C.; Becher, P.G.; Von Reuß, S.H.; Jüttner, F. Sesquiterpenes of the Geosmin-Producing Cyanobacterium *Calothrix* PCC 7507 and Their Toxicity to Invertebrates. *Z. Naturforsch. Sect. C-A J. Biosci.* **2009**, *64*, 49–55. [CrossRef]
165. Delaney, J.M.; Wilkins, R.M. Toxicity of microcystin-LR, isolated from *Microcystis aeruginosa*, against various insect species. *Toxicon* **1995**, *33*, 771–778. [CrossRef]
166. Aguilar-Marcelino, L.; Pineda-Alegría, J.A.; Salinas-Sánchez, D.O.; Hernández-Velázquez, V.M.; Silva-Aguayo, G.I.; Navarro-Tito, N.; Sotelo-Leyva, C. In Vitro Insecticidal Effect of Commercial Fatty Acids, β-Sitosterol, and Rutin against the Sugarcane Aphid, *Melanaphis sacchari* Zehntner (Hemiptera: Aphididae). *J. Food Prot.* **2022**, *85*, 671–675. [CrossRef] [PubMed]
167. Perumalsamy, H.; Jang, M.J.; Kim, J.R.; Kadarkarai, M.; Ahn, Y.J. Larvicidal Activity and Possible Mode of Action of Four Flavonoids and Two Fatty Acids Identified in *Millettia pinnata* Seed toward Three Mosquito Species. *Parasit. Vectors* **2015**, *8*, 1–14. [CrossRef]

168. El-Mougy, N.S.; Abdel-Kader, M.M. Effect of Commercial Cyanobacteria Products on the Growth and Antagonistic Ability of Some Bioagents Under Laboratory Conditions. *J. Pathog.* **2013**, *2013*, 1–11. [CrossRef] [PubMed]
169. Medeiros, D.L.; Moreira, Í.T.A. Microalgae Biomass Production from Cultivation in Availability and Limitation of Nutrients: The Technical, Environmental and Economic Performance. *J. Clean. Prod.* **2022**, *370*, 133538. [CrossRef]

agriculture

MDPI

Article

Priestia megaterium KW16: A Novel Plant Growth-Promoting and Biocontrol Agent Against *Rhizoctonia solani* in Oilseed Rape (*Brassica napus* L.)—Functional and Genomic Insights

Bożena Nowak, Daria Chlebek and Katarzyna Hupert-Kocurek *

Institute of Biology, Biotechnology and Environmental Protection, Faculty of Natural Sciences, University of Silesia, Jagiellońska 28, 40-032 Katowice, Poland; bozena.d.nowak@us.edu.pl (B.N.)
* Correspondence: katarzyna.hupert-kocurek@us.edu.pl; Tel.: +48-32-2009-462

Abstract: Plant diseases caused by *Rhizoctonia solani* present a significant challenge in agriculture. While chemical pesticides remain a common control strategy, their use leads to health and environmental problems. In contrast, endophytic bacteria with plant growth-promoting (PGP) activity offer a promising, sustainable alternative. In this context, a novel endophytic *Priestia* megaterium strain, KW16, originated from the bluegrass (*Poa pratensis* L.), demonstrated distinct biocontrol potential against *R. solani.* in vitro assays showed that KW16 inhibited *R. solani* growth by up to 58%, primarily by releasing volatile compounds. In planta experiments further highlighted KW16′s ability to colonize oilseed rape internal tissues, significantly enhancing its growth and development. In the presence of the pathogen, KW16 abolished the negative impact of *R. solani* and promoted plant growth, increasing shoot and root biomass by 216% and 1737%, respectively, when compared to the plants grown in fungal-infested soil. Biochemical and genome analyses confirmed the strain's metabolic versatility, resistance to biotic and abiotic factors, and a whole spectrum of PGP and biocontrol traits such as biofilm formation, production of phytohormones, and synthesis of lytic enzymes, siderophores, and volatiles, alongside its ability to survive in the presence of autochthonous soil microflora. These findings position KW16 as a potent biological alternative to synthetic fungicides, with significant potential for sustainable crop protection.

Keywords: biocontrol; plant growth promotion; *Priestia*; *Rhizoctonia*; genome

1. Introduction

Pathogenic fungi are the most numerous plant pathogens, using plants as a habitat and source of nutrients and posing a threat to their growth and development. Plant diseases caused by these pathogens are a serious problem in agriculture, resulting in yield losses in all globally important crops, such as wheat, rice, corn, potato, soybean, and oilseed rape, and a significant threat to food security [1,2]. A ubiquitous soil necrotroph, *Rhizoctonia solani,* poses a serious risk [3]. In the pathogenesis of the fungus infection, the secretion of reactive oxygen species, detoxifying enzymes, lignocellulosic enzymes, and host-selective phytotoxins plays an essential role. For example, a specific toxin against rice (*Oryza* L.) is a carbohydrate consisting of mannose, N-acetylgalactosamine, N-acetylglucosamine, and glucose. Other biologically active molecules synthesized by *R. solani* are oxalic acid or phenylacetic acid and their derivatives [4]. Characteristic symptoms of *R. solani* infection are gray water-soaked spots on leaves and leaf sheaths, and rotting of hypocotyls, leaves, young seedlings, and roots. Chlorosis and leaf deformation are also frequently observed.

The earliest lesions are observed 24–72 h after infection. Rhizoctoniosis is very difficult to control due to the broad host range of *R. solani* and the transmission of mycelial hyphae in soil, water, and infected plants [4–6].

Plant protection against this pathogen is overwhelmingly based on chemical pesticides. However, the widespread use of such compounds is increasingly disputed, as it leads to serious health and environmental problems, including pollution, destruction of ecosystems, and ecological imbalances [7]. Chemical residues in food are also a problem. For example, a report prepared on behalf of Greenpeace indicated that as many as 50% of tested rapeseed honey samples had exceeded limits for the pesticides used to cultivate this crop [8]. Pesticide contamination of many other bee products has also been demonstrated by Swiatly-Blaszkiewicz et al. [9]. In addition, regular and long-term use of pesticides results in increased resistance in target organisms and the spread of genes that determine resistance to the applied pesticide into the environment [10].

A valuable option to reduce the use of pesticides and their detrimental impacts on humans and benefit microbial diversity, soil chemistry, and underground water lies in biological methods. Therefore, there is a constant search for microorganisms that effectively control pests and plant pathogens and could be used as active ingredients in biopesticides as well as biofertilizers [11,12]. In recent years, special attention has been paid to endophytic bacteria, which inhabit the same ecological niche as phytopathogens. Their constant direct contact with plant cells facilitates their beneficial effects on the host and contributes to systemic plant immunity [13].

Among various microbial candidates, *Priestia megaterium* (formerly *Bacillus megaterium*), a Gram-positive, spore-forming bacterium, holds significant promise as a potential biocontrol agent and has been reported to inhibit bacterial and fungal diseases in various plants. For example, *P. megaterium* KD7, is a known antagonist of *Erwinia amylovora*, a plant pathogen that causes fire blight disease in Rosaceous plants [14]. *P. megaterium T3* reduces leaf spot disease symptoms induced by *Xanthomonas vesicatoria* [15], and the *B. megaterium* strain NBAII-63 suppressed bacterial wilt caused by *Ralstonia solanacearum* in tomato plants [16]. In addition, *B. megaterium* AB4 controlled the disease symptoms caused by *Alternaria japonica* [17], while *B. megaterium* BM344-1 inhibited growth and mycotoxin production in *Aspergillus flavus*, *A. carbonarius*, *Penicillium verrucosum*, and *Fusarium verticillioides* [18]. However, only a few studies deal with combating *R. solani* infections by *Priestia* strains. Zheng et al. [19] described the positive effect of *B. megaterium* strain B153-2-2 on suppressing *Rhizoctonia* root rot in soybeans, and Solanki et al. [20] reported the protective effect of *P. megaterium* on root rot in tomatoes. Nevertheless, the antagonistic efficacy of different *Priestia* representatives against pathogens varies, making exploring their in-depth disease-inhibitory potential crucial.

In addition to its ability to inhibit the growth and development of pathogens, *P. megaterium* was also reported as a plant growth promoter. For instance, Nascimento et al. [21] demonstrated the significant impact of the STB1 strain on the development of roots, shoots, and leaves in tomatoes, which resulted in an expressive increase in plant total dry biomass (295.8 mg) relative to that of the non-inoculated control (81.9 mg). In turn, the *P. megaterium* JR48 strain described by Li et al. [22] notably promoted the growth of *Arabidopsis* (0.25-fold increase in fresh weight), Chinese cabbage (0.29-fold increase in fresh weight), and tomato plants (increase in fresh weight of 17.29%), when compared to the control.

Biological activity, colonization, and promotion of plant growth are determined by the presence of many genes involved in PGP and biocontrol mechanisms, including nitrogen, phosphorus, and sulfur metabolism; biofilm formation; and the production of phytohormones, siderophores, volatile compounds, lytic enzymes, and antibiotics, which may participate in one or both of these activities [23–26]. Understanding the molecular basis of these mechanisms, and thus the potential of the studied strain along with its response to the pathogen at the level of gene expression and revealed activity, provides a more comprehensive view, allowing for future application of the microorganism in agriculture.

Due to the lack of data related to control rhizoctonia-related diseases in oilseed rape by *P. megaterium* and the continuous search for new strains with high efficiency, a wide range of biocontrol characteristics, modes of action, and plant growth promotion (PGP) potential, the aims of our study are as follows: (i) examine biological activity of a new endophytic *Priestia megaterium* strain KW16 isolated from the bluegrass (*Poa pratensis* L.) against *R. solani*, (ii) evaluate the effect of the strain on the growth and protection of oilseed rape (*Brassica napus* L.) against the fungal pathogen, (iii) characterize the KW16 strain in terms of traits necessary for plant colonization, endophytic–host interactions, and biocontrol mechanisms, (iv) identify genes crucial for plant growth promotion and antifungal activity in the genome of the strain, and (v) to determine the effect of *R. solani* on the expression of genes potentially involved in the biocontrol process in the KW16 strain, and its ability to form a biofilm and produce siderophores, the main traits engaged in effective plant colonization and biological activity.

2. Materials and Methods

2.1. Bacterial and Fungal Strains, Growth Media, and Culture Conditions

Priestia megaterium KW16 was isolated from the surface-sterilized roots of the bluegrass (*Poa pratensis* L.) overgrowing soils in the vicinity of Kalina Pond in Świętochłowice, Poland (50.278758 N 18.928032 E), according to the standard protocol [27], and deposited in the Polish Collection of Microorganisms (Wrocław, Poland) under deposit no. B/00441. Bacteria were cultivated routinely in Luria–Bertani broth (LB Broth) at 30 °C with shaking (140 rpm), or on LB agar (LBA) at 30 °C.

Rhizoctonia solani W70, isolated from the tissues of the grapevine (*Vitis vinifera* L.), was derived from the Microbial Culture Collection of the Institute of Biology, Biotechnology, and Environmental Protection (Faculty of Natural Sciences, University of Silesia in Katowice, Poland) and incubated on PDA (A&A Biotechnology, Gdańsk, Poland) at 28 °C.

2.2. In Vitro Antifungal Activity of KW16

The antifungal activity of the KW16 strain towards *R. solani* was tested by a dual-culture in vitro assay on PDA and LBA medium, according to the method described by Chlebek et al. [27]. The 5 mm agar disks of actively growing fungal mycelium were inoculated separately on one side of the Petri dishes with PDA or the LBA medium. A loop of overnight KW16 culture was then streaked 30 mm away from the disk of the pathogen. As a control, plates inoculated only with the fungus were used. Samples were incubated for 3 days at 28 °C. After incubation, the distance between the center of the agar disk and the edge of the actively growing mycelium was measured in both the control plate and those inoculated with the fungus and the bacterium. The percent growth inhibition (PGI) was calculated using the following formula:

$$\mathrm{PGI} = \frac{\mathrm{KR} - \mathrm{R1}}{\mathrm{KR}} \times 100\% \tag{1}$$

where KR represents the distance (in mm) from the point of fungal inoculation to the mycelium growing edges on control dishes; and R1 represents the distance of fungal growth towards the antagonist, from the point of fungal inoculation to the fungal colony margin on plates inoculated with bacteria [28]. The experiment was performed in triplicate.

2.3. Effect of VOCs Produced by KW16 on the Growth of *R. solani*

100 µL of overnight bacterial culture was spread on one half of a BI plate with LBA medium, and a 5 mm agar disk of *R. solani* mycelium was placed at the center of the other half with PDA medium. BI plates inoculated only with the fungus were used as controls. The plates were immediately tightly sealed with Parafilm and incubated for 3 days at 28 °C. The inhibition of fungal growth was calculated as in Section 2.2. The experiment was performed in triplicate.

2.4. In Planta Growth Promotion and Antifungal Activity of KW16

2.4.1. Soil Samples

The soil for planting was collected from an oilseed rape field near Pszczyna, Poland (49.988546 N, 18.848113 E), from the top (0–20 cm) surface layer. Physico-chemical analysis of the sampled soil was performed by the external certified laboratory i2 Analytical Ltd., Poland (Table S1 Supplementary Materials).

2.4.2. Generation of a Rifampicin-Resistant Mutant of KW16

Spontaneous rifampicin-resistant mutants of the strain were obtained to monitor its presence and survival in soil and plant tissues. The KW16 strain was inoculated into LB and grown overnight with shaking (130 rpm) at 30 °C. Later, 100 µL of the overnight culture was plated onto an LBA medium supplemented with 2 µg mL^{-1} of rifampicin and incubated at 30 °C for 24 h. The grown colonies were transferred to an LBA medium with 5 µg mL^{-1} of rifampicin and incubated at 30 °C for 24 h. In each subsequent round, the colonies were transferred onto an LBA medium with increasing antibiotic concentration until colonies growing in the presence of 150 µg mL^{-1} rifampicin were obtained. To confirm the persistence of the acquired resistance, obtained mutants grown in the presence of the highest antibiotic concentration were cultured five times in LB medium without antibiotic selection and then re-cultured on LBA medium with rifampicin at a concentration of 150 µg/mL.

2.4.3. The Inoculum Preparation and Experiment Design

The rifampicin-resistant mutant $KW16^{RIF}$ was inoculated into LB medium supplemented with rifampicin (150 µg/mL) and grown overnight with shaking (130 rpm) at 30 °C. The cells were harvested by centrifugation at 4415× *g* for 15 min, and the pellet was washed three times with sterile 0.9% NaCl, following the cell suspension with a concentration of 10^9 CFU mL^{-1} preparation. Subsequently, four different treatments were designed as follows:

(1) non-inoculated soil (control);
(2) soil inoculated with the $KW16^{RIF}$ strain ($KW16^{RIF}$),
(3) soil inoculated with *R. solani* (RS);
(4) soil inoculated with the $KW16^{RIF}$ strain and *R. solani* (RS + $KW16^{RIF}$).

The study was conducted in a randomized design. Before starting the experiment, the soil was sieved (2 mm) to remove clods and stones. Then, 750 g of the soil was weighed into separate pots, and five seeds of oilseed rape (*B. napus* L.) were randomly sawn in every pot at a depth of 1 cm. In the *R. solani* system, the soil was infested with fungus by introducing 5 mm agar disks of actively growing fungal mycelium. Then, 50 mL of the prepared inoculum was introduced into the pots except for the control and soil inoculated only with *R. solani* (RS). Cultures were carried out in the culture room at a temperature of 21 °C, constant soil moisture, and a 16 h of light photoperiod. On the 14th, 28th, and 35th day of culture, 10 plants were taken from each of the test systems, and the length and weight of the roots and shoots were measured. Additionally, the survival rate of the bacterium in plants (roots and shoots) and their presence in soil were determined. The chlorophyll (Chl), NB, flavonol (Flav), and anthocyanin (Anth) indices were measured on the 35th day of the experiment using a Dualex Scientific+^{TM} sensor (Force-A, Orsay, France) to analyze plants physiological status.

2.4.4. Evaluation of KW16^{RIF} Strain Survival in Plant Tissues and Soil

To verify the endophytic colonization of the oilseed rape with strain KW16^{RIF}, 1 g of plant mass was taken, and isolation of the bacteria was carried out under sterile conditions. Root and shoot fragments of selected plants were surface sterilized by successively immersing in 70% ethanol (roots—30 s, shoots—20 s) and 5% ACE (4–5 min), followed by washing three times with sterile distilled water for 5 min. Sterile samples were homogenized in sterile mortars, and 9 mL of sterile saline (0.9% NaCl) was added. A series of decimal dilutions (10^{-1}–10^{-5}) of the resulting macerate were prepared, and 20 µL of each dilution was plated, in duplicate, onto an LBA medium supplemented with rifampicin (150 µg/mL). The plates were incubated at 30 °C for 48 h. A sterility control was also prepared by inoculating 100 µL of water from the last wash of the plant organ onto an LBA medium with rifampicin (150 µg/mL).

To assess the survival of the KW16^{RIF} strain in soil, 10 g of soil from every treatment was weighed, suspended in 90 mL of 0.9% NaCl with 1% Tween 80, and shaken for 30 min (120 rpm, 28 °C). From the resulting soil solutions, a series of decimal dilutions (10^{-1}–10^{-5}) were prepared, and 20 µL of each dilution was plated, in duplicate, onto LBA medium supplemented with rifampicin (150 µg/mL). Plates were incubated for 24 h at 30 °C.

The pot experiments were carried out from March to April 2022.

2.5. Physiological and Biochemical Characterization of KW16

2.5.1. Colonization and Plant Growth-Promoting Features

Swimming, swarming, and twitching motility assays and oxidase and catalase activity were tested as described by Naveed et al. [29]. Production of exopolysaccharides (EPS) was assessed by cultivating the KW16 on Congo Red Agar (CRA) [30], while solubilization of phosphate was detected on the Pikovskay'a medium containing insoluble $Ca_3(PO_4)_2$ [31]. The Phosphate Solubilization Index (PSI) was calculated according to Edi-Premono [32]. The ability of the KW16 strain to produce indole-3-acetic acid (IAA), ammonia, and 1-aminocyclopropane-1-carboxylate (ACC) deaminase was determined according to the method of Szilagyi-Zecchin et al. [33], Cappuccino and Sherman [34], and Sandhya et al. [35], respectively. Biofilm formation was analyzed using the crystal violet (CV) method in 96-well microtiter plates by staining bacteria with 0.1% CV for 20 min, following a triple wash with 200 µL of PBS buffer and fixation with 200 µL of methanol. Further, 150 µL of ethanol-detained CV was transferred into a new microtiter plate, and the absorbance was measured at 590 nm (modified from Naveed et al. [29]). The ability of the KW16 strain to utilize 1 mM glucose, arabinose, rhamnose, mannose, trehalose, succinic acid, 4-hydroxyphenylacetic acid, fumaric acid, benzoic acid, mannitol, citric acid, or p-coumaric acid as the sole source of carbon and energy was studied using the 96-well microplates as described by Chlebek et al. [27]. All experiments were performed in three biological replicates.

2.5.2. Biocontrol Traits

Production of proteases, cellulases, and lipases was verified, as described by Vijayalakshmi et al. [36], and chitinase production was verified following the method of Kuddus and Ahmad [37]. The ability of the KW16 strain to produce acetoin, 2,3-butanediol, and hydrogen cyanide (HCN) was analyzed according to Johnston-Monje and Raizada [38], Syamala and Sivaji [39], and Ahmad et al. [40], respectively. All experiments were performed in three biological replicates.

2.5.3. Siderophore Production

The siderophore production was evaluated using both qualitative and quantitative methods. For the qualitative assay, the Chrome Azurol Blue (CAS) agar plate method was used [41]. The microtiter plate method was used to quantify the strain's siderophore activity. Briefly, the KW16 strain was inoculated into LB and grown overnight with shaking (130 rpm) at 30 °C. The cells were harvested by centrifugation at 4415× *g* for 15 min. Then, 100 µL of the culture supernatant was added into the separate wells of the 96-well microplates, followed by the addition of 100 µL CAS reagent. After 30 min incubation, the absorbance of each sample was measured at 630 nm using TECAN SPARK 10M Multimode Microplate Reader, and the percentage of siderophore was calculated according to Kumar et al. [42]. Eight replicates for each biological treatment were performed.

Biochemical assays and microbial tests were conducted between 2022 and 2025.

2.6. R. solani Filtrates Preparation

The liquid medium was inoculated with plugs taken from actively growing fungus cultures and incubated for 14 days at 28 °C in darkness. The fungal biomass was collected by centrifugation (4415× *g*, 20 min), and the obtained supernatant was filtered through a 0.22 µm pore size syringe filter to remove fungal cells.

2.7. Influence of R. solani *on the Expression Level of Selected Genes of the KW16 Strain*

In detail, 25 mL of LB medium was inoculated with 1 mL of overnight LB culture of the KW16 strain, and 5 mL of fungal filtrate was added. The KW16 strain grown in LB medium without the filtrate served as a control. Cultures were incubated with shaking (130 rpm) at 30 °C for 72 h. After 24 h, 48 h, and 72 h of incubation, total RNA was isolated from the control and the filtrate-treated cultures using a GeneMATRIX Universal RNA Purification Kit (EURx, Gdansk, Poland) and purified with TURBO DNA-free™ DNase (Invitrogen, ThermoFisher Scientific, Waltham, MA, USA). The concentration and purity of obtained RNA were assessed using an ND-1000 NanoDrop spectrophotometer (ThermoFisher Scientific, Waltham, MA, USA). The synthesis of a single-stranded cDNA was carried out using a RevertAid First Strand cDNA Synthesis Kit (ThermoFisher Scientific, Waltham, MA, USA) with 1 µg of the total RNA and random hexamer primers. The generated cDNA was used as a template in qPCR reactions performed in a 10 µL reaction volume with 5 µL LightCycler® 480SYBR Green I Master, 2 µL PCR-grade water, 0.5 µM of each gene-specific primer, and 2 µL cDNA as a template, in LightCycler®480 Multiwell Plates 96 under the following conditions: 10 min at 95 °C and 45 cycles of 15 s at 94 °C, 30 s at 60 °C, and 30 s at 72 °C (LightCycler® 480 Real-Time PCR System; Roche, Basel, Switzerland). Two biological and two technical replicates were performed for each treatment. The expression levels of the following genes were analyzed: *gltB*, *iucC*, *katA*, *ilvB*, *bdhA*, and *bglA*. In order to increase the stability of the measurements, two genes whose expression was stable across the treatments, *pyk* and *ftsZ*, were used as the internal controls [43]. Each gene-specific primer was designed using Geneious Prime (version 2022.0.1; Table S2—Supplementary Materials). The amplification efficiency of the primers was checked according to Taylor et al. [44]. The relative expression level was calculated according to Livak and Schmittgen [45].

Gene expression studies were conducted in 2023.

2.8. Influence of R. solani on Auto-Aggregation, Biofilm Formation and Siderophore Production by KW16 Strain

Auto-aggregation assays were conducted following the method of Chlebek et al. [27]. Two biological and four technical replicates were performed.

The effect of *R. solani* on the biofilm formation activity of the KW16 strain was analyzed in 96-well microtiter plates. 180 μL of LB medium mixed with the fungal filtrate in the ratio of 5:1 (*v:v*) was introduced into separate wells of the plates and inoculated with 20 μL of the overnight LB bacterial culture. As a control, 180 μL of LB medium inoculated with 20 μL of the bacterial culture was used. The abiotic control was 200 μL of sterile LB medium. The plates were protected from evaporation and incubated for 24, 48, and 72 h at 30 °C. Thereafter, biofilm formation was evaluated, as described in Section 2.5.1. Two biological and eight technical replicates were performed.

To determine the effect of *R. solani* on the siderophore activity of the KW16 strain, 1 mL of the overnight LB bacterial culture was introduced into 25 mL of fresh LB medium, and 5 mL of the phytopathogen filtrate was added. The KW16 strain grown in LB medium without the filtrate served as a control. Cultures were incubated with shaking (130 rpm) at 30 °C for 72 h. After 24 h, 48 h, and 72 h of incubation, cells were harvested by centrifugation at 4415× *g* for 15 min, and siderophore production was quantified as described in Section 2.5.3. Two biological and eight technical replicates were performed.

2.9. Genome Sequencing and Sequence Analysis

Genomic DNA of *P.megaterium* KW16 was extracted using the GeneMatrix Bacterial and Yeast Genomic Purification Kit (EURx). The sequencing was performed with MicrobesNG using the Illumina MiSeq platform with 2 × 250 bp paired-end reads. The results of the sequencing were put through a standard MicrobesNG analysis pipeline and were submitted to the GenBank database under the accession number JAHTKR000000000.1. Functional annotation of the genes was performed using a multitude of tools and databases, such as the NCBI Prokaryotic Genome Annotation Pipeline (PGAP) (www.ncbi.nlm.nih.gov/genome/annotation_prok/, accessed on 24 November 2020) and the eggNOG orthology prediction pipeline (http://eggnog5.embl.de, accessed on 24 November 2020) [46]. The genes that were assigned to multiple COG (clusters of orthologous groups) categories were counted as being present in each of these categories. For the annotation of gene function, genes were compared to the KEGG database (Functional Kyoto Encyclopedia of Genes and Genomes database) (http://www.genome.jp/kegg/, accessed on 24 November 2020) [47]. The identification of gene clusters responsible for the biosynthesis of secondary metabolites was performed using antiSMASH (https://antismash.secondarymetabolites.org/, accessed on 17 December 2020) [48]. The CAZy database (Carbohydrate Active Enzymes database, http://www.cazy.org/, 15 June 2021) was used to classify cell wall degrading enzymes (CWDEs) and divide them into different families. CAZy families were identified with dbCAN2 according to the DIAMOND database [49,50]. Sequencing data and assembly were submitted to a public NCBI database under the BioProject accession number PRJNA529642.

2.10. Statistical Analyses

All data presented in this study are expressed as the mean values and were analyzed using Statistica®13.3 PL (StatSoft® Inc., Tulsa, OK, USA). A Student's *t*-test or a one-way analysis of variance (ANOVA) was performed to evaluate the effect of KW16 and/or RS application on bacteria or plants. For samples with normal distributions, homogeneity of variance was assessed using Levene's or Brown Forsythe test, depending on sample size. Tukey Honestly Significant Difference (HSD) test at the 5% level of significance ($p \leq 0.05$) was used for the comparisons of means. The correlation between plant fitness indices (shoots and roots length, mass, and pigments) and the application of microorganisms was determined *via* Principal Component Analysis (PCA). For data without a normal distribution, the non-parametric U Mann–Whitney test was used. For all data, the different lowercase letters indicate statistical significance between the samples at $p \leq 0.05$ level.

3. Results and Discussion

Microorganisms capable of degrading xenobiotics have been studied at our institute for years. The *P. megaterium* strain KW16 was isolated from the roots of the bluegrass (*Poa pratensis* L.) growing in the soil around the Kalina reservoir contaminated with various toxic compounds, including aromatic hydrocarbons and heavy metals. This strain showed the ability to degrade phenol and its chlorinated derivatives, as well as benzoate and catechol. It also showed resistance to high concentrations of zinc, copper, cadmium, and nickel. Nowadays, there is increasing scientific interest in searching for microorganisms that combine the characteristics of a good plant growth promoter and biocontrol agent and show survival ability in extreme and contaminated environments. Therefore, we attempted to find out whether strain KW16, as an endophyte, is capable of colonizing plants other than grasses, whether it significantly promotes their growth and development, and whether it can assist plants in their fight against pathogens.

3.1. *Priestia megaterium KW16 as an Effective Biocontrol Agent*

In the first stage of the study, we tested the ability of strain KW16 to inhibit the growth of *R. solani*, a common pathogen of oilseed rape. Many studies indicate that vast genera of strains with confirmed PGP traits may lack the ability to control the growth of phytopathogenic fungi in culture, depending on the identity of competing fungi, the mode of transmission of antifungal substances in the environment, and the composition of the growth medium, which may affect the interaction outcomes by modulating antifungal molecules [51]. That is why experiments were conducted on PDA and LBA solid media to assess the effect of diffusible compounds secreted by bacteria. In addition, the impact of microbial volatile compounds was studied on bipartite plates where bacteria and fungi and the media in which they grew did not contact each other.

When microorganisms were grown separately on the LBA or PDA medium, a significant 30% difference was found in fungal growth (Figure 1a,d) in favor of the PDA (Figure 1a). The media also affected mycelial morphology. Similarly, differences in bacterial growth were observed, with the ability of the bacteria to move (motility) and enhanced expansion visible solely on LBA (Figure 1e). Next, dual-culture assays were performed, and it was revealed that the endophyte inhibited the fungus growth in each system relative to the control (Figure 1a,d). This effect was evident to the greatest extent on the LBA medium (58%) and to the least on the PDA medium (21%), as shown in Figure 1b and Figure 1e, respectively.

The almost three-times more potent inhibition on LBA was most likely caused by the fact that the KW16 growing on this medium secreted more antifungal substances and/or they had a different composition than those produced on the PDA medium. The bacteria not only reduced the growth of the fungus but also induced changes in mycelial morphology and color. A lower (50.5%), though statistically insignificant, inhibition than that observed for LBA was exerted by the volatile compounds secreted by the bacteria (Figure 1c). It can, therefore, be assumed that strain KW16 controls *R. solani* mainly through the secretion of volatile compounds. The ability to produce such substances gives the bacteria a significant advantage as microbial volatile compounds are effective even at low concentrations and do not require physical contact with pathogens since they can be transferred over long distances.

In contrast, the lesser inhibition of pathogen expansion observed on the PDA medium may have been due to several factors. First, the production of volatile organic compounds by bacteria may require specific ingredients in the medium, as many studies indicate. For example, Huang et al. [52] observed significantly higher (up to 90%) growth inhibition of *R. solani* and *Pythium phanidermatum* (Edson) by culturing *B. mycoides* on protein-rich media, while on PDA, the bacteria showed no biocontrol activity. A similar relationship was observed by Prigigallo et al. [53] in a study in which bacteria did not show biocontrol ability when cultured on media without peptone or tryptone. These components were necessary to convert amino acids into the antifungal biogenic ammonia, which can act directly on the fungal pathogen or indirectly by increasing the pH of the agar, a phenomenon also observed in other studies [54]. The outstanding antifungal activity of KW16 cultured on LBA medium may also be related to the production of specific volatile organic compounds (VOCs) dependent on the composition of the medium, e.g., when growing on M9 medium, different strains of *B. amyloliquefaciens* produced 2,3-butanediol and acetoin. In contrast, TSA and LBA media favored the synthesis of the antifungal 5-methylheptanone, 2-methylpyridine, and 2-pentanone [55].

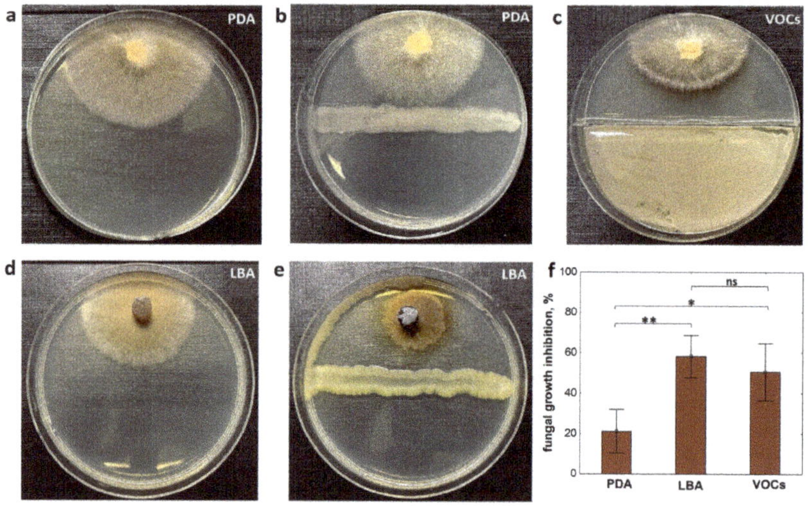

Figure 1. Biocontrol activity of strain KW16 in dual culture tests. Growth of RS on PDA (**a**) and LBA (**d**) medium; and the antifungal effect of diffusing substances secreted by KW16 in PDA (**b**) and LBA (**e**) medium. Inhibition of RS growth in the presence of volatile compounds secreted by KW16 (**c**). Percentage of fungal growth inhibition on different media (**f**). The values are means ± SD of three replicates, and significant differences were assessed with a one-way ANOVA test and Tukey's post hoc comparisons: ns—not significant, * $p < 0.05$, and ** $p < 0.01$.

The inhibition of fungal growth observed on PDA may be related to the ability of *P. megaterium* KW16 to secrete diffusible antifungal agents (e.g., enzymes, soluble bioactive secondary metabolites—see later sections), which can be produced on PDA, the standard medium used to test biocontrol activity against fungi. However, the effect may be less pronounced than on LBA, a medium that favors the synthesis of volatile compounds, because diffusible compounds are only transported over short distances between organisms, also requiring higher concentrations of signaling molecules [56]. Significant differences in the degree of pathogen inhibition on different media are illustrated in Figure 1f.

3.2. *Priestia megaterium KW16 as Pronounced Plant Growth Promoter*

The next step was to evaluate the effect of inoculation of the KW16 strain into the soil on plant growth and development. For the pot study, we chose oilseed rape as a crop of great economic importance and with higher nitrogen, phosphorus, and sulfur requirements than other crops [57]. The increase in the length and weight of the shoots and roots of this plant was studied over 35 days. To further assess whether the effect of KW16 introduction into the soil on plants is due to colonization of plant internal tissues, it was necessary to use spontaneous rifampicin-resistant mutants, as this approach allowed their selective isolation from soil and plants [58]. The colonization of plant internal tissues had to be confirmed because, according to Burns et al. [59], plant species and specific root exudates are among the most critical factors for successful PGPB colonization, and KW16 was isolated from bluegrass, not oilseed rape.

On day 14, after sowing, no significant effect of the bacterial strain on the tested plant parameters was observed. This effect became apparent in the following weeks when all parameters of the control and inoculated plants differed significantly ($p < 0.05$). The most significant differences, even at $p < 0.01$, were observed after 35 days in plant weight when the shoots and roots of inoculated plants were about 30 and 60% longer and about 130% heavier than the control plants (Figure 2c–f). As can be seen from Figure 2a,b, after 35 days of the experiment, leaf numbers, total leaf areas, and development of lateral roots (LR) were significantly higher than the control plants.

The positive effect of bacteria of the genera *Bacillus* and *Priestia* on plant length and weight has long been indicated. Among others, *Priestia* sp. LWS1 increased the biomass of rice shoots and roots by 63% and 47%, respectively [60]. Li et al. [22] also showed that plant inoculation with the *P. megaterium* strain JR48 significantly increased the fresh weight of *Arabidopsis*, Chinese cabbage (*Brassica rapa* ssp. *chinensis*), and tomato (*Solanum lycopersicum*) seedlings in a dose-dependent manner. Up to a 0.3-fold increase in the fresh weight of plant seedlings was observed. However, a clear answer regarding what factor induces the changes is still being sought. It is known that auxin synthesis is responsible for plants' morphogenesis and root/shoots development. A study by Hwang et al. [61] showed that inoculation of *Arabidopsis* and pak choi plants with IAA producing *P. megaterium* strain BP-R2 resulted in heavier, taller, and larger plants. This was also confirmed in studies on the *B. megaterium* strain CDK25, which induced the growth of *Capsicum annum* L. (chili plant) [62]. In addition to modulating phytohormones, PGPB increases external plant nutrient availability and improves plant biomass yield. As the main bacterial target of PGP, the root is the first organ to show morphological and functional changes after bacterial infection. The most prominent effect of PGPB is the increase in lateral root development, which facilitates gas diffusion, water, ion, and nutrient uptake, thus contributing significantly to plant growth [63]. For example, readily available iron can stimulate LR emergence and elongation when locally available to plants [64] through bacterial siderophore production. In the case of nitrogen supply, root length increases under mild conditions and decreases under severe nitrogen deficiency. Interestingly, deficiency in other nutrients results in a progressive reduction in total root length [65,66]. In our study, a more than 100% increase in the roots' biomass, with a significantly higher number of additional lateral roots in oilseed rape, was observed as compared to the control. Such changes in the overall root morphology can also result from altered nutrient balance induced, e.g., by biogenic ammonium, which stimulates branching [66,67]. The role of VOCs produced by *Bacillus* resulting in an increase in lateral roots and root hair was also reported by Fincheira et al. [68] and Gutiérrez-Luna et al. [69].

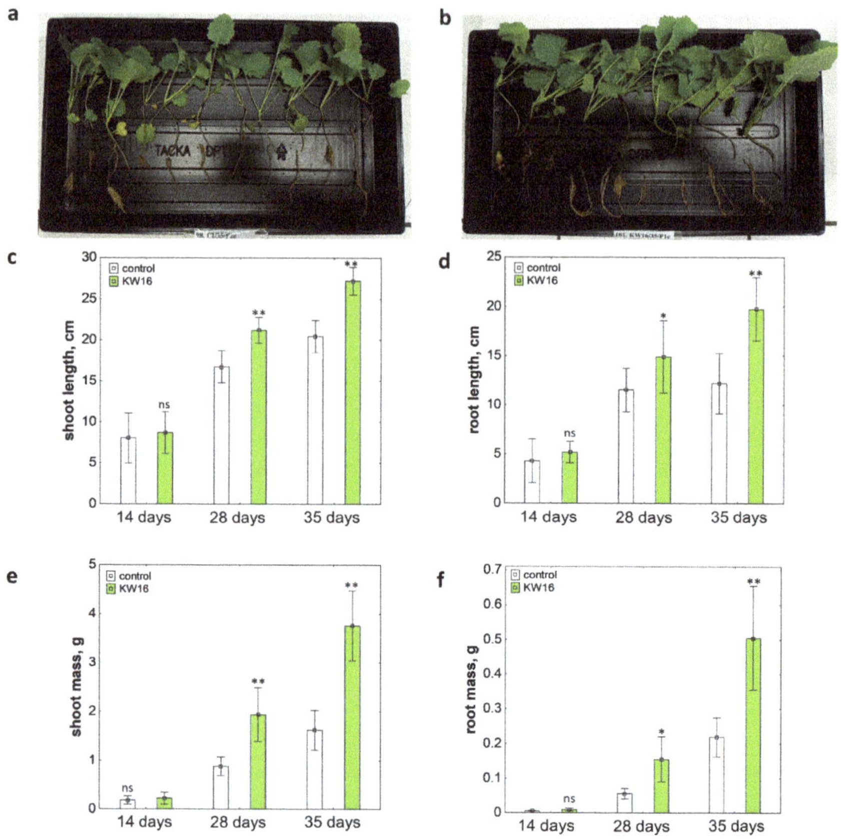

Figure 2. The effect of the KW16 on oilseed rape growth. Control plants (**a**) and plants after 35 days of inoculation with the rifampicin-resistant mutant KW16^{RIF} (**b**). Effect of bacteria on length of shoots (**c**) and roots (**d**), and mass of shoots (**e**) and roots (**f**) after 14, 28, and 35 days of the experiment. The values are means ± SD of ten replicates, and significant differences between corresponding control and inoculated plants were assessed with the *t*-test: ns—not significant, * $p < 0.05$, and ** $p < 0.01$.

3.3. *Priestia megaterium KW16 as an Efficient Plant Protector*

The search for ever-new strains beneficial in reducing plant fungal diseases has led to the selection of many strains from the genera *Bacillus* and *Priestia*. Among them are *B. amyloliquefaciens* LMR2, *B. halotolerans* (SF3 and SF4), and *B. mojarvensis* SF16, with high efficacy in reducing fire blight of apples [70]; *B. velezensis* FZB42 (commercial product), inhibiting lower lettuce rot caused by *R. solani* or *B. amyloliquefaciens* SF14; and *B. amyloliquefaciens* SP10, inhibiting *Monilinia fructigena* and *M. laxa*, the efficacy of which was only slightly lower than that of commercial synthetic fungicides [71–73]. The effectiveness of endophytes against various diseases depends on many factors. For example, Liu et al. [74] found that the growth of endophytic bacteria depends on the source plants and their secondary metabolites and activities; in this study, *P. megaterium* P-NA14 and *P. megaterium* D-HT207 showed higher efficacy against pathogens associated with the host plant and much lower efficacy against fungal pathogens associated with other plant species. Therefore, further studies assessed whether the endophytic strain KW16 derived from the bluegrass has a protective effect on oilseed rape sown in *R. solani*-infested soil, since in vitro assays do not necessarily reflect the strength of antagonistic interactions of bacteria with fungi in planta [75,76]. One of the reasons might be that the amount of secondary metabolites produced by microorganisms in the in vitro systems is likely much higher than that reached in natural habitats [77].

Our previous work confirmed the ability of *R. solani* W70 to damage oilseed rape seedlings from the very early stages of plant development in sterile, soilless cultures [27]. In this study, statistically significant differences between control plants (sown in RS-infested soil) and plants sown in RS-infested soil and inoculated with KW16^{RIF} were observed in the second half of the experiment. The most significant differences (at $p < 0.01$) for both groups were shown in the shoot and root weights on the 35th day of cultivation, which are also visualized in the photo (Figure 3a,b).

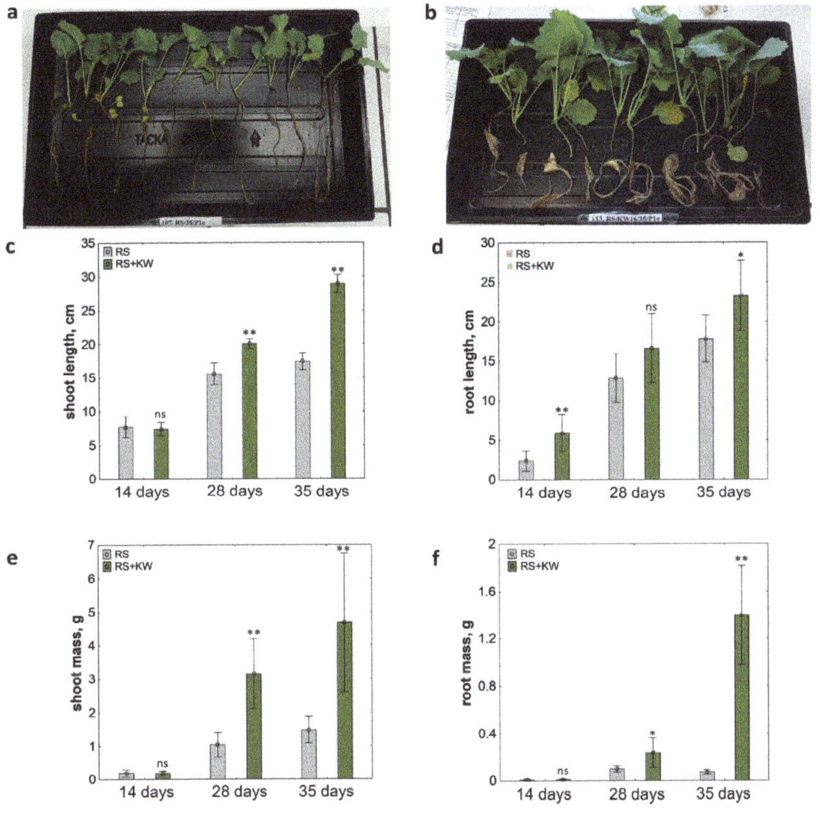

Figure 3. Effect of the KW16 strain on the protection of oilseed rape against *R. solani*. Plants after 35 days of culture in soil infested with the fungus (**a**) and soil co-inoculated with RS and rifampicin-resistant mutant KW16^{RIF} (**b**). Length of shoots (**c**) and roots (**d**), and mass of shoots (**e**) and roots (**f**) of plants after 14, 28, and 35 days of growth in the presence of fungus (RS) or fungus and bacteria (RS + KW16^{RIF}). The values are means ± SD of ten replicates, and significant differences between plants growing in fungus-infested soil and corresponding plants co-inoculated with RS and KW16^{RIF} mutants were assessed with the *t*-test: ns—not significant, * $p < 0.05$, and ** $p < 0.01$.

In addition, a comparison of the shoot and root lengths of the control plants (C) (Figure 2a) and plants grown in *R. solani*-infested soil (RS), as visible in Figure 3a (both groups not inoculated with bacteria), showed that oilseed rape infected with RS had 15% shorter shoots but 46% longer roots.

There are some potential explanations for this phenomenon. The first is based on the "cry for help" hypothesis, by which plants recruit beneficial microorganisms from the soil that can help fight the pathogen after initial contact and disease outbreak. Higher amounts of beneficial microorganisms improve nutrient uptake and activate the plant defense system, which may, at that moment, influence infected plants' growth [78,79]. Thus, in our experiment, root growth-stimulating, yet unknown, microorganisms that plants recruited in response to the presence of the pathogen may have been present in the soil from the oilseed rape fields used in pot experiments. The second explanation is based on the theory that some pathogens manipulate auxin signaling, encouraging plants to develop wounds with emerging lateral roots that can be an entry point for pathogens [80]. An interesting experiment on *Brassica rapa* showed that pathogenic fungi *R. solani* produce VOCs that attract the roots of plants growing in soil, gaining an advantage over them [81]. Also, others reported that fungal volatiles promoted root biomass in vitro. Research by Cordovez et al. [82] revealed that VOCs released from *R. solani* mycelium and sclerotia enhance *A. thaliana* root biomass growth, thus predisposing plants to infection. As a result, the authors observed significant increase in shoot weight of 96%.

Inoculation with KW16^{RIF} abolished the adverse effect of *R. solani* on the oilseed rape shoot length and enhanced root growth. A comparison of the average shoot and root lengths of RS-infested plants with those inoculated with RS + KW16^{RIF} showed that the addition of the bacteria supported shoot length growth by 66% and root growth by 30% (Figure 3c,d), while shoot and root weights increased by 216% and 1737%, respectively (Figure 3e,f). The most likely explanation is related to the increased recruitment of inoculated bacteria from the soil under the influence of signals sent by the fungus, whose increased 'rescue' signal significantly stimulated the growth of plant roots and shoots.

3.4. *Priestia megaterium KW16 as a Facultative Endophyte*

Despite many years of work on the effects of endophytic bacteria on plants, only a few papers have dealt with tracking bacteria in the rhizosphere and plants, mainly because of the complex procedures involved in the study [83]. In this work, we used stable rifampicin mutants, whose isolation on a selective medium allowed us to trace the effect of the presence of *R. solani* on KW16^{RIF} behavior related to the protective impacts on oilseed rape. The addition of rifampicin into the LBA medium effectively inhibited the growth of soil and endophytic bacteria other than the KW16^{RIF} mutant strain, allowing only rifampicin-resistant KW16 to be recovered from the inoculated soil and plants. To assess whether the endophytic strain introduced into the soil can survive directly in the soil and/or whether it can efficiently colonize plant tissues to exhibit long-term effective action, its survival in soil, as well as shoots and roots of oilseed rape, was tested. These relationships were tested for both the system in which the soil was inoculated solely with bacteria (KW16^{RIF}) and the system infested with *R. solani* and inoculated with bacteria (RS + KW16^{RIF}). It was found that on day 14 after sowing, in the KW16^{RIF} system, bacteria were only found in the soil (0.9 log CFU/g soil), whereas, in the presence of the fungus, bacteria colonized not only the soil (0.94 log CFU/g soil) but above all the interior of plant root tissues, where they multiplied to 37 log CFU/g roots fw (Figure 4a).

On day 28 of the experiment in the bacterium system, microorganisms colonized both the soil and the roots of the oilseed rape to amounts of 0.72 CFU/g soil and 2.65 log CFU/g roots fw, respectively. In the system with the fungus (RS + KW16^{RIF}), the presence of bacteria was recorded in quantities similar not only in the soil and roots but also in the shoots (2.77 log CFU/g shoots fw) (Figure 4b). On day 35, the number of microorganisms and their location in the bacterial system did not change significantly. The presence of RS did not affect the number of bacteria in the soil but stimulated bacterial proliferation in the roots, where the number of bacteria increased to 3.5 log CFU/g roots fw. No bacteria were isolated from oilseed rape shoots (Figure 4c).

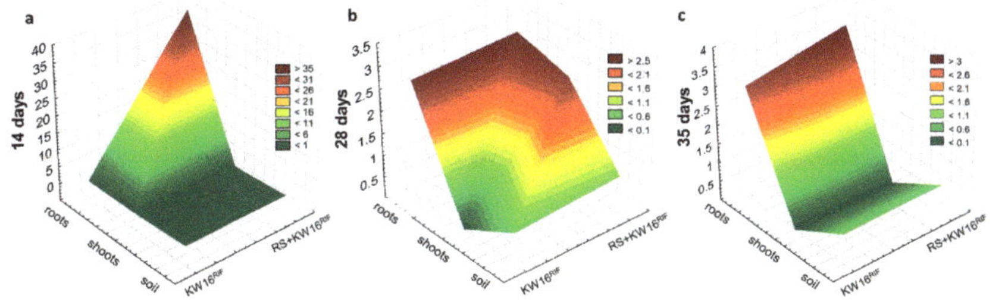

Figure 4. Changes in KW16^{RIF} abundance in soil, shoots and roots of oilseed rape after 14 (**a**), 28 (**b**), and 35 (**c**) days from inoculation in the absence (KW16^{RIF}) and presence of the fungus (RS + KW16^{RIF}). The values are means of at least three replicates.

In our experiment, a distinct translocation of bacteria from the soil to the roots and shoots of plants was observed in a fungus–bacteria system. The number of bacteria in the soil did not change significantly, irrespective of whether the soil was infested with fungus, which meant that after 2 weeks, the equilibrium of the amount of introduced strain in the soil had been established. In contrast, a fascinating pattern was observed when colonizing the interior of plants. For plants grown in non-infested soil, a constant equilibrium of the amount of KW16^{RIF} inside the roots was established. However, the number of bacteria was lower than indicated in the literature for sub-populations inhabiting roots ranging from 5 to 7 log CFU/g fresh weight [84]. In contrast, the significantly higher number of bacteria in the interior of plants growing in the presence of *R. solani* was probably related, as mentioned earlier, to their increased recruitment by the plants to protect themselves from the pathogen. The intensified recruitment on day 14 of cultivation probably also resulted in the entry of some bacteria, e.g., via the transpiration stream or intercellular spaces to aerial parts, which was only observed after 28 days of cultivation. According to Compant et al. [85] and Afzal et al. [86], PGPB can establish stem and leaf population densities between 3 and 4 log CFU/g fw under natural conditions. The decrease in bacterial counts in plants on day 35 of the experiment to amounts comparable to the uncontaminated system was most likely due to the activation of systemic plant resistance (ISR), which not only allowed plants to fight off the pathogen but also prevented over-colonization by *P. megaterium* KW16. Zinniel et al. [87] reported that contrary to pathogenic bacteria found in plant tissues in amounts around 7–10 log CFU/g fw, maintaining low cell densities (between 2 and 6 log CFU/g fw) by endophytic bacteria is crucial for avoiding being detected by the plant.

3.5. *Priestia megaterium KW16 as an Active Colonizer*

The active penetration of plants by bacteria requires a set of complex machinery involving, as the first step, biofilm formation and the secretion of cell wall-degrading enzymes at the level that prevent triggering the plant defense system. Insufficient rhizosphere colonization by PGPB is commonly related to variable and inadequate biocontrol activity in field tests [12]. In our work, when assessing the effect of *P. megaterium* KW16 on oilseed rape growth, it was noted that the bacterium has a promising ability to colonize its roots and shoots, particularly in the presence of *R. solani* (Figure 4), which may be due to its ability to move towards plant-specific exudates. The ability of the KW16 strain to form a biofilm and characteristics related to this process, such as aggregation and siderophore synthesis, were further investigated. Since substances secreted by pathogenic fungi can affect the mentioned features, the effect of filtrates containing *R. solani* exudates after 24 h, 48 h, and 72 h of contact was also assessed (Figure 5).

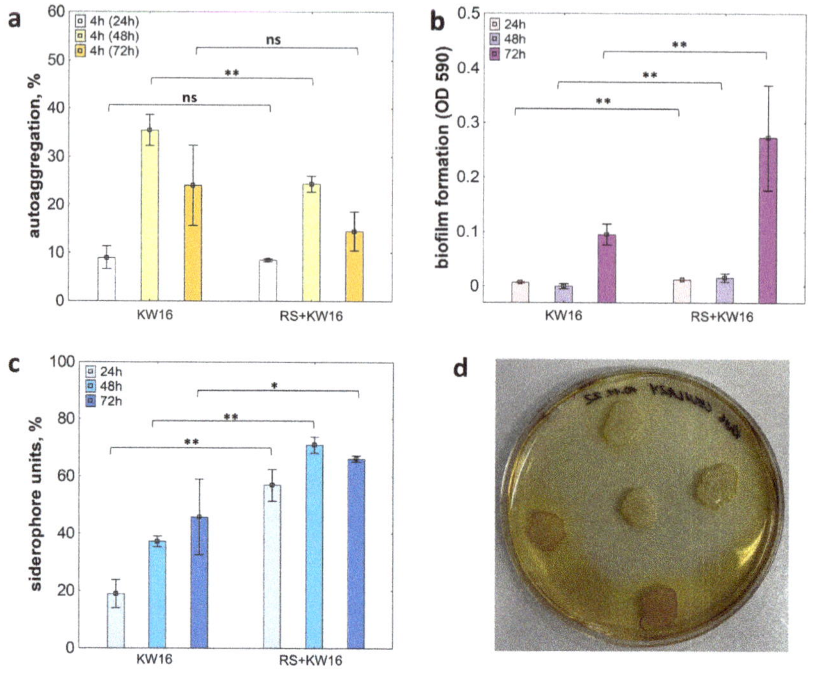

Figure 5. Traits of strain KW16 determining plant colonization. Auto-aggregation (**a**), biofilm formation (**b**), and siderophore production (**c**) after 24 h, 48 h, and 72 h of incubation without and in the presence of fungus (RS). Detection of cellulolytic activity (**d**). The values are means ± SD of two replicates, and significant differences between KW16 and RS + KW16 auto-aggregation capacity (**a**) and siderophore production (**c**) were assessed with the *t*-test, while the Mann–Whitney U test was used to evaluate biofilm formation (**b**): ns—not significant, * $p < 0.05$, and ** $p < 0.01$.

Generally, the ability to auto aggregate may be fixed or reveal itself phenotypically under the influence of an inducer, which may be an environmental abiotic or biotic factor or the need to occupy and persist in a new ecological niche [88], since rapidly aggregating bacteria are more resistant to adverse environmental conditions [89]. After the first 24 h of the experiment, the aggregation of the bacteria recorded after 4 h from the start of the measurement did not change significantly in the control system (KW16) compared to the bacteria in contact with the fungal filtrate. In contrast, after 48 h, a sharp 3-fold increase in aggregation was observed for the control system and a 2-fold increase for the bacteria incubated with the filtrate. After 72 h, a proportional reduction in this capacity was observed in both systems, with the control system still showing faster auto-aggregation (Figure 5a). The reduced auto-aggregation capacity of strain KW16 under the influence of fungal filtrates and the substances contained therein may have been due to the presence of biosurfactants, which, by adhering to bacterial cells, may alter their physicochemical properties, including hydrophobicity. It may have favored cell retention in suspension, limiting the capacity for aggregation [90]. The biosurfactants may have been the metabolic products of both *R. solani* and the bacterium, since *P. megaterium* is known for its efficient production [91,92]. It was also reported that through intercellular communication, bacterial biosurfactants facilitate plant colonization by affecting motility, virulence, and biofilm formation [93].

Biofilm formation on plant roots has several functions. Firstly, by remaining in close contact with the plant, bacteria reduce losses in the mutual exchange of nutrients and hormones. Secondly, by enveloping plant roots, bacteria protect them from invasion by other organisms through antimicrobial compounds secreted in various effective local concentrations. Thirdly, bacteria in the biofilm are more resistant to biotic and abiotic factors, which provides an advantage when occupying a new niche. Checking this ability is, therefore, crucial in identifying a strain with PGP properties. The tests that were conducted showed that KW16 was capable of biofilm formation. The ability of the culture to form a biofilm increased with time. In the control culture, this was because the bacteria entered a metabolic state in which cells produced more extracellular polymeric substances, which are crucial for establishing the bacterial biofilm. The bacteria incubated for 72 h with the fungal filtrate produced a significantly more abundant biofilm than the control cultures (Figure 5b). This may indicate that the previously observed inhibition of planktonic cell auto-aggregation did not reduce the ability to form a biofilm.

Root exudates drive another factor that influences plant colonization. Siderophores secreted by PGPB recruited in the root zone contribute to changes in the social motility of bacterial cells and promote biofilm formation. In addition, they exhibit indirect antagonism against pathogens via competition for nutrients and for the occupation of niches [12]. Our results regarding the secretion of siderophores by KW16 (control) agree with the work of Santos et al. [94], where *B. megaterium* continuously produced siderophores throughout the stationary growth phase. In turn, in the presence of fungal filtrates (RS + KW16), a significant ($p < 0.01$) increase of more than 200% in their concentration relative to the control, was already observed after the first 24 h of incubation. Over the next two days, we observed fluctuations in their concentration, with a maximum after 48 h (Figure 5c). This capability for detecting fungi and their metabolites and the overproduction of certain compounds, including siderophores, can affect the pathogen by destroying it. Bacteria of the *Bacillaceae* family can synthesize a mixture of siderophores and other secondary metabolites, the composition and action of which depend on the fungal competitors [95]. An example is bacillibactin—a siderophore secreted by all members of the *B. subtilis* species complex, whose synthesis seems dependent on a fungal competitor and is linked with direct antifungal activity. One of many examples might be the reaction of *B. amyloliquefaciens,* for which *R. solani* triggered the overproduction of bacillomycin, while the opposite response was observed upon contact with *Rhizomucor variabilis* [96,97].

Endophytic bacteria may enter plants passively through root tips, cracks, or lesions. However, the determining factor for the effective colonization and spread in plant tissues is the production of CWDEs, including cellulase. Tests carried out on plates containing cellulose derivative showed that KW16 was capable of producing and secreting this enzyme directly to the medium. In addition, the enzyme showed high activity, as can be seen by the significant clarity visible in the image (Figure 5d). The presence and activity of cellulose-hydrolyzing enzymes in endophytic strains belonging to different genera, including *Bacillus* and *Priestia,* have been confirmed by numerous studies [98,99].

3.6. *Priestia megaterium KW16 as a Plant Metabolism Booster*

It is known from the literature that the quantity and quality of pigments synthesized by plants in leaves vary with environmental conditions and the physiological state of the plants. Since the colonization of plants by growth-promoting bacteria and the presence and attack of pathogens can affect their synthesis or degradation, plant metabolite parameters such as NBI, Chl, Flav, and Anth indices were studied after 35 days of plant growth in the tested systems (Figure 6).

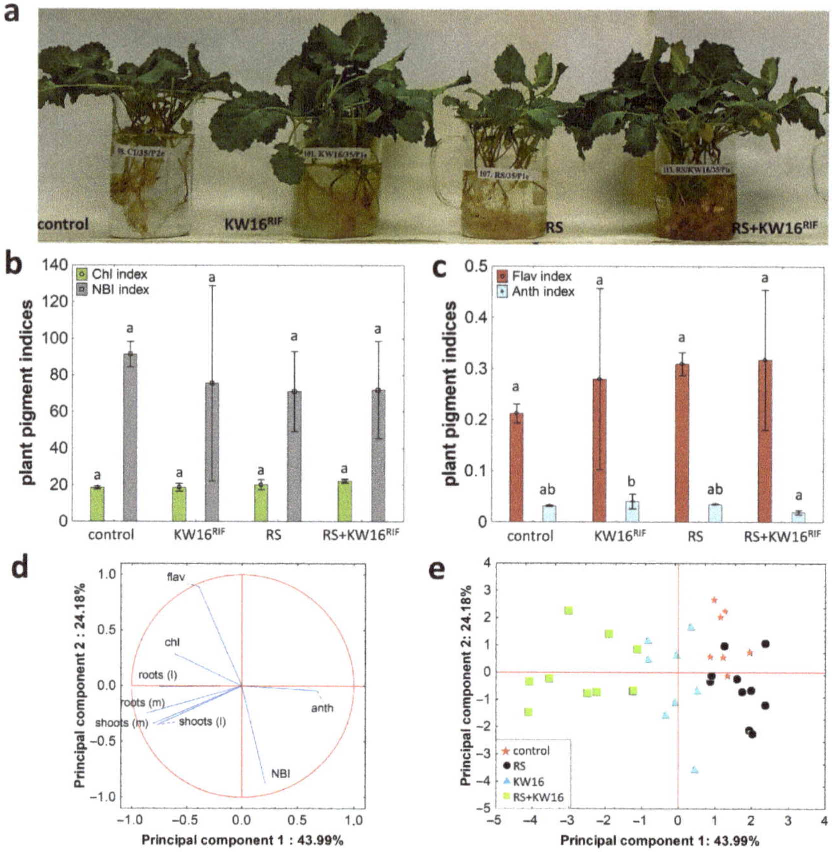

Figure 6. The oilseed rape grown for 35 days in the non-inoculated soil (control), in soil inoculated with bacteria (KW16^{RIF}), with *R. solani* (RS) or the pathogen and bacteria (RS + KW16^{RIF}) (**a**). Plant metabolite parameters: chlorophyll and NBI (**b**), flavonols and anthocyanins (**c**). The values are means ± SD of ten replicates, and significant differences ($p < 0.05$) were assessed with a one-way ANOVA test and Tukey's post hoc comparisons. Different lowercase letters indicate significant differences between treatments. PCA of all examined plant parameters (**d**,**e**).

There were no statistically significant differences observed in the chlorophyll index in any of the experimental setups (Figure 6b). It has been reported that *R. solani*, as a necrotrophic fungus, can hijack some plant components, known as susceptible genes, to induce host susceptibility to fungal infection. Upon infection, plants often respond by cell death and production of ROS, a strategy effective against biotrophic pathogens [100,101]. Unfortunately, for necrotrophs producing additional phytotoxic compounds to exacerbate cell death further, such signals might promote infection [102]. Therefore, most studies on fungus-infected plants indicate that the amount of chlorophyll in infected plants decreases significantly after contact with pathogens [3,103,104]. In a study by Cao et al. [100] on rice and maize with greater resistance to *R. solani*, it was shown that their ability to inhibit chlorophyll degradation induced by contact with the pathogen prevented senescence in these plants and thus limited the spread of fungal infection. Since our previous study [27] showed a destructive effect of *R. solani* on *B. napus* seedlings, we were confident that the plants were not resistant to the fungus. In our current study, therefore, there must have been a factor that prevented chlorophyll degradation in plants incubated in *R. solani*-infected soil. Thus, the only reasonable explanation for the lack of chlorophyll degradation in oilseed rape leaves could be the presence of other indigenous microorganisms in the soil, as previously mentioned, which had a protective effect on germinating and growing plants.

Flavonoids are another group of plant pigments important in the context of pathogen invasion. They control plant cell wall synthesis, participate in mutual interactions between plants and symbiotic microorganisms, and also act as phytoalexins that modulate plant defensive mechanisms, e.g., sakuranetin, which improved the plant's resistance to *R. solani* [105]. Moreover, catechins and rutin directly inhibit pathogen invasion by preventing biofilm formation, disrupting the fungal cell wall, or uncoupling the electron transport chain [106]. In our study, a higher flavonol index was observed in all experimental setups compared to the control. However, these differences were not statistically significant ($p > 0.05$) (Figure 6c). Interestingly, the Anth index was significantly lower in the plants treated with the fungus and bacterium than in the other plants. Reduced anthocyanin concentrations with increased flavonoid synthesis in tomato plants after *R. solani* contamination were also observed in previous work [3]. Confirmation of the protective role of flavonoids can be found in a study conducted by [107], which showed a correlation between plant infection and increased expression of genes coding for isoflavonoid-specific enzymes involved in the biosynthesis of phytoalexins. Also, PGP bacteria may induce flavonoid production in plants, increasing the plant defense system [105]. For example, Li et al. [107] inoculated Nanguo pear fruit with *P. megaterium* strain PH3 and observed an enhanced and prolonged production of phenolics and flavonoids. A similar trend was observed in their previous study [22] on the protective role of *P. megaterium* strain JR48 upon infection of *A. thaliana* by the pathogenic bacterium *Xanthomonas campestris*. Strain JR48 induced the biosynthesis of lignin and secondary metabolites such as flavonoids and salicylic acid in planta. As a result, the enforced plant cell wall more efficiently hinders pathogen entry, while phenolic compounds can be converted to quinones with antimicrobial activity.

Nitrogen balance index (NBI) reflects the ratio of chlorophyll to epidermal flavonoids (Chl/Flav) [108]. It is widely used for diagnosing plants' N nutrition, crucial for better crop yield prediction. Our study showed no statistically significant differences between the NBI in all four systems (Figure 6b).

In order to assess the mutual correlations between the determined plant parameters, Principal Component Analysis (PCA) was used. Three main components described 85.5% of the variability, with the most significant contribution from principal component 1 (PCA1), which accounted for 43.99% of the total variability, and second principal component (PCA2), which accounted for 24.18% of the total variability. In total, the two components explained 68.17% of the variability. PCA1 can be characterized as a factor depicting plant growth promotion, and PCA2 as a factor representing plant metabolic status.

As one can see from Figure 6d,e, plants can be divided into four clusters depending on the experimental setup. The group of plants treated with RS emerged as a separate cluster due to differences in metabolic status relative to the control group. On the other hand, the two groups treated with bacteria were separated primarily due to the intensity of the plant growth promotion effect, although some plants manifested changes similar to the pathogen-treated group. This was probably because the presence of *R. solani* in the soil did not induce plant disease, despite the detrimental effect previously found on oilseed rape [27]. However, the fungus and its diffusible and volatile secondary metabolites may have stimulated the plant defense system by acting similarly to vaccines. A growing body of evidence suggests that adequately severe disruption of plant homeostasis can result in increased plant fitness and resistance to a wide range of pathogens [109]. Ethylene, salicylic acid, and jasmonic acid—phytohormones playing a central role in plants' defense systems—are orchestrated by such disturbances triggered by abiotic and biotic factors.

The resultant resistance is determined by the interplay between these pathways [110]. For example, research conducted by Trapet et al. [111] shows that Fe deficiency arising from the activity of fungal siderophores induces plants' defense against *B. cinerea* through the ethylene-signaling pathway. In turn, *B. velezensis* VOCs triggered a salicylic acid-mediated defense signaling pathway, resulting in improved plant biomass, chlorophyll content, and biocontrol effectiveness in pepper seedlings [112]. Sharifi and Ryu [113] have found that the application of *B. subtilis* GB03 volatiles elicited 90% of ISR and prevented the invasion of *B. cinerea* into *A. thaliana* plants, even though the bacteria themselves had not shown any antifungal activity in vitro.

3.7. Genetic Potential and Biochemical Features of KW16 as a Comprehensive PGP and Biocontrol Factor

The ability of bacteria to invade plant internal tissues and fight pathogens is determined by chance features and bacterial genetic determinants that enable mutual crosstalk between plants and bacteria and between bacteria and pathogens [83,86]. Therefore, upon demonstrating the positive effect of the endophytic *P. megaterium* KW16 on plant growth and protection against the fungal pathogen, its genome was explored, and biochemical tests were carried out to shed light on the potential of the strain for PGP and biocontrol activity at the molecular level.

The general properties of the sequenced genome are presented in Table S3 (Supplementary Materials). As a result, 6.167.380 bp were assembled into 141 contigs with an average G + C content of 64%. In total, 6402 genes, of which 6231 were annotated as coding DNA sequences (CDSs), 123 tRNAs, and 40 rRNAs, have been predicted. Additionally, 5323 out of the 6231 predicted CDSs (85.40%) were, assigned to a cluster of orthologous groups (COG). The genome of the KW16 strain was enriched with genes belonging to five COG categories: transcription (K category, 10.05%), amino acid transport, and metabolism (E category, 10.84%), inorganic ion transport and metabolism (P category, 6.73%), carbohydrate transport and metabolism (G category, 7.50%), cell wall, membrane, envelope biogenesis (M category, 4.72%), and energy production and conversion (C category, 5.56%). Specifically, the number of genes participating in the G and E categories indicates the ability of KW16 to metabolize the wide carbon and nitrogen sources. It is worth noting that the metabolism class was the most significant functional group, which assigned 2262 predicted genes, followed by the poorly characterized class (1381, 25.90%), information storage and processing class (943, 17.70%), and cellular process and signaling class (737, 13.82%). The KEGG functional analysis identified 3125 genes (50.20% of predicted CDS). The most significant fraction of these genes were involved in the metabolism of carbohydrates (343; 10.98%), amino acids (282, 9.09%), cofactors and vitamins (171, 5.47%), and energy metabolism (137, 4.38%). Figure S1 and Table S4 (Supplementary Materials) present the CDS numbers allocated to the KEGG and COG. Furthermore, we analyzed the potential production of secondary metabolites using the antiSMASH tool, finding possible gene clusters involved in the biosynthesis of a siderophore, terpene, lassopeptide, phosphonate, and lanthipeptide (Table S5—Supplementary Materials). The functional annotation of proteins based on the eggNOG and KEGG database revealed crucial genes contributing to plant-beneficial and biocontrol activity of the KW16 strain, and were grouped according to their general roles.

It is well documented that the first step of effective plant root colonization involves the coordinated expression of many genes engaged in bacterium motility, chemotaxis, and adherence [114]. Bacterial motility, which depends mainly on the presence of flagella or pili, is one of the key factors promoting adhesion to various surfaces and biofilm formation [115]. The genome of the KW16 strain showed the presence of a number of genes involved in flagella biosynthesis and folding, such as genes of the *flb*, *flg*, and *fli* operons (Table S6—Supplementary Materials).

The results of tests on solid media with increasing agar content confirmed its ability to swim and express swarm motility (Table S7—Supplementary Materials), which is powered by rotating flagella. Twitching motility, a slow surface movement not requiring the locomotion structure [116], was also observed.

Swarming is described as a social form of motility that allows flagellated bacteria to travel rapidly toward a nutrient-rich environment, and it is supposed to be crucial for plant colonization by bacteria [117]. This is confirmed by the results of Gao et al. [118], among others, on tomato root colonization by mutants of *B. subtilis* SWR01 defective in motility and chemotaxis. The active movement of bacteria oriented by chemotaxis is determined by the presence of genes responsible for transmitting external environmental signals, modulating the activity and direction of rotation of the bacterial flagellum. An essential role in this process is played by methyl-accepting chemotaxis proteins (MCPs) belonging to transmembrane receptors found in the inner cell membrane of many bacteria [119]. In silico analyses confirmed the presence of genes crucial for chemotaxis in the KW16 genome (Table S6—Supplementary Materials). Apart from genes for the MCPs, *che*, and *mot* genes encoding response regulators, chemotaxis, signal transduction, and flagellar motor proteins were identified. The importance of chemotaxis in plant colonization was reported for *B. subtilis* FB17 in *A. thaliana* [120] and *B. subtilis* N11 in cucumber and banana roots [121]. The studies on *B. subtilis* mutants also confirmed the significant role of genes involved in chemotaxis. Allard-Massicotte et al. [122] showed that mutants in 10 genes encoding chemoreceptors exhibited a reduced ability to colonize *Arabidopsis* roots compared to the wild strain.

The auto-aggregation and biofilm formation ability confirmed by biochemical tests and genetic analysis (Figure 5, Table S6 and S7—Supplementary Materials) highlights KW16's potential to colonize plants and thus express its capacity to promote and protect plant hosts against biotic and abiotic stresses. One crucial component of biofilms are EPSs, affecting bacteria aggregation efficiency and attachment to the root surface [123]. These polymeric compounds also contribute to the survival of bacteria in the plant or soil by acting as a barrier to plant defense mechanisms or protecting against environmental factors such as drought or predation [124]. The production of polysaccharides by KW16 was confirmed by plating the strain on the CRA medium, while the genome analyses revealed the presence of *algA* involved in the biosynthesis of alginate and the *hxlBAR* cluster engaged in the biosynthesis of polysaccharides (Table S6—Supplementary Materials).

Plant host colonization is associated with osmotic stress, which bacteria can cope with by synthesizing and accumulating osmoprotectants. Similarly to *P. megaterium* ZS-3 [24], in the KW16 genome, we identified key genes involved in the mitigation of osmotic stress, including genes for the biosynthesis of highly soluble molecules such as proline (*pro*) and glutamine (*gln*), along with genes involved in glycine betaine/proline transport (*proPVWX*). In addition, the *des* genes encoding fatty acid desaturase and the *clsB* gene for cardiolipin synthase, and enzymes modulating lipid and fatty acid composition in the lipid membrane, thus playing a central role in the tolerance to salt and to cold stress, were also found (Table S6—Supplementary Materials).

To successfully colonize a plant, bacterial endophytes must be adapted to survive in an environment rich in reactive oxygen or nitrogen species, which plants produce as a mechanism protecting against viral, bacterial, or fungal infections, or during colonization by endophytic organisms [98,123]. Therefore, plant-colonizing organisms are equipped with mechanisms to overcome oxidative stress. A wide range of genes responsible for the oxidative/nitrosative stress response have been identified in the genome of *P. megaterium* KW16, including catalase (*catE*, *catX*, *catA*), superoxide dismutase (*sodA*), thiol peroxidase (*tpx*), or nitric oxide dioxygenase (*hmp*) genes (Table S6—Supplementary Materials).

Similarly to other known plant colonizers and biocontrol agents [125], KW16 produced catalases and oxidases (Table S7—Supplementary Materials), enzymes capturing toxic free radicals generated by abiotic and biotic stresses and enabling them to enter and settle in *B. napus* tissues and protect the plant from the oxidative damage caused by *R. solani* infection. Since ROS are essential for vegetative hyphal growth and the differentiation of conidial anastomosis tubes, in addition to their protective activity, catalases may also have a crucial role in the biocontrol process. For example, Srikhong et al. [126] observed that the *Bacillus* sp. strain M10 produced a protein similar to vegetative catalase, the activity of which induced abnormal elongation of the shoots and swelling and cracking of *C. capsici* conidia.

The ability of KW16 to produce phytohormones, including IAA, is also a vital characteristic of the strain, positively affecting the growth and development of oilseed rape roots (Figure 1). The results of the biochemical tests show that strain KW16 produced this hormone at a concentration of 17.42 ± 0.9 µg/mL (Table S7—Supplementary Materials). However, since the strain's ability to produce IAA was evaluated in a tryptophan-supplemented medium, it is impossible to assess the amount of IAA secreted by this strain in the plant. IAA production by endophytic strains in vitro is widely discussed in the literature. In a study by Kukla et al. [127], all strains tested were able to produce between 0.7 ± 0.1 and 132.0 ± 1.3 µg/mL of IAA, while a study conducted by Amaresan et al. [128] showed that endophytic bacteria isolated from tomatoes and chilies produced IAA at concentrations ranging from 15.0 to 59.2 µg/mL. However, since plant growth is determined by plant-synthesized auxin, IAA produced by the bacteria can only enhance shoot and root development in cases of low plant auxin levels. Therefore, it is difficult to determine at what concentration bacterial IAA can positively affect the plant. Analysis of the KW16 genome revealed genes encoding proteins related to auxin biosynthesis and modulation, including genes for tryptophan biosynthesis (*trpABCDE*) and conversion to IAA intermediates (*yod*, *yug*), indicating trp-dependent IAA production. The presence of the *patB*, *yclB/yclC*, and *dhaS* encoding tryptophan transaminase, indole-3-pyruvate decarboxylase, and indole-3-acetaldehyde dehydrogenase, respectively, suggests that in the studied strain, IAA synthesis proceeds through the indole-3-pyruvate (IPyA) pathway. However, the *ycl* genes might also be engaged in the transformation of tryptophan to tryptamine, a substrate characteristic of the TAM pathway of IAA biosynthesis [129]. Genes involved in IAA biosynthesis were also identified in the *B. megaterium* strain STB1, which increased the biomass of tomato plants [21], and the *B. megaterium* strain RmBm31, which significantly improved *Arabidopsis* seedling growth [25]. Apart from the genes engaged in IAA synthesis, several genes for polyamine metabolism (*put*, *spd*, *spm*) and transport (*puuP*, *potABCD*) were also identified in the genome of the strain. Polyamines such as putrescine, spermidine, and spermine play essential roles in plant growth promotion as they are among the synthetic precursors of GABA [130]. As an example, the GABA-producing *B. velezensis* strain FX-6 increased biomass in tomato plants by improving nutrient absorption and influencing hormone levels, promoting the elongation and differentiation of plant cells [131].

Another important mechanism that promotes plant growth is the ability of endophytes to produce ACC deaminase, reducing ethylene levels in the plant [132]. At low concentrations, ethylene regulates plant growth and various metabolic processes. However, under stress conditions, the ethylene concentration in plants increases significantly, adversely affecting root growth and the metabolism of the entire plant. The strain KW16 grown on DF medium, supplemented with 3 mM ACC as a nitrogen source, confirmed its ability to produce the enzyme (Table S7—Supplementary Materials) and suggested its role in reducing ethylene levels during rapeseed growth in soil infested with *R. solani* and in alleviating the adverse effects of the pathogen (Figure 3).

Genome analysis also confirmed the potential of the KW16 strain for ammonia production and acquisition. The presence of genes for urea degradation and transport (*ureABCDEFGH*), the *amtB* gene for ammonia transport, and a number of genes encoding enzymes involved in ammonia synthesis (Table S6—Supplementary Materials) were found. We hypothesize that ammonium produced by the strain might have contributed to promoting the growth of oilseed rape roots and shoots, thus increasing the fresh weight of the inoculated plants (Figure 2), and also helped the plants to minimize the effect of *R. solani* (Figure 3). Moreover, this inorganic volatile compound could negatively affect the growth of the pathogen (Figure 1).

Phosphorus is the primary nutrient mandated for plant growth and increased plant immune and defense mechanisms. The presence of genes of the *pstABSC* operon involved in phosphate transport, genes *phoA*, *phoD* encoding phosphatases, and *phoR* regulating phosphate uptake, together with genes involved in phosphate metabolism, e.g., *phoQ*, *pqq*, *murF*, *pepM*, *phnW*, confirms the ability of the strain KW16 to manage phosphate efficiently (Table S6). Indeed, the bacterium showed the ability to solubilize phosphate (PSI 1.3 $\pm$ 0.08) (Table S7), as evidenced by clear zones around the bacterial colonies on a medium containing tricalcium phosphate as the sole source of phosphorus (Figure S1a—Supplementary Materials). However, it is suggested that in vitro P solubilization by microorganisms might not be associated with the promotion of plant growth [133], and tricalcium phosphate is supposed to be unreliable as a universal selection factor for isolating phosphate-solubilizing bacteria (PSB) for enhancing plant growth [134]. Since PSB-gathering phosphates in the plant tissues deplete the nutrient pool for pathogens, phosphate solubilization may also be considered a pathogen control mechanism [135]. Therefore, the genes discussed above are also included in Table S7 (see further paragraphs).

Genes involved in sulfur metabolism in the KW16 strain are also noteworthy. Many bacteria can use sulfate as a primary sulfur source under sulfur starvation [136]. The assimilatory reduction in sulfate and the formation of cysteine comprise a sequence of enzymatic reactions catalyzed by adenosine 5′-phosphosulfate kinase, 3′-phosphoadenosine 5′-phosphosulfate (PAPS) sulfotransferase, and sulfite reductase (among others), encoded by *cysC*, *sat*, and *cysIJ*, respectively (Table S6). These genes were identified in the KW16 genome along with *cysPW*, which is responsible for the transport of sulfonates. In turn, the presence of *ssuD* and *ssuE* genes (Table S6) encoding alkanesulfonate monooxygenase indicates its ability to obtain sulfur from the environment by organosulfonate or sulfonate-derived compound utilization. Moreover, sulfur-oxidizing bacteria improve sulfur availability for many plants, including oilseed rape promoting their growth and inducing resistance against various pathogens such as *R. solani* [137].

Root exudates abundant in sugars, organic acids, amino acids, polyamines, fatty acids, purines, vitamins, and other potent sources of nutrients for PGPB have a massive impact on shaping the plant microbiome, especially since their composition varies according to the plant genotype and growth stage and can be modified in the presence of pathogens [98,138]. The positive chemotaxis toward plant exudates and the ability of bacteria to utilize them are crucial for the successful colonization of the root surface and competition with other microorganisms [139]. Biochemical tests showed that the strain KW16 exhibits metabolic activity and nutrient diversity, utilizing a wide range of organic compounds such as organic acids, phenolic compounds, and sugars (Table S7—Supplementary Materials). Furthermore, the COG and KEGG analyses of the KW16 genome (Figure S2 and Table S4—Supplementary Materials) confirmed the presence of many genes involved in the metabolic transformation of these compounds. We speculate that this metabolic diversity supported strain adaptability to the new environment during plant colonization and increased its potential to compete with microorganisms present in the soil. Moreover, this may explain the fact that the KW16 strain originated from the bluegrass was able to colonize the roots of the oilseed rape.

Except for the above-described potential of the KW16 strain for plant promotion, the tested strain holds tremendous potential as a biocontrol agent. The biochemical assays and genome analyses revealed many traits and mechanisms engaged in combating pathogens. An exciting feature is a spectrum of genes for hydrolytic enzymes (*cel*, *yqeZ*, *amyY*, *ypmRS*) (Table S8—Supplementary Materials) and the activity of cellulases, proteases, amylases, and lipases confirmed by biochemical tests (Table S7, and Figure S2b—Supplementary Materials). A wide array of hydrolytic enzymes produced by the strain may have the significant advantage of suppressing multiple pathogens present in the host rhizosphere, since it is believed that the hydrolysis of the cell wall of pathogenic fungi is the result of the coordinated action of the lytic enzyme complex. For example, the production of chitinase, β-1,3-glucanase, and protease by four strains of the genus *Bacillus* was an important mechanism responsible for inhibiting the growth of *R. solani* [19]. Similarly, the *B. subtilis* strain 330-2 isolated from rapeseed, which produced β-1,3-glucanase, β-1,4-glucanase, and proteases, strongly inhibited the growth of fungi from *A. alternata*, *B. cinerea*, *C. heterostrophus*, *F. oxysporum*, *R. solani* AG1-IA, and *Nigrospora oryzae* [140], and the cellulase along with extracellular pectinase and chitinase from the *B. subtilis* strain EG21 inhibited the growth of *P. infestans* and lowered zoospore germination and infection. In turn, in *S. marcescens* B2 and *S. proteamaculans* 568, the antagonistic effect correlated with high chitinase activity [141,142]. Strain KW16 did not exhibit chitinolytic activity. However, as shown in studies by Kamenski et al. [143], chitinase was not essential for the biocontrol of *B. cinerea* and *Sclerotinia sclerotiorum* by *Serratia plymuthica* IC14 in synthesis of other enzymes, such as proteases. Similarly, in *Paenibacillus* sp. B2 activity against *Phytophthora parasitica* among various hydrolytic enzymes, only proteases were responsible for inhibition of the mycelial growth [144].

Competition for nutrients, especially iron, is an essential antagonistic trait against plant pathogens. It has been frequently described that siderophore-producing bacteria can reduce iron availability to phytopathogens, inhibiting their growth and development. In addition, they can sequester it around plant roots to toxic levels, preventing pathogens from surviving in such an environment. The vital role of siderophores produced by various antagonistic microorganisms has been proven to inhibit the growth of fungi from the genera *Colletotrichum*, *Fusarium*, *Rhizoctonia*, and *Sclerotinia* [27,145]. Many genes involved in iron transport and the production and transport of siderophores have also been identified in the genome of the KW16 strain. The analyses revealed the presence of *feoA* and *feoB* genes in a Feo system, which are common, especially for Gram-negative bacteria, and dedicated to iron [146]; *yusV*, *yfhA*, and *yfiZ* genes involved in the import of iron-hydroxamate and catecholate siderophores [147]; and genes involved in siderophore synthesis (*rhbA*, *iucC*). KW16's ability to produce siderophores (Table S7—Supplementary Materials) was proved by the change in CAS medium color from blue to orange (Figure S2c—Supplementary Materials). The *P. megaterium* species is known to be a producer of hydroxamate siderophores. One is schizokinen and its derivatives, with a citrate backbone and a high affinity for iron, aluminum, and gallium ions [145]. Recently, Zhu et al. [148] isolated and characterized another, Ferrioxamine E [M + Fe-2 H], with the structure of a hexadentate octahedral complex. The siderophore engaged in the biological activity of KW16 could be rhizobactin 1021, a siderophore similar to the aerobactin commonly found in *Enterobacteriaceae*, as indicated by the presence of the *rhbA* and *iucC* genes.

Another important biocontrol feature identified in the KW16 genome is the presence of genes determining VOCs production. For instance, Nagrale et al. [149] showed the biocontrol potential of benzene, 1,3-diethyl- and 1,4-diethylbenzene, naphthalene, methylacetophenone, and ethanone, 1-(4-ethylphenyl), produced by three rhizobacterial strains, *B. cereus* CICR-D3, *B. aryabhattai* CICR-D5, and *B. tequilensis* CICR-H3 against *Macrophomina phaseolina*. In contrast, VOCs secreted by *B. megaterium* BM344-1 not only inhibited the growth of fungal pathogens of *Aspergillus*, *Penicillium*, and *Fusarium* but also caused complete inhibition of the synthesis of aflatoxins, ochratoxin A, and fumonisin B1 by these pathogens [18]. Notably, in the genome of KW16, *ilvB*, *ilvD*, *ilvE*, *ilvC*, *ilvA*, *ilvN*, and *bdhA* genes involved in the synthesis of acetoin and butanediol, the two most studied VOCs, were present. As evidenced by many studies, VOCs act not only directly on pathogens but also enhance plant resistance to them and stimulate plant growth by inducing the expression of genes essential in SA, jasmonic acid, and ethylene signaling pathways [150]. Therefore, it can be speculated that the protective effect of the KW16 strain, which significantly inhibited the growth of *R. solani* by VOCs production (Figure 1c,f), may also be exerted indirectly by the positive influence of these compounds on oilseed rape fitness and the induction of ISR in the colonized plant. Moreover, as the KW16 strain synthesized ammonium (inorganic volatile compound), the pathogen inhibition and spectacular increase in the number of lateral roots in plants inoculated with the strain that was observed in the pot experiments (Figure 3) could be a result of the synergistic action of different volatile compounds.

Antibiosis is considered an essential mechanism for combating plant pathogens. According to many studies, bacteria of the *Bacillus* and *Priestia* genera are excellent producers of a wide variety of antimicrobial agents [14,150,151]. Genes determining the synthesis of biologically active secondary metabolites, including lanthionine-containing antibiotics, lipopeptides, polyketides, and phenazines, were also identified in the genome of strain KW16 (Table S8—Supplementary Materials). Of these compounds, documented antifungal activity has been shown for bacillomycin D of the iturin family, responsible for causing damage to the cell wall and membrane of the hyphae and spores of *C. gloeosporioides* [152], and for polyketides synthesized from carboxylic acids and phenazines—compounds with redox activity, which, apart from their antifungal activity, are known to enable the survival and persistence of bacteria in soil through a key role in biofilm formation and iron reduction [151]. Although the presence of genes for the above-mentioned antibiotics indicates the potential of the strain to synthesize them, confirmation of their contribution to the biological activity of the KW16 strain against *R. solani* requires further research.

3.8. Rhizoctonia Solani Affects Expression of Selected Genes in Priestia Megaterium KW16

Upon their growth, fungal pathogens produce and release a mixture of secondary metabolites, enzymes, phytotoxins, and volatile compounds that affect plants and nearby microorganisms in the soil. Sequencing and genome analysis of *P. megaterium* KW16 revealed the presence of many genes potentially involved in plant colonization and biocontrol mechanisms. To verify if the presence of *R. solani* filtrates influence the KW16 activity, the relative expressions of selected genes, *gltB*, *iucC*, *katA*, *ilvB*, *bdhA*, and *bglA*, were measured using RT-qPCR after 24, 48, and 72 h of culturing bacteria in the presence of fungal filtrates. Changes of at least 2-fold in the expression of a given gene were considered significant.

It was found that the strain KW16 showed significant differences in the expression of most of the genes in response to the presence of the phytopathogen. It was also observed that the expression profiles of individual genes varied significantly depending on the exposure time (Figure 7).

After 24 h of incubation in the presence of *R. solani*, significant changes were observed in the expression levels of *iucC*, *gltB*, and *katA* genes. The expression of the *iucC* gene increased by about 13-fold; a 10-fold increase in expression was observed for the *gltB* gene, while the expression of the *katA* gene was 2-fold higher compared to the control. In contrast,

48 h incubation of the KW16 strain in the presence of the fungal filtrate resulted in the overexpression of *ilvB* by about 3.5-fold, while the expression levels of the *iucC* and *gltB* genes were significantly lower when compared to the 24 h incubation. After 72 h of culturing the KW16 strain in the presence of fungal filtrates, the expression level of the *katA* gene was almost 3 times higher compared to the control. Longer incubation with the pathogen did not affect the expression of *ilvB*. However, the *gltB* gene was significantly down-regulated (Figure 7). The presence of *R. solani* has no significant impact on the expression of *bglA* gene encoding β-glucosidase (GH1), an enzyme involved in the degradation of β-glucosides, and the *bdhA* gene, encoding an acetoin reductase that catalyzes the conversion of acetoin to 2,3-butanediol.

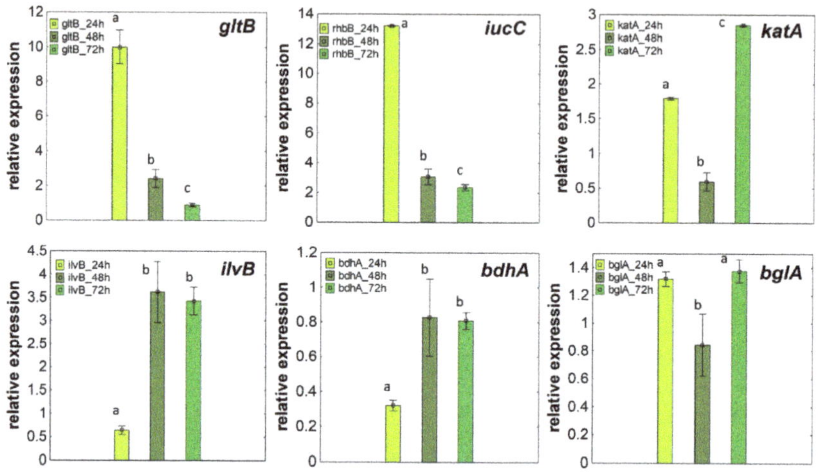

Figure 7. Changes in expression levels of selected genes of strain KW16 after 24 h, 48 h, and 72 h of incubation in the presence of fungal filtrate. The values are means ± SD of two biological and two technical replicates, and significant differences ($p < 0.05$) were assessed with a one-way ANOVA test and Tukey's post hoc comparisons. Different lowercase letters indicate significant differences between treatments.

The *gltB* gene plays a pivotal role in biofilm formation. As shown in a study by Zhou et al. [153] on *B. subtilis* Bs916 mutants, the *gltB* gene product regulates the production of γ-polyglutamate (γ-PGA), which is one of the polysaccharides involved in biofilm formation as well as the synthesis of the lipopeptide antibiotics bacillomycin L and surfactin, thereby affecting the colonization of plant tissues by bacteria. The study also confirmed the importance of the *gltB* gene in the biocontrol of *R. solani*. The importance of γ-polyglutamate synthesis in biofilm development was also confirmed in the study by Liu et al. [154], in which a mutant strain of *B. amyloliquefaciens* C06 defective in γ -PGA production showed lower efficiency in biofilm formation, surface adhesion, and swarming ability, which impaired colonization of apple surfaces. However, the results of this work did not answer the question of whether changes in biofilm synthesis, as a physical barrier against the adverse effects of the pathogen, are induced by the pathogen. Our study clearly indicates the effect of *R. solani* filtrates on the activity of the *gltB* gene, as its expression significantly increased after 24 h of the strain's growth in the presence of fungus filtrate and was maintained on the following day of incubation (Figure 7). This was confirmed by in vitro observations indicating an increase in the biofilm formed by strain KW16 in the presence of the fungus (Figure 5b). The drastic decrease in the level of *gltB* gene expression after 72 h incubation with KW16 in the presence of the fungus may result from the tight regulation of γ -PGA synthesis in the tested strain, the interaction of other genes involved in biofilm formation, and the metabolic state of the cell as well as substrate availability. The explanation of this phenomenon, however, requires further detailed studies.

On the first day after the exposure of strain KW16 to the phytopathogen, an almost 13-fold increase in expression of the *iucC* gene was also observed. Overexpression of the gene persisted, but at a lower level for the next 48 h, and biochemical tests conducted in parallel indicated an increased efficiency in the production of siderophores by strain KW16 over time (Figure 5c). The gene *iucC*, encoding a protein of the IucA/IucC protein family, is associated with iron acquisition, mainly in *Enterobacteriaceae*. However, the cluster containing three ORFs, identified as the iron-binding IucA/IucC family siderophore biosynthesis proteins, were recently described in *P. megaterium* AB-S79 [26] and a novel *P. megaterium* strain MARUCO02 [155]. The siderophore biosynthetic gene clusters are tightly regulated to balance iron acquisition with iron toxicity, and the primary signal for inducing their synthesis is a low iron concentration. To explain our findings, we speculate that the siderophore gene overexpression in the KW16 strain was connected to the rapid decrease in iron concentration due to *R. solani* siderophores present in its filtrates [156]. The fast response of the bacterium to changing environmental conditions may indicate its significant role in fighting pathogens through siderophore release. In some bacteria, siderophore production is also influenced by population density *via* quorum sensing signals (QS) [157], which may explain the continuing overexpression and production of siderophores while the culture grows.

KW16 showed a significant increase in *katA* gene expression in response to treatment with *R. solani* filtrate. As noted by Faulkner et al., 2012 [158], the increased expression of the major vegetative catalase gene (*katA*) in *Bacillus subtilis* depends on the global regulator PerR, which is highly sensitive to H_2O_2. In addition, PerR is involved in the regulation of the Fur protein that controls siderophore synthesis. Therefore, we speculate that the overexpression of the *katA* gene and the increased siderophore synthesis by the KW16 strain in the presence of a pathogen could be the answer to the presence of fungus elicitors like ROS. The production of these molecules is an essential feature for the development and successful pathogenesis of various necrotrophic fungi [159].

Our study also shows the effect of *R. solani* on the expression profile of the *ilvB* gene involved in the synthesis of acetoin and butanediol. The overexpression of *ilvB* in response to fungal pathogens and the results obtained in the dual-culture assays, where secreted VOCs significantly inhibited the growth of the fungus (Figure 1) by more than 50%, confirm the relevant role of VOCs in the control of *R. solani* by the KW16 strain.

4. Conclusions

P. megaterium KW16 isolated from the plant environment and devoid of plasmids—major vectors of gene dissemination in the environment—can be used as a replacement for fungicides used in oilseed rape protection. By applying properties such as ability to survive in the presence of soil autochthonous microflora, eligibility to enter plant internal tissues, and distinct biocontrol traits against *R. solani*, oilseed rape treated with both biotic factors (KW16 and RS) showed more intensive growth and better metabolic fitness than plants in the single-factor systems. This was most likely due to the blockade of pathogen invasion into the plant interior resulting from KW16's prompt response at the molecular level, with simultaneous stimulation of oilseed rape root density under the influence of *R. solani* metabolites. This so-called 'chemical priming' of biotic stress responses might be used in agriculture in the near future. However, the long-term study and development of formulation for field application require further investigation.

5. Patents

Patent no. PL 246324 B1 "Endophytic strain of *Priestia megaterium* bacteria and its applications."

Supplementary Materials: The following supporting information can be downloaded at https://www.mdpi.com/article/10.3390/agriculture15131435/s1, Figure S1: Biochemical characterization of the KW16 strain; Figure S2: COG classification of predicted genes in the KW16 strain. Colored bars indicate the CDS assigned to each COG category; Table S1: Physico-chemical characteristic of the sampled soil; Table S2: The designed specific primers for selected genes used in RT-qPCR reactions; Table S3: General genome features of the KW16 strain; Table S4: KEGG pathway classification of predicted genes in the tested strain using KEGG BlastKoala; Table. S5: Secondary metabolite gene clusters identified in the tested strain using antiSMASH; Table S6: Genes potentially involved in plant colonization and growth promotion identified in the KW16 genome; Table S7: Physiological and biochemical characteristic of the strain KW16; Table S8: Genes potentially involved in biocontrol activity localized in the KW16 genome.

Author Contributions: Conceptualization, B.N., D.C. and K.H.-K.; methodology, B.N., D.C. and K.H.-K.; software, B.N., D.C. and K.H.-K.; validation, B.N. and K.H.-K.; formal analysis, B.N. and K.H.-K.; investigation, B.N., D.C. and K.H.-K.; data curation, B.N. and K.H.-K.; writing—original draft preparation, B.N., D.C. and K.H.-K.; writing—review and editing, B.N. and K.H.-K.; visualization, B.N. and K.H.-K.; supervision, K.H.-K.; funding acquisition, K.H.-K. All authors have read and agreed to the published version of the manuscript.

Funding: This research was funded in part by the National Science Centre, Poland (grant number UMO-2020/39/B/NZ9/00491), and in part by the Ministry of Education and Science "Innovation Incubator 4.0" (UŚ/5/II 4.0/2021). For the purpose of Open Access, the author has applied a CC-BY public copyright license to any Author Accepted Manuscript (AAM) version arising from this submission.

Institutional Review Board Statement: Not applicable.

Data Availability Statement: Sequencing data and assembly are available at the NCBI database under the BioProject accession number PRJNA529642. The data are also included in the Supplementary Materials, which are available online or can be provided upon request.

Acknowledgments: We wish to thank Luis A. J. Mur (Aberystwyth University) for his assistance with the whole-genome sequencing of *Priestia megaterium* KW16. The genome sequencing was provided by MicrobesNG (http://www.microbesng.uk/, accessed on 5 October 2018), which is supported by the BBSRC (grant number BB/L024209/1). The authors are very grateful to Karolina Kaniuch for taking some of the pictures in the Supplementary Materials.

Conflicts of Interest: The authors declare no conflicts of interest. The funders had no role in the design of the study, in the collection, analyses, or interpretation of data, in the writing of the manuscript, or in the decision to publish the results.

Abbreviations

The following abbreviations are used in this manuscript:

Anth	Anthocyanins
CFU	Colony-Forming Unit
Chl	Chlorophyll
CRA	Congo Red Agar
CV	Crystal Violet
CWDEs	Cell Wall Degrading Enzymes
EPS	Exopolysaccharides
Flav	Flavonols
LR	Latheral Roots

NBI	Nitrogen Balance Index
PGI	Percent Growth Inhibition
PGP	Plant Growth Promotion
PGPR	Plant Growth Promoting Rhizobacteria
PSI	Phosphate Solubilization Index

References

1. Savary, S.; Willocquet, L.; Pethybridge, S.J.; Esker, P.; McRoberts, N.; Nelson, A. Assessing the global burden of pathogens and pests on major food crops. *Nat. Ecol. Evol.* **2019**, *3*, 430–439. [CrossRef] [PubMed]
2. Kashyap, A.S.; Manzar, N.; Meshram, S.; Sharma, P.K. Screening microbial inoculants and their interventions for cross kingdom management of wilt disease of solanaceous crop—A step towards sustainable agriculture. *Front. Microbiol.* **2023**, *14*, 1174532. [CrossRef]
3. Al-Surhanee, A.A.; Afzal, M.; Bouqellah, N.A.; Ouf, S.A.; Muhammad, S.; Jan, M.; Kaleem, S.; Hashem, M.; Alamri, S.; Abdel Latef, A.A.H.; et al. The Antifungal Activity of Ag/CHI NPs against *Rhizoctonia solani* Linked with Tomato Plant Health. *Plants* **2021**, *10*, 2283. [CrossRef]
4. Li, D.; Li, S.; Wei, S.; Sun, W. Strategies to manage rice sheath blight: Lessons from interactions between rice and *Rhizoctonia solani*. *Rice* **2021**, *14*, 21. [CrossRef]
5. Senapati, M.; Tiwari, A.; Sharma, N.; Chandra, P.; Bashyal, B.M.; Ellur, R.K.; Bhowmik, P.K.; Ballinadi, H.; Vinod, K.K.; Singh, A.K.; et al. *Rhizoctonia solani* Kühn pathophysiology: Status and prospects of sheath blight disease management in rice. *Front. Plant Sci.* **2022**, *13*, 1–22. [CrossRef]
6. Ajayi-Oyetunde, O.O.; Bradley, C.A. *Rhizoctonia solani*: Taxonomy, population biology and management of rhizoctonia seedling disease of soybean. *Plant Pathol.* **2017**, *67*, 3–17. [CrossRef]
7. Ons, L.; Bylemans, D.; Thevissen, K.; Cammue, B.P.A. Combining biocontrol agents with chemical fungicides for integrated plant fungal disease control. *Microorganisms* **2020**, *8*, 1930. [CrossRef] [PubMed]
8. Greenpeace Polska. Available online: https://tinyurl.com/bdz2ma6m (accessed on 9 May 2025).
9. Swiatly-Blaszkiewicz, A.; Klupczynska-Gabryszak, A.; Matuszewska-Mach, E.; Matysiak, J.; Attard, E.; Kowalczyk, D.; Adamkiewicz, A.; Kupcewicz, B.; Matysiak, J. Pesticides in Honeybee Products—Determination of Pesticides in Bee Pollen, Propolis, and Royal Jelly from Polish Apiary. *Molecules* **2025**, *30*, 275. [CrossRef] [PubMed]
10. Aktar, M.W.; Sengupta, D.; Chowdhury, A. Impact of pesticides use in agriculture: Their benefits and hazards. *Interdiscip. Toxicol.* **2009**, *2*, 1–12. [CrossRef]
11. El Saadony, M.T.; Saad, A.M.; Soliman, S.M.; Salem, M.; Ahmed, A.I.; Mahmood, M.; El-Tahan, A.M.; Ebrahim, A.A.M.; El-Mageed, T.A.A.; Negm, S.H.; et al. Plant growth promoting microorganisms as biocontrol agents of plant diseases: Mechanisms, challenges and future perspectives. *Front. Plant Sci.* **2022**, *13*, 923880. [CrossRef]
12. Wang, H.; Liu, R.; You, M.P.; Barbetti, M.J.; Chen, Y. Pathogen Biocontrol Using Plant Growth-Promoting Bacteria (PGPR): Role of Bacterial Diversity. *Microorganisms* **2021**, *9*, 1988. [CrossRef] [PubMed]
13. Xia, Y.; Liu, J.; Chen, C.; Mo, X.; Tan, Q.; He, Y.; Wang, Z.; Yin, J.; Zhou, G. The multifunctions and future prospects of endophytes and their metabolites in plant disease management. *Microorganisms* **2022**, *10*, 1072. [CrossRef]
14. Cui, Z.; Hu, L.; Zeng, L.; Meng, W.; Guo, D.; Sun, L. Isolation and characterization of *Priestia megaterium* KD7 for the biological control of pear fire blight. *Front. Microbiol.* **2023**, *14*, 1099664. [CrossRef] [PubMed]
15. Gupta, S.; Pandey, S.; Nandi, S.P.; Singh, M. Modulation of ethylene and ROS-scavenging enzymes by multifarious plant growth-promoting endophytes in tomato (*Solanum lycopersicum*) plants to combat *Xanthomonas*-induced stress. *Plant Physiol. Biochem.* **2023**, *202*, 107982. [CrossRef]
16. Sivakumar, G.; Rangeswaran, R.; Sriram, S.; Yandigeri, M.S. *Bacillus megaterium* strain NBAII 63 a potential biocontrol agent for the management of bacterial wilt of tomato caused by *Ralstonia solanacearum*. *Indian J. Agri. Sci.* **2012**, *84*, 102. [CrossRef]
17. Vásconez, R.D.A.; Moya, E.M.T.; Yépez, L.A.C.; Chiluisa-Utreras, V.P.; Suquillo, I.d.L.Á.V. Evaluation of *Bacillus megaterium* strain AB4 as a potential biocontrol agent of *Alternaria japonica*, a mycopathogen of *Brassica oleracea* var. *italica*. *Biotechnol. Rep.* **2020**, *26*, e00454. [CrossRef] [PubMed]
18. Saleh, A.E.; Ul-Hassan, Z.; Zeidan, R.; Al-Shamary, N.; Al-Yafei, T.; Alnaimi, H.; Salah Higazy, N.; Migheli, Q.; Jaoua, S. Biocontrol Activity of *Bacillus megaterium* BM344-1 against Toxigenic Fungi. *ACS Omega* **2021**, *6*, 10984–10990. [CrossRef] [PubMed]
19. Zheng, X.Y.; Sinclair, J.B. The effects of traits of *Bacillus megaterium* on seed and root colonization and their correlation with the suppression of Rhizoctonia root rot of soybean. *BioControl* **2000**, *45*, 223–243. [CrossRef]
20. Solanki, M.K.; Robert, A.S.; Singh, R.K.; Kumar, S.; Pandey, A.K.; Srivastava, A.K.; Arora, D.K. Characterization of Mycolytic Enzymes of *Bacillus* Strains and Their Bio-Protection Role Against *Rhizoctonia solani* in Tomato. *Curr. Microbiol.* **2012**, *65*, 330–336. [CrossRef]

21. Nascimento, F.X.; Hernández, A.G.; Glick, B.R.; Rossi, M.J. Plant growth-promoting activities and genomic analysis of the stress-resistant *Bacillus megaterium* STB1, a bacterium of agricultural and biotechnological interest. *Biotechnol. Rep.* **2020**, *25*, e00406. [CrossRef]
22. Li, Q.; Hou, Z.; Zhou, D.; Jia, M.; Lu, S.; Yu, J. A plant growth-promoting bacteria *Priestia megaterium* JR48 induces plant resistance to the crucifer black rot via a salicylic acid-dependent signaling pathway. *Front. Plant Sci.* **2022**, *13*, 1046181. [CrossRef]
23. Shehata, H.R.; Ettinger, C.L.; Eisen, J.A.; Raizada, M.N. Genes required for the anti-fungal activity of a bacterial endophyte isolated from a corn landrace grown continuously by subsistence farmers since 1000 BC. *Front. Microbiol.* **2016**, *7*, 1548. [CrossRef] [PubMed]
24. Shi, L.; Zhu, X.; Qian, T.; Du, J.; Du, Y.; Ye, J. Mechanism of Salt Tolerance and Plant Growth Promotion in *Priestia megaterium* ZS-3 Revealed by Cellular Metabolism and Whole-Genome Studies. *Int. J. Mol. Sci.* **2023**, *24*, 15751. [CrossRef] [PubMed]
25. Dahmani, M.A.; Desrut, A.; Moumen, B.; Verdon, J.; Mermouri, L.; Kacem, M.; Coutos-Thevenot, P.; Kaid-Harche, M.; Berges, T.; Vriet, C. Unearthing the plant growth-promoting traits of *Bacillus megaterium* RmBm31, an endophytic bacterium isolated from root nodules of *Retama monosperma*. *Front. Plant Sci.* **2020**, *11*, 124. [CrossRef]
26. Adeniji, A.A.; Chukwuneme, C.F.; Conceição, E.C.; Ayangbenro, A.S.; Wilkinson, E.; Maasdorp, E.; de Oliveira, T.; Babalola, O.O. Unveiling novel features and phylogenomic assessment of indigenous *Priestia megaterium* AB-S79 using comparative genomics. *Microbiol. Spectr.* **2025**, *13*, e0146624. [CrossRef]
27. Chlebek, D.; Pinski, A.; Żur, J.; Michalska, J.; Hupert-Kocurek, K. Genome Mining and Evaluation of the Biocontrol Potential of *Pseudomonas fluorescens* BRZ63, a New Endophyte of Oilseed Rape (*Brassica napus* L.) against Fungal Pathogens. *Int. J. Mol. Sci.* **2020**, *21*, 8740. [CrossRef]
28. Siala, R.; Chobba, I.B.; Vallaeys, T.; Triki, M.A.; Jrad, M.; Cheffi, M.; Ayedi, I.; Elleuch, A.; Nemsi, A.; Cerqueira, F.; et al. Analysis of the cultivable endophytic bacterial diversity in the date palm (*Phoenix dactylifera* L.) and evaluation of its antagonistic potential against pathogenic *Fusarium* species that cause date palm bayound disease. *J. App. Environ. Microbiol.* **2016**, *4*, 93–104.
29. Naveed, M.; Matter, B.; Yosuaf, S.; Pastar, M.; Afzal, M.; Sessitsch, A. The endophyte *Enterobacter* sp. FD17: A maize growth enhancer selected based on rigorous testing of plant beneficial traits and colonization characteristics. *Biol. Fert. Soils* **2014**, *50*, 249–262. [CrossRef]
30. Freeman, D.J.; Falkiner, F.R.; Keane, C.T. New method for detecting slime production by coagulase negative staphylococci. *J. Clin. Pathol.* **1989**, *42*, 872–874. [CrossRef]
31. Pikovskaya, R.I. Mobilization of phosphorus in soil in connection with vital activity of some microbial species. *Microbiology* **1948**, *17*, 362–370.
32. Edi-Premono, M.; Moawad, M.A.; Vleck, P.L.G. Effect of phosphate solubilizing *Pseudomonas putida* on the growth of maize and its survival in the rhizosphere. *Ind. J. Crop Sci.* **1996**, *11*, 13–23.
33. Szilagyi-Zecchin, V.J.; Ikeda, A.C.; Hungria, M.; Adamoski, D.; Kava-Cordeiro, V.; Glienke, C.; Galli-Terasawa, L.V. Identification and characterization of endophytic bacteria from corn (*Zea mays* L.) roots with biotechnological potential in agriculture. *AMB Express* **2014**, *4*, 26. [CrossRef]
34. Cappuccino, J.G.; Sherman, N. *Biochemical Activities of Microorganisms. Microbiology, A Laboratory Manual*; The Benjamin/Cummings Publishing Co.: California, CA, USA, 1992; pp. 188–247.
35. Sandhya, V.; Shrivastava, M.; Ali, S.Z.; Prasad, V.S.S.K. Endophytes from maize with plant growth promotion and biocontrol activity under drought stress. *Russ. Agric. Sci.* **2017**, *43*, 22–34. [CrossRef]
36. Vijayalakshmi, R.; Kairunnisa, K.; Sivvaswamy, S.N.; Dharan, S.S.; Nataranjan, S. Enzyme production and antimicrobial activity of endophytic bacteria isolated from medicinal plants. *Indian J. Sci. Technol.* **2016**, *9*, 1–8. [CrossRef]
37. Kuddus, S.M.; Ahmad, R.I.Z. Isolation of novel chitinolytic bacteria and production optimization of extracellular chitinase. *J. Genet. Eng. Biotechnol.* **2013**, *11*, 39–46. [CrossRef]
38. Johnston-Monje, D.; Raizada, M.N. Conservation and diversity of seed associated endophytes in *Zea* across boundaries of evolution, ethnography and ecology. *PLoS ONE* **2013**, *6*, e20396. [CrossRef] [PubMed]
39. Syamala, M.; Sivaji, M. Functional characterization of various plant growth promoting activity of *Pseudomonas fluorescens* and *Bacillus subtilis* from *Aloe vera* rhizosphere. *J. Pharmaco. Phytochem.* **2017**, *6*, 120–122.
40. Ahmad, F.; Ahmad, I.; Khan, M.S. Screening of free-living rhizospheric bacteria for their multiple growth promoting activities. *Microbiol. Res.* **2008**, *163*, 173–181. [CrossRef]
41. Schwyn, B.; Neilands, J.B. Universal chemical assay for the detection and determination of siderophores. *Anal. Biochem.* **1987**, *160*, 47–56. [CrossRef]
42. Kumar, V.S.; Menon, S.; Agarwal, H.; Gopalakrishnan, D. Characterization and optimization of bacterium isolated from soil samples for the production of siderophores. *Resour.-Effic. Technol.* **2017**, *3*, 434–439. [CrossRef]
43. Pfaffl, M.W.; Tichopad, A.; Prgomet, C.; Neuvians, T.P. Determination of stable housekeeping genes, differentially regulated target genes and sample integrity: Bestkeeper-excel-based tool using pair-wise correlations. *Biotechnol. Lett.* **2004**, *26*, 509–515. [CrossRef]

44. Taylor, S.; Wakem, M.; Dijkman, G.; Alsarraj, M.; Nguyen, M. A practical approach to RT-qPCR-Publishing data that conform to the MIQE guidelines. *Methods* **2010**, *50*, S1–S5. [CrossRef] [PubMed]
45. Livak, K.J.; Schmittgen, T.D. Analysis of relative gene expression data using real-time quantitative PCR and the $^{2-\Delta\Delta CT}$ method. *Methods* **2001**, *25*, 402–408. [CrossRef] [PubMed]
46. Huerta-Cepas, J.; Szklarczyk, D.; Heller, D.; Hernández-Plaza, A.; Forslund, S.K.; Cook, H.; Mende, D.R.; Letunic, I.; Rattei, T.; Jensen, L.J.; et al. EggNOG 5.0: A hierarchical, functionally and phylogenetically annotated orthology resource based on 5090 organisms and 2502 viruses. *Nucleic Acid Res.* **2019**, *47*, 309–314. [CrossRef] [PubMed]
47. Kanehisa, M.; Sato, Y.; Kawashima, M.; Furumichi, M.; Tanabe, M. KEGG as a reference resource for gene and protein annotation. *Nucleic Acids Res.* **2016**, *44*, D457–D462. [CrossRef]
48. Blin, K.; Shaw, S.; Steinke, K.; Villebro, R.; Ziemert, N.; Lee, S.Y.; Medema, M.H.; Weber, T. AntiSMASH 5.0: Updates to the Secondary Metabolite Genome Mining Pipeline. *Nucleic Acids Res.* **2019**, *47*, W81–W87. [CrossRef]
49. Yin, Y.; Mao, X.; Yang, J.; Chen, X.; Mao, F.; Xu, Y. dbCAN: A web resource for automated carbohydrate-active enzyme annotation. *Nucleic Acids Res.* **2012**, *40*, W445–W451. [CrossRef]
50. Zhang, H.; Yohe, T.; Huang, L.; Entwistle, S.; Wu, P.; Yang, Z.; Busk, P.K.; Xu, Y.; Yin, Y. dbCAN2: A meta server for automated carbohydrate-active enzyme annotation. *Nucleic Acids Res.* **2018**, *46*, W95–W101. [CrossRef]
51. Mezghanni, H.; Khedher, S.B.; Tounsi, S.; Zouari, N. Medium optimization of antifungal activity production by *Bacillus amyloliquefaciens* using statistical experimental design. *Prep. Biochem. Biotechnol.* **2012**, *42*, 267–278. [CrossRef]
52. Huang, J.-S.; Peng, Y.-H.; Chung, K.-R.; Huang, J.-W. Suppressive efficacy of volatile compounds produced by *Bacillus mycoides* on damping-off pathogens of cabbage seedlings. *J. Agric. Sci.* **2018**, *156*, 795–809. [CrossRef]
53. Prigigallo, M.I.; De Stradis, A.; Anand, A.; Mannerucci, F.; L'Haridon, F.; Weisskopf, L.; Bubici, G. Basidiomycetes Are Particularly Sensitive to Bacterial Volatile Compounds: Mechanistic Insight Into the Case Study of *Pseudomonas protegens* Volatilome Against *Heterobasidion abietinum*. *Front. Microbiol.* **2021**, *12*, 684664. [CrossRef]
54. Avalos, M.; Garbeva, P.; Raaijmakers, J.M.; van Wezel, G.P. Production of ammonia as a low-cost and long-distance antibiotic strategy by *Streptomyces* species. *ISME J.* **2020**, *14*, 569–583. [CrossRef] [PubMed]
55. Asari, S.; Matzén, S.; Petersen, M.A.; Bejai, S.; Meijer, J. Multiple effects of *Bacillus amyloliquefaciens* volatile compounds: Plant growth promotion and growth inhibition of phytopathogens. *FEMS Microbiol. Ecol.* **2016**, *92*, fiw070. [CrossRef] [PubMed]
56. Chandrasekaran, M.; Paramasivan, M.; Sahayarayan, J.J. Microbial volatile organic compounds: An alternative for chemical fertilizers in sustainable agriculture development. *Microorganisms* **2023**, *11*, 42. [CrossRef]
57. Friedt, W.; Snowdon, R. Oilseed Rape. In *Oil Crops. Handbook of Plant Breeding*; Vollmann, J., Rajcan, I., Eds.; Springer: New York, NY, USA, 2009; Volume 4, pp. 91–126. ISBN 978-0-387-77594-4. [CrossRef]
58. Liu, J.; Ridgway, H.J.; Jones, E.E. The use of rifampicin mutants and ERIC-PCR to track plant colonization and *in planta* efficacy of bacterial biocontrol agents against *Neonectria ditissima*. *J. Appl. Microbiol.* **2025**, *136*, lxaf086. [CrossRef]
59. Burns, J.H.; Anacker, B.L.; Strauss, S.Y.; Burke, D.J. Soil microbial community variation correlates most strongly with plant species identity, followed by soil chemistry, spatial location and plant genus. *AoB PLANTS* **2015**, *7*, plv030. [CrossRef] [PubMed]
60. Lin, X.-R.; Chen, H.-B.; Li, Y.-X.; Zhou, Z.-H.; Li, J.-B.; Wang, Y.-Q.; Zhang, H.; Zhang, Y.; Han, Y.-H.; Wang, S.-S. *Priestia* sp. LWS1 Is a Selenium-Resistant Plant Growth-Promoting Bacterium That Can Enhance Plant Growth and Selenium Accumulation in *Oryza sativa* L. *Agronomy* **2022**, *12*, 1301. [CrossRef]
61. Hwang, H.-H.; Chien, P.-R.; Huang, F.-C.; Yeh, P.-H.; Hung, S.-H.W.; Deng, W.-L.; Huang, C.-C. A Plant Endophytic Bacterium *Priestia megaterium* Strain BP-R2 Isolated from the Halophyte *Bolboschoenus planiculmis* Enhances Plant Growth under Salt and Drought Stresses. *Microorganisms* **2022**, *10*, 2047. [CrossRef]
62. Bhatt, K.; Maheshwari, D.K. *Bacillus megaterium* strain CDK25, a novel plant growth promoting bacterium enhances proximate chemical and nutritional composition of *Capsicum annuum* L. *Front. Plant Sci.* **2020**, *11*, 1147. [CrossRef]
63. Colmer, T.D. Long-distance transport of gases in plants: A perspective on internal aeration and radial oxygen loss from roots. *Plant Cell Environ.* **2003**, *26*, 17–36. [CrossRef]
64. Giehl, R.F.H.; Lima, J.E.; von Wirén, N. Localized iron supply triggers lateral root elongation in *Arabidopsis* by altering the AUX1-mediated auxin distribution. *Plant Cell* **2012**, *24*, 33–49. [CrossRef] [PubMed]
65. Gruber, B.D.; Giehl, R.F.H.; Friedel, S.; von Wirén, N. Plasticity of the *Arabidopsis* root system under nutrient deficiencies. *Plant Physiol.* **2013**, *163*, 161–179. [CrossRef] [PubMed]
66. Giehl, R.F.; von Wirén, N. Root nutrient foraging. *Plant Physiol.* **2014**, *166*, 509–517. [CrossRef]
67. Lima, J.E.; Kojima, S.; Takahashi, H.; von Wirén, N. Ammonium triggers lateral root branching in *Arabidopsis* in an AMMONIUM TRANSPORTER1; 3-dependent manner. *Plant Cell* **2010**, *22*, 3621–3633. [CrossRef]
68. Fincheira, P.; Venthur, H.; Mutis, A.; Parada, M.; Quiroz, A. Growth promotion of *Lactuca sativa* in response to volatile organic compounds emitted from diverse bacterial species. *Microbiol. Res.* **2016**, *193*, 39–47. [CrossRef]

69. Gutiérrez-Luna, F.M.; López-Bucio, J.; Altamirano-Hernández, J.; Valencia-Cantero, E.; de la Cruz, H.R.; Macías-Rodríguez, L. Plant growth promoting rhizobacteria modulate root-system architecture in *Arabidopsis thaliana* through volatile organic compound emission. *Symbiosis* **2010**, *51*, 75–83. [CrossRef]
70. Ait Bahadou, S.; Ouijja, A.; Karfach, A.; Tahiri, A.; Lahlali, R. New potential bacterial antagonists for the biocontrol of fire blight disease (*Erwinia amylovora*) in Morocco. *Microb. Pathog.* **2018**, *117*, 7–15. [CrossRef]
71. Lahlali, R.; Mchachti, O.; Radouane, N.; Ezrari, S.; Belabess, Z.; Khayi, S.; Mentag, R.; Tahiri, A.; Ait Barka, E. The Potential of Novel Bacterial Isolates from Natural Soil for the Control of Brown Rot Disease (*Monilinia fructigena*) on Apple Fruits. *Agronomy* **2020**, *10*, 1814. [CrossRef]
72. De Curtis, F.; Ianiri, G.; Raiola, A.; Ritieni, A.; Succi, M.; Tremonte, P.; Castoria, R. Integration of biological and chemical control of brown rot of stone fruits to reduce disease incidence on fruits and minimize fungicide residues in juice. *Crop Prot.* **2019**, *119*, 158–165. [CrossRef]
73. Lahlali, R.; Ezrari, S.; Radouane, N.; Kenfaoui, J.; Esmaeel, Q.; El Hamss, H.; Belabess, Z.; Barka, E.A. Biological control of plant pathogens: A global perspective. *Microorganisms* **2022**, *10*, 596. [CrossRef]
74. Liu, J.M.; Liang, Y.T.; Wang, S.S.; Jin, W.; Sun, J.; Lu, C.; Sun, Y.F.; Li, S.Y.; Fan, B.; Wang, F.Z. Antimicrobial activity and comparative metabolomic analysis of *Priestia megaterium* strains derived from potato and dendrobium. *Sci. Rep.* **2023**, *13*, 5272. [CrossRef]
75. Pérez-Martínez, J.; Ploetz, R.; Konkol, J. Significant in vitro antagonism of the laurel wilt pathogen by endophytic fungi from the xylem of avocado does not predict their ability to control the disease. *Plant. Pathol.* **2018**, *67*, 1768–1776. [CrossRef]
76. Whitaker, B.K.; Bakker, M.G. Bacterial endophyte antagonism toward a fungal pathogen in vitro does not predict protection in live plant tissue. *FEMS Microbiol. Ecol.* **2019**, *95*, fiy237. [CrossRef]
77. Köhl, J.; Kolnaar, R.; Ravensberg, W.J. Mode of action of microbial biological control agents against plant diseases: Relevance beyond efficacy. *Front. Plant. Sci.* **2019**, *10*, 845. [CrossRef]
78. Rudrappa, T.; Czymmek, K.J.; Pare, P.W.; Bais, H.P. Root-secreted malic acid recruits beneficial soil bacteria. *Plant Physiol.* **2008**, *148*, 1547–1556. [CrossRef]
79. Lee, B.; Lee, S.; Ryu, M.R. Foliar aphid feeding recruits rhizosphere bacteria and primes plant immunity against pathogenic and non-pathogenic bacteria in pepper. *Ann. Bot.* **2012**, *110*, 281–290. [CrossRef]
80. Kong, F.; Yang, L. Pathogen-triggered changes in plant development: Virulence strategies or host defense mechanism? *Front. Microbiol.* **2023**, *14*, 1122947. [CrossRef] [PubMed]
81. Moisan, K.; Raaijmakers, J.M.; Dicke, M.; Lucas-Barbosa, D.; Cordovez, V. Volatiles from soil-borne fungi affect directional growth of roots. *Plant Cell Environ.* **2021**, *44*, 339–345. [CrossRef] [PubMed]
82. Cordovez, V.; Mommer, L.; Moisan, K.; Lucas-Barbosa, D.; Pierik, R.; Mumm, R.; Carrion, V.J.; Raaijmakers, J.M. Plant Phenotypic and Transcriptional Changes Induced by Volatiles from the Fungal Root Pathogen *Rhizoctonia solani*. *Front. Plant Sci.* **2017**, *8*, 1262. [CrossRef]
83. Rilling, J.I.; Acuña, J.J.; Nannipieri, P.; Cassan, F.; Maruyama, F.; Jorquera, M.A. Current opinion and perspectives on the methods for tracking and monitoring plant growth—promoting bacteria. *Soil Biol. Biochem.* **2019**, *130*, 205–219. [CrossRef]
84. Hallmann, J.; Berg, G. Spectrum and Population Dynamics of Bacterial Root Endophytes. In *Microbial Root Endophytes. Soil Biology*; Schulz, B.J.E., Boyle, C.J.C., Sieber, T.N., Eds.; Springer: Berlin/Heidelberg, Germany, 2006; Volume 9. [CrossRef]
85. Compant, S.; Clément, C.; Sessitsch, A. Plant growth-promoting bacteria in the rhizo-and endosphere of plants: Their role, colonization, mechanisms involved and prospects for utilization. *Soil Biol. Biochem.* **2010**, *42*, 669–678. [CrossRef]
86. Afzal, I.; Shinwari, Z.K.; Sikandar, S.; Shahzad, S. Plant beneficial endophytic bacteria: Mechanisms, diversity, host range and genetic determinants. *Microbiol. Res.* **2019**, *221*, 36–49. [CrossRef] [PubMed]
87. Zinniel, D.K.; Lambrecht, P.; Harris, N.B.; Feng, Z.; Kuczmarski, D.; Higley, P.; Ishimaru, C.A.; Arunakumari, A.; Barletta, R.G.; Vidaver, A.K. Isolation and characterization of endophytic colonizing bacteria from agronomic crops and prairie plants. *Appl. Environ. Microbiol.* **2002**, *68*, 2198–2208. [CrossRef]
88. Nwoko, E.Q.A.; Okeke, I.N. Bacteria autoaggregation: How and why bacteria stick together. *Biochem. Soc. Trans.* **2021**, *49*, 1147–1157. [CrossRef]
89. Trunk, T.; Khalil, H.S.; Leo, J.C. Bacterial autoaggregation. *AIMS Microbiol.* **2018**, *4*, 140–164. [CrossRef] [PubMed]
90. Shen, G.; Yang, L.; Lv, X.; Zhang, Y.; Hou, X.; Li, M.; Zhou, M.; Pan, L.; Chen, A.; Zhang, Z. Antibiofilm Activity and Mechanism of Linalool against Food Spoilage *Bacillus amyloliquefaciens*. *Int. J. Mol. Sci.* **2023**, *24*, 10980. [CrossRef] [PubMed]
91. Ighilahriz, K.; Benchouk, A.; Belkebir, Y.; Seghir, N.; Yahi, L. Production of biosurfactant by *Bacillus megaterieum* using agro-food wastes and its application in petroleum sludge oil recovery. *J. Environ. Health. Sci. Eng.* **2024**, *22*, 413–424. [CrossRef]
92. Stancu, M.M. Biosurfactant production by a *Bacillus megaterium* strain. *Open Life Sci.* **2020**, *15*, 629–637. [CrossRef]
93. Sharma, J.; Sundar, D.; Srivastava, P. Biosurfactants: Potential Agents for Controlling Cellular Communication, Motility, and Antagonism. *Front. Mol. Biosci.* **2021**, *8*, 727070. [CrossRef]

94. Santos, S.; Neto, I.F.F.; Machado, M.D.; Soares, H.M.V.M.; Soares, E.V. Siderophore production by *Bacillus megaterium*: Effect of growth phase and cultural conditions. *Appl. Biochem. Biotechnol.* **2014**, *172*, 549–560. [CrossRef]
95. Andrić, S.; Meyer, T.; Ongena, M. *Bacillus* Responses to Plant-Associated Fungal and Bacterial Communities. *Front. Microbiol.* **2020**, *11*, 1350. [CrossRef] [PubMed]
96. Chowdhury, S.P.; Uhl, J.; Grosch, R.; Alquéres, S.; Pittroff, S.; Dietel, K.; Schmitt-Kopplin, P.; Borriss, R.; Hartmann, A. Cyclic Lipopeptides of *Bacillus amyloliquefaciens* subsp. *plantarum* Colonizing the Lettuce Rhizosphere Enhance Plant Defense Responses Toward the Bottom Rot Pathogen *Rhizoctonia solani*. *Mol. Plant Microbe Interact.* **2015**, *28*, 984–995. [CrossRef] [PubMed]
97. Kulimushi, P.Z.; Arias, A.A.; Franzil, L.; Steels, S.; Ongena, M. Stimulation of fengycin-type antifungal lipopeptides in *Bacillus amyloliquefaciens* in the presence of the maize fungal pathogen *Rhizomucor variabilis*. *Front. Microbiol.* **2017**, *8*, 850. [CrossRef]
98. Liu, H.; Carvalhais, L.C.; Crawford, M.; Singh, E.; Dennis, P.G.; Pieterse, C.M.J.; Schenke, P.M. Inner plant values: Diversity, colonization and benefits from endophytic bacteria. *Front. Microbiol.* **2017**, *8*, 2552. [CrossRef]
99. Rios-Ruiz, W.F.; Tuanama-Reátegui, C.; Huamán-Córdova, G.; Valdez-Nuñez, R.A. Co-Inoculation of Endophytes *Bacillus siamensis* TUR07-02b and *Priestia megaterium* SMBH14-02 Promotes Growth in Rice with Low Doses of Nitrogen Fertilizer. *Plants* **2023**, *12*, 524. [CrossRef]
100. Cao, W.; Zhang, H.; Zhou, Y.; Zhao, J.; Lu, S.; Wang, X.; Chen, X.; Yuan, L.; Guan, H.; Wang, G.; et al. Suppressing chlorophyll degradation by silencing OsNYC3 improves rice resistance to *Rhizoctonia solani*, the causal agent of sheath blight. *Plant Biotechnol. J.* **2022**, *20*, 335–349. [CrossRef]
101. Kashyap, A.S.; Manzar, N.; Nebapure, S.M.; Rajawat, M.V.S.; Deo, M.M.; Singh, J.P.; Kesharwani, A.K.; Singh, R.P.; Dubey, S.C.; Singh, D. Unraveling Microbial Volatile Elicitors Using a Transparent Methodology for Induction of Systemic Resistance and Regulation of Antioxidant Genes at Expression Levels in Chili against Bacterial Wilt Disease. *Antioxidants* **2022**, *11*, 404. [CrossRef]
102. Macioszek, V.K.; Jęcz, T.; Ciereszko, I.; Kononowicz, A.K. Jasmonic Acid as a Mediator in Plant Response to Necrotrophic Fungi. *Cells* **2023**, *12*, 1027. [CrossRef]
103. Kaniyassery, A.; Hegde, M.; Sathish, S.B.; Thorat, S.A.; Udupa, S.; Murali, T.S.; Muthusamy, A. Leaf spot-associated pathogenic fungi alter photosynthetic, biochemical, and metabolic responses in eggplant during the early stages of infection. *Physiol. Mol. Plant Pathol.* **2024**, *133*, 102320. [CrossRef]
104. Rivas, S.; Fincheira, P.; González, F.; Santander, C.; Meier, S.; Santos, C.; Contreras, B.; Ruiz, A. Assessment of the Photosynthetic Response of Potato Plants Inoculated with *Rhizoctonia solani* and Treated with Flesh-Colored Potato Extracts Nanoencapsulated with Solid Lipid Nanoparticles. *Plants* **2025**, *14*, 156. [CrossRef]
105. Wang, L.; Chen, M.; Lam, P.Y.; Dini-Andreote, F.; Dai, L.; Wei, Z. Multifaceted roles of flavonoids mediating plant-microbe interactions. *Microbiome* **2022**, *10*, 233. [CrossRef] [PubMed]
106. Patil, J.R.; Mhatre, K.J.; Yadav, K.; Yadav, L.S.; Srivastava, S.; Nikalje, G.C. Flavonoids in plant-environment interactions and stress responses. *Discov. Plants* **2024**, *1*, 68. [CrossRef]
107. Li, Z.; Jiang, J.; Sun, K.; Ye, S. Mechanisms of *Priestia megaterium* PH3 in alleviating postharvest disease caused by *Penicillium expansum* in Nanguo pear fruit. *Food Microbiol.* **2025**, *131*, 104784. [CrossRef] [PubMed]
108. Fan, K.; Li, F.; Chen, X.; Li, Z.; Mulla, D.J. Nitrogen Balance Index Prediction of Winter Wheat by Canopy Hyperspectral Transformation and Machine Learning. *Remote Sens.* **2022**, *14*, 3504. [CrossRef]
109. Pieterse, C.M.J.; Zamioudis, C.; Berendsen, R.L.; Weller, D.M.; Van Wees, S.C.M.; Bakker, P.A.H.M. Induced systemic resistance by beneficial microbes. *Ann. Rev. Phytopathol.* **2014**, *52*, 347–375. [CrossRef]
110. Hönig, M.; Roeber, V.M.; Schmülling, T.; Cortleven, A. Chemical priming of plant defense responses to pathogen attacks. *Front. Plant Sci.* **2023**, *14*, 1146577. [CrossRef]
111. Trapet, P.L.; Verbon, E.H.; Bosma, R.R.; Voordendag, K.; Van Pelt, J.A.; Pieterse, C.M.J. Mechanisms underlying iron deficiency-induced resistance against pathogens with different lifestyles. *J. Exp. Bot.* **2021**, *72*, 2231–2241. [CrossRef]
112. Jiang, C.H.; Liao, M.J.; Wang, H.K.; Zheng, M.Z.; Xu, J.J.; Guo, J.H. *Bacillus velezensis*, a potential and efficient biocontrol agent in control of pepper gray mold caused by Botrytis cinerea. *Biol. Control* **2018**, *126*, 147–157. [CrossRef]
113. Sharifi, R.; Ryu, C.M. Are bacterial volatile compounds poisonous odors to a fungal pathogen *Botrytis cinerea*, alarm signals to *Arabidopsis* seedlings for eliciting induced resistance, or both? *Front. Microbiol.* **2016**, *7*, 196. [CrossRef]
114. Fernández-Llamosas, H.; Díaz, E.; Carmona, M. Motility, adhesion and c-di-GMP influence the endophytic colonization of rice by *Azoarcus* sp. CIB. *Microorganisms* **2021**, *8*, 554. [CrossRef]
115. Pankievicz, V.C.; Camilios-Neto, D.; Bonato, P.; Balsanelli, E.; Tadra-Sfeir, M.Z.; Faoro, H.; Monteiro, R.A. RNA-seq transcriptional profiling of *Herbaspirillum seropedicae* colonizing wheat (*Triticum aestivum*) roots. *Plant. Mol. Biol.* **2016**, *90*, 589–603. [CrossRef]
116. Kearns, D.B. A field guide to bacterial swarming motility. *Nat. Rev. Microbiol.* **2010**, *8*, 634–644. [CrossRef] [PubMed]
117. Fan, H.; Zhang, Z.; Li, Y.; Zhang, X.; Duan, Y.; Wang, Q. Biocontrol of Bacterial Fruit Blotch by *Bacillus* subtilis 9407 via Surfactin-Mediated Antibacterial Activity and Colonization. *Front. Microbiol.* **2017**, *8*, 1973. [CrossRef]
118. Gao, S.; Wu, H.; Yu, X.; Qian, L.; Gao, X. Swarming motility plays the major role in migration during tomato root colonization by *Bacillus subtilis* SWR01. *Biological Control* **2016**, *98*, 11–17. [CrossRef]

119. Feng, H.; Fu, R.; Hou, X.; Lv, Y.; Zhang, N.; Liu, Y.; Xu, Z.; Miao, Y.; Krell, T.; Shen, Q.; et al. Chemotaxis of beneficial rhizobacteria to root exudates: The first step towards root–microbe rhizosphere interactions. *Int. J. Mol. Sci.* **2021**, *22*, 6655. [CrossRef]
120. Rudrappa, T.; Splaine, R.E.; Biedrzycki, M.L.; Bais, H.P. Cyanogenic pseudomonads influence multitrophic interactions in the rhizosphere. *PLoS ONE* **2008**, *3*, e2073. [CrossRef] [PubMed]
121. Zhang, N.; Wang, D.; Liu, Y.; Li, S.; Shen, Q.; Zhang, R. Effects of different plant root exudates and their organic acid components on chemotaxis, biofilm formation and colonization by beneficial rhizosphere-associated bacterial strains. *Plant Soil* **2014**, *374*, 689–700. [CrossRef]
122. Allard-Massicotte, R.; Tessier, L.; Lécuyer, F.; Lakshmanan, V.; Lucier, J.-F.; Garneau, D.; Caudwell, L.; Vlamakis, H.; Bais, H.P.; Beauregard, P.B. *Bacillus subtilis* early colonization of *Arabidopsis thaliana* roots involves multiple chemotaxis receptors. *MBio* **2016**, *7*, e01664-16. [CrossRef] [PubMed]
123. Kandel, S.L.; Joubert, P.M.; Doty, S.L. Bacterial endophyte colonization and distribution within plants. *Microorganisms* **2017**, *5*, 77. [CrossRef]
124. Costa, O.Y.A.; Raaijmakers, J.M.; Kuramae, E.E. Microbial extracellular polymeric substances: Ecological function and impact on soil aggregation. *Front. Microbiol.* **2018**, *9*, 1636. [CrossRef]
125. Fouda, A.; Eid, A.M.; Elsaied, A.; El-Belely, E.F.; Barghoth, M.G.; Azab, E.; Gobouri, A.A.; Hassan, S.E. Plant growth-promoting endophytic bacterial community inhabiting the leaves of *Pulicaria incisa* (Lam.) DC Inherent to Arid Regions. *Plants* **2021**, *10*, 76. [CrossRef]
126. Srikhong, P.; Lertmongkonthum, K.; Sowanpreecha, R.; Rerngsamran, P. *Bacillus* sp. strain M10 as a potential biocontrol agent protecting chili pepper and tomato fruits from anthracnose disease caused by *Colletotrichum capsici*. *BioControl* **2018**, *63*, 833–842. [CrossRef]
127. Kukla, M.; Płociniczak, T.; Piotrowska-Seget, Z. Diversity of endophytic bacteria in *Lolium perenne* and their potential to degrade petroleum hydrocarbons and promote plant growth. *Chemosphere* **2014**, *117*, 40–46. [CrossRef]
128. Amaresan, N.; Jayakumar, V.; Thajuddin, N. Isolation and characterization of endophytic bacteria associated with chilli (*Capsicum annuum*) grown in coastal agricultural ecosystem. *Indian J. Biotechnol.* **2014**, *13*, 247–255.
129. Shao, J.; Li, S.; Zhang, N.; Cui, X.; Zhou, X.; Zhang, G.; Shen, Q.; Zhang, R. Analysis and cloning of the synthetic pathway of the phytohormone indole-3-acetic acid in the plant-beneficial *Bacillus amyloliquefaciens* SQR9. *Microb. Cell Fact.* **2015**, *14*, 130. [CrossRef]
130. Khan, M.I.R.; Jalil, S.U.; Chopra, P.; Chhillar, H.; Ferrante, A.; Khan, N.A.; Ansari, M.I. Role of GABA in plant growth, development and senescence. *Plant Gene* **2021**, *26*, 100283. [CrossRef]
131. Smith, J.; Johnson, A. Effects of *B. velezensis* FX-6 on GABA production and biomass increase in tomato plants. *J. Agric. Sci.* **2021**, *18*, 350–365.
132. Gamalero, E.; Lingua, G.; Glick, B.R. Ethylene, ACC, and the plant growth-promoting enzyme ACC deaminase. *Biology* **2023**, *12*, 1043. [CrossRef] [PubMed]
133. Collavino, M.M.; Sansberro, P.A.; Mroginski, L.A.; Aguilar, O.M. Comparison of in vitro solubilization activity of diverse phosphate-solubilizing bacteria native to acid soil and their ability to promote *Phaseolus vulgaris* growth. *Biol. Fertil. Soils* **2010**, *46*, 727–738. [CrossRef]
134. Bashan, Y.; Kamnev, A.A.; de-Bashan, L.E. A proposal for isolating and testing phosphate-solubilizing bacteria that enhance plant growth. *Biol. Fertil. Soils* **2013**, *49*, 1–2. [CrossRef]
135. Mitra, D.; Anđelković, S.; Panneerselvam, P.; Senapati, A.; Vasić, T.; Ganeshamurthy, A.N.; Chauhan, M.; Uniyal, N.; Mahakur, B.; Radha, T.K. Phosphate-Solubilizing Microbes and Biocontrol Agent for Plant Nutrition and Protection: Current Perspective. *Commun. Soil Sci. Plant Anal.* **2020**, *51*, 645–657. [CrossRef]
136. Mansilla, M.C.; Albanesi, D.; de Mendoza, D. Transcriptional control of the sulfur-regulated *cysH* operon, containing genes involved in L-cysteine biosynthesis in *Bacillus subtilis*. *J. Bacteriol.* **2000**, *182*, 5885–5892. [CrossRef]
137. Ranadev, P.; Revanna, A.; Bagyaraj, D.J.; Shinde, A.H. Sulfur oxidizing bacteria in agro ecosystem and its role in plant productivity-a review. *J. Appl. Microbiol.* **2023**, *134*, lxad161. [CrossRef]
138. Sharma, M.; Saleh, D.; Charron, J.-B.; Jabaji, S. A crosstalk between *Brachypodium* root exudates, organic acids, and *Bacillus velezensis* B26, a growth promoting bacterium. *Front. Microbiol.* **2020**, *11*, 575578. [CrossRef]
139. Oso, S.; Walters, M.; Schlechter, R.O.; Remus-Emsermann, M.N.P. Utilisation of hydrocarbons and production of surfactants by bacteria isolated from plant leaf surfaces. *FEMS Microbiol. Lett.* **2019**, *366*, fnz061. [CrossRef]
140. Ahmad, Z.; Wu, J.; Chen, L.; Dong, W. Isolated *Bacillus subtilis* strain 330-2 and its antagonistic genes identified by the removing PCR. *Sci. Rep.* **2017**, *7*, 1777. [CrossRef] [PubMed]
141. Someya, N.; Nakajima, M.; Watanabe, K.; Hibi, T.; Akutsu, K. Potential of *Serratia marcescens* strain B2 for biological control of rice sheath blight. *Biocontrol Sci. Technol.* **2005**, *15*, 105–109. [CrossRef]
142. Purushotham, P.; Arun, P.V.; Prakash, J.S.; Podile, A.R. Chitin binding proteins act synergistically with chitinases in *Serratia proteamaculans* 568. *PLoS ONE* **2012**, *7*, e36714. [CrossRef]

143. Kamensky, M.; Ovadis, M.; Chet, I.; Chernin, L. Soil-borne strain IC14 of *Serratia plymuthica* with multiple mechanisms of antifungal activity provides biocontrol of *Botrytis cinerea* and *Sclerotinia sclerotiorum* diseases. *Soil Biol. Biochem.* **2003**, *35*, 323–331. [CrossRef]
144. Budi, S.W.; van Tuinen, D.; Arnould, C.; Dumas-Gaudot, E.; Gianinazzi-Pearson, V.; Gianinazzi, S. Hydrolytic enzyme activity of *Paenibacillus* sp. strain B2 and effects of the antagonistic bacterium on cell integrity of two soil-borne pathogenic fungi. *Appl. Soil Ecol.* **2000**, *15*, 191–199. [CrossRef]
145. Timofeeva, A.M.; Galyamova, M.R.; Sedykh, S.E. Bacterial Siderophores: Classification, Biosynthesis, Perspectives of Use in Agriculture. *Plants* **2022**, *11*, 3065. [CrossRef] [PubMed]
146. Lau, C.K.; Krewulak, K.D.; Vogel, H.J. Bacterial ferrous iron transport: The Feo system. *FEMS Microbiol. Rev.* **2016**, *40*, 273–298. [CrossRef]
147. Ollinger, J.; Song, K.B.; Antelmann, H.; Hecker, M.; Helmann, J.D. Role of the Fur regulon in iron transport in *Bacillus subtilis*. *J. Bacteriol.* **2006**, *188*, 3664–3673. [CrossRef] [PubMed]
148. Zhu, X.X.; Shi, L.N.; Shi, H.M.; Ye, J.R. Characterization of the *Priestia megaterium* ZS-3 siderophore and studies on its growth-promoting effects. *BMC Microbiol.* **2025**, *25*, 133. [CrossRef] [PubMed]
149. Nagrale, D.T.; Gawande, S.P.; Shah, V.; Verma, P.; Hiremani, N.S.; Prabhulinga, T.; Gokte-Narkhedkar, N. Biocontrol potential of volatile organic compounds (VOCs) produced by cotton endophytic rhizobacteria against *Macrophomina phaseolina*. *Eur. J. Plant Pathol.* **2022**, *163*, 467–482. [CrossRef]
150. Dimkić, I.; Janakiev, T.; Petrović, M.; Degrassi, G.; Fira, D. Plant-associated *Bacillus* and *Pseudomonas* antimicrobial activities in plant disease suppression *via* biological control mechanisms—A review. *Physiol. Mol. Plant Pathol.* **2022**, *117*, 101754. [CrossRef]
151. Khalifa, A.; Alsowayeh, N. Whole-Genome Sequence Insight into the Plant-Growth-Promoting Bacterium *Priestia filamentosa* Strain AZC66Obtained from *Zygophyllum coccineum* Rhizosphere. *Plants* **2023**, *12*, 1944. [CrossRef]
152. Jin, P.; Wang, H.; Tan, Z.; Xuan, Z.; Dahar, G.Y.; Li, Q.X.; Miao, W.; Liu, W. Antifungal mechanism of bacillomycin D from *Bacillus velezensis* HN-2 against *Colletotrichum gloeosporioides* Penz. *Pestic. Biochem. Physiol.* **2020**, *163*, 102–107. [CrossRef]
153. Zhou, H.; Luo, C.; Fang, X.; Xiang, Y.; Wang, X.; Zhang, R.; Chen, Z. Loss of GltB Inhibits Biofilm Formation and Biocontrol Efficiency of *Bacillus subtilis* Bs916 by Altering the Production of γ-Polyglutamate and Three Lipopeptides. *PLoS ONE* **2016**, *11*, e0156247. [CrossRef]
154. Liu, J.; He, D.; Li, X.; Gao, S.; Wu, H.; Liu, W.; Gao, X.; Zhou, T. γ-Polyglutamic acid (γ-PGA) produced by *Bacillus amyloliquefaciens* C06 promoting its colonization on fruit surface. *Int. J. Food Microbiol.* **2010**, *142*, 190–197. [CrossRef]
155. Maghembe, R.S.; Mdoe, F.P.; Makaranga, A.; Mpemba, J.A.; Mark, D.; Mlay, C.; Moto, E.A.; Mtewa, A.G. Complete genome sequence data of *Priestia megaterium* strain MARUCO02 isolated from marine mangrove-inhabited sediments of the Indian Ocean in the Bagamoyo Coast. *Data Brief.* **2023**, *48*, 109119. [CrossRef] [PubMed]
156. Xu, S.; Shen, C.; Li, C.; Dong, W.; Yang, G. Genome sequencing and comparative genome analysis of *Rhizoctonia solani* AG-3. *Front. Microbiol.* **2024**, *15*, 1360524. [CrossRef] [PubMed]
157. Stintzi, A.; Evans, K.; Meyer, J.M.; Poole, K. Quorum-sensing and siderophore biosynthesis in *Pseudomonas aeruginosa*: LasR/lasI mutants exhibit reduced pyoverdine biosynthesis. *FEMS Microbiol. Lett.* **1998**, *166*, 341–345. [CrossRef] [PubMed]
158. Faulkner, M.J.; Ma, Z.; Fuangthong, M.; Helmann, J.D. Derepression of the *Bacillus subtilis* PerR peroxide stress response leads to iron deficiency. *J. Bacteriol.* **2012**, *194*, 1226–1235. [CrossRef]
159. Foley, R.C.; Kidd, B.N.; Hane, J.K.; Anderson, J.P.; Singh, K.B. Reactive Oxygen Species Play a Role in the Infection of the Necrotrophic Fungi, *Rhizoctonia solani* in Wheat. *PLoS ONE* **2016**, *11*, e0152548. [CrossRef]

agriculture

MDPI

Article

Utilization of *Gluconacetobacter diazotrophicus* in Tomato Crop: Interaction with Nitrogen and Phosphorus Fertilization

Nelson Ceballos-Aguirre [1], Gloria M. Restrepo [2,*], Sergio Patiño [1], Jorge A. Cuéllar [2] and Óscar J. Sánchez [3]

1 Faculty of Agricultural Sciences, Universidad de Caldas, Manizales 170004, Colombia; nelson.ceballos@ucaldas.edu.co (N.C.-A.); sergio.501211011@ucaldas.edu.co (S.P.)
2 Faculty of Health Sciences, Research Institute in Microbiology and Agro-Industrial Biotechnology, Universidad Católica de Manizales, Manizales 170002, Colombia; jorge.cuellar@ucm.edu.co
3 Center for Technological Development—Bioprocess and Agro-Industry Plant, Department of Engineering, Universidad de Caldas, Manizales 170004, Colombia; osanchez@ucaldas.edu.co
* Correspondence: grestrepo@ucm.edu.co; Tel.: +57-606-8933050 (ext. 3142)

Abstract: *Gluconacetobacter diazotrophicus* is a nitrogen-fixing bacterium capable of colonizing different host plants. This work evaluated the impact of Colombian native isolate *G. diazotrophicus* GIBI029 on tomato yield and quality in response to nitrogen and phosphorus fertilization levels. The experiment was conducted under semi-controlled conditions using a split-plot design with four replicates and ten plants per treatment. Variables assessed included fruit count per plant, fruit weight, average fruit weight, production per plant, and yield. Application of GIBI029 without fertilization resulted in a higher number and weight of fruits per plant across harvests (7.1 fruits, 509.2 g) compared to both the unfertilized control (4.8 fruits, 271.7 g) and with complete nitrogen and phosphorus fertilization (5.2 fruits, 288.8 g). The behavior of these variables were similar for GIBI029 and complete fertilization (7.0 fruits and 510.7 g per harvest). The highest yields were obtained with GIBI029 without fertilization (106.1 t ha^{-1}) and with full nitrogen and phosphorus fertilization (106.4 t ha^{-1}). Under the evaluated conditions, native *G. diazotrophicus* GIBI029 isolate could effectively improve tomato growth and yield in contrast to the controls. Based on these findings, the reproducibility of this behavior should be confirmed, and the mechanisms involved in the plant–bacteria interaction should be determined.

Keywords: Colombian native isolate; fruit count per plant; fruit weight per plant; greenhouse cultivation; nitrogen fixation; plant growth-promoting bacteria; *Solanum lycopersicum*; yield

1. Introduction

Nitrogen is an indispensable nutrient for plant growth. Despite the substantial amount of nitrogen present in the atmosphere and soil, its availability to plants is limited, necessitating the application of this nutrient to crops [1]. Over the past 50 years, agricultural production has witnessed a tenfold increase, with a corresponding surge in the utilization of nitrogen fertilizers [2]. Concurrently, agricultural practices account for approximately 80% of the nitrogen released into the environment, contributing to adverse effects such as eutrophication [3], in addition to escalating production costs. Furthermore, the industrial processes employed in the production of nitrogenous chemical fertilizers consume significant quantities of fossil fuels, which are non-renewable resources. This consumption results in the release of carbon dioxide, a potent greenhouse gas, thereby contributing to global warming [2].

Phosphorus exists within the soil in both organic and mineral compound forms. However, the quantity of phosphorus readily available for plant uptake is significantly lower than the total phosphorus content present in the soil matrix. Consequently, the application of phosphorus fertilizers becomes essential to fulfill the nutritional requirements of crops. Nevertheless, the efficiency of phosphorus recovery from chemically synthesized fertilizers is notably limited. Only a fraction, ranging between 10% and 20%, of the applied phosphorus is utilized by plants during the year of application, as the majority of phosphorus is rapidly immobilized or fixed within the soil [4].

Gluconacetobacter diazotrophicus is recognized as a plant growth-promoting bacterium due to its inherent capabilities for biological fixation of atmospheric nitrogen, solubilization of phosphorus that is fixed within soils, and the production of indolic compounds [5]. This bacterium was initially isolated and characterized by Cavalcante and Döbereiner [6] from sugarcane, and it has since become one of the primary endophytic microorganisms employed as a model organism to investigate plant–bacteria interactions in non-leguminous plant species [7]. In the context of cultivation of economically important crops, it has been suggested that *G. diazotrophicus* holds substantial potential, largely attributed to the unique combination of its physiological characteristics, which are effectively expressed under conditions that closely resemble those found within the fluids of sugarcane, namely acidic pH and elevated sucrose concentrations [8]. While several studies have documented the presence of *G. diazotrophicus* and related growth-promoting effects in various crops, including sugar beet, carrot [9,10], radish, papaya, coffee [11], pineapple [12], sweet potato, taro, and cassava [13], as well as its general occurrence in many tropical and subtropical plants [14,15], further research is needed to elucidate its impact on commercially significant crops such as tomatoes.

Tomato (*Solanum lycopersicum*) is a dicotyledonous plant belonging to the Solanaceae family and the *Solanum* genus. This vegetable is the most extensively cultivated species and has nine related wild species. In 2023, global tomato production was estimated at 192,317,973 metric tons of fruit, harvested from approximately 5,412,458 hectares, resulting in an average global yield of approximately 35.53 metric tons per hectare [16]. Tomato cultivation is characterized by relatively high production costs. In middle-income countries like Colombia, tomato production represents a particularly risky agricultural activity, characterized by several key aspects: generally small and dispersed cultivation areas, intensive capital and labor requirements, substantial post-harvest losses, and significant phytosanitary challenges, among other factors. In particular, tomato cultivation in Colombia exhibits relatively low yields and limited technological adoption in production practices [17,18], such as the underutilization of environmentally friendly fertilizers, namely biofertilizers. Furthermore, a considerable portion of production expenses is allocated to fertility inputs, including costly chemical fertilizers, which can lead to severe environmental disturbances if applied excessively [19].

To reduce the production costs associated with tomato cultivation and to alleviate its negative environmental impacts, research is increasingly focused on exploring novel alternatives, such as microbial inoculants derived from plant growth-promoting bacteria [20] or plant growth-promoting rhizobacteria (PGPR) [21]. These bacteria possess the capacity to adapt, colonize, and persist within the rhizosphere, thereby promoting plant growth and development through their metabolic activities. Furthermore, they can significantly influence the bioavailability of essential nutrients, as exemplified by their ability to facilitate phosphorus solubilization and biological nitrogen fixation [22].

Studies have been conducted to assess the growth-promoting potential of *G. diazotrophicus* in tomato, such as that by Cocking et al. [23], who demonstrated that root meristem cells of tomato were intracellularly colonized by *G. diazotrophicus* strain 5541 UAP under gnotobiotic conditions. On the other hand, Luna et al. [24] studied the colonization and yield promotion of *G. diazotrophicus* PAL5 (also known as the ATCC 49037 strain) in tomato seedlings under gnotobiotic conditions, where it was evidenced that the bacterium colonizes lateral root emergence sites, root hairs, and stomata. In the same study, under greenhouse conditions, they found that plants inoculated with the bacterium significantly increased both the number and weight of fruit production compared to uninoculated controls. Botta et al. [25] evaluated *G. diazotrophicus* PAL5 under gnotobiotic conditions as well as in tomato seedlings, both individually and in a mixture, with *Azospirillum brasilense*, *Herbaspirillum seropedicae*, and *Burkholderia ambifaria*. During the in vitro experiment, *G. diazotrophicus* was the best isolate that colonized the entire tomato plant. A high number of bacteria was recovered in both roots (10^8 CFU/g dry weight) and in the stem and leaves (10^5 CFU/g dry weight). However, in seedlings during the in vivo experiment, *G. diazotrophicus* did not show good results in stimulating plant growth when applied individually. In response, Pallucchini et al. [26] investigated the effect of biological nitrogen fixation on growth promotion in a hydroponic culture of tomato plants by inoculating two isolates of *G. diazotrophicus*: a mutant with impaired nitrogen fixation (Gd *nifD*-) and a wild-type strain (Gd WT) under nitrogen-rich and limiting conditions. It was evidenced that the functional *nifD* gene is crucial for the optimal promotion of plant growth by *G. diazotrophicus*.

The application of high levels of fertilizers can significantly impact the interaction between bacteria and their host plants. This aspect has been evaluated in the case of *G. diazotrophicus* under different nitrogen and phosphorus fertilization regimes. In particular, nitrogen seems to act as a selective factor, as suggested by Caballero-Mellado et al. [27] and Kirchhof et al. [28]. Muthukumarasamy et al. [29] showed that nitrogen fertilization limited the isolation of *G. diazotrophicus*, with higher bacterial counts (from 10^4 to 10^7) resulting in plants without additional nitrogen fertilization. Likewise, Bueno dos Reis Junior et al. [30] noted that, in certain sugarcane genotypes, higher nitrogen doses reduced the population of *G. diazotrophicus*, indicating a genotype-dependent response. Moreover, several studies [31,32] have reported differential responses among sugarcane varieties and diazotrophic bacterial isolates regarding yield, nitrogen content, and nitrogenase activity under different nitrogen levels. Recent research has expanded these findings to include other crops. For instance, Tufail et al. [33] highlighted the role of *G. diazotrophicus* in mitigating combined drought and nitrogen stress in maize by enhancing biomass, chlorophyll content, and nitrogen uptake, while Van Long et al. [34] demonstrated that combining *G. diazotrophicus* inoculation with reduced nitrogen fertilizer maintained rice yield and improved soil quality in a triple rice cropping system. These studies indicate the potential of *G. diazotrophicus* to reduce nitrogen fertilizer inputs while sustaining crop productivity and soil health under diverse environmental conditions. Therefore, the study of the interaction between this bacterium, in conjunction with nitrogen fertilizer, and crops like tomato represents a relevant and ongoing research topic.

On the other hand, the interaction between *G. diazotrophicus* and phosphorus fertilization is a critical factor influencing plant–microbe dynamics and nutrient acquisition. Several studies have highlighted that phosphorus availability can significantly affect the colonization and activity of plant growth-promoting bacteria, including *G. diazotrophicus*. For instance, Delaporte-Quintana et al. [35] reported that phosphorus deficiency in strawberry plants was alleviated by inoculation with *G. diazotrophicus*, which increased phosphorus uptake and biomass by solubilizing inorganic phosphate. Similarly, Delaporte-Quintana et al. [36] showed that the PAL5 strain effectively solubilized different phosphate sources, supporting its functional role as a phosphate-solubilizing bacterium. Additionally, Idogawa et al. [37] demonstrated that high phosphate concentrations stimulated the production of levan, an exopolysaccharide that enhances stress resistance and colonization. These findings suggest that combining *G. diazotrophicus* inoculation with the addition of appropriate levels of phosphorus fertilizers may optimize nutrient cycling and contribute to sustainable crop production. Therefore, research on the role of this bacterium in relation to phosphorus fertilization, particularly in crops such as tomato, represents significant agronomic value.

Our research group has recovered a Colombian wild strain of *G. diazotrophicus* GIBI029, which has demonstrated notable plant growth-promoting properties, including nitrogenase activity, phosphate solubilization, and the synthesis of indolic compounds [38]. Furthermore, its potential in stimulating the growth of tomato seedlings (*S. lycopersicum* genotypes Santa Clara and Torrano) has been identified. Consequently, this study aimed to evaluate the interactive effects of inoculation with the *G. diazotrophicus* GIBI029 native isolate and varying doses of nitrogen- and phosphorus-based chemical fertilizers on tomato crop quality and yield. We hypothesized that the application of *G. diazotrophicus* GIBI029 native isolate will have a differential effect on tomato crop quality and yield compared to varying doses of nitrogen and phosphorus chemical fertilizers; it is expected that the bacterium can enhance yield even under reduced fertilizer input. This information will contribute to further establishing the behavior of this bacterium in tomato cultivation.

2. Materials and Methods

2.1. Location

The study was conducted at the Tesorito farm, Universidad de Caldas, situated in Manizales city (latitude 5°01′49″ N, longitude 75°26′13″ W), Colombia, in 2022 between February and September. The location experiences an average temperature of 17 °C, at an elevation of 2340 m above sea level (m a.s.l.), with an annual average rainfall of 1800 mm and a relative humidity of 78% [39]. The evaluation site was selected because it is one of the experimental farms of the Universidad de Caldas, which allowed for ensuring local control of the evaluated treatments. Additionally, the agroecological and climatic conditions of the geographical zone where the farm is located, above 2000 m a.s.l., have become ideal for the development of tomato cultivation under semi-controlled conditions, with neighboring associations of tomato producers identified as potential beneficiaries of extrapolatable and promising results.

2.2. Plant Material

A commercial Chonto-type hybrid tomato (Roble 956 F1 Hybrid Tomato, ImpulSemillas, Tocancipá, Colombia) was used, which is characterized by its good vigor and size, with strong stems and medium internodes. It is a material with indeterminate growth and a fruit weight between 140 and 160 g. It adapts to production areas with cold and medium climates (1800 to 2200 m a.s.l.), is tolerant to low temperatures, and resistant to pathogens such as nematodes, mosaic virus, and *Verticillium* or *Fusarium* 1, 2, and 3 [40]. The seeds used are certified and endorsed by a competent Colombian authority (Colombian Agricultural Institute) from the commercial company ImpulSemillas, under the commercial name "tomate Híbrido Roble" [41].

Planting was conducted in 128-cell nursery trays. Seedling production began with the disinfection of trays using agricultural iodine Agrodyne® (West, La Estrella, Colombia) at a concentration of 5 mL per liter of water, in which the trays were submerged and then allowed to dry in the sun. Subsequently, the cells were filled with grade 4 peat-type substrate, avoiding compaction, and drilling holes were made with double the size of the seed for planting. Tomato seeds were then sown, one per cell, and the trays were placed in a germinator under semi-controlled conditions for roughly 30 days. Seedlings were kept in the germinator until they reached an average of 4 true leaves, indicating they were ready for transplanting. During the process, irrigation, fertilization, and pest control tasks were performed.

2.3. Microbial Cultures

Indigenous *G. diazotrophicus* isolation obtained in previous work [38], identified in the collection of microorganisms of the Universidad Católica de Manizales as GIBI029 (Manizales, Colombia), and the ATCC 49037 reference strain of *G. diazotrophicus* (ATCC, Manassas, VA, USA) were used. The bacteria were recovered from cryopreserved vials at −80 °C by reactivation at 37 °C for 5 min and subsequent sowing by depletion in Potato Dextrose Agar (PDA) (Oxoid®, Waltham, MA, USA). The plates were incubated for 5 days at 30 °C until isolated colonies were obtained. From these cultures, the inoculations were prepared to be applied in the field.

2.4. Preparation of Greenhouses

The production system used was semi-controlled, which presented an indirect influence of external climatic variables. For this, two experimental semi-greenhouses were built. Total control was maintained only for the climatic variable of precipitation, managing the amount of water required by the plants through a drip irrigation system with emitters spaced at 20 cm. The temperature inside the greenhouses showed increases of between 2 °C and 4 °C in relation to the outside temperature. Black-black plastic mulch installed on the furrows was used to ensure weed control throughout the crop production cycle, thus avoiding a critical weed period that could alter the results. Additionally, the plastic mulch allows for moisture retention in the root zone because the evaporation process prevents the loss of water and nutrients. Constant humidity also provides an adequate microclimate for the inoculated bacteria, allowing for a more reliable measurement of the effect of soil application.

The soil in both greenhouses was prepared for tomato cultivation by adjusting the pH based on initial soil analysis (Table A1, Appendix A). Agricultural lime (240 g per linear meter) was manually incorporated using a hoe. Additionally, 360 g of organic matter was also incorporated into the soil. The land was leveled with a rake, and the irrigation hoses were further installed (see Figure 1).

(a) (b)

Figure 1. Preparation of the greenhouses for the experiment setup, illustrating (**a**) the initial soil preparation and (**b**) the installation of irrigation hoses at 20 cm spacing and black plastic mulch over the planting rows to control weeds and retain moisture, which created semi-controlled conditions for the tomato cultivation and bacterial inoculation study.

2.5. Experimental Design

The experiment employed a split-plot design arranged in four randomized, completely replicated blocks. The main plot (main factor) corresponded to the application of *G. diazotrophicus* reference strain (ATCC 49037), the Colombian native *G. diazotrophicus* isolate (GIBI029), and the control for a total of 3 main plots (the control main plot was included in both greenhouses, as shown in Figure 2; for the statistical analysis, an equal and appropriate number of effective plants corresponding to this control main plot were taken from both greenhouses). Thus, the subplot (subfactor) comprised four levels of nitrogen and phosphorus fertilization. For this, each main plot was further divided into four smaller plots or blocks (n = 4 replicates), to which the four fertilization treatments were randomly assigned (see Table 1).

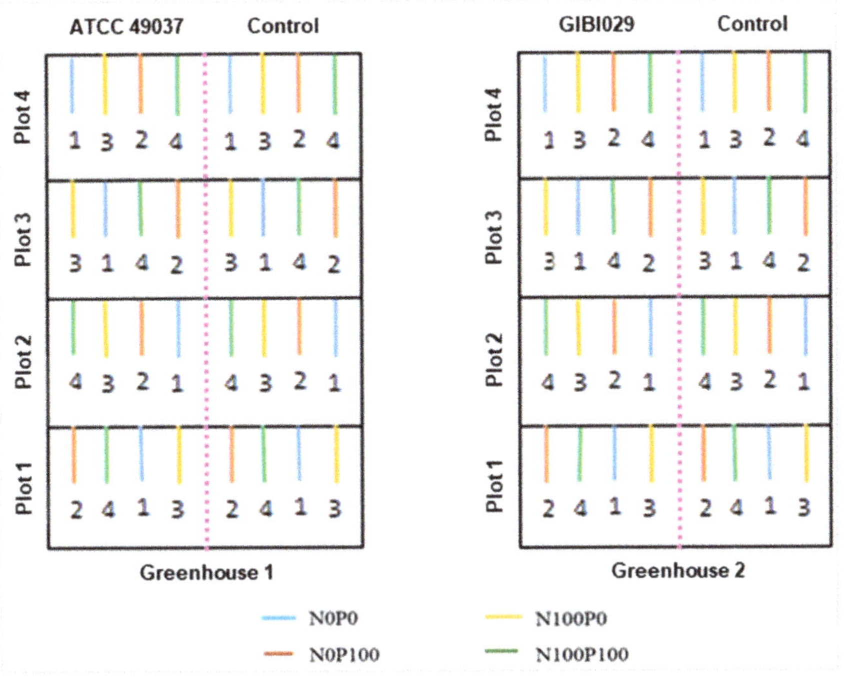

Figure 2. Schematic diagram of the split-plot experimental design conducted within two greenhouses, demonstrating the arrangement of bacterial treatments (main plots) and fertilization levels (subplots or smaller plots) to assess their interactive effects on tomato yield. The design facilitated the comparison of GIBI029 and ATCC 49037 against controls under varying nutrient conditions. Solid colored lines represent rows receiving different edaphic fertilization treatments. The dotted plum-colored line depicts the bamboo stalks used for structural support, dividing each greenhouse into two main plots. Codes corresponding to the percentages of nitrogen and phosphorus fertilization are deciphered in Table 1.

Each treatment combination (main plot × subplot) was applied to 10 individual plants. Each main plot, with its four replicates (blocks), contained a total of 160 plants (4 nitrogen and phosphorus fertilization levels × 10 plants × 4 blocks), i.e., a total of 480 plants in the 3 main plots. Consequently, each block consisted of 40 plants, representing a given bacterial treatment and all combinations of nitrogen and phosphorus fertilization. Each treatment was applied directly to 10 plants per combination per block.

Table 1. Scheme of edaphic fertilization during the crop cycle according to the treatment, detailing the specific treatments, fertilization sources, and application rates.

Treatments	Degree of Fertilization	Code	Fertilization Source	Days After Transplanting [g per Plant]					Dosage [kg ha^{-1}]
				0	30	60	90	150	
1	0% N + 0% P	N0P0	-	-	-	-	-	-	0
2	0% N + 100% P	N0P100	TSP	8.0	4.0	4.2	-	-	338
3	100% N + 0% P	N100P0	Urea	8.0	4.0	4.0	-	-	335
4	100% N + 100% P	N100P100	TSP	8.0	4.0	4.2	-	-	338
			Urea	8.0	4.0	4.0	-	-	335
1–4			Micronutrients	2.0	2.0	2.0	2.0	2.0	205
1–4			KCl	6.0	6.0	12.0	12.0	9.0	937
1–4			$MgSO_4$	5.0	5.0	5.0	5.0	5.0	520

TSP: triple superphosphate.

The plants were established at a distance of 1.2 m between rows, spaced at 0.4 m between plants. In the corresponding main plot, 75 mL of a bacterial suspension of strain ATCC 49037 or isolate GIBI029 was applied as a drench to the base of each plant in the experimental units. The bacterial suspension was prepared from a bacterial culture with a concentration of 18×10^7 CFU mL^{-1}, at a dose of 5 mL L^{-1}, and was inoculated 20 days after transplanting, as determined in previous works [42]. Four fertilization treatments were applied, arranged as illustrated in Figure 2 and Table 1.

2.6. Transplant and Crop Management

At 30 days post-germination, plants were transplanted from the nursery to the two experimental greenhouses. Seedlings were established at a spacing of 40 cm intra-row and 120 cm inter-row, resulting in a planting density of 20,833 plants ha^{-1}. Nutrient supplementation was tailored to the crop based on a comprehensive soil analysis of the experimental site (see Table A1 in Appendix A). All essential elements were supplied according to the species' requirements, with the exception of nitrogen and phosphorus. These macronutrients were administered at either 100% or 0% of the recommended levels, interacting with the evaluated microorganisms as defined by the fertilization treatments outlined in the experimental design (Table 1). Additional micronutrients required by the crop were applied according to the soil analysis: $MgSO_4$ (IFFCO Kisan Suvidha Ltd., New Delhi, India) at a rate of 520 kg ha^{-1}, KCl (Lvfeng Fertilizer Co., Ltd., Zibo, China) at a rate of 937 kg ha^{-1}, and a commercial micronutrient mixture containing boron, copper, manganese, zinc, and iron (Yara International ASA, Oslo, Norway) at a rate of 205 kg ha^{-1}. A split application of fertilizers was carried out, starting at transplanting and continuing for 150 days, as presented in Table 1. All fertilizations were applied in a ring around each plant at the corresponding times and at the doses described in Table 1. To facilitate fertilizer dilution, continuous irrigation was ensured through drip irrigation lines with emitters spaced at 15 cm, one line per row, with a discharge rate of 25 mL min^{-1}. The amount of water applied per plant was 500 mL $\times$ plant^{-1} $\times$ day^{-1} from transplanting up to 30 days, 1000 mL $\times$ plant^{-1} $\times$ day^{-1} from 30 to 60 days, and 1500 mL $\times$ plant^{-1} $\times$ day^{-1} from 60 to 180 days (end of the cultivation period).

Crop architecture was managed by maintaining two stems per plant across all experimental units and standard pruning practices were implemented [43,44]. Upon reaching a total of 12 floral clusters on the main axis, apical bud pruning was performed to terminate vegetative growth and prevent further cluster development.

2.7. Evaluation of Yield and Quality

For each experimental unit of every replicate, the following variables were quantified: number of fruits per plant, weight of fruits per plant (in g/plant), mean fruit weight (in g), and total production per plant (in g plant^{-1}). Crop yield (in ton ha^{-1}) was subsequently estimated based on the determined production per plant and the established planting density. Fruit weight per harvest was recorded over a 10-week period, encompassing 10 distinct harvests throughout the crop cycle, with one harvest conducted per week. Fruits harvested weekly were those attaining a ripening index corresponding to either green-yellow (scale 2) or greenish-yellow (scale 3), as determined by visual fruit coloration according to the United States' standards for grades of fresh tomatoes [45]. The summation of all partial harvests within each treatment provided the total production per plant (g plant^{-1}). The time leading up to the first harvest, specifically 90 days post-transplantation, was recorded, noting that the initial harvest and the first week of harvest coincided.

2.8. Statistical Analysis

Statistical analysis of the data was conducted using analysis of variance (ANOVA), employing the General Linear Model (GLM) procedure with SAS software version 9.4 (SAS Institute, Cary, NC, USA) to assess the effects of each experimental factor and their interactions. This analysis was performed to determine statistically significant differences among the various treatments. Where significant differences were detected, post hoc mean comparisons were carried out using Duncan's multiple-range test at a significance level of $p < 0.05$.

3. Results

3.1. Number of Fruits per Plant

The impact of bacterial type (native isolation or reference strain) on tomato fruit number revealed statistically significant differences ($p < 0.05$) between the standard strain and the native *G. diazotrophicus* isolate in relation to varying nitrogen and phosphorus fertilization regimes throughout the crop cycle. This outcome is depicted in Figure 3, where the plants applied with the native isolate exhibited higher number of fruits per plant.

The GIBI029 isolate demonstrated superior performance across all interactions, exhibiting peak production during the second and fourth harvests (see Figure 4). Specifically, it yielded 11.2 fruits per plant in the second harvest under 100% nitrogen and 100% phosphorus fertilization, and it yielded 12.85 fruits per plant in the fourth harvest under nitrogen-free fertilization with 100% phosphorus. For the case when the native GIBI029 isolate was applied, statistical analysis indicated no significant differences ($p < 0.05$) in fruit number per harvest between varying nitrogen and phosphorus doses. Significantly, across all treatments and interactions, the highest mean fruit per plant value throughout the ten harvest cycles was observed in the N0P0 control interaction (7.1 fruits per plant) and in the N100P100 complete nitrogen and phosphorus fertilization treatment (7.0 fruits per plant), under the influence of the GIBI029 isolate. This suggests that the bacterium can function effectively utilizing inherent soil nutrient reserves, obviating the need for chemical fertilization when soil nutrient availability aligns with the requirements of the cultivated species.

(**a**)

(**b**)

Figure 3. Main plots within the two experimental greenhouses used in this study. (**a**) The main plot treated with the *G. diazotrophicus* ATCC 49037 reference strain (**left**) and the corresponding control main plot (**right**). (**b**) The main plot treated with the native *G. diazotrophicus* GIBI029 isolate (**left**) and the corresponding control main plot (**right**). GIBI029-treated plots visibly show denser and more productive plants than the controls, suggesting enhanced growth performance from the native isolate. Images were captured at the time of the fourth harvest.

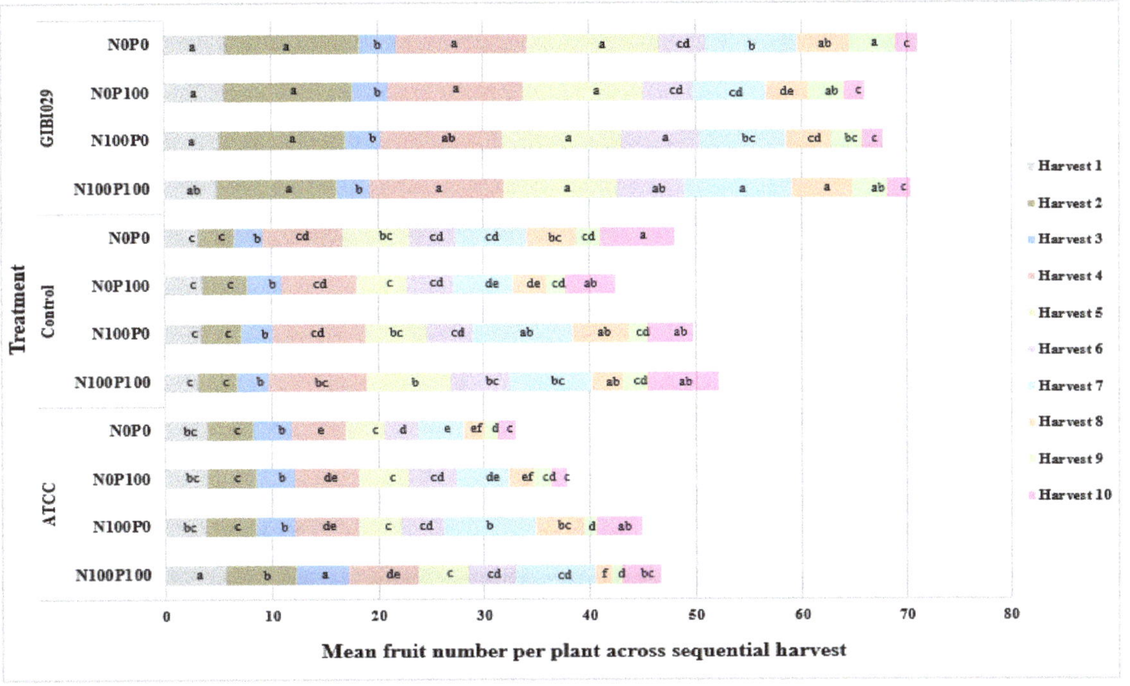

Figure 4. Effect of *G. diazotrophicus* and four types of fertilization on the mean fruit number per plant across sequential harvests of the greenhouse tomato crop. Tomato plants treated with the native isolate GIBI029 consistently produced a significantly higher number of fruits across all fertilization regimes, with peaks during the 2nd and 4th harvests. This indicates a strong early-to-mid-cycle effect of GIBI029 on fruit set, independent of chemical fertilization. The codes corresponding to the percentages of nitrogen and phosphorus fertilization are deciphered in Table 1. ATCC: *G. diazotrophicus* reference strain (ATCC 49037); control: conventional control treatment; GIBI029: native *G. diazotrophicus* isolation; Harvest 1–10: number of tomato fruits per plant in each harvest (the numerical value indicates the harvest number). Treatments with different letters are significantly different ($p < 0.05$) according to Duncan's multiple-range test ($n = 4$ replicates or blocks).

The ATCC 49037 strain exhibited the lowest fruit number values across the experiments compared to the native isolate's performance. The values recorded for this strain were consistently lower than those of the absolute control (nitrogen and phosphorus combinations without bacterial inoculation, see Figure 4). The interaction between nitrogen and phosphorus fertilization levels, along with the application of the ATCC 49037 strain, yielded peak fruit numbers when 100% of both phosphorus and nitrogen fertilizers were administered, exceeding the performance of the absolute control (N0P0). This observation underscores the dependence of this strain on external nitrogen and phosphorus sources to enhance fruit number production (Figure 4). For this variable, the efficacy of the GIBI029 isolate was evident in enhancing production outcomes within systems employing technical fertilization, thereby capitalizing on the utilization of native isolates.

3.2. Weight of Fruits per Plant

Concerning fruit production per harvest over time (Figure 5), statistically significant differences ($p < 0.05$) were observed across the majority of harvest periods for the two bacterial treatments and the controls. Notably, the tenth harvest exhibited no statistically significant differences between the two bacterial treatments. The peak harvest production was achieved 118 days post-sowing (fourth harvest). This harvest yielded values of 997 g plant^{-1} and 356 g plant^{-1} when utilizing the GIBI029 isolate and the ATCC 49037 strain, respectively. It was also found that the highest values were reported in the harvests 2, 4, and 5, respectively, and the lowest values were observed in the last three harvests 8, 9, and 10, respectively, for all the evaluated combinations, as depicted in Figure 5.

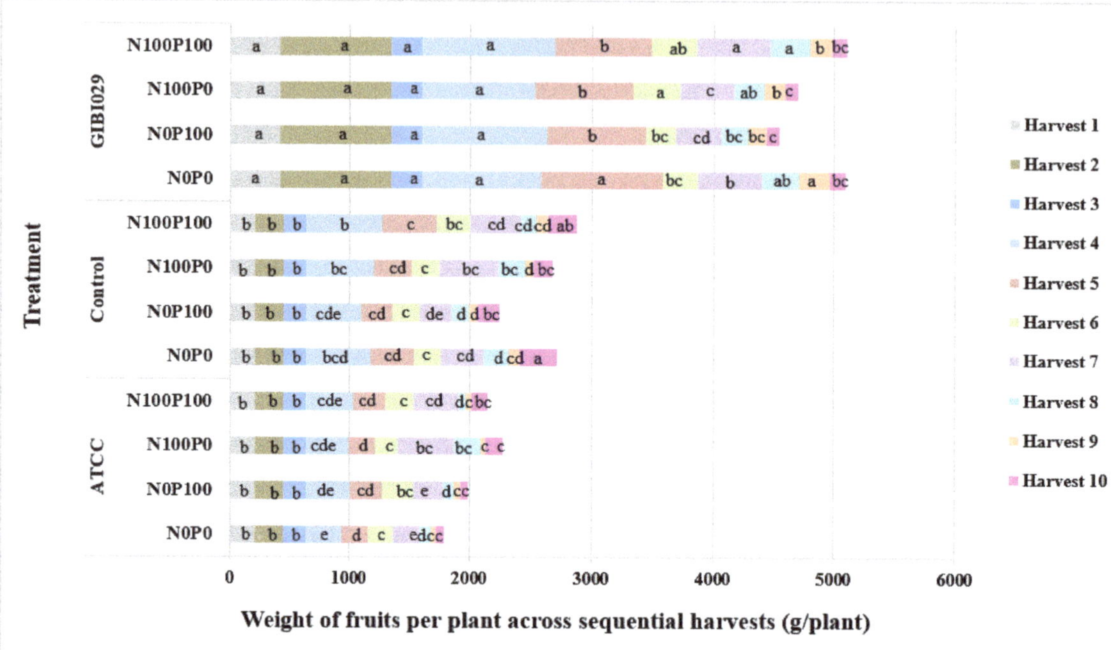

Figure 5. Effect of *G. diazotrophicus* and four types of fertilization on the weight of fruits per plant across sequential harvests of the greenhouse tomato crop. GIBI029-treated plants showed higher fruit weights in early and mid-harvests (notably harvests 2, 4, and 5), outperforming both the control and ATCC 49037 treatments. The effect was consistent even in the absence of nitrogen and phosphorus fertilization, highlighting the isolate's growth-promoting potential. The codes corresponding to the percentages of nitrogen and phosphorus fertilization are deciphered in Table 1. ATCC: *G. diazotrophicus* reference strain (ATCC 49037); control: conventional control treatment; GIBI029: native *G. diazotrophicus* isolation. Treatments with different letters are significantly different ($p < 0.05$) according to Duncan's multiple-range test ($n = 4$ replicates or blocks).

During harvests 4, 5, and 6, GIBI029 reported values of 997 g plant^{-1}, 846 g plant^{-1}, and 331 g plant^{-1}, respectively, while the control exhibited values of 555 g plant^{-1}, 354 g plant^{-1}, and 238 g plant^{-1}, respectively. Reference ATCC 49037 strain also showed statistically significant differences ($p < 0.05$) with values of 356 g plant^{-1}, 245 g plant^{-1}, and 231 g plant^{-1} in harvests 4, 5, and 6, respectively (Table 2).

Table 2. Comparative Duncan's multiple-range tests for production indexes over time during the evaluation of *G. diazotrophicus* addition to the tomato culture.

Variable	Factor		Harvest									
			1	2	3	4	5	6	7	8	9	10
Number of fruits per plant	Type of bacteria	GIBI029	5.25 ^{a}	11.94 ^{a}	3.32 ^{b}	12.38^{a}	11.38 ^{a}	5.72 ^{a}	8.35 ^{a}	4.78 ^{a}	3.49 ^{a}	2.18 ^{b}
		ATCC	4.31 ^{b}	4.98 ^{b}	3.97 ^{a}	5.95^{c}	4.23 ^{c}	4.13 ^{a}	6.33 ^{b}	2.65 ^{b}	1.24 ^{c}	2.77 ^{b}
		Control	3.28 ^{c}	3.65 ^{c}	2.97 ^{b}	8.17^{b}	6.19 ^{b}	4.73 ^{b}	7.33 ^{ab}	4.10 ^{a}	1.99 ^{b}	5.65 ^{a}
	Type of fertilization	N0P0	3.93 ^{a}	5.85 ^{a}	3.11 ^{a}	8.13^{a}	7.27 ^{ab}	5.62 ^{a}	6.41 ^{b}	4.20 ^{ab}	2.57 ^{a}	4.37 ^{a}
		N0P100	4.11 ^{a}	6.19 ^{a}	3.35 ^{a}	8.29^{a}	6.30 ^{b}	4.54 ^{ab}	5.79 ^{b}	3.22 ^{b}	2.05 ^{a}	3.32 ^{a}
		N100P0	3.93 ^{a}	5.95 ^{a}	3.23 ^{a}	8.75^{a}	6.69 ^{ab}	5.03 ^{ab}	8.92 ^{a}	4.87 ^{a}	1.930 ^{a}	3.73 ^{a}
		N100P100	4.15 ^{a}	6.23 ^{a}	3.52 ^{a}	9.49^{a}	7.73 ^{a}	5.62 ^{a}	8.22 ^{a}	3.35 ^{b}	2.17 ^{a}	4.84 ^{a}
Weight of fruits [g per plant]	Type of bacteria	GIBI029	418.0 ^{a}	940.0 ^{a}	260.0 ^{a}	996.8^{a}	846.2 ^{a}	331.0 ^{a}	487.2 ^{a}	275.4 ^{a}	189.9 ^{a}	117.9 ^{b}
		ATCC	207.0 ^{b}	240.0 ^{b}	191.0 ^{b}	356.1^{c}	244.9 ^{c}	230.9 ^{b}	321.8 ^{b}	114.0 ^{c}	43.6 ^{c}	101.7 ^{b}
		Control	207.0 ^{b}	240.0 ^{b}	191.0 ^{b}	554.8^{b}	354.0 ^{b}	238.8 ^{b}	372.5 ^{b}	177.4 ^{b}	81.3 ^{b}	218.8 ^{a}
	Type of fertilization	N0P0	259.8 ^{a}	415.0 ^{a}	208.3 ^{a}	591.8^{a}	492.1 ^{a}	238.6 ^{a}	353.6 ^{b}	209.8 ^{ab}	116.8 ^{b}	195.4 ^{a}
		N0P100	259.8 ^{a}	415.0 ^{a}	208.3 ^{a}	580.4^{a}	396.9 ^{a}	246.9 ^{a}	286.4 ^{b}	146.2 ^{c}	87.1 ^{a}	131.1 ^{a}
		N100P0	259.8 ^{a}	415.0 ^{a}	208.3 ^{a}	602.2^{a}	419.3 ^{a}	260.3 ^{a}	462.9 ^{a}	229.3 ^{a}	87.4 ^{a}	142.6 ^{a}
		N100P100	259.8 ^{a}	415.0 ^{a}	208.3 ^{a}	688.0^{a}	490.8 ^{a}	293.6 ^{a}	451.0 ^{a}	158.8 ^{bc}	104.4 ^{a}	187.9 ^{a}

Remarks: Within each column, means followed by different letters indicate statistically significant differences ($p < 0.05$). Differences may be attributed to either fertilization treatments (subplot effect; $n = 4$ replicates or blocks) or bacterial treatments (main plot effect). The codes corresponding to the percentages of nitrogen and phosphorus fertilization are deciphered in Table 1. ATCC: *G. diazotrophicus* reference strain (ATCC 49037).

A similar trend to that observed for the number of fruits per plant was found for fruit weight over time (Figure 5). *G. diazotrophicus* GIBI029 isolate (in its different interactions with nitrogen and phosphorus fertilization levels) showed the highest values compared to the control and the reference ATCC 49037 strain. Nitrogen and phosphorus doses (0% and 100%) did not show statistically significant differences ($p < 0.05$) within each larger plot.

Throughout all harvest periods, the GIBI029 isolate consistently demonstrated a greater effect than the ATCC 49037 strain (Table 2). This phenomenon can be attributed to the fact that GIBI029 was isolated under agroclimatic conditions closely resembling those of the experimental site. This similarity likely facilitated the adaptation of the bacterium to the experimental conditions, thereby maximizing its potential for production enhancement.

The differential impact of four nitrogenous and phosphorous fertilization regimes on the per plant weight of fruits across the ten harvests in the tomato cultivation was observed at the seventh harvest, revealing statistically significant differences ($p < 0.05$) for the 0% nitrogen/100% phosphorus and 100% nitrogen/100% phosphorus treatments. The critical role of nitrogen application to reach the highest production was demonstrated by the superior production observed in treatments receiving 100% nitrogen. Specifically, at the seventh harvest, per plant weight of 451 g and 463 g were recorded for the N100P100 and N100P0 treatments, respectively, compared to 354 g for the unfertilized control (N0P0) and 286 g for the N0P100 treatment (Table 2). Evaluation of the interaction between the GIBI029 isolate, the ATCC 49037 strain, and the varied nitrogen and phosphorus fertilization levels yielded results consistent with those observed for fruit numbers across the progression of harvests. Notably, the GIBI029 isolate consistently exhibited the highest fruit weight across all nitrogen and phosphorus combinations throughout the crop's lifecycle. In particular, the GIBI029 isolate demonstrated peak performance at the second, fourth, and fifth harvests, followed by the control and the ATCC 49037 strain treatments, with statistically significant differences observed ($p < 0.05$).

3.3. *Mean Fruit Weight*

Regarding the influence of bacterial type on the temporal dynamics of mean fruit weight, statistical significance was established ($p < 0.05$) across a majority of harvests, with the sole exception being the fourth harvest, where no significant divergence was observed between the two bacterial types applied. Throughout the sequential harvests, the GIBI029 isolate consistently exhibited a superior effect on mean fruit weight compared to the standard ATCC 49037 strain (Figure 6a).

The mean fruit weight, measured between 90 days after harvest and 167 days after transplanting, exhibited statistically significant differences ($p < 0.05$) for the GIBI029 isolate when compared to both the control group and the reference strain ATCC. While the initial harvests for GIBI029 yielded mean fruit weights exceeding 70 g and 75 g, respectively, a gradual decrease in weight was observed over time. Notably, the overall mean fruit weight for the GIBI029 treatment remained consistently higher than that of the control and the reference strain throughout the experimental period. It is important to clarify that the mean fruit weight reported here represents the mean of all harvested fruits, irrespective of fruit quality.

Analogously, during the second week (harvest), corresponding to 101 days post-transplantation (dpt), an increase in the mean fruit weight was observed. This augmentation was succeeded by a period of stabilization spanning the fourth and fifth harvests (123 dpt and 134 dpt), after which a progressive decline persisted until the termination of the cultivation cycle. This pattern evidenced the inherent tendency of the crop to exhibit a reduction in mean fruit weight over the duration of the harvest period (Figure 6a). Conversely, the influence of the four distinct nitrogen and phosphorus fertilization treatments on the mean fruit weight of the tomato crop, assessed throughout the eight harvests, did not show statistically significant variations ($p < 0.05$), except for the seventh and eighth harvests, specifically at 156 dpt and 167 dpt (Figure 6b).

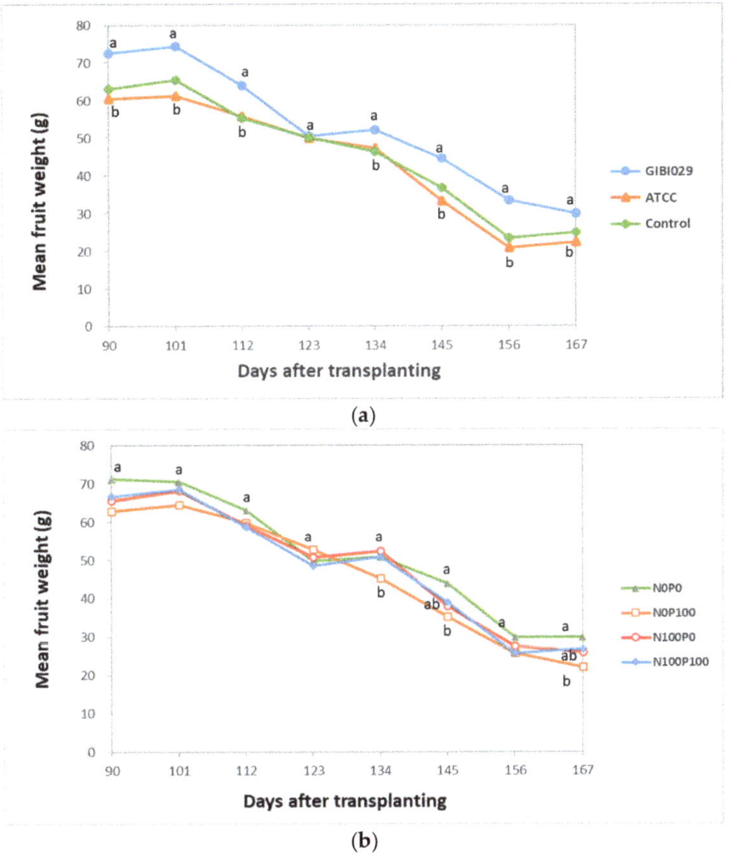

Figure 6. (**a**) Mean fruit weight over time by bacterial treatment; GIBI029 application resulted in higher average fruit weights, especially in early harvests, with a gradual decline over time, mirroring normal crop senescence. (**b**) Effect of four types of fertilization on the mean fruit weight (g) over time in the greenhouse tomato crop. Differences were minor across treatments, except during later harvests (7th and 8th), suggesting bacterial influence outweighed fertilization effects on this variable. ATCC: *G. diazotrophicus* reference strain (ATCC 49037); control: conventional control treatment; GIBI029: native *G. diazotrophicus* isolation. The codes corresponding to the percentages of nitrogen and phosphorus fertilization are deciphered in Table 1. Different letters indicate significant differences between treatments ($p < 0.05$) according to Duncan's multiple-range test, with the figure showing both the main plot effect (**a**) and the subplot effect (**b**), where $n = 4$ replicates or blocks.

3.4. Total Number of Fruits per Plant and Total Production per Plant

In general, the total number of fruits per plant across the 10 harvests (see Table 3) showed statistically significant differences ($p < 0.05$) for the bacterial application factor. The highest values were observed with the GIBI029 isolate, exceeding 65 fruits per plant by the end of the 10 harvests. In contrast, no significant differences ($p < 0.05$) were found among the smaller plots within each main plot, indicating that the bacterial treatment had a greater effect than the nitrogen and phosphorus fertilization levels. Specifically, the GIBI029 isolate outperformed both the control and the reference strain ATCC 49037. Furthermore, nitrogen and phosphorus fertilization at 100% levels, based on soil analysis, did not differentially affect fruit number in the smaller plots.

Table 3. Cumulative values of total number of fruits per plant, total production per plant, and overall yield at the end of the tenth harvest, according to the bacterial treatment and nitrogen and phosphorus fertilization levels.

Variable	Factor	Treatment	Total Value
Total number of fruits per plant	Type of bacteria	GIBI029	69.0 ^{a} ± 16.5
		ATCC	40.5 ^{c} ± 15.5
		Control	48.0 ^{b} ± 16.1
	Type of fertilization	N0P0	49.9 ^{bc} ± 19.0
		N0P100	47.1 ^{c} ± 18.0
		N100P0	53.1 ^{ab} ± 17.2
		N100P100	55.3 ^{a} ± 21.6
Total production [g per plant]	Type of bacteria	GIBI029	4862.4 ^{a} ± 1268.8
		ATCC	2051.1 ^{c} ± 854.6
		Control	2635.6 ^{b} ± 937.5
	Type of fertilization	N0P0	3081.3 ^{a} ± 1579.4
		N0P100	2758.2 ^{b} ± 1365.7
		N100P0	3087.2 ^{a} ± 1294.4
		N100P100	3258.1 ^{a} ± 1620.5
Yield [t ha^{-1}]	Type of bacteria	GIBI029	101.3 ^{a} ± 26.4
		ATCC	42.7 ^{c} ± 17.8
		Control	54.9 ^{b} ± 19.5
	Type of fertilization	N0P0	64.2 ^{a} ± 32.9
		N0P100	57.5 ^{b} ± 28.4
		N100P0	64.3 ^{a} ± 26.9
		N100P100	67.9 ^{a} ± 33.7

Remarks: Mean values with different letters in the column are significantly different ($p < 0.05$). Standard deviation is presented following the total values. Fertilization treatment comparisons (subplot effect), and bacterial treatment comparisons (main plot effect) are both represented ($n = 4$ replicates or blocks). The codes corresponding to the percentages of nitrogen and phosphorus fertilization are deciphered in Table 1. ATCC: *G. diazotrophicus* reference strain (ATCC 49037).

The cumulative mean value reached by the end of the last harvest (also called total production per plant) was 4862.4 ± 1268.8 g per plant for the GIBI029 isolate, followed by 2635.6 ± 937.5 g per plant for the controls (without bacterial application), and finally 2051.1 ± 854.6 g per plant for the ATCC 49037 strain. These results indicate an 84.5% increase of the total production per plant reached by the GIBI029 isolate related to the control mean value. In contrast, a 22.2% decrease of the total production per plant related to the controls was verified for the reference strain (Table 3).

Consistent with the preceding analysis of yield components (number of fruits per plant and fruit weight per plant), the best results of the total production per plant were found with the GIBI029 isolate that always exhibited values above 4500 g per plant, followed by the production achieved in the control treatments (without bacteria) and by the ATCC 49037 strain, with statistical differences ($p < 0.05$). Quantitatively, the GIBI029 isolate achieved production levels of 5107 g per plant under full nitrogen and phosphorus fertilization and 5092 g per plant in the absence of chemical fertilization, contrasting sharply with the ATCC 49037 strain under unfertilized conditions (N0P0), which yielded 1792 g plant^{-1} (Figure 7a). The bacterium's capacity to utilize soil phosphorus reserves, as reported in the soil analysis (see Appendix A), and its ability as a nitrogen fixer to ensure yields similar to those achieved with complete fertilization (100% nitrogen supplementation) according to the crop's nutritional demand are evident.

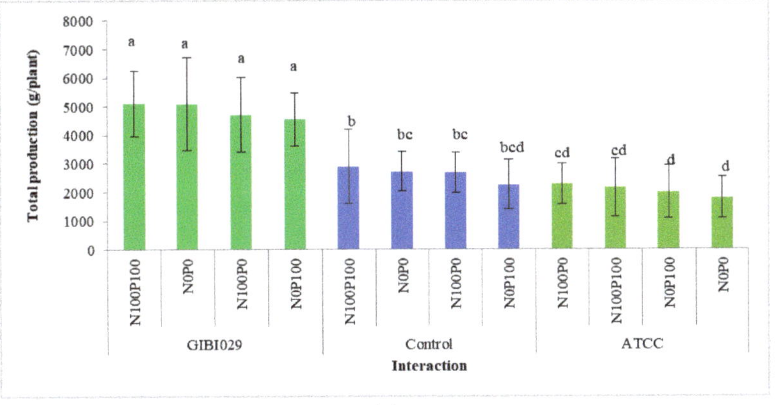

(**a**)

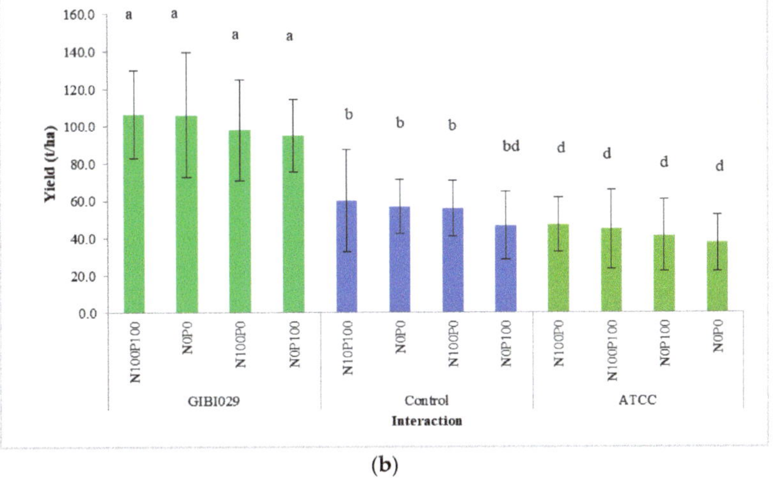

(**b**)

Figure 7. (**a**) Effect of the type of *G. diazotrophicus* bacteria applied and four types of fertilization on the total production per plant in the greenhouse tomato crop. GIBI029 treatments achieved the highest cumulative fruit production, exceeding 5000 g per plant even without nitrogen and phosphorus fertilization, demonstrating its robust effect on overall productivity. (**b**) Effect of the type of *G. diazotrophicus* bacteria applied and four types of fertilization on the yield in the greenhouse tomato crop. Yields exceeded 100 t ha^{-1} with GIBI029, matching or surpassing those achieved with full fertilization, indicating potential for reducing chemical inputs without compromising productivity. ATCC: *G. diazotrophicus* reference strain (ATCC 49037); control: conventional control treatment; GIBI029: native *G. diazotrophicus* isolation. The codes corresponding to the percentages of nitrogen and phosphorus fertilization are deciphered in Table 1. Error bars show standard deviation (n = 4 replicates or blocks). Significant differences between treatments ($p < 0.05$) are indicated by different letters, as determined by Duncan's multiple-range test.

In cases where one of these two elements was absent, such as the 0% phosphorus or 0% nitrogen treatments, production showed a slight downward trend, falling below 5000 g per plant in the presence of the GIBI029 bacterium. However, no statistically significant differences were observed at a 95% confidence level among the different interactions within this bacterium. In contrast, the values achieved for the control group and the reference bacterium ATCC fluctuated at around 3000 g per plant, showing statistically significant differences at a 95% confidence level compared to the GIBI029 bacterium across the different groups of the control and the ATCC strain in their respective interactions.

In the control treatment, the inherent response of the soil's native nutrient reserves was evident. However, these reserves alone did not ensure increased production, highlighting the necessity of interventions to enhance nutrient availability. Within the control treatment group, nitrogen and phosphorus sources showed no statistically significant differences ($p < 0.05$), with values ranging from 2248.5 g plant^{-1} (0% nitrogen, 100% phosphorus) to 2888.3 g plant^{-1} (100% nitrogen, 100% phosphorus). In contrast, the ATCC reference strain, across its various nitrogen and phosphorus interactions, exhibited values below 2273.9 g plant^{-1} (100% nitrogen, 0% phosphorus), reaching a minimum of 1792.4 g plant^{-1} in the absence of both nitrogen and phosphorus (Figure 7a).

3.5. Crop Yield

To estimate the hectare-based yield (t ha^{-1}), we extrapolated from the plant density (20,833 plants ha^{-1}) and the mean production per plant (g plant^{-1}). This calculation revealed substantial differences among treatments, highlighting the potential impact on the production system's sustainability. Specifically, the GIBI029 isolate, under varying nitrogen and phosphorus levels, resulted in yields exceeding 94.7 t ha^{-1} (N0P100) and surpassing 100 t ha^{-1} in the N0P0 and N100P100 treatments. In contrast, the ATCC reference strain exhibited significantly lower yields, ranging from a minimum of 37.3 t ha^{-1} (N0P0) to a maximum of 47.4 t ha^{-1} (N100P0) (Figure 7b).

Correspondingly, in the evaluation of crop yield per treatment, statistically significant differences ($p < 0.05$) were observed between the reference strain and the native isolate. Specifically, the GIBI029 isolate exhibited a yield of 101.3 tons per hectare, while the ATCC 49037 strain yielded 42.7 tons per hectare. This disparity signifies an augmentation of approximately 58.6 t ha^{-1}, representing a 137% increase in crop yield attributable to the application of the GIBI029 isolate (refer to Figure 7b and Table 3).

Regarding crop yield as a function of the four distinct fertilization regimes (Table 3), analysis revealed that the maximum yield was attained with complete nitrogen and phosphorus fertilization, registering a mean value of 67.9 tons per hectare. Subsequently, treatments featuring 100% nitrogen fertilization without phosphorus fertilization yielded 64.3 t ha^{-1}, closely followed by the treatment lacking both nitrogen and phosphorus fertilization, which yielded 64.2 t ha^{-1}. Conversely, the minimum yield was associated with the N0P100 fertilization regime, producing 57.5 t ha^{-1}.

During the examination of the interactive effects of the evaluated factors, namely bacterial type and nitrogen–phosphorus fertilization regimes, it was evident that combined treatments yielded values exceeding those obtained with individual factors. Specifically, the GIBI029 isolate, in conjunction with both N100P100 and N0P0 fertilization regimes, achieved a crop yield of 106 t ha^{-1}. These results demonstrated statistically significant differences ($p < 0.05$) when compared to the control treatments. The control treatments, which lacked bacterial inoculation, included the following variations in inorganic fertilization: no nitrogen and phosphorus (N0P0), full nitrogen fertilization without phosphorus (N100P0), full phosphorus fertilization without nitrogen (N0P100), and full nitrogen and phosphorus application (N100P100). These controls exhibited yields ranging from 46.8 t ha^{-1} (N0P100) to 60.2 t ha^{-1} (N100P100) when compared to the ATCC 49037 strain, which produced yields between 37.3 t ha^{-1} (N0P0) and 47.4 t ha^{-1} (N100P0). This observation underscores the synergistic potential of the GIBI029 isolate in combination with specific nitrogen and phosphorus fertilization regimes to substantially enhance crop yield.

These findings indicate the necessity of developing robust biological tools optimized for target environments to ensure beneficial interactions. The synergistic effects of such tools are crucial for achieving sustainable yields that surpass conventional agricultural practices, as demonstrated by the superior performance of the native GIBI029 isolate compared to a market reference strain. While reference strains may exhibit significant efficacy in their native habitats, their effectiveness can be diminished when transferred to different environments. Consequently, their inherent growth-promoting potential and cumulative impact on yield components may not be fully realized in productive systems, a limitation observed in this study with the market reference strain in contrast to the native GIBI029 isolation.

4. Discussion

4.1. Soil Analysis

The soil analysis reported a high phosphorus value of 299 ppm and low levels of calcium, magnesium, and potassium at 2.38 cmol kg^{-1}, 0.65 cmol kg^{-1}, and 0.19 cmol kg^{-1}, respectively. In contrast, the microelement values were within optimal average levels according to soil interpretation values reported by Molina [46]. The nutritional supplementation was tailored to the crop based on the soil analysis of the experimental site, the characteristics of which are detailed in Appendix A. According to IGAC and CORPOCALDAS [47], the soils in the Bellavista forest area, where the farm for the experiment is located, are composed of volcanic ash and are moderately weathered. They are soils with low-to-medium acidity, rich in organic matter, and have a medium-to-very high cation exchange capacity.

In general, the native GIBI029 isolate demonstrated the capacity to function effectively using the existing nitrogen and phosphorus levels indicated by the soil analysis (without supplemental nitrogen and phosphorus). Moreover, applying 100% nitrogen and 100% phosphorus did not change the bacterium's effect, implying that nitrogen and phosphorus fertilization is not required when the soil contains sufficient reserves of these elements. In such cases, these elements can be mobilized by nitrogen-fixing and phosphorus-solubilizing bacteria, such as the native *Gluconacetobacter* isolate (GIBI029).

4.2. Fruit Number and Weight

Concerning fruit number, the GIBI029 isolate exhibited the most favorable performance among the bacterial treatments. In particular, across the ten harvests, the highest number of fruits per plant was consistently observed in treatments inoculated with the GIBI029 bacterium across the various combinations of nitrogen and phosphorus sources. This suggests a positive influence of GIBI029 on fruit number and, consequently, yield.

In the context of microbial inoculant evaluations, Hernández and Chailloux [48] and Luna et al. [24] documented 17.73 fruits per plant when inoculating tomato cultures with *Glomus mosseae* and *Pseudomonas fluorescens* under optimal treatment conditions. This observation contrasts with the results obtained in the present study, where the GIBI029 isolate yielded a mean value of 68.9 fruits per plant across the ten harvests. Luna et al. [24] demonstrated increased tomato fruit number and weight under gnotobiotic and greenhouse conditions using the *G. diazotrophicus* ATCC 49037 strain. However, unlike the results obtained in this work, their experiment did not explore responses across nutrient stress gradients. In the present study, the GIBI029 isolate exhibited a higher performance than the ATCC 49037 strain and showed significant yield benefits under both full and zero fertilization, underscoring the superior adaptability of the native isolate under local agroecological conditions.

Aguilar and Sánchez [49], in their study involving Santa Clara tomato plants inoculated with *Azotobacter* sp. at the time of transplantation and subjected to 50% of the recommended fertilization, reported 17 fruits per plant in their most effective treatment. This indicates the enhanced performance of *G. diazotrophicus* in tomato plants as well as the conditions selected for its application compared to the application of *Azotobacter* sp.

Reis et al. [50] have reported that fruit number in tomato plants can fluctuate in response to endogenous phytohormone production, particularly auxins such as 3-indoleacetic acid. In this regard, *G. diazotrophicus* is recognized for its capacity to fix atmospheric nitrogen and enhance the synthesis of phytohormones, including auxins [5,50].

4.3. Per Plant Production and Mean Fruit Weight

Concerning per plant production, a 156% increase was observed relative to the commercial control (N100P100 without bacteria) when utilizing the GIBI029 isolate (5107 g plant^{-1}). This finding aligns with the previously cited report by Aguilar and Sánchez [49], which documented a production of 2000 g plant^{-1}. In general, within each of the main factors or main plots (application of native isolate, reference strain, or control), a positive interaction was evident between the bacteria and the fertilization source.

The interaction that promoted fruit production the most was observed with the GIBI029 isolate. The highest values for this interaction occurred at the 100% nitrogen and phosphorus fertilization levels, as well as in the absence of added nitrogen and phosphorus. It is important to clarify that the zero level of nitrogen and phosphorus refers to the absence of added fertilizer sources, but these plots still contained the residual nutrient levels reported in the soil analysis (see Appendix A). The control treatments, representing conditions without bacterial inoculation and varying levels of inorganic fertilization, yielded intermediate plant weights exceeding 2500 g. This observation demonstrates the validity of the control treatments selected for the experimental setup disclosed in this work since this intermediate value is consistent with the generally reported fruit production range of 2000 to 3000 g per plant in conventional tomato production systems [17]. This suggests that even without the introduction of bacterial treatment, standard fertilization practices in our experimental setup were capable of achieving yields within the typical range for traditionally cultivated tomatoes.

Specifically, for the GIBI029 isolate, the positive impact of *G. diazotrophicus* was confirmed through an observed increase in mean fruit weight. This augmentation in mean fruit weight has been documented in other studies involving tomato [24] and papaya [51]. The results obtained revealed no statistically significant differences in mean fruit weight among treatments with varying fertilization levels ($p < 0.05$), which aligns with findings reported by Aguilar and Sánchez [49] in tomato crops subjected to moderate and complete fertilization regimes.

4.4. Crop Yield

The application of a suspension of *G. diazotrophicus* GIBI029 isolation to the tomato crop demonstrated a positive effect on both fruit production and yield. This beneficial influence is also observed in the inoculation of other crops, including sugarcane, carrot [9,10], sugar beet [10], radish, coffee [11], and pineapple [12], utilizing this microbial species, where production increases ranging from 30% to 80% above control treatments have been documented. Likewise, the introduction of various bacteria from *Gluconacetobacter* genus to diverse crops has shown the potential to enhance yield, as evidenced by yield increases in taro (38%) [13], papaya (38%) [14], and cassava (45%) [13].

In this study, treatments with 100% nitrogenous fertilization levels exhibited the highest mean fruit number per plant. This observation aligns with the findings of Aujla et al. [52] and Direkvandi et al. [53] in their respective studies on tomato plants, where increasing nitrogen levels from synthetic sources resulted in an elevated fruit count per plant. Furthermore, Jaramillo et al. [44] reported that nitrogen is a fundamental factor in flower and fruit production and in the regulation of maturation. Similarly, the highest production per harvest during the production cycle was achieved with treatments receiving 100% nitrogen fertilization. In this context, Rojas et al. [5] emphasized the importance of nitrogen in tomato crop yield, as elevated nitrogen concentrations promote flower and fruit development. However, regarding final production, the maximum yield for the tomato crop was obtained with a balanced fertilization regimen (100% nitrogen and 100% phosphorus). Jaramillo et al. [44] asserted that balanced fertilization is crucial for successful greenhouse tomato cultivation. Direkvandi et al. [53] also observed the highest yields in the treatment with the highest nitrogen fertilization rate (225 kg N ha^{-1}), reaching approximately 4796 g tomato plant^{-1}. Additionally, Hernández and Chailloux [48] reported a yield increase of 300 kg ha^{-1} when implementing a balanced fertilization strategy (100 kg nitrogen per hectare, 25 kg phosphorus per hectare, and 50 kg potassium per hectare) compared to fertilization with only 100 kg nitrogen per hectare.

In comparison to the overall Colombian yield (96 tons per hectare) and the regional yield for Caldas region (70 t ha^{-1}) in tomato crops cultivated under semi-controlled conditions [54], the application of the GIBI029 isolate enables an increase exceeding 31 t ha^{-1} above the regional yield, thus approaching the national value.

The yield performance of the tomato crop was enhanced by the application of the *G. diazotrophicus* GIBI029 isolate. The yield exhibited a 168.3% increase related to the control with balanced fertilization (N100P100 with no bacteria), corresponding to a mean value of 101.3 t ha^{-1}, which can be compared to the findings of Alfonso and Galán [55], who achieved 32.7 t ha^{-1} in their optimal treatment utilizing co-inoculation of mycorrhizae and rhizobacteria. This treatment involved the application of a mixture of *Azospirillum brasilense* and *Glomus clarum*, supplemented with 30 kg nitrogen per hectare (in the seedbed) and 60 kg nitrogen per hectare (in the field). Hernández and Chailloux [48], in their study, attained 34.3 t ha^{-1} in their most effective treatment by inoculating the tomato crop with *Glomus mosseae*. Consequently, the singular application of *G. diazotrophicus* demonstrates the potential to yield superior outcomes for the growth promotion of tomato crops.

4.5. Harvest-Stage Analysis

The graph depicting the total production achieved in grams per plant over time, across different treatments (Figure 7a), illustrates a positive cumulative trend throughout the 10 harvests. This trend was evident in both the number of fruits and the weight of fruits obtained with each harvest. Specifically, the values reported for the bacterium GIBI029 across its various interactions reached up to 5000 g per plant for the 100% nitrogen and 100% phosphorus treatments, as well as the 0% nitrogen and 0% phosphorus treatments.

It is worth highlighting that fructification reaches its maximum production period between 90 and 100 days after transplanting. Subsequently, it begins to decrease until the end of the crop cycle. The genotype used has a harvest initiation time between 75 and 90 days after transplanting. This is a normal phenomenon over time. The initial harvests have the highest average fruit weight values, which decline as the number of harvests progresses. From the results obtained, it was evidenced that the GIBI029 isolate consistently exhibited a higher performance across the sequential harvests on mean fruit weight compared to the standard ATCC 49037 strain (Figure 6a).

This study reports statistically significant improvements in yield components (fruit number, total fruit weight, average fruit weight, and yield) when using the GIBI029 isolate. In addition to the cumulative yield improvements observed, the results obtained suggest that the influence of the native GIBI029 isolate varies across the crop cycle. For instance, peak fruit production and weight were observed with the GIBI029 isolate in the earlier harvests (specifically, harvests 2, 4, and 5), with a gradual decline in fruit weight in later harvests. During this period, plants treated with the native GIBI029 isolate consistently outperformed both the uninoculated controls and the plants inoculated with the reference strain ATCC 49037. This temporal behavior indicates that the beneficial impact of GIBI029 is closely associated with the early and mid-reproductive phases of tomato development, likely mediated by peak biological nitrogen fixation, enhanced phosphate solubilization, and increased production of indolic compounds (e.g., auxin) during critical stages of fruit set and early fruit development. This can be explained by the fact that the native *G. diazotrophicus* GIBI029 isolate may enhance nutrient mobilization (particularly, of nitrogen and phosphorus) early in the reproductive phase, supporting the development of a high number of fruits and increased fruit weight. In fact, other diazotrophs such as *Azospirillium* sp. have the ability to enhance nutrient availability in the rhizosphere [56]. As the plant matures and the demand for nutrients increases, the isolate's capacity to sustain the same level of nutrient supply may become limiting; this may contribute to the decline in fruit weight in later harvests. On the other hand, the plant's sink strength (the capacity of fruit to attract resources) changes over time. Early in development, there are fewer fruits, and they act as strong sinks. Later, with a higher number of fruits, the sink strength of individual fruits may decrease, naturally leading to smaller fruit size [57]. Finally, the activity of GIBI029 itself might be influenced by the plant's physiological stage, root exudate composition [58], or changing environmental conditions within the rhizosphere over time [59].

While our study demonstrates the positive impact of GIBI029 on overall yield, the complex interactions between the plant and GIBI029 likely involve stage-specific dynamics that justify further investigation. Future studies could incorporate detailed plant physiological measurements (e.g., nutrient uptake, photosynthesis rates, hormone levels) at different growth stages, rhizosphere dynamics (e.g., root exudate analysis, microbial community changes, nutrient availability), and gene expression analysis to obtain an understanding of the mechanisms underlying the temporal patterns observed in the data presented in this work.

4.6. Overall Comparison of Results on Yield Components

The significant improvements in yield components observed with the native *G. diazotrophicus* isolate GIBI029 align with previous reports of growth promotion by endophytic diazotrophs but also present notable contrasts. Compared to Cocking et al. [23], our results demonstrate tangible agronomic benefits under semi-controlled greenhouse conditions, advancing from the earlier focus on colonization mechanisms. Likewise, Luna et al. [24] found that inoculation with the ATCC 49037 strain increased fruit number and weight under greenhouse conditions; however, their study did not test nutrient-limited conditions. In contrast, our findings show that GIBI029 outperformed ATCC 49037 and even full nitrogen and phosphorus fertilization, suggesting that strain origin and soil adaptation are critical. This indicates that native strains like GIBI029 may unlock greater yield potential in low-input systems than previously reported with standard reference strains. Botta et al. [25] found that the ATCC 49037 strain was ineffective alone in promoting tomato seedling growth in vivo. This sharply contrasts with GIBI029's robust performance as a standalone inoculant, suggesting that local environmental adaptation is key to microbial success in real-world applications. Our results underscore that effectiveness is strain- and context-specific. Further contrast arises with Hernández and Chailloux [48] and with Aguilar and Sánchez [49], who reported lower maximum tomato yields (32.7 t ha^{-1} and

34.3 t ha^{-1}, respectively) with co-inoculation of mycorrhizae and rhizobacteria, compared to the 106.1 t ha^{-1} achieved with GIBI029 isolate even without synthetic fertilization. This suggests GIBI029 may offer greater efficiency in nutrient mobilization, particularly under conditions with moderate soil fertility and high organic matter. This superior performance may also relate to GIBI029's higher nitrogenase activity, phosphate solubilization index, and indole compound production compared to the ATCC 49037 strain. These differences indicate that microbial inoculants can be effective alternatives to chemical fertilizers if properly selected for environmental and crop compatibility. Notably, GIBI029's ability to maintain yield in nutrient-limited soils highlights its potential for reducing costs and environmental impacts, especially for smallholders and organic farming.

4.7. Comparison of Native Isolate and Reference Strain

This study successfully highlighted the differential effects observed when utilizing a regional isolation compared to a foreign reference strain. *G. diazotrophicus* ATCC 49037 strain (the same PAL5 strain) was used in this experiment since it has become a key endophytic bacterial strain employed as a model organism to investigate plant–bacteria interactions in non-leguminous plant species [7]. Furthermore, this strain has been evaluated in tomato seedlings in studies such as Luna et al. [24], where its plant growth-promoting potential has been demonstrated. Moreover, in seedbed studies, the application of strain ATCC 49037 at a dose of 2.5 mL L^{-1} with added phosphorus to Santa Clara tomato seeds at sowing has been identified as a promising strategy for enhancing seedling growth. These assessments, focused on the seedbed stage, evaluated the vegetative development of the tomato plants [42]. For this reason, it was selected for inclusion as the reference strain of the same species as the native isolate used in the experimental design of this study.

On the other hand, the Colombian wild isolate of *G. diazotrophicus* GIBI029 exhibited a nitrogenase activity of 0.010 nmol of reduced acetylene $\times$ mL^{-1} $\times$ h^{-1}, production of indole compounds of 78.5 μg $\times$ mL^{-1}, and a solubilization index for tricalcium phosphate of 3.619, as demonstrated in earlier in vitro assays [38]. The physiological capabilities of the GIBI029 isolate, which were reinforced by this study, suggest a more efficient mobilization of soil nutrients and greater auxin production, potentially contributing to its superior performance compared to other strains. This contrasts with the lower nitrogenase activity and phosphate solubilization capacity reported for the reference ATCC 49037 strain, as also observed in the comparative assays performed.

In this study, the GIBI029 isolation demonstrated superior performance compared to both the uninoculated control and the reference ATCC 49037 strain. This phenomenon should be analyzed in light of the results obtained during comparative in vitro assays on plant growth-promoting properties conducted in previous work [38]. The native GIBI029 isolate was recovered from the stem tissue of sugarcane plants cultivated in soil characterized by a substantial content of organic matter (ranging between 10.0% and 20.4%), acidic to slightly acidic pH values (between 4.4 and 5.9), and considerable nitrogen levels (between 0.40 and 0.68%). These inherent soil properties of the GIBI029 origin site are quite similar to the soil conditions in the experimental greenhouses used in this study (0.53% nitrogen, pH of 4.8, and 14.1% organic matter, as presented in Table A1 of Appendix A). The performance of this isolation can be considered as promising, taking into account that the bacterium was successfully applied to a dicotyledonous vegetable crop (tomato) different from the grass species (sugarcane) from which it originally was recovered.

The reasons for the low performance of this reference strain remain unclear. Several factors could explain this reduced performance. The ATCC 49037 strain, which was originally isolated from Brazilian sugarcane cultivars [6], may not be optimally adapted to the specific and unique agroecological conditions prevalent in the West Central region of Colombia, where this study was conducted. This can potentially limit its colonization and plant growth-promoting properties. In this sense, the best results observed with GIBI029 correlate strongly with the comparatively lower performance exhibited by the reference ATCC 49037 strain during the aforementioned in vitro assays [38], which showed a nitrogenase activity of 0.005 nmol of reduced acetylene $\times$ mL^{-1} $\times$ h^{-1} (half of the activity displayed by GIBI029), a production of indole compounds of 64.3 μg $\times$ mL^{-1}, (about 82% of the production observed with GIBI029), and a solubilization index for tricalcium phosphate of 3.097 (approximately 86% of the solubilization capacity of GIBI029). Although standard procedures were used for strain revival, there exists the possibility of some degree of deterioration in the ATCC 49037 strain during storage or subculturing, which might have affected its viability or effectiveness. In this regard, it is worth highlighting that this strain has been preserved by cryopreservation at −80 °C through a monitoring and custody process by the collection in which it is stored. In addition, differences in specific plant–microbe interactions between the tomato cultivar used and the two types of *G. diazotrophicus* applied (native isolate or reference strain) could also contribute to the observed differences in performance. It is worth highlighting that this study did not specifically investigate these factors, and therefore, these explanations remain speculative. Further research is needed to elucidate the specific reasons for the differential performance of GIBI029 and ATCC 49037.

Consequently, these elevated values of plant growth-promoting indices explain why the native GIBI029 isolate exhibited the most pronounced effects on the yield component variables of the tomato crop assessed in the present work, irrespective of nitrogenous or phosphorous fertilization.

4.8. Advantages of Native Colombian GIBI029 Isolate

The presented findings on the native Colombian *G. diazotrophicus* GIBI029 isolate, which significantly improved tomato yield and fruit production align with and extend prior research on the plant growth-promoting effects of this bacterium in different crops. The results obtained corroborate the colonization capacity and positive growth effects of this species observed under gnotobiotic conditions in the studies by Cocking et al. [23] and Luna et al. [24], both of which demonstrated the ability of *G. diazotrophicus* to enter tomato root tissue and enhance fruit yield components. However, our study extends this understanding by confirming the performance of the native Colombian GIBI029 isolate under semi-controlled greenhouse conditions, showing consistent fruit and yield increases even in the absence of synthetic nitrogen and phosphorus fertilization, a context not explored in those earlier works.

In contrast to Botta et al. [25], who found that *G. diazotrophicus* performed poorly under in vivo conditions when applied alone to tomato seedlings, this study provides robust evidence that the native GIBI029 isolate performs strongly when applied independently in soil systems. This highlights the importance of local adaptation and suggests that strain origin plays a critical role in determining inoculant efficacy.

The findings disclosed in this work expand upon the work of Ríos Rocafull et al. [10], who demonstrated that the plant growth-promoting effects of *G. diazotrophicus* on crops such as carrot and sugar beet are highly dependent on the type of culture medium used for inoculum preparation. While they observed variable responses depending on the carbon and nitrogen sources in the medium, our study used a standardized inoculum preparation (PDA-based) and still achieved consistent and robust effects on tomato yield across all fertilization regimes. This suggests that the native GIBI029 isolate may have enhanced adaptability or a more efficient interaction mechanism with tomato plants under semi-controlled conditions. Additionally, while these authors reported moderate biomass gains in short-cycle crops, the results of this study indicate that GIBI029 isolate significantly boosts yield components such as fruit number and fruit weight, achieving values over 100 t ha^{-1} even in the absence of nitrogen and phosphorus fertilization. This contrast highlights the importance of host specificity and environmental context, indicating that GIBI029 may be particularly suited to Solanaceous crops and the agroecological conditions of the West Central Colombian highlands.

On the other hand, Pallucchini et al. [26] demonstrated the importance of functional nitrogen fixation genes in *G. diazotrophicus* for growth promotion in hydroponic tomato, aligning with our observation that the native GIBI029 isolate, which has shown a high nitrogenase activity, significantly improved fruit yield even without nitrogen and phosphorus fertilization.

4.9. Limitations of This Study

Although the present study clearly demonstrated that inoculation with the native *G. diazotrophicus* GIBI029 isolate significantly improved tomato yield and fruit production, it is important to recognize certain experimental limitations. Specifically, the rhizosphere microbial community structure and the colonization patterns of the bacterial isolates within the plant roots were not directly assessed. Consequently, while the results suggest that the observed benefits can largely be attributed to the strain-specific advantages of GIBI029, the possibility that unmeasured biotic interactions or shifts in the native microbiome contributed to these outcomes cannot be entirely excluded.

However, several elements within this study reinforce the hypothesis that the beneficial effects observed are predominantly due to intrinsic characteristics of the GIBI029 isolate. As noted above, previous in vitro characterization revealed that GIBI029 exhibits a higher nitrogenase activity, a greater capacity for phosphate solubilization, and enhanced production of indolic compounds relative to the reference ATCC 49037 strain. Furthermore, GIBI029 was isolated from agroecological conditions highly similar to those of the experimental site, including soil pH, organic matter content, and nitrogen availability. This environmental congruence likely contributed to the higher adaptation and performance of the native isolate under the conditions evaluated. Moreover, the consistent superiority of GIBI029 across all fertilization regimes, including treatments without nitrogen and phosphorus supplementation, indicates a robust growth-promoting effect that appears independent of external chemical inputs. This suggests that the advantages conferred by GIBI029 are primarily linked to its physiological capabilities rather than potential indirect effects mediated by shifts in the soil microbial community.

The semi-controlled conditions within the greenhouses where the experiment was conducted may have influenced why some treatments, such as the reference strain, did not exhibit the expected responses. The region's average temperatures of 17 °C could have affected the bacteria and, consequently, their behavior. Furthermore, this study has areas for improvement, including the need to integrate physiological, molecular, and ecological analyses to clarify how the GIBI029 isolate promotes growth and whether its performance remains stable across different environments and cultivation systems. These considerations will be taken into account in the design of future research.

Nevertheless, as recognized in this study, future research efforts should incorporate a more comprehensive evaluation of the plant–microbe interaction mechanisms as well. Specifically, integrated approaches involving rhizosphere microbiome profiling through metagenomics, visualization of colonization patterns using fluorescently labeled strains, and transcriptomic analyses of both plant and bacterial gene expression during the interaction are necessary. These investigations will be critical to elucidate the functional basis of the observed plant growth promotion and to definitively exclude the contribution of unmeasured biotic interactions.

4.10. Comparison with Findings from Previous Studies

The present study confirms that the *G. diazotrophicus* GIBI029 isolate, combined with nitrogen and phosphorus fertilization (or without additional fertilization of these two macronutrients), enhances tomato yield and quality of the fruits. This aligns with prior research under macro-tunnel systems [60], where tomato yields increased by 28% and profitability improved by 35% with inoculation under 50–75% fertilization regimes.

The microbial effectiveness observed here is supported by prior work on the design of the culture medium for the production of *G. diazotrophicus* cell suspension, where the influence of pH and sucrose-based media was modeled to improve viability and bacterial cultivation performance [61]. That study demonstrated biomass yields up to 5.3 g L^{-1} and stable nitrogenase activity under bench-scale conditions, ensuring viability and consistency for field application.

Comparable findings were reported in carrot cultivation, where inoculation with *G. diazotrophicus* under 75% nitrogen and phosphorus fertilization led to yield increases of 29% and higher root quality compared to full fertilization without inoculation [9]. The economic evaluation of that system showed profitability improvements of up to 22% [62], consistent with our results for tomato under open-field conditions.

In tomato seedlings, earlier results revealed significant interactions between *G. diazotrophicus*, plant genotype, and phosphorus fertilization levels, with shoot biomass increases of 30–45% under low-phosphorus conditions in the cultivars [42]. These findings support the results obtained in the present work and highlight the importance of genotype-specific responses to microbial inoculants.

Furthermore, plant growth-promoting properties such as nitrogen fixation, phosphate solubilization, and production of indole compounds have been previously identified in isolates of *G. diazotrophicus* (including GIBI029) and *G. sacchari*, recovered from Colombian sugarcane cultivars [38]. Such physiological traits provide a mechanistic basis for the improved nutrient efficiency and yield observed in this and the previous studies.

These findings collectively strengthen the evidence that *G. diazotrophicus* can sustainably enhance tomato production while reducing chemical fertilizer use.

4.11. Future Research

The superior performance of the native *G. diazotrophicus* GIBI029 isolate in this study undoubtedly suggests its adaptation to the specific agroecological conditions of the West Central region of Colombia. To fully elucidate the underlying physiological and molecular mechanisms responsible for these beneficial effects, further research is required. Specifically, our ongoing work includes a genomic analysis aimed at assessing biological nitrogen fixation through the identification of *nif* genes within the GIBI029 genome. Additionally, transcriptomic studies will be essential to identify the genes activated during the establishment of the interaction between this bacterium and the tomato plant. Another promising avenue for future investigation involves classical biochemical assays utilizing marked bacterial mutants, enabling the visualization of the bacterium's colonization patterns within the plant through fluorescence microscopy. The comprehensive insights gained from these future research efforts will be crucial for strengthening the scientific evidence required to definitively define and recommend the native GIBI029 isolate as a valuable microbial inoculant, not only for vegetables like tomatoes but also potentially for carrots and other economically important crops in similar agroecological zones.

Results obtained in this work indicate that the native GIBI029 isolate holds promise as a potential commercial inoculant. However, it is crucial to recognize that this work did not consider its formulation stability, field reproducibility, and regulatory feasibility. The current study was conducted under semi-controlled greenhouse conditions, and several key aspects need to be addressed before moving towards commercial application. Research is required to determine appropriate formulations (e.g., liquid, solid, encapsulation) that ensure the stability and viability of GIBI029 during storage and application. Multi-location field trials under diverse agroclimatic conditions are necessary to validate the reproducibility of our results and the consistency of the effects of GIBI029 isolate in real-world agricultural settings. A thorough cost–benefit analysis is essential to evaluate the economic viability of using GIBI029 as a bioinoculant compared to traditional fertilization practices. The regulatory requirements for registering and commercializing a microbial inoculant in different regions must be carefully considered.

Therefore, while this study provides a strong foundation for further research, claims regarding its immediate commercial application would be premature. Future work should focus on addressing the points mentioned above and determining the precise mechanisms by which GIBI029 promotes plant growth to optimize its application and ensure consistent results.

5. Conclusions

Under sandy loam soil conditions, the application of the native Colombian *G. diazotrophicus* isolate GIBI029 to tomato cultivars exhibited a positive effect on plant growth compared to unfertilized controls. Using GIBI029 without nitrogen and phosphorus fertilization led to higher yields than those obtained with full conventional fertilization. This study revealed that *G. diazotrophicus* GIBI029 positively influenced tomato development throughout the growing cycle, significantly affecting key yield components, including fruit count per plant, fruit weight per plant, mean fruit weight, and total production per plant. The resulting final yields were substantial, reaching 5092 g per plant and 106 t ha^{-1}.

To fully understand the mechanisms by which GIBI029 enhances tomato plant growth and to assess the consistency of its performance across diverse environments and cropping systems, future research should incorporate integrated physiological, molecular, and ecological analyses. Based on the findings of the present study, these future investigations will be crucial for advancing our understanding of *G. diazotrophicus*'s potential in promoting economically significant crops like tomato. Consequently, a thorough understanding of these aspects could enable the formulation of an effective microbial inoculant for tomato production in tropical agricultural systems.

Author Contributions: Conceptualization, N.C.-A.; methodology, S.P. and J.A.C.; software, N.C.-A.; validation, G.M.R. and N.C.-A.; formal analysis, G.M.R., N.C.-A., and Ó.J.S.; investigation, S.P. and J.A.C.; resources, G.M.R., N.C.-A., and Ó.J.S.; data curation, N.C.-A. and S.P.; writing—original draft preparation, S.P., J.A.C., and N.C.-A.; writing—review and editing, Ó.J.S.; visualization, S.P. and J.A.C.; supervision, N.C.-A. and G.M.R.; project administration, G.M.R.; funding acquisition, G.M.R. and Ó.J.S. All authors have read and agreed to the published version of the manuscript.

Funding: This research was funded by the Colombian Ministry of Science, Technology, and Innovation, the Universidad de Caldas, and the Universidad Católica de Manizales through the research initiative "Production of a biofertilizer based on *Gluconacetobacter diazotrophicus* on the pilot scale and its evaluation in promoting the growth of vegetable crops" of the research project "Development of biotechnology-based products for the production chaining of biofactories" (code 87055) in the framework of the research program "Biofactories: A bioeconomic development opportunity for Caldas through biotechnology" (grant 1235-903-86957). The APC was funded by the Vice-rectorate of Research and Graduate Studies and the Food and Agro-industry research group at the Universidad de Caldas.

Institutional Review Board Statement: Not applicable.

Data Availability Statement: The data presented in this study are available on request from the corresponding author. The data are not publicly available due to privacy restrictions.

Acknowledgments: The authors thank the technical support of the Tesorito Farm at the Universidad de Caldas and the Research Institute in Microbiology and Agro-industrial Biotechnology at the Universidad Católica de Manizales.

Conflicts of Interest: The authors declare no conflicts of interest.

Appendix A

Table A1. Analysis of the initial soil sample of the Tesorito farm at the Universidad de Caldas.

Parameter	Units	Value
N	%	0.53
pH		4.8
Na	cmol/kg	0.43
Ca	cmol/kg	2.38
Mg	cmol/kg	0.65
K	cmol/kg	0.19
P	ppm	299
Fe	ppm	669
Cu	ppm	7.97
Zn	ppm	28.65
Mn	ppm	38.35
B	ppm	1.9
S	ppm	25.54
OM	%	14.08
Texture	Sand—76%; Silt—13%; Clay—11%	

OM: organic matter.

Table A2. Mean comparison tests for the effect of nitrogen and phosphorus fertilization and two types of *G. diazotrophicus* on the content of nitrogen and phosphorus in a tomato-cultivated soil 180 days after transplanting under controlled conditions.

Type of Treatment	Nitrogen (%)	Phosphorus (mg/kg)
N0P0	0.61 ^{a}	220.33 ^{b}
N0P100	0.59 ^{a}	223.67 ^{b}
N100P0	0.58 ^{a}	273.67 ^{a}
N100P100	0.60 ^{ab}	259.22 ^{ab}
ATCC 49037	0.60 ^{b}	227.33 ^{b}
GIBI029	0.64 ^{a}	284.75 ^{a}
No bacteria	0.56 ^{c}	220.58 ^{b}

Remarks: Values with different letters in the column are significantly different ($p < 0.05$).

References

1. Anas, M.; Liao, F.; Verma, K.K.; Sarwar, M.A.; Mahmood, A.; Chen, Z.-L.; Li, Q.; Zeng, X.-P.; Liu, Y.; Li, Y.-R. Fate of nitrogen in agriculture and environment: Agronomic, eco-physiological and molecular approaches to improve nitrogen use efficiency. *Biol. Res.* **2020**, *53*, 47. [CrossRef] [PubMed]
2. Ren, K.; Xu, M.; Li, R.; Zheng, L.; Liu, S.; Reis, S.; Wang, H.; Lu, C.; Zhang, W.; Gao, H. Optimizing nitrogen fertilizer use for more grain and less pollution. *J. Clean. Prod.* **2022**, *360*, 132180. [CrossRef]
3. Pöyry, J.; Carvalheiro, L.G.; Heikkinen, R.K.; Kühn, I.; Kuussaari, M.; Schweiger, O.; Valtonen, A.; van Bodegom, P.M.; Franzén, M. The effects of soil eutrophication propagate to higher trophic levels. *Glob. Ecol. Biogeogr.* **2017**, *26*, 18–30. [CrossRef]
4. Choudhary, A.K.; Pooniya, V.; Bana, R.; Kumar, A.; Singh, U. Mitigating pulse productivity constraints through phosphorus fertilization—A review. *Agric. Rev.* **2014**, *35*, 314–319. [CrossRef]

5. Rojas, M.M.; Rodríguez, A.J.; Trujillo, I.D.; Heydrich, M. Relación de la fijación de nitrógeno y la producción de auxinas en cepas de *Gluconacetobacter diazotrophicus* procedentes de diferentes cultivos (Relationships between nitrogen fixation and auxin production in *Gluconacetobacter diazotrophicus* strains from different crops). *Rev. Colomb. Biotecnol.* **2009**, *11*, 84–93. (In Spanish)
6. Cavalcante, V.A.; Döbereiner, J. A new acid-tolerant nitrogen-fixing bacterium associated with sugar cane. *Plant Soil* **1988**, *108*, 23–31. [CrossRef]
7. Cocking, E.C. Endophytic colonization of plant roots by nitrogen-fixing bacteria. *Plant Soil* **2003**, *252*, 169–175. [CrossRef]
8. Leandro, M.; Andrade, L.; De Souza Vespoli, L.; Soares, F.; Moreira, J.; Pimentel, V.; Barbosa, R.; De Oliveira, M.; Silveira, V.; De Souza Filho, G. Combination of osmotic stress and sugar stress response mechanisms is essential for *Gluconacetobacter diazotrophicus* tolerance to high-sucrose environments. *Appl. Microbiol. Biotechnol.* **2021**, *105*, 7463–7473. [CrossRef]
9. Ceballos-Aguirre, N.; Cuellar, J.A.; Restrepo, G.M.; Sánchez, Ó.J. Effect of the application of *Gluconacetobacter diazotrophicus* and its interaction with nitrogen and phosphorus fertilization on carrot yield in the field. *Int. J. Agron.* **2023**, *2023*, 6899532. [CrossRef]
10. Ríos Rocafull, Y.; Sánchez López, M.; Dibut Álvarez, B.; Ortega García, M.; Tejeda González, G.; Rodríguez Sánchez, J.; Rojas Badía, M. The culture medium effect in plant growth promotion activity of *Gluconacetobacter diazotrophicus* in carrot and sugar beet. *Rev. Bio Cienc.* **2019**, *6*, e470. [CrossRef]
11. Jiménez, T.; Fuentes, L.E.; Tapia, A.; Macarua, M.; Martínez, E.; Caballero, J. *Coffea arabica* L., a new host plant for *Acetobacter diazotrophicus*, and isolation of other nitrogen-fixing acetobacteria. *Appl. Environ. Microbiol.* **1997**, *63*, 3676–3683. [CrossRef] [PubMed]
12. Tapia, A.; Bustillos, M.R.; Jiménez, T.; Caballero, J.; Fuentes, L.E. Natural endophytic occurrence of *Acetobacter diazotrophicus* in pineapple plants. *Microb. Ecol.* **2000**, *39*, 49–55. [CrossRef] [PubMed]
13. Dibut, B.; Martínez, R.; Ríos, Y.; Plana, L.; Rodríguez, J.; Ortega, M.; Tejada, G. Estudio de la asociación *Gluconacetobacter diazotrophicus*-viandas tropicales en suelo ferralítico rojo. I. selección de cepas efectivas para la biofertilización de boniato, yuca y malanga (Study of the association *Gluconacetobacter diazotrophicus*-tropical foods in red ferralite Soil I. Selection of effective strains for the biofertilization of sweet potato, cassava and taro). *Cultiv. Tropic.* **2010**, *31*, 51–57. (In Spanish)
14. Madhaiyan, M.; Saravanan, V.; Jovi, D.B.S.S.; Lee, H.; Thenmozhi, R.; Hari, K.; Sa, T. Occurrence of *Gluconacetobacter diazotrophicus* in tropical and subtropical plants of Western Ghats, India. *Microbiol. Res.* **2004**, *159*, 233–243. [CrossRef] [PubMed]
15. Saravanan, V.; Madhaiyan, M.; Osborne, J.; Thangaraju, M.; Sa, T. Ecological occurrence of *Gluconacetobacter diazotrophicus* and nitrogen-fixing *Acetobacteraceae* members: Their possible role in plant growth promotion. *Microb. Ecol.* **2008**, *55*, 130–140. [CrossRef]
16. Food and Agriculture Organization of the United Nations. Crops and Livestock Products. Available online: https://www.fao.org/faostat/en/#data/QCL (accessed on 5 March 2025).
17. Gil, R.; Bojacá, C.R.; Schrevens, E. Understanding the heterogeneity of smallholder production systems in the Andean tropics—The case of Colombian tomato growers. *NJAS Wagening J. Life Sci.* **2019**, *88*, 1–9. [CrossRef]
18. Herrera, H.J.; Hurtado-Salazar, A.; Ceballos-Aguirre, N. Estudio técnico y económico del tomate tipo cereza élite (*Solanum lycopersicum* L. var. *cerasiforme*) bajo condiciones semicontroladas (Economic study of the elite cherry tomato type (*Solanum lycopersicum* L. var. *cerasiforme*) under semicontrolled conditions). *Rev. Colomb. Cienc. Hortic.* **2015**, *9*, 290–300. (In Spanish) [CrossRef]
19. Rahman, K.A.; Zhang, D. Effects of fertilizer broadcasting on the excessive use of inorganic fertilizers and environmental sustainability. *Sustainability* **2018**, *10*, 759. [CrossRef]
20. Baez-Rogelio, A.; Morales-García, Y.E.; Quintero-Hernández, V.; Muñoz-Rojas, J. Next generation of microbial inoculants for agriculture and bioremediation. *Microb. Biotechnol.* **2017**, *10*, 19–21. [CrossRef]
21. De Andrade, L.A.; Santos, C.; Frezarin, E.T.; Sales, L.; Rigobelo, E. Plant growth-promoting rhizobacteria for sustainable agricultural production. *Microorganisms* **2023**, *11*, 1088. [CrossRef]
22. Ladha, J.K.; Peoples, M.B.; Reddy, P.M.; Biswas, J.C.; Bennett, A.; Jat, M.L.; Krupnik, T.J. Biological nitrogen fixation and prospects for ecological intensification in cereal-based cropping systems. *Field Crops Res.* **2022**, *283*, 108541. [CrossRef] [PubMed]
23. Cocking, E.C.; Stone, P.J.; Davey, M.R. Intracellular colonization of roots of *Arabidopsis* and crop plants by *Gluconacetobacter diazotrophicus*. *Vitr. Cell. Dev. Biol.-Plant* **2006**, *42*, 74–82. [CrossRef]
24. Luna, M.F.; Aprea, J.; Crespo, J.M.; Boiardi, J.L. Colonization and yield promotion of tomato by *Gluconacetobacter diazotrophicus*. *Appl. Soil Ecol.* **2012**, *61*, 225–229. [CrossRef]
25. Botta, A.L.; Santacecilia, A.; Ercole, C.; Cacchio, P.; Del Gallo, M. In vitro and in vivo inoculation of four endophytic bacteria on *Lycopersicon esculentum*. *N. Biotechnol.* **2013**, *30*, 666–674. [CrossRef]
26. Pallucchini, M.; Franchini, M.; El-Ballat, E.M.; Narraidoo, N.; Pointer-Gleadhill, B.; Palframan, M.J.; Hayes, C.J.; Dent, D.; Cocking, E.C.; Perazzolli, M. *Gluconacetobacter diazotrophicus* AZ0019 requires functional *nifD* gene for optimal plant growth promotion in tomato plants. *Front. Plant Sci.* **2024**, *15*, 1469676. [CrossRef]
27. Caballero-Mellado, J.; Fuentes-Ramirez, L.E.; Reis, V.M.; Martinez-Romero, E. Genetic structure of *Acetobacter diazotrophicus* populations and identification of a new genetically distant group. *Appl. Environ. Microbiol.* **1995**, *61*, 3008–3013. [CrossRef]

28. Kirchhof, G.; Reis, V.M.; Baldani, J.I.; Eckert, B.; Döbereiner, J.; Hartmann, A. Occurrence, physiological and molecular analysis of endophytic diazotrophic bacteria in gramineous energy plants. *Plant Soil* **1997**, *194*, 45–55. [CrossRef]
29. Muthukumarasamy, R.; Revathi, G.; Lakshminarasimhan, C. Influence of N fertilisation on the isolation of *Acetobacter diazotrophicus* and *Herbaspirillum* spp. from Indian sugarcane varieties. *Biol. Fertil. Soils* **1999**, *29*, 157–164. [CrossRef]
30. Bueno dos Reis Junior, F.; Massena Reis, V.; Urquiaga, S.; Döbereiner, J. Influence of nitrogen fertilisation on the population of diazotrophic bacteria *Herbaspirillum* spp. and *Acetobacter diazotrophicus* in sugar cane (*Saccharum* spp.). *Plant Soil* **2000**, *219*, 153–159. [CrossRef]
31. Hari, K.; Srinivasan, T. Response of sugarcane varieties to application of nitrogen fixing bacteria under different nitrogen levels. *Sugar Tech.* **2005**, *7*, 28–31. [CrossRef]
32. Medeiros, A.; Polidoro, J.; Reis, V. Nitrogen source effect on *Gluconacetobacter diazotrophicus* colonization of sugarcane (*Saccharum* spp.). *Plant Soil* **2006**, *279*, 141–152. [CrossRef]
33. Tufail, M.A.; Touceda-González, M.; Pertot, I.; Ehlers, R.-U. *Gluconacetobacter diazotrophicus* Pal5 enhances plant robustness status under the combination of moderate drought and low nitrogen stress in *Zea mays* L. *Microorganisms* **2021**, *9*, 870. [CrossRef] [PubMed]
34. Van Long, V.; Duong, V.T.; Van Dung, T.; Nghia, N.K.; Nam, T.S.; Viet, N.Q.; Van Cuong, L.; Van Tien, H. *Gluconacetobacter diazotrophicus* bacteria combined nitrogen fertilization promotes rice yield and soil quality in the paddy field in the Vietnamese Mekong Delta Region. *Malays. J. Soil Sci.* **2024**, *28*, 359–368.
35. Delaporte-Quintana, P.; Grillo-Puertas, M.; Lovaisa, N.C.; Teixeira, K.R.; Rapisarda, V.A.; Pedraza, R.O. Contribution of *Gluconacetobacter diazotrophicus* to phosphorus nutrition in strawberry plants. *Plant Soil* **2017**, *419*, 335–347. [CrossRef]
36. Delaporte-Quintana, P.; Grillo-Puertas, M.; Lovaisa, N.C.; Rapisarda, V.A.; Teixeira, K.R.; Pedraza, R.O. Solubilization of different sources of insoluble phosphate of *Gluconacetobacter diazotrophicus*. *Rev. Agron. Noreste Argent.* **2016**, *36*, 33–35.
37. Idogawa, N.; Ryuta, A.; Kousaku, M.; Kawai, S. Phosphate enhances levan production in the endophytic bacterium *Gluconacetobacter diazotrophicus* Pal5. *Bioengineered* **2014**, *5*, 173–179. [CrossRef]
38. Restrepo, G.M.; Sánchez, Ó.J.; Marulanda, S.M.; Galeano, N.F.; Taborda, G. Evaluation of plant-growth promoting properties of *Gluconacetobacter diazotrophicus* and *Gluconacetobacter sacchari* isolated from sugarcane and tomato in West Central region of Colombia. *Afr. J. Biotechnol.* **2017**, *16*, 1619–1629. [CrossRef]
39. Universidad de Caldas. Sistema de Granjas (System of Farms). Available online: http://www.ucaldas.edu.co/portal/sistema-de-granjas/ (accessed on 12 October 2020). (In Spanish).
40. ImpulSemillas. Tomate Híbrido Roble 956 F1 (Roble 956 F1 Hybrid Tomato). Available online: https://www.impulsemillas.com/documentos/fichas/Toamate_Roble_compressed.pdf (accessed on 4 June 2024). (In Spanish).
41. ImpulSemillas. Tomate Híbrido Roble 956 F1 (Roble 956 F1 Hybrid Tomato). Available online: https://www.impulsemillas.com/producto/tomate-roble-956-f1/ (accessed on 1 February 2022). (In Spanish).
42. Restrepo, G.M.; Ceballos, N.; Valencia, L.F.; Sánchez, Ó.J. Plant growth promotion by *Gluconacetobacter diazotrophicus* and its interaction with genotype and phosphorus availability in tomato seedlings. *Org. Agr.* **2021**, *11*, 601–614. [CrossRef]
43. Jaramillo, J.; Rodríguez, V.P.; Guzmán, M.; Zapata, M.; Rengifo, T. *Buenas Prácticas Agrícolas (BPA) en la Producción de Tomate bajo Condiciones Protegidas (Good Agricultural Practices (GAP) in Tomato Production Under Protected Conditions)*; Organización de las Naciones Unidas para la Agricultura y la Alimentación (FAO): Rome, Italy, 2007. (In Spanish)
44. Jaramillo, J.E.; Rodríguez, V.P.; Gil, L.F.; García, M.C.; Clímaco, J.; Quevedo, D.; Sánchez, G.D.; Aguilar, P.A.; Pinzón, L.M.; Zapata, M.Á.; et al. *Tecnología para el Cultivo de Tomate bajo Condiciones Protegidas (Technology for Growing Tomatoes Under Protected Conditions)*; Corpoica: Bogotá, Colombia, 2012. (In Spanish)
45. United States Standards for Grades of Fresh Tomatoes. Available online: https://www.ams.usda.gov/sites/default/files/media/Tomato_Standard[1].pdf (accessed on 1 October 2022).
46. Análisis de Suelos y su Interpretación (Soil Analysis and Its Interpretation). Available online: www.infoagro.go.cr/Inforegiones/RegionCentralOriental/Documents/Suelos/SUELOS-AMINOGROWanalisiseinterpretacion.pdf (accessed on 1 February 2024). (In Spanish).
47. IGAC; CORPOCALDAS. *Estudio Semidetallado de Suelos de los Municipios de Manizales, Chinchiná, Palestina, Neira y Villamaría (Semi-Detailed Soil Study of the Municipalities of Manizales, Chinchiná, Palestina, Neira, and Villamaría)*; IGAC & CORPOCALDAS: Manizales, Colombia, 2013. (In Spanish)
48. Hernández, M.I.; Chailloux, M. Las micorrizas arbusculares y las bacterias rizosfericas como alternativa a la nutricion mineral del tomate (Arbuscular mycorrhizae and rhizosphere bacteria as an alternative to mineral nutrition of tomato). *Cultiv. Tropic.* **2004**, *25*, 5–12. (In Spanish)
49. Aguilar, J.E.; Sánchez, M. Efecto de una rizobacteria nitrofijadora y niveles de fertilizante en el comportamiento agronómico del tomate *Lycopersicon esculentum* var (Effect of nitrogen fixing rhizobacteria and chemical fertilization on yield of *Lycopersicon esculentum* Mill var Santa Clara). *Acta Agron.* **1998**, *48*, 60–70. (In Spanish)

50. Reis, V.; Lee, S.; Kennedy, C. Biological nitrogen fixation in sugarcane. In *Associative and Endophytic Nitrogen-Fixing Bacteria and Cyanobacterial Associations*; Elmerich, C., Newton, W., Eds.; Springer: Dordrecht, The Netherlands, 2007; pp. 213–232.
51. Dibut, B.; Martínez-Viera, R.; Ortega, M.; Ríos, Y.; Planas, L.; Rodríguez, J.; Tejeda, G. Situación actual y perspectiva de las relaciones endófitas planta-bacteria. Estudio de caso *Gluconacetobacter diazotrophicus*-cultivos de importancia económica (Current situation and perspective of plant-bacteria endophytic relationships. *Gluconacetobacter diazotrophicus*-cultures of economic importance case study). *Cultiv. Tropic.* **2009**, *30*, 16–23. (In Spanish)
52. Aujla, M.; Thind, H.; Buttar, G. Fruit yield and water use efficiency of eggplant (*Solanum melongena* L.) as influenced by different quantities of nitrogen and water applied through drip and furrow irrigation. *Sci. Hortic.* **2007**, *112*, 142–148. [CrossRef]
53. Direkvandi, S.N.; Ansari, N.A.; Dehcordie, F.S. Effect of different levels of nitrogen fertilizer with two types of bio-fertilizers on growth and yield of two cultivars of tomato (*Lycopersicon esculentum* Mill). *Asian J. Plant Sci.* **2008**, *7*, 757–761. [CrossRef]
54. Cifras Agropecuarias (Agricultural Figures). Available online: http://www.agronet.gov.co/estadistica/Paginas/Precios.aspx (accessed on 1 March 2016). (In Spanish)
55. Alfonso, E.T.; Galán, A.L. Evaluación agrobiológica de la coinoculación micorrizas-rizobacterias en tomate (Agrobiological evaluation of coinoculation micorrhyza-rhyzobacteria in tomato). *Agron. Costarric.* **2006**, *30*, 65–73. (In Spanish)
56. Bashan, Y.; Holguin, G. *Azospirillum*—Plant relationships: Environmental and physiological advances (1990–1996). *Can. J. Microbiol.* **1997**, *43*, 103–121. [CrossRef]
57. Marcelis, L. Sink strength as a determinant of dry matter partitioning in the whole plant. *J. Exp. Bot.* **1996**, *47*, 1281–1291. [CrossRef]
58. Bais, H.P.; Weir, T.L.; Perry, L.G.; Gilroy, S.; Vivanco, J.M. The role of root exudates in rhizosphere interactions with plants and other organisms. *Annu. Rev. Plant Biol.* **2006**, *57*, 233–266. [CrossRef]
59. Zhou, W.; Shen, W.; Li, Y.; Hui, D. Interactive effects of temperature and moisture on composition of the soil microbial community. *Eur. J. Soil Sci.* **2017**, *68*, 909–918. [CrossRef]
60. Ceballos-Aguirre, N.; Hurtado-Salazar, A.; Restrepo, G.M.; Sánchez, Ó.J.; Hernández, M.C.; Montoya, M. Technical and economic assessment of tomato cultivation through a macro-tunnel production system with the application of *Gluconacetobacter diazotrophicus*. *Horticulturae* **2024**, *10*, 1110. [CrossRef]
61. Restrepo, G.M.; Rincón, A.; Sánchez, Ó.J. Kinetic analysis of *Gluconacetobacter diazotrophicus* cultivated on a bench scale: Modeling the effect of pH and design of a sucrose-based medium. *Fermentation* **2023**, *9*, 705. [CrossRef]
62. Ceballos-Aguirre, N.; Restrepo, G.M.; Hurtado-Salazar, A.; Cuellar, J.A.; Sánchez, Ó.J. Economic feasibility of *Gluconacetobacter diazotrophicus* in carrot cultivation. *Rev. Ceres* **2022**, *69*, 40–47. [CrossRef]

agriculture

MDPI

Article

The Effect of New Zeolite Composites from Fly Ashes Mixed with Leonardite and Lignite in Enhancing Soil Organic Matter

Renata Jarosz [1,*], Joanna Beata Kowalska [2], Krzysztof Gondek [3], Romualda Bejger [4], Lilla Mielnik [4], Altaf Hussain Lahori [5] and Monika Mierzwa-Hersztek [3]

1 Department of Mineralogy, Petrography and Geochemistry, Faculty of Geology, Geophysics and Environmental Protection, AGH University of Krakow, Mickiewicza 30 Av., 30-059 Krakow, Poland
2 Institute of Soil Science, Plant Nutrition and Environmental Protection, Wroclaw University of Environmental and Life Sciences, Grunwaldzka 53, 50-357 Wroclaw, Poland
3 Department of Agricultural and Environmental Chemistry, University of Agriculture in Krakow, Mickiewicza 21 Av., 31-120 Krakow, Poland
4 Department of Bioengineering, West Pomeranian University of Technology in Szczecin, Papieza Pawla VI 3, 71-459 Szczecin, Poland
5 Department of Environmental Sciences, Sindh Madressatul Islam University, Karachi 74000, Pakistan
* Correspondence: renata.jarosz@agh.edu.pl

Abstract: The aim of this study was to evaluate the influence of innovative mineral–organic mixtures containing zeolite composites produced from fly ashes and lignite or leonardite on the fractional composition of soil organic matter in sandy loam soil under two-year pot experiments with maize. The fractional composition of soil organic matter (SOM) was analyzed and changes in the functional properties of soil groups were identified using the ATR-FTIR method. Changes in the content of phenolic compounds were assessed, and the potential impact of fertilizer mixtures on soil carbon stocks was investigated. The addition of these mixtures improved the stability of SOM. The application of mineral–organic mixtures significantly increased the total organic carbon (TOC) by 18% after the 2nd year of the experiment. The maximum TOC content in the soil was observed by 33% with the addition of MC3%Leo3% amendment. Nitrogen content in soil was increased by 62% with MV9%Leo6% additive, indicating increased soil fertility. The study highlighted an increase in fulvic acid carbon relative to humic acid carbon, signaling positive changes in organic matter quality. The new mineral–organic mixtures influence changes in specific functional groups (ATR-FTIR) present in the soil matrix, compared to mineral fertilization alone. The additive mixtures also contributed to an increase in soil carbon stocks, highlighting their potential for long-term improvement of soil fertility and carbon sequestration.

Keywords: carbon stocks; contaminated soil; exogenous organic matter; fertilizer mixtures; maize; soil additives

1. Introduction

Soil health and nutrient management are critical elements of sustainable agriculture. The physical, chemical, and biological properties of healthy soil promote optimal plant growth as well as water and air quality [1]. Most parameters used to assess soil condition are interrelated and their interactions can be predicted. The Soil Health Institute identified organic carbon, soil carbon mineralization potential, and aggregate stability as key indicators based on analysis of over 30 indicators at 124 long-term agricultural research centers in North America [2,3]. Healthy soil contains high levels of organic matter and a diverse biological population [4].

Fertilizers play a critical role in preserving soil fertility and thereby increasing crop yields. Their use is essential for the production of high-quality food. Consequently, the significance of fertilizers in sustaining the global food supply is undeniable [5,6]. The combination of organic and inorganic fertilizers can help to increase the organic matter content of the soil and enhance the quantity and diversity of soil microorganisms [7,8].

In order to limit the decline in soil organic matter, it is necessary to implement sustainable, environmentally friendly and economically viable production systems. Zeolites as a soil additive offer numerous benefits, as they increase or maintain soil pH compared to mineral fertilizers [9]. Due to their ability to exchange ions, zeolites improve nutrient retention. One of the main applications of zeolites in agriculture is the absorption, storage, and gradual release of nitrogen [9–12]. Megías-Sayago et al. [13] observed that in addition to nitrogen retention, zeolites can also capture CO_2. The effectiveness of CO_2 capture relies heavily on the concentration of Al atoms in the zeolite structure and its cationic character. Zeolites present a promising solution to achieve balance in the agricultural ecosystem. In this regard, zeolites can potentially aid farmers in achieving fertilizer savings, improved water management, soil remediation, and groundwater purification [11,12]. The results of studies by various authors show that zeolites can be successfully synthesized from fly ash. Therefore, it can be considered that the synthesis of zeolites from fly ash is an effective way of managing these industrial wastes, contributing to their utilization and use in practical applications [14,15].

Lignite represents a natural organic reservoir that is of significant value to agricultural soils because it is a rich source of humic acids [16–19]. According to Dubey and Mailapalli [20] and Nan et al. [21], the use of lignite or lignite-based fertilizers can improve the properties of soil organic matter, thereby increasing the soil's ability to retain water and nutrients. The addition of lignite to the soil can affect the soil microflora quantitatively and qualitatively and the enzymatic activity [8,16,22,23]. Moreover, the study conducted by Chen et al. [24] indicates that the use of lignite-based fertilizer had a positive impact on the reduction of CH_4 and CO_2 emissions.

Leonardite is a deposit rich in humic acids, making it a valuable addition to agricultural soils [25]. The chemical composition of leonardite indicates a great potential for its use as a soil improver [18,19,26]. The decline in soil organic matter significantly impacts soil biochemical processes, highlighting the importance of preserving and enhancing soil organic carbon for the long-term stability of agricultural and environmental ecosystems. Consequently, the restoration of adequate soil organic matter content is a prevalent objective in soil science research [16].

Despite extensive research on zeolites and organic amendments, the combined effects of these materials on long-term SOM stability and nutrient cycling remain poorly understood. This study aims to fill this gap by analyzing the transformation of soil organic matter after the application of mineral–organic mixtures in a two-year pot experiment.

We hypothesize that the addition of mineral–organic mixtures containing zeolite composites, lignite, or leonardite would possess heterogeneous dissimilarities to increase soil carbon stocks, stabilize soil organic matter content, and positively influence its quality. In order to verify these assumptions, we analyzed (i) the fractional composition of soil organic matter, (ii) changes in soil carbon stocks, (iii) ATR-FTIR characterization of soil, and (iv) the content of phenolic compounds.

2. Materials and Methods

2.1. Description of the Soil Sampling Site

Soil samples were collected from an agricultural area with grass mixtures adjacent to a coniferous forest in South Malopolska (coordinates 50°09′36″N 19°66′53″E, 377.6 m a.s.l.).

Samples were taken from the upper soil layer at a depth of 0–30 cm using a hydraulic soil sampler (Auto-Field Sampler Wintex 1000 adapted for ATV Polaris, Wintex Agro, Thisted, Denmark). According to the World Reference Base for Soil Resources [27], the soil identified as Eutric Cambisol (CM-eu) comprised sand 850 $g \cdot kg^{-1}$, silt 120 $g \cdot kg^{-1}$, clay 30 $g \cdot kg^{-1}$, which categorizes it as loamy sand. The physical composition of the soil was as follows: particle density 2.65 $g \cdot cm^{-3}$, bulk density 1.45 $g \cdot cm^{-3}$. The initial soil for the research had a pH value of 5.24 and an electrical conductivity (EC) value of 850 $\mu S \cdot cm^{-3}$. The total organic carbon (TOC) and total nitrogen (TN) contents were 5.74 $g \cdot kg^{-1}$ and 0.40 $g \cdot kg^{-1}$, respectively, with a C:N ratio of 14.4 [8,19,23].

2.2. Materials Used in the Experiment

The components of the mineral–organic mixtures used in the research were zeolite composites (zeolite–carbon composite (NaX-C) or zeolite–vermiculite composite (NaX-V)) and organic materials (lignite or leonardite), in two doses. The doses of zeolite composites were determined based on the analysis of previous literature data. Zeolite composites were applied at rates of 3% (0.043 $g \cdot kg^{-1}$) and 9% (0.129 $g \cdot kg^{-1}$). Zeolite composites were obtained through alkaline synthesis of fly ash from the "Kozienice" Power Plant (Poland), which underwent hydrothermal treatment. A detailed description of the zeolite composites used has been presented in a previous article [28,29]. The rates of lignite and leonardite were 3% (0.043 $g \cdot kg^{-1}$) and 6% (0.086 $g \cdot kg^{-1}$). Lignite was sourced from the "Sieniawa" lignite mine located in Poland. Leonardite was supplied by the producer—Energy Investment Company Ltd., Kyiv, Ukraine. The properties of lignite and leonardite are given in Table S1 in the Supplementary Material. The reference treatment of the experiment was the control soil without any additives and the soil with only mineral fertilization. Chemical salts were applied as a solution prior to planting at the following rates: 0.20 $g \cdot kg^{-1}$ NH_4NO_3; 0.10 $g \cdot kg^{-1}$ $Ca(H_2PO_4)_2$ H_2O; 0.25 $g \cdot kg^{-1}$ KCl. The chemical composition of each mixture was determined by considering the percentage share and chemical composition of individual ingredients.

2.3. Description of the Pot Experiment Design

The two-year pot experiment was conducted in 2020 and 2021 during the summer season in the vegetation hall located in Krakow. Details of the experimental design are presented in Table 1. All treatments were replicated four times ($n = 4$). Soil was air-dried in the shade for 3–4 days, then ground to <2 mm sieve size and homogenized prior to pot experiments. The pots, with dimensions of 0.25 m in height, 0.22 m in diameter, and a volume of 0.009 cubic meters, were filled with soil blended together with mineral–organic mixtures. Each pot held 9 kg of soil and mineral–organic mixture, and the Kosynier variety of maize (*Zea mays* L.) was sown in each pot. Maize is an excellent crop for testing due to its long growing season and high nutrient requirements. For this reason, information on the accuracy or justification of the fertilization applied can be obtained relatively quickly [30,31]. During the growing season, the plants exhibited symptoms indicative of N deficiency. Consequently, supplementary NH_4NO_3 was applied to each pot (except the control) at a rate of 0.02 $g N \cdot kg^{-1}$ DM soil. Providing the optimal amount of water and nutrients resulted in a reduction in the length of the growing season of the plants. In a field experiment, this period is usually extended by 2–3 months. The doses of NPK fertilizer and microelements were established based on the recommended doses for maize. Soil moisture levels were maintained between 40 and 60% of the maximum water capacity of soil, with adjustments made according to the stage of plant development. Soil moisture levels were regulated using a portable probe equipped with an ECH_2O EC5 sensor (Decagon Devices, Inc., Pullman, WA, Washington, USA). At the end of the growing season, the roots were removed from the pots and, after the necessary preparation, their morphological parameters were determined. These findings are presented in more detail in our previous paper [19]. In the 2nd year, the mixtures were reapplied to the same pots and, after manual mixing, new maize seeds were sown in each pot.

Table 1. Description of experimental treatments.

Symbol	Treatment Description
C	Control soil without any additivities
MF	Soil with addition of only mineral (NPK) fertilization
MV3%L3%	Soil with 94% NPK, 3% zeolite–vermiculite composite and 3% lignite
MV9%L6%	Soil with 85% NPK, 9% zeolite–vermiculite composite and 6% lignite
MV3%Leo3%	Soil with 94% NPK, 3% zeolite–vermiculite composite and 3% leonardite
MV9%Leo6%	Soil with 85% NPK, 9% zeolite–vermiculite composite and 6% leonardite
MC3%L3%	Soil with 94% NPK, 3% zeolite–carbon composite and 3% lignite
MC9%L6%	Soil with 85% NPK, 9% zeolite–carbon composite and 6% lignite
MC3%Leo3%	Soil with 94% NPK, 3% zeolite–carbon composite and 3% leonardite
MC9%Leo6%	Soil with 85% NPK, 9% zeolite–carbon composite and 6% leonardite

2.4. Description of the Experimental Area

The experimental vegetation hall where the pot experiment was conducted is located at the research facility of the University of Agriculture in Krakow. The experimental region, situated in the south of Krakow in Poland, is characterized by a moderate climate zone and is dominated by four distinct seasons. Data on mean monthly temperature and humidity are shown in the Supplementary Materials, Table S2. The vegetation hall was constructed with a transparent roof to allow natural light to enter, and the use of netting instead of walls permitted unrestricted airflow while providing protection from birds. The pots were placed on horizontal carts to provide equal access to light and air.

2.5. Soil Sampling

Soil samples were collected from each pot at the end of the plant growth cycle (126 days in year one and 118 days in year two). Soil samples were obtained using the standard soil sampler probe (Egner's staff) from the depth of the entire pot. The soil was then manually homogenized and placed on PVC trays to dry. Soil samples were subsequently dried at room temperature, 2 mm sieved, cleaned of plant residues, and stored in plastic bags for subsequent physicochemical analyses. Soil samples for carbon fraction determination were additionally ground in an agate mortar immediately before analysis.

2.6. Laboratory Analysis

All analyses were conducted in four replicates for each treatment.

2.6.1. Basal Soil Analysis

Soil pH (1:2.5 w/v) was determined electrochemically, while electrical conductivity values were obtained conductometrically (soil/water ratio, 1:2.5 w/v) using an Elmetron Multifunction Meter CX-502. Total nitrogen (TN) content in soil was determined using a Vario EL Cube (Elementar Analysensysteme GmbH, Langenselbold, Germany). Total sorption capacity (T) was determined as the sum of base cations (S) and hydrolytic acidity (Hh). S was determined by the Kappen method by extracting the cations from the soil with 0.1 M HCL, and Hh was determined by titration after extraction with 1 M sodium acetate. The T value was calculated according to the formula

$$T = S + Hh\ [\mathrm{mmol}(+) \cdot \mathrm{kg}^{-1}].$$

2.6.2. Carbon Fractions Determination

The obtained solutions of humic substances and fulvic acids (after humic acid precipitation) were then analyzed by the oxidation titration method [32] to determine TOC (total organic carbon content), including the following carbon fractions: C Ext—$NaOH+Na_4P_2O_7$ separated carbon; CHA—carbon of the humic acid fraction; CFA—carbon of the fulvic acid fraction (calculated as CFA = C Ext—CFA); CNH—the amount of carbon remaining in the soil after extraction (calculated as CNH = TOC—C Ext). An important indicator for evaluating organic matter quality is the humification index, which is measured by the CHA:CFA ratio.

Organic carbon stocks (ZC) were calculated based on the TOC content (in %) and the volumetric density of soil (ρc in $Mg \cdot m^{-3}$) [33] using the following formula:

$$ZC = TOC \cdot \rho c \cdot 10$$

where ZC represents the soil carbon stock (in $Mg\ C \cdot ha^{-1}$), TOC stands for the carbon content ($mg\ C \cdot g\ soil^{-1}$), and ρc is the soil density (bulk density in $Mg \cdot m^{-3}$).

2.6.3. ATR-FTIR Spectroscopy Method

ATR-FTIR spectroscopy is a valuable tool for analyzing the organic and mineral composition of soil mixtures. Individual spectra of soil samples were measured using an IR300 FT-IR spectrometer (Thermo Mattson, Madison, WI, USA) fitted with a Quest Single Reflection Diamond ATR (Specac, Cranston, RI, USA). Spectral analysis was performed in the wave number range of 4000 to 400 cm^{-1}. Recordings were made with a resolution of 4 cm^{-1} and 32 scans per sample. To correct for background interference, each spectrum was adjusted using ambient air as a reference spectrum. The ATR-FTIR spectra were then smoothed to avoid shifting of the maximum band. All spectra were baseline-corrected and normalized to ensure comparability of spectral characteristics between treatments.

2.6.4. Water-Soluble Phenolic Compounds (WPC)

Water-soluble phenolic compounds were extracted from the soil with redistilled water at a soil/water ratio of 1:10 (*w*/*v*). The extraction (using a rotary mixer) was carried out for 4 h. The solution was then filtered through a Büchner funnel and 10 cm^3 of the solution was transferred to 25 cm^3 test tubes. Subsequently, 3 cm^3 of Na_2CO_3 (20%) and 1 cm^3 of Folin–Ciocalteu reagent were added and the solution was topped-up with redistilled water. After mixing, the samples were left at room temperature (20–25 °C) for 1 h. The absorbance was then read at a wavelength of 750 nm using a UV-VIS spectrophotometer (HITACHI U-5100, Tokyo, Japan).

2.6.5. Statistical Analysis

A two-way analysis of variance (ANOVA) with Duncan's test ($p \leq 0.05$) was performed using Statistica® PL 13.3 TIBCO Software Inc (StatSoft Inc., Tulsa, OK, USA), with both treatments and year of experiment as factors. The standard deviation (SD) was computed to assess the variability within each mixture. Pearson linear correlation coefficients were employed to calculate the correlation between carbon fractions, treatment, year, and other soil properties (with significance levels: $p \leq 0.05$, $p \leq 0.01$, and $p \leq 0.001$). Graphical visualizations of the data sets were generated using OriginPro® 2022b software (OriginLab®).

3. Results and Discussion

3.1. The pH and Electrical Conductivity (EC) in Soil

The pH values in the soil after the first and second year of the experiment did not differ significantly within each treatment (Table 2). Significant differences between the years were recorded between the soil from the control treatments and the soil from the fertilized treatments. It should be noted that regardless of the year of the study, fertilization led to a decrease in the soil pH value.

Table 2. The pH and EC values in the soil following the first and second year of the experiment.

Treatment	pH 1st Year	pH 2nd Year	EC 1st Year $\mu S \cdot cm^{-1}$	EC 2nd Year $\mu S \cdot cm^{-1}$	T 1st Year $mmol(+) \cdot kg^{-1}$	T 2nd Year $mmol(+) \cdot kg^{-1}$
C	5.91 d ± 0.10	5.97 d ± 0.12	365.8 a ± 58.6	341.3 a ± 9.2	106.5 c ± 1.8	129.8 d ± 5.1
MF	5.28 abc ± 0.06	5.34 c ± 0.16	426.7 b ± 46.9	762.7 j ± 20.8	93.7 a ± 3.6	137.6 ef ± 3.2
MV3%L3%	5.22 abc ± 0.12	5.17 abc ± 0.13	564.7 g ± 53.3	709.5 i ± 15.1	94.6 a ± 1.3	133.3 def ± 1.8
MV9%L6%	5.30 abc ± 0.07	5.25 abc ± 0.18	330.3 a ± 14.6	640.7 h ± 49.1	100.7 abc ± 3.8	131.1 de ± 0.8
MV3%Leo3%	5.31 bc ± 0.16	5.24 abc ± 0.13	525.5 efg ± 28.2	699.3 i ± 52.9	99.6 abc ± 2.9	131.9 def ± 1.8
MV9%Leo6%	5.23 abc ± 0.05	5.09 a ± 0.14	499.5 def ± 55.1	1111.7 l ± 33.0	101.9 bc ± 3.9	132.0 def ± 1.0
MC3%L3%	5.12 ab ± 0.12	5.15 abc ± 0.13	544.3 fg ± 11.8	708.3 ij ± 23.8	94.4 a ± 3.7	131.7 def ± 5.2
MC9%L6%	5.24 abc ± 0.06	5.27 abc ± 0.18	448.7 bcd ± 4.2	698.0 i ± 43.6	99.0 ab ± 4.0	138.5 f ± 5.6
MC3%Leo3%	5.25 abc ± 0.16	5.10 a ± 0.05	441.0 bcd ± 9.1	1263.3 m ± 62.7	104.8 bc ± 3.4	138.4 f ± 4.0
MC9%Leo6%	5.27 abc ± 0.05	5.15 abc ± 0.13	484.0 cde ± 18.0	824.5 k ± 19.6	100.4 abc ± 9.1	126.6 d ± 9.0

EC—electric conductivity; T—total cation sorption capacity. ±—standard deviation (SD); $n = 4$; means labeled with the same letters do not differ significantly according to Duncan's test at significance level of $p \leq 0.05$. Factors: treatment × year.

In the soil with treatments, an increase in electrical conductivity values was observed after the second year of the experiment in all treatments except the control soil (Table 2). The highest EC values after the 2nd year of the experiment were determined in the MV9%Leo6% treatment—1111.7 $\mu S \cdot cm^{-1}$ and in the MC3%Leo3% treatment—1263.3 $\mu S \cdot cm^{-1}$. The majority of these treatments showed a considerable elevation in EC over time, indicating an increase in soil EC. The impact of the different treatments varies, with certain combinations (e.g., MV9%Leo6%) demonstrating a notable increase in EC, potentially attributable to the elevated concentrations of NaX-V or NaX-C and lignite.

The results of the total sorption capacity (T) determination show significant differences between the tested variants in the first and second years of the study. All fertilized variants showed an increase in T values in the second year of the study, confirming the positive effect of the applied substances on the soil's ability to retain cations. The highest T values in the second year were observed in the variants MC9%L6% (138.5 $mmol(+) \cdot kg^{-1}$) and MC3%Leo3% (138.4 $mmol(+) \cdot kg^{-1}$), indicating that these combinations had the greatest effect on improving soil sorption properties.

In all fertilized variants, an increase in T values was observed in the second year of the study, suggesting that the effect of fertilization builds up over time and that the improvement in soil properties is not immediate. Particularly significant increases in sorption capacity were observed in the MV3%L3% and MC3%L3% treatments, indicating that these combinations are more effective in improving the soil's ability to retain cations.

3.2. Total Carbon and Total Nitrogen Contents in Soil

The total soil organic carbon (TOC) content after the first year of the study ranged from 5.13 $g \cdot kg^{-1}$ in the soil with MV3%L3% to 6.17 $g \cdot kg^{-1}$ in the soil of the MV9%Leo6% addition. The TOC content after the second year of the study ranged from 5.57 $g \cdot kg^{-1}$ in the soil with MV3%L3% to 6.87 $g \cdot kg^{-1}$ in the soil with the MC9%L6% addition. Soil organic carbon (TOC) content was higher in each treatment after the second year of the experiment compared to the first year (Table 3). The addition of mineral–organic mixtures resulted in an average 18% increase in TOC content after the second year of the experiment. The soil containing MC3%Leo3% showed the highest increase in TOC content (33%). The ANOVA results indicate that the different fertilization treatments had no significant effect on the TOC content in the soil compared to the control ($p \leq 0.05$). However, an analysis of the results obtained after the first and second years of testing revealed a statistically significant increase in the TOC content following the application of the MV3%Leo3%, MC3%L3%, MC9%L6%, and MC3%Leo3% mixtures. These results suggest that the use of mineral–organic mixtures significantly changed the TOC content, which is consistent with previous literature reports [34,35]. The study by Spaccini et al. [36] proved that the introduction of hydrophobic humic acid from an external source into the soil resulted in a notable increase in organic carbon sequestration in the soil. A portion of the microbially oxidized carbon derived from organic material was successfully incorporated into the stable fraction of soil organic matter, which was represented by humic substances. The cited authors contended that the retention of carbon in the soil results in a subsequent reduction in CO_2 emissions. It is estimated that up to 60% of total applied nitrogen is lost from poor, low-yielding soils. In addition, low-yielding soils exhibit reduced sorption and ion-exchange capacity, resulting in additional nitrogen losses through leaching and surface runoff. The application of zeolites to arable soils has the potential to enhance nitrogen efficiency and significantly reduce possible losses. This may contribute to increased yields [37,38]. The soil TN content is presented in Table 3. The results of the ANOVA analysis for TN content revealed significant differences between the various fertilization treatments. After the first year of the experiment, the highest TN content (0.477 $g \cdot kg^{-1}$) was identified in the control soil, exhibiting a notable elevation compared to all other treatments. The lowest TN content was observed in the MV3%Leo3% treatment (0.376 $g \cdot kg^{-1}$). However, after the second year of the experiment, the significantly highest TN content was recorded in the MV9%Leo6% treatment (0.609 $g \cdot kg^{-1}$), indicating that the addition of leonardite and NaX-V increased the nitrogen content in the soil. The control treatment (C) had a lower TN content (0.512 $g \cdot kg^{-1}$) compared to several treatments that included the addition of mineral–organic mixtures. The greatest increase (62%) in TN content was determined in the soil with the MV9%Leo6% addition. This was probably due to the lower amount of biomass and, consequently, reduced nitrogen uptake. In the treatment amended solely with mineral fertilization, the total nitrogen content increased by a mere 3%. The use of mineral–organic mixtures led to an average 9% increase in the total nitrogen content in the soil after the second year of the study, compared to the control. However, the addition of mineral–organic mixtures helped in nitrogen retention, probably due to enhanced sorption and ion-exchange capacity provided by the mixtures, as supported by the findings of Aziz et al. [37] and Naz et al. [38].

The carbon to nitrogen ratio (TOC:TN) of soil organic matter reflects the amount of carbon relative to the amount of nitrogen present. A TOC:TN ratio between 1 and 15 results in rapid mineralization and release of N available for plant uptake. A lower TOC:TN ratio results in accelerated release of nitrogen into the soil, thereby facilitating its direct use by the crop. A ratio of 20–30 provides a balance between mineralization and immobilization [39]. It is assumed that the TOC:TN ratio in soil organic matter is in the range of 10–13:1 and that it decreases significantly after the removal of plant residues from the soil. On the other hand, in clayey agricultural soils, the TOC:TN ratio can be as low as 6:1 [40]. The results of the TOC:TN ratio (Table 3) obtained after the 1st year of the experiment ranged from 12.8 $\pm$ 0.7 (MV3%L3% and MC3%Leo3%) to 16.6 $\pm$ 2.7 in the MV9%Leo6% treatment. The TOC:TN ratio results for the analyzed treatments indicate a higher TOC:TN ratio value after the second year of the experiment in the control treatment and in the treatment with mineral fertilization, compared to the values obtained after the first year. However, in the remaining treatments with mineral–organic additives, the TOC:TN values obtained after the second year were 5 to 35% lower than those obtained after the first year. The statistical analysis revealed no significant differences in the TOC:TN ratios across treatments analyzed. However, when comparing the values obtained from year to year, a notable decline was observed only in the MV9%Leo6% treatment. The variation in TOC:TN ratios suggests a balanced release of nitrogen for plant uptake, supporting crop growth.

Table 3. The content of soil organic carbon (TOC), total nitrogen (TN), and the TOC:TN ratio in soil after the first and the second year of the experiment.

Treatment	TOC 1st Year	TOC 2nd Year	TN 1st Year	TN 2nd Year	TOC:TN 1st Year	TOC:TN 2nd Year
	$g \cdot kg^{-1}$ DM	$g \cdot kg^{-1}$ DM	$g \cdot kg^{-1}$ DM	$g \cdot kg^{-1}$ DM		
C	6.12 abcdefgh ± 0.44	6.23 gh ± 0.81	0.477 bc ± 0.050	0.512 cd ± 0.041	12.9 abcde ± 1.5	13.4 abcde ± 1.5
MF	5.61 abcd ± 0.36	6.42 defgh ± 0.41	0.435 abc ± 0.082	0.449 abc ± 0.093	14.3 bcde ± 2.3	15.0 cde ± 4.6
MV3%L3%	5.13 a ± 0.28	5.57 abcd ± 0.13	0.402 ab ± 0.062	0.585 de ± 0.048	12.8 abcd ± 0.7	9.6 a ± 1.0
MV9%L6%	5.87 abcdef ± 0.29	6.72 efg ± 0.33	0.401 ab ± 0.082	0.524 cde ± 0.055	15.0 cde ± 2.8	13.0 abcde ± 2.1
MV3%Leo3%	5.27 abc ± 0.73	6.82 gh ± 0.24	0.376 a ± 0.051	0.509 cd ± 0.063	14.2 bcde ± 2.7	13.5 bcde ± 1.7
MV9%Leo6%	6.17 cdefgh ± 0.34	6.52 efgh ± 0.89	0.377 a ± 0.050	0.609 e ± 0.062	16.6 e ± 2.7	10.7 ab ± 0.5
MC3%L3%	5.73 abcde ± 0.56	6.84 gh ± 0.40	0.377 a ± 0.096	0.525 cde ± 0.050	15.7 de ± 2.8	13.1 abcde ± 1.2
MC9%L6%	5.94 abcdefg ± 0.64	6.87 h ± 0.19	0.452 abc ± 0.058	0.595 de ± 0.032	13.2 abcde ± 0.5	11.6 abc ± 0.8
MC3%Leo3%	5.14 a ± 0.35	6.86 h ± 0.66	0.401 ab ± 0.004	0.595 de ± 0.069	12.8 abcd ± 0.9	11.7 abc ± 2.4
MC9%Leo6%	5.24 ab ± 0.81	6.06 bcdefgh ± 0.94	0.402 ab ± 0.050	0.515 cd ± 0.029	14.3 bcde ± 4.3	11.8 abc ± 1.6

±standard deviation (SD), n = 4; means labeled with the same letters do not differ significantly according to Duncan's test at significance level of $p \leq 0.05$. Factor: treatment × year.

3.3. Content of Carbon Fractions in Soil

Humic acids, the principal constituent of SOM, can form an enzymatically active complex that initiates various reactions commonly attributed to the metabolic activity of microorganisms [25]. Fulvic acids have a higher abundance of carboxylic, phenolic, and ketone groups than humic acids, resulting in greater solubility at all pH levels. Humic acids, which are characterized by their aromatic properties, become insoluble when their carboxylate groups are protonated at low pH. This structural characteristic of humic materials influences their ability to bind both hydrophobic and hydrophilic substances [41]. The contents of humic acid carbon (CHA) and fulvic acid carbon (CFA) after the first and second year of the experiment are depicted in Figure 1. The CHA content identified after the first year of the experiment ranged from 0.33 ± 0.01 $g \cdot kg^{-1}$ in the soil with only mineral fertilizer addition to 0.61 ± 0.01 $g \cdot kg^{-1}$ in the soil with MC3%L3% addition. After the second year of the experiment, the CHA content increased in all treatments analyzed. The greatest increases in CHA content were found in the MF treatment—89%—and in the MC9%L6% treatment—63%. In contrast, the results showed a decrease in CFA content in the soil after the second year of the study compared to the first year. The CFA content determined after the first year of the experiment ranged from 1.23 ± 0.19 $g \cdot kg^{-1}$ in MC3%Leo3%, to 2.09 ± 0.34 $g \cdot kg^{-1}$ in the soil with MV9%Leo6% addition. After the second year of the experiment, the highest CFA content of 1.93 ± 0.19 $g \cdot kg^{-1}$ was found in the control treatment, while the lowest of 1.27 ± 0.23 $g \cdot kg^{-1}$ in the treatment with the addition of MV3%L3%. In general, the CFA content was higher than the CHA content in all the soils analyzed. The study of Truc and Yoshida [42] demonstrated that the carbon accumulation of humic fractions and the degree of humification and aromaticity of humic acids increased as a result of Ca-zeolite application to soil. This proves that the decomposition process of humic acids in soil without zeolites exceeds the rates of oxidation, decarboxylation, and demethanation. This suggests that the presence of certain amendments, such as zeolites, can significantly influence the balance between the decomposition and stabilization processes of humic substances in soil [43]. The observed changes in CHA and CFA levels across treatments underscore the dynamic nature of SOM and its sensitivity to external inputs and soil management practices [44,45]. The results of our study indicate an increase in CHA content in soils containing mineral and organic amendments, highlighting the potential of these treatments to enhance soil carbon sequestration and improve soil condition by promoting the formation of stable humic substances.

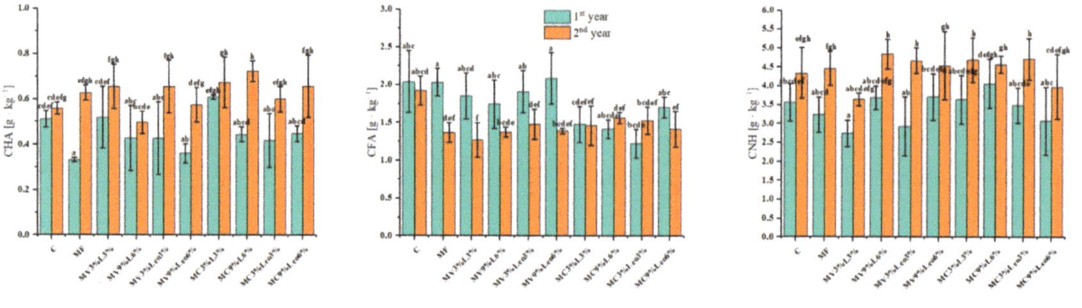

Figure 1. Humic compound fractions in the soil after the first and second years of the experiment; ±standard deviation (SD), n = 4; means labeled with the same letters do not differ significantly according to Duncan's test at significance level of $p \leq 0.05$. Factor: treatment × year.

The content range of non-hydrolyzable humic compounds (CNH) (Figure 1) varied from 2.75 ± 0.35 g · kg^{-1} in soil treated with MV3%L3% to 4.07 ± 0.65 g · kg^{-1} in soil with the addition of MC9%L6% after the first year of the experiment. On the other hand, after the second year of the experiment, the non-hydrolyzing fraction of humic compounds showed varying contents across treatments, ranging from 3.65 ± 0.17 g · kg^{-1} in MV3%L3% to 4.85 ± 0.39 g · kg^{-1} in MV9%L6%. All treatments showed an increase in CNH content after the second year, with the highest increase of 59% in the MV3%Leo3% treatment. The content and composition of humic substances indicate enhanced soil microbial activity and carbon stabilization. The increase in CNH content, especially the significant rise in the MV3%Leo3% treatment, indicates that these amendments can effectively promote the formation of stable carbon compounds in the soil. Compiling these results can be helpful in understanding which fertilizer combinations are most effective in improving soil quality. Overall, the findings support the use of mineral–organic mixtures as a sustainable soil management practice, with implications for soil health and environmental sustainability.

Soil organic carbon (ZC) stocks are influenced by various factors, including soil properties, cultivation and fertilization methods, climate, irrigation, and the type and frequency of agrotechnical practices [34,35]. The ZC values (Figure 2) obtained after the first year of the experiment ranged from 7.45 ± 0.52 Mg · ha^{-1} in the soil treated with MC3%Leo3% to 9.08 ± 0.32 Mg · ha^{-1} in the soil treated with MV9%Leo6%. ANOVA for ZC content in the soil after the first and second years of the experiment showed significant differences between various fertilization treatments and between years. The study findings revealed that all the fertilization variants used resulted in an increase in ZC after the second year of the experiment, with an average increase of 16%. The greatest increase in ZC (39%) after the second year of the experiment was determined in the soil with MC3%Leo3% addition. The fertilized treatments exhibited either lower or slightly higher ZC values compared to the control after the first year of the experiment. However, after the second year, the ZC values were lower in the MV3%L3% and MC9%Leo6% treatments compared to the control. Moreover, the values determined in the other treatments were notably higher. Additives such as NaX-V, NaX-C, lignite, and leonardite were found to significantly influence the increase in ZC in the soil.

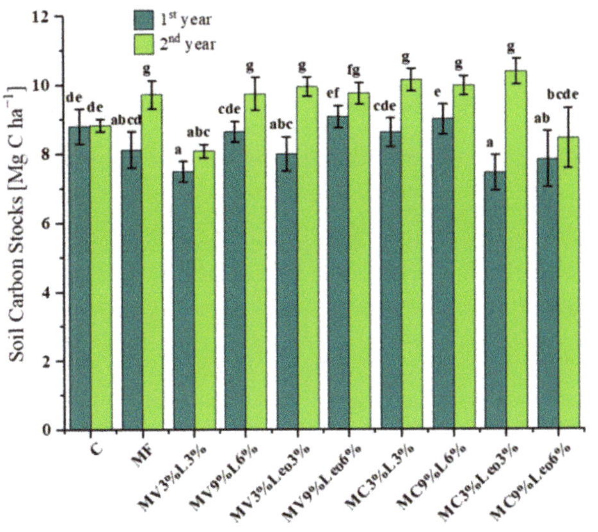

Figure 2. Soil carbon stocks (ZC) in the soil after the first and second years of the experiment; ±standard deviation (SD), n = 4; means labeled with the same letters do not differ significantly according to Duncan's test at significance level of $p \leq 0.05$. Factor: treatment × year.

The enhancement of soil carbon stocks has the potential to reduce carbon dioxide emissions into the atmosphere. The simultaneous mitigation of climate change impacts and improvement of soil health and fertility can ultimately lead to higher-quality crops [46]. Optimizing soil organic carbon stocks provides many benefits, including carbon sequestration, improved water retention, enhanced soil structure, and support for microbiological and enzymatic activities [47]. In a study by Zhang et al. [48], the use of organic amendments compared to mineral fertilization practices had a positive effect on the soil carbon sequestration in rice crops, capturing carbon at a rate of 0.20–0.88 Mg $\cdot$ ha^{-1} $\cdot$ year^{-1}. However, the soil carbon content did not increase proportionally with increasing dose of organic amendment. Therefore, the cited authors advised against using excessive amounts of organic amendments to enhance carbon sequestration. Policymakers, farmers, and scientists need to promote changes in C sequestration practices that provide additional benefits, e.g., climate change mitigation, improved soil health, food security [49].

3.4. Humic Acid Carbon: Fulvic Acid Carbon Ratio

The CHA:CFA ratios (Table 4) calculated after the first year of the experiment ranged from 0.165 ± 0.018 in the MF treatment to 0.420 ± 0.069 in the treatment with the addition of MC3%L3%. After the second year, the lowest CHA:CFA ratio was obtained in treatment C—0.293 ± 0.039—and the highest in the MV3%L3% treatment—0.536 ± 0.163. Based on the average CHA:CFA ratio over two years, the soil treated with MC3%L3% exhibited the highest value (0.448 ± 0.101), suggesting superior humification compared to the control soil, which showed the lowest value (0.277 ± 0.049). The quality of soil organic matter can be assessed by calculating the CHA:CFA ratio. It is generally assumed that fertile soils are characterized by a higher concentration of humic substances and a CHA:CFA ratio greater than 1. On agricultural land, the humic properties of soils are mainly influenced by post-harvest residues [50]. In our study, all the CHA:CFA values obtained were less than 1. Nevertheless, the results of the second year of the study demonstrated a significant increase in the CHA:CFA ratio following the application of mineral–organic mixtures.

Table 4. The CHA:CFA ratio in the soil after the first and second years of the experiment.

	CHA:CFA	
Treatment	**1st year**	**2nd year**
C	0.260 abc ± 0.060	0.293 abcd ± 0.039
MF	0.165 a ± 0.018	0.463 def ± 0.055
MV3%L3%	0.294 abcd ± 0.126	0.536 f ± 0.163
MV9%L6%	0.264 abc ± 0.147	0.362 bcde ± 0.033
MV3%Leo3%	0.239 ab ± 0.136	0.458 def ± 0.152
MV9%Leo6%	0.176 a ± 0.030	0.413 bcdef ± 0.060
MC3%L3%	0.420 cdef ± 0.069	0.475 ef ± 0.132
MC9%L6%	0.316 abcde ± 0.038	0.464 def ± 0.048
MC3%Leo3%	0.355 bcde ± 0.145	0.399 bcdef ± 0.081
MC9%Leo6%	0.263 abc ± 0.021	0.489 ef ± 0.207

±standard deviation (SD), $n = 4$; means labeled with the same letters do not differ significantly according to Duncan's test at significance level of $p \leq 0.05$. Factor: treatment × year.

3.5. Analysis of ATR-FTIR Spectra of Soil Samples

In the tested soil samples amended with mineral–organic mixtures (MV3%L3%—MV9%Leo6%) (Figure 3), a small band emerged, centered at 1520 cm^{-1}, attributed to the asymmetric stretching vibrations of carboxylate (COO^-) [51,52]. The appearance of a band at 1520 cm^{-1} in samples containing mineral–organic additives suggests the formation of carboxyl (COO-) complexes with metal cations present in the soil. The presence of this band in samples with organic additives, but not in control (C) or mineral fertilization (MF), suggests that organic matter introduces additional functional groups capable of interacting with soil minerals. The most intense bands in this range were observed in the MC3%Leo3% treatment.

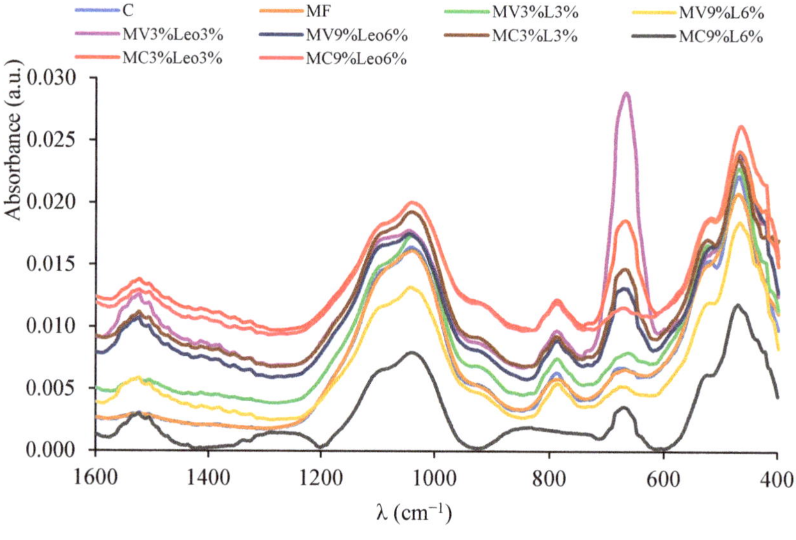

Figure 3. ATR-FTIR spectra of soil samples from individual treatments in the range of 1600–400 cm^{-1}.

The control soil samples, both without additives and with the addition of mineral fertilization (C and MF, respectively), did not show the aforementioned band. The 1400–1100 cm^{-1} region is difficult to analyze without data from other IR techniques to evaluate both the organic matter and mineral composition of the soil matrix, but differences in band intensities between variants may be due to changes in the content of humic substances and their interactions with soil minerals [53,54]. All soil samples showed well-defined peaks of high or medium intensity between 1100 and 450 cm^{-1}, which may correspond to the overtones and combination bands of bending vibrations involving O–Si–O bonds in quartz and hydrosilicates; clay minerals; or oxides such as hematite, magnetite, and rutile anions [53–55].

Our results indicate that mineral–organic mixtures add distinct functional groups to the soil matrix that are absent when only mineral fertilizers are used. The peaks with a maximum around 1080 cm^{-1} are attributed to the asymmetric stretching vibration of Si-O groups, symmetric stretching at 780 cm^{-1}, and symmetric and asymmetric Si-O bending modes at 695, 520, and 450 cm^{-1}, respectively [56]. In general, C and MF treatments were found to have relatively weak band intensities in most key regions. These samples contain less hydrogen, carbonyl, and C-O groups. Average band intensities compared to other soil samples are shown by MV9%L6%, MV3%Leo3%, and MV9%Leo6% samples. They are characterized by a higher number of functional groups such as -OH, C=O, and C-O. In contrast, MC3%L3%, MC9%L6%, MC3%Leo3%, and MC9%Leo6% samples are characterized by the highest intensity of bands, especially in the C-H range and -OH and C=O groups. This indicates a higher content of hydrocarbon compounds and functional groups such as alcohol and carbonyl groups.

These spectral modifications highlight the role of mineral–organic mixtures in altering soil chemistry by introducing distinct functional groups and facilitating organic-mineral interactions absent in mineral-only fertilized soils. The formation of carboxylate-metal complexes and increased humic substance associations with silicates suggest improved soil structure and nutrient retention, demonstrating the agronomic benefits of these amendments.

3.6. Water-Soluble Phenolic Compounds (WPC) in Soil

Phenolic compounds are crucial for the humification process of organic matter and are considered to be the major precursors of humic substances [57,58]. They occur in soil in soluble, absorbed, and polymerized forms and influence nutrient cycling and soil health. The results of our study indicate that mineral–organic mixtures significantly increased WPC levels, likely due to their acidifying effect on the soil. These findings are consistent with those reported by Min et al. [59]. The degradation and reactivity of phenols depend on their chemical structure and forms. Many phenolic compounds, including phenolic acids and tannins, are soluble in water. Microbial condensation and polymerization reactions of phenolic compounds with amino acids and proteins in the soil matrix result in the formation of higher molecular weight soil organic acids such as fulvic acids, humic acids, and humin. This humification process alters physicochemical processes in soils through modifications in soil health and qualitative characteristics [60]. Despite extensive research on phenols, there is ongoing debate regarding their transformations in soils and their impact on the rate of the soil organic matter decomposition. It is generally accepted that an elevated concentration of phenolic compounds in the soil solution reduces the intensity of soil organic matter decomposition, thereby promoting carbon sequestration and reducing carbon dioxide emissions. Our study revealed that the mineral–organic mixtures utilized had a substantial impact on the quantity of water-soluble phenolic compounds (WPC) (Figure 4). The WPC content determined after the first year of the experiment ranged from 1038 ± 40 $\mu g \cdot kg^{-1}$ in the control soil to 1722 ± 91 $\mu g \cdot kg^{-1}$ in the soil with the addition of MC3%L3%. However, after the second year, the determined values ranged from 899 ± 39 $\mu g \cdot kg^{-1}$ (treatment C) to 1489 ± 164 $\mu g \cdot kg^{-1}$ in the treatment with the addition of MC3%L3%. A reduction in WPC content after the second year was found in the treatments with the following additives: MV9%Leo6%, MC3%L3%, MV3%L3%, MC9%Leo6%, and C. In the remaining cases, the WPC content was higher or the same as in MC9%L6%.

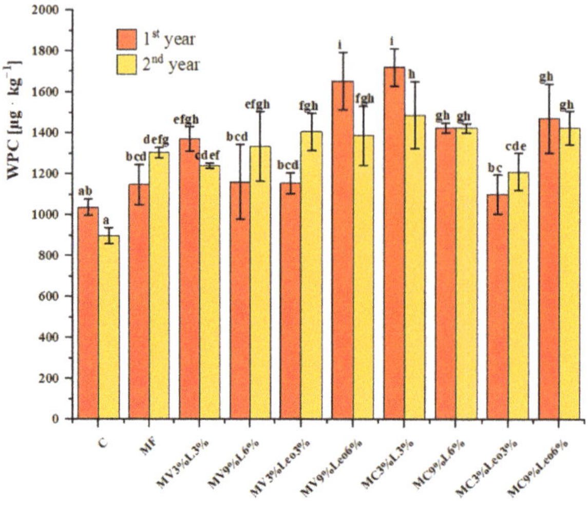

Figure 4. Content of water-soluble phenolic compounds (WPC) in the soil after the 1first and second years of the experiment; ±standard deviation (SD), n = 4; means labeled with the same letters do not differ significantly according to Duncan's test at significance level of $p \leq 0.05$. Factor: treatment × year.

The results showed that fertilization boosted the WPC content in both years of the study. The MC3%L3% and MV9%Leo6% treatments had the highest WPC content in both years, most likely due to the acidifying effect of these mixtures on the soil. Min et al. [59] reported that reducing soil pH increases the solubility of phenolic compounds and this relationship was confirmed in our study (Table 2, Figure 5). A positive correlation of WPC with CHA content and CHA:CFA ratio was also demonstrated (Figure 5). The conducted study is in agreement with literature reports where a decreasing proportion of phenolic compounds is observed with increasing degree of humification [58]. The content of phenolic compounds is also important in the context of heavy metal mobility [60]. An increase in WPC content compared to the control may suggest that the fertilizer mixtures used stimulated the production of phenols by plants or microorganisms.

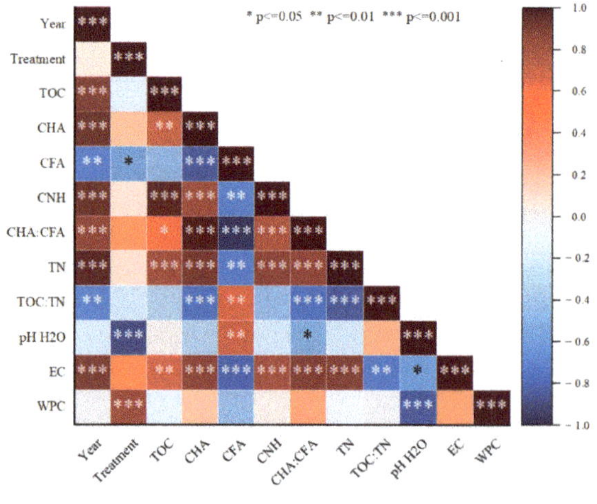

Figure 5. Pearson's correlation between carbon fractions, soil chemical properties, treatment, and year of the experiment. Positive correlations are represented by brown color and negative correlation

are represented by navy blue color. * indicates $0.01 < p \leq 0.05$, ** indicates $0.001 < p \leq 0.01$, *** indicates $p \leq 0.001$; TOC—total organic carbon; CHA—humic acid carbon; CFA—fulvic acid carbon; CNH—non-hydrolyzing carbon fraction; CHA:CFA—ratio; TN—total nitrogen content; TOC:TN—ratio; pH H_2O—pH determined in water; EC—electrical conductivity; WPC—water-soluble phenolic compounds.

3.7. Pearson's Correlation and Redundancy Analysis Among the Studied Parameters

The results of the statistical analysis, conducted using Pearson's linear correlation, are presented in Figure 5. The analysis examined the relationship between the soil properties and the parameters under investigation, taking into account the year of the experiment and the type of treatment employed. A positive correlation was found between the year of the experiment and the determined values of TN, EC, TOC, CHA, and CNH. A notable negative relationship was observed between pH and EC. The Pearson's correlation demonstrated the significant influence of mineral–organic mixtures on water-soluble phenolic compounds, emphasizing the role of fertilization and its impact on soil characteristics. The pH and EC values have important but different correlations with other parameters that may affect the mobility and availability of nutrients and contaminants in the soil. The chemical properties of the soil are significantly dependent on both the year of the experiment and the type of treatment used, which emphasizes the significant influence of fertilization on soil characteristics.

4. Future Perspective

Further studies should investigate the long-term effects of mineral–organic mixtures on soil health, crop yields, and greenhouse gas emissions. Assessing the economic feasibility of implementing these practices on a larger scale and under different environmental conditions would be valuable for promoting sustainable agriculture. In addition, understanding the mechanisms behind the observed changes in carbon fractions and humic substances will help to refine soil management strategies [61,62]. Further studies should address the specific mechanisms behind the observed interactions, providing deeper insights into the transformations of soil-phenolic compounds and their implications for soil health and fertility [60]. Moreover, it is recommended that the effects of mixtures incorporating zeolite composites, lignite, and leonardite on the properties, composition, and stability of soil organic matter under diverse climatic conditions be investigated, taking into account different soil types. Clearly, new and expanded research is needed to identify improvements in agricultural management practices and technologies that will contribute to increasing soil carbon stocks [49]. The results suggest that a combination of zeolite composites and organic additives (lignite/leonardite) may be a promising strategy for improving sandy soils. Under field conditions, their application would require doses adapted to the specific soil and cultivation system. Farmers are encouraged to integrate nutrient management by combining organic additives with inorganic fertilizers to improve the soil carbon pool and increase crop productivity in the long term. One of the main factors limiting the use of these additives on an industrial scale is their availability and the cost of production and transportation. Zeolites derived from fly ash can vary in composition depending on the source and require careful analysis before their agricultural use. The next recommended step is to carry out field experiments on different soil types to determine optimal application rates and methods. Further research should focus on the effects of additives on soil microbial composition, organic matter stability, and nutrient bioavailability.

5. Conclusions

Research on soil organic matter is important not only from a cognitive point of view, but also from the point of view of assessing soil health, fertility, and productivity. Our study showed that the introduction of new innovative mineral–organic mixtures into the soil resulted in an increase in the level of non-hydrolysable carbon, which suggests a greater stabilization of humic compounds and, at the same time, lower CO_2 emissions. The mineral–organic mixtures used increased the organic carbon content of the soil after the second year of the experiment. In each of the treatments, regardless of the type of mixture used, the carbon content of fulvic acids was higher than the carbon content of humic acids. The values of CHA:CFA ratios obtained in all treatments after the second year of the study were higher compared to those obtained after the first year, with the greatest difference found in the soil with the addition of mixtures containing a double dose of organic additives. It was observed that the application of mineral and organic materials to the soil significantly increased the content of carbon and organic matter precursors in the soil, including phenolic compounds. After the application of mineral–organic mixtures, a positive trend in the qualitative composition of SOM was observed, as evidenced by changes in various functional groups identified by ATR-FTIR analysis. The results suggest that mineral–organic mixtures influence changes in specific functional groups present in the soil matrix compared to mineral fertilization alone. Analysis of the composition and quality of soil organic matter can be an important indicator of soil health. Further correlations of the obtained results with other soil properties will contribute to a comprehensive understanding of the interaction of various soil parameters. Our study showed that zeolites synthesized from fly ash can be a valuable addition to fertilizer mixtures and contribute to the improvement of soil properties.

Supplementary Materials: The following supporting information can be downloaded at: https://www.mdpi.com/article/10.3390/agriculture15070786/s1. Table S1. Chemical properties of components used for organic fertilizers formulation. Table S2. Monthly average temperature and humidity during the experiment.

Author Contributions: Conceptualization, R.J. and M.M.-H.; data curation, R.J. and R.B.; formal analysis, R.J., K.G., and M.M.-H.; funding acquisition, M.M.-H.; investigation, R.J., K.G., and M.M.-H.; methodology, R.J., K.G., R.B., and M.M.-H.; project administration, M.M.-H.; resources, R.J. and M.M.-H.; software, R.J., J.B.K., and A.H.L.; supervision, R.J. and M.M.-H.; visualization, R.J., J.B.K., and A.H.L.; writing—original draft, R.J. and M.M.-H.; writing—review and editing, R.J., J.B.K., K.G., R.B., L.M., A.H.L., and M.M.-H. All authors have read and agreed to the published version of the manuscript.

Funding: The "Fly ash as the precursors of functionalized materials for applications in environmental engineering, civil engineering and agriculture" no. POIR.04.04.00-00-14E6/18-00 project is carried out within the TEAM-NET program of the Foundation for Polish Science co-financed by the European Union under the European Regional Development Fund.

Institutional Review Board Statement: Not applicable.

Data Availability Statement: The original contributions presented in this study are included in the article/Supplementary Material. Further inquiries can be directed to the corresponding author.

Conflicts of Interest: The authors have no relevant financial or non-financial interests to disclose.

References

1. Pandao, M.R.; Thakare, A.A.; Choudhari, R.J.; Navghare, N.R.; Sirsat, D.D.; Rathod, S.R. Soil Health and Nutrient Management. *Int. J. Plant Soil Sci.* **2024**, *36*, 873–883. [CrossRef]
2. Bagnall, D.K.; Rieke, E.L.; Morgan, C.L.S.; Liptzin, D.L.; Cappellazzi, S.B.; Honeycutt, C.W. A Minimum Suite of Soil Health Indicators for North American Agriculture. *Soil Secur.* **2023**, *10*, 100084. [CrossRef]

3. Maharjan, B.; Das, S.; Thapa, V.R.; Sharma Acharya, B. Soil Health Cycle. *Agrosystems Geosci. Environ.* **2024**, *7*, e20504. [CrossRef]
4. Arriaga, F.J.; Guzman, J.; Lowery, B. *Conventional Agricultural Production Systems and Soil Functions*; Elsevier Inc.: Amsterdam, The Netherlands, 2017; ISBN 9780128054017.
5. Wollenberg, E.; Vermeulen, S.J.; Girvetz, E.; Loboguerrero, A.M.; Ramirez-Villegas, J. Reducing Risks to Food Security from Climate Change. *Glob. Food Secur.* **2016**, *11*, 34–43. [CrossRef]
6. Fróna, D.; Szenderák, J.; Harangi-Rákos, M. The Challenge of Feeding Theworld. *Sustainability* **2019**, *11*, 5816. [CrossRef]
7. Yang, Q.; Zheng, F.; Jia, X.; Liu, P.; Dong, S.; Zhang, J.; Zhao, B. The Combined Application of Organic and Inorganic Fertilizers Increases Soil Organic Matter and Improves Soil Microenvironment in Wheat-Maize Field. *J. Soils Sediments* **2020**, *20*, 2395–2404. [CrossRef]
8. Wolny-Koładka, K.; Jarosz, R.; Juda, M.; Mierzwa-Hersztek, M. Distinct Changes in Abundance of Culturable Microbial Community and Respiration Activities in Response to Mineral–Organic Mixture Application in Contaminated Soil. *Sustainability* **2022**, *14*, 15004. [CrossRef]
9. Mondal, M.; Biswas, B.; Garai, S.; Sarkar, S.; Banerjee, H.; Brahmachari, K.; Bandyopadhyay, P.K.; Maitra, S.; Brestic, M.; Skalicky, M.; et al. Zeolites Enhance Soil Health, Crop Productivity and Environmental Safety. *Agronomy* **2021**, *11*, 448. [CrossRef]
10. Ramesh, K.; Reddy, D.D. Zeolites and Their Potential Uses in Agriculture. In *Advances in Agronomy*; Elsevier: Amsterdam, The Netherlands, 2011; Volume 113, pp. 219–241. ISBN 9780123864734.
11. Cataldo, E.; Salvi, L.; Paoli, F.; Fucile, M.; Masciandaro, G.; Manzi, D.; Masini, C.M.; Mattii, G.B. Application of Zeolites in Agriculture and Other Potential Uses: A Review. *Agronomy* **2021**, *11*, 1547. [CrossRef]
12. Jarosz, R.; Szerement, J.; Gondek, K.; Mierzwa-Hersztek, M. The Use of Zeolites as an Addition to Fertilisers—A Review. *Catena* **2022**, *213*, 106125. [CrossRef]
13. Megías-Sayago, C.; Bingre, R.; Huang, L.; Lutzweiler, G.; Wang, Q.; Louis, B. CO_2 Adsorption Capacities in Zeolites and Layered Double Hydroxide Materials. *Front. Chem.* **2019**, *7*, 551. [CrossRef] [PubMed]
14. Fan, Y.; Huang, R.; Liu, Q.; Cao, Q.; Guo, R. Synthesis of Zeolite A from Fly Ash and Its Application in the Slow Release of Urea. *Waste Manag.* **2023**, *158*, 47–55. [CrossRef] [PubMed]
15. Liang, Z.; Liu, Z.; Yu, L.; Wang, W. Fly Ash-Based Zeolites: From Waste to Value—A Comprehensive Overview of Synthesis, Properties, and Applications. *Chem. Eng. Res. Des.* **2024**, *212*, 240–260. [CrossRef]
16. Kim Thi Tran, C.; Rose, M.T.; Cavagnaro, T.R.; Patti, A.F. Lignite Amendment Has Limited Impacts on Soil Microbial Communities and Mineral Nitrogen Availability. *Appl. Soil Ecol.* **2015**, *95*, 140–150. [CrossRef]
17. Akimbekov, N.S.; Digel, I.; Tastambek, K.T.; Sherelkhan, D.K.; Jussupova, D.B.; Altynbay, N.P. Low-Rank Coal as a Source of Humic Substances for Soil Amendment and Fertility Management. *Agriculture* **2021**, *11*, 1261. [CrossRef]
18. Głąb, T.; Gondek, K.; Marcińska-Mazur, L.; Jarosz, R.; Mierzwa–Hersztek, M. Effect of Organic/Inorganic Composites as Soil Amendments on the Biomass Productivity and Root Architecture of Spring Wheat and Rapeseed. *J. Environ. Manag.* **2023**, *344*, 118628. [CrossRef]
19. Głąb, T.; Jarosz, R.; Gondek, K.; Mierzwa-Hersztek, M. Maize Root Architecture and Biomass Productivity after Application of Organic and Inorganic Additives in Contaminated Soil. *Ecol. Chem. Eng.* **2024**, *31*, 75–87. [CrossRef]
20. Dubey, A.; Mailapalli, D.R. Zeolite Coated Urea Fertilizer Using Different Binders: Fabrication, Material Properties and Nitrogen Release Studies. *Environ. Technol. Innov.* **2019**, *16*, 100452. [CrossRef]
21. Nan, J.; Chen, X.; Chen, C.; Lashari, M.S.; Deng, J.; Du, Z. Impact of Flue Gas Desulfurization Gypsum and Lignite Humic Acid Application on Soil Organic Matter and Physical Properties of a Saline-Sodic Farmland Soil in Eastern China. *J. Soils Sediments* **2016**, *16*, 2175–2185. [CrossRef]
22. Wolny-Koładka, K.; Marcińska-Mazur, L.; Jarosz, R.; Juda, M.; Lošák, T.; Mierzwa-Hersztek, M. Effect of Soil Application of Zeolite-Carbon Composite, Leonardite and Lignite on the Microorganisms. *Ecol. Chem. Eng.* **2022**, *29*, 553–563. [CrossRef]
23. Wolny-Koładka, K.; Jarosz, R.; Marcińska-Mazur, L.; Gondek, K.; Lahori, A.H.; Szara-Bąk, M.; Lošák, T.; Szerement, J.; Mokrzycki, J.; Karcz, R.; et al. The Impact of Mineral and Organic Supplements on the Abundance of Selected Groups of Culturable Microorganisms in Soil Contaminated with Heavy Metals. *J. Elem.* **2023**, *28*, 595–617. [CrossRef]
24. Chen, Z.; Huang, G.; Li, Y.; Zhang, X.; Xiong, Y.; Huang, Q.; Jin, S. Effects of the Lignite Bioorganic Fertilizer on Greenhouse Gas Emissions and Pathways of Nitrogen and Carbon Cycling in Saline-Sodic Farmlands at Northwest China. *J. Clean. Prod.* **2022**, *334*, 130080. [CrossRef]
25. Akimbekov, N.; Qiao, X.; Digel, I.; Abdieva, G.; Ualieva, P.; Zhubanova, A. The Effect of Leonardite-Derived Amendments on Soil Microbiome Structure and Potato Yield. *Agriculture* **2020**, *10*, 147. [CrossRef]
26. Ratanaprommanee, C.; Chinachanta, K.; Chaiwan, F.; Shutsrirung, A. Chemical Characterization of Leonardite and Its Potential Use as Soil Conditioner and Plant Growth Enhancement. *Asia-Pac. J. Sci. Technol.* **2017**, *22*, 1–10.
27. IUSS Working Group WRB IUSS Working Group WRB. World Reference Base for Soil Resources. In *International Soil Classification System for Naming Soils and Creating Legends for Soil Maps*, 4th ed.; International Union of Soil Sciences: Vienna, Austria, 2022; Volume 13, ISBN 9798986245119.

28. Mokrzycki, J.; Franus, W.; Panek, R.; Sobczyk, M.; Rusiniak, P.; Szerement, J.; Jarosz, R.; Marcińska-Mazur, L.; Bajda, T.; Mierzwa-Hersztek, M. Zeolite Composite Materials from Fly Ash: An Assessment of Physicochemical and Adsorption Properties. *Materials* **2023**, *16*, 2142. [CrossRef]
29. Szerement, J.; Jurek, K.; Mokrzycki, J.; Jarosz, R.; Oleszczuk, P.; Mierzwa-Hersztek, M. Zeolite Composites from Fly Ashes Mixed with Leonardite as a Useful Addition to Fertilizer for Accelerating the PAHs Degradation in Soil. *Soil Tillage Res.* **2023**, *230*, 105701. [CrossRef]
30. Zhang, H.; Dang, Z.; Zheng, L.C.; Yi, X.Y. Remediation of Soil Co-Contaminated with Pyrene and Cadmium by Growing Maize (*Zea Mays* L.). *Int. J. Environ. Sci. Technol.* **2009**, *6*, 249–258. [CrossRef]
31. Salam, A.K.; Rizki, D.O.; Santa, I.T.D.; Supriatin, S.; Septiana, L.M.; Sarno, S.; Niswati, A. The Biochar-Improved Growth-Characteristics of Corn (*Zea Mays* L.) in a 22-Years Old Heavy-Metal Contaminated Tropical Soil. In *IOP Conference Series: Earth and Environmental Science*; IOP Publishing: Bristol, UK, 2022; Volume 1034. [CrossRef]
32. Kalembasa, S. Quick Method of Determination of Organic Carbon in Soil. *Pol. J. Soil Sci.* **1991**, *24*, 17–22.
33. Guo, L.B.; Gifford, R.M. Soil Carbon Stocks and Land Use Change: A Meta Analysis. *Glob. Change Biol.* **2002**, *8*, 345–360. [CrossRef]
34. Garcia-Pausas, J.; Rabissi, A.; Rovira, P.; Romanyà, J. Organic Fertilisation Increases C and N Stocks and Reduces Soil Organic Matter Stability in Mediterranean Vegetable Gardens. *Land Degrad. Dev.* **2017**, *28*, 691–698. [CrossRef]
35. Jarosz, R.; Mierzwa-Hersztek, M.; Gondek, K.; Kopeć, M.; Lošák, T.; Marcińska-Mazur, L. Changes in Quantity and Quality of Organic Matter in Soil after Application of Poultry Litter and Poultry Litter Biochar—5-Year Field Experiment. *Biomass Convers. Biorefinery* **2022**, *12*, 2925–2934. [CrossRef]
36. Spaccini, R.; Piccolo, A.; Conte, P.; Haberhauer, G.; Gerzabek, M.H. Increased Soil Organic Carbon Sequestration through Hydrophobic Protection by Humic Substances. *Soil Biol. Biochem.* **2002**, *34*, 1839–1851. [CrossRef]
37. Aziz, Y.; Shah, G.A.; Rashid, M.I. ZnO Nanoparticles and Zeolite Influence Soil Nutrient Availability but Do Not Affect Herbage Nitrogen Uptake from Biogas Slurry. *Chemosphere* **2019**, *216*, 564–575. [CrossRef] [PubMed]
38. Naz, M.; Akhtar, K.; Khan, A.; Nizamani, G.S. Agricultural Practices Can Reduce Soil Greenhouse Gas Emissions: Challenges and Future Perspectives. In *Engineering Tolerance in Crop Plants Against Abiotic Stress*; CRC Press: Boca Raton, FL, USA, 2021; pp. 263–274. ISBN 9781000462135.
39. Brust, G.E. *Management Strategies for Organic Vegetable Fertility*; Elsevier Inc.: Amsterdam, The Netherlands, 2019; ISBN 9780128120606.
40. Paul, E.A. The Nature and Dynamics of Soil Organic Matter: Plant Inputs, Microbial Transformations, and Organic Matter Stabilization. *Soil Biol. Biochem.* **2016**, *98*, 109–126. [CrossRef]
41. Gaffney, J.S.; Marley, N.A.; Clark, S.B. Humic and Fulvic Acids and Organic Colloidal Materials in the Environment. In *ACS Symposium Series*; American Chemical Society: Washington, DC, USA, 1996; Volume 651. [CrossRef]
42. Truc, M.T.; Yoshida, M. Effect of Zeolite on the Decomposition Resistance of of Organic Matter in Tropical Soils under Global Warming. *Int. J. Environ. Chem. Ecol. Geol. Geophys. Eng.* **2011**, *5*, 664–668.
43. Doni, S.; Gispert, M.; Peruzzi, E.; Macci, C.; Mattii, G.B.; Manzi, D.; Masini, C.M.; Grazia, M. Impact of Natural Zeolite on Chemical and Biochemical Properties of Vineyard Soils. *Soil Use Manag.* **2021**, *37*, 832–842. [CrossRef]
44. Zhang, B.; Xu, C.; Zhang, Z.; Hu, C.; He, Y.; Huang, K.; Pang, Q.; Hu, G. Response of Soil Organic Carbon and Its Fractions to Natural Vegetation Restoration in a Tropical Karst Area, Southwest China. *Front. For. Glob. Change* **2023**, *6*, 1172062. [CrossRef]
45. Murindangabo, Y.T.; Kopecký, M.; Konvalina, P.; Ghorbani, M.; Perná, K.; Nguyen, T.G.; Bernas, J.; Baloch, S.B.; Hoang, T.N.; Eze, F.O.; et al. Quantitative Approaches in Assessing Soil Organic Matter Dynamics for Sustainable Management. *Agronomy* **2023**, *13*, 1776. [CrossRef]
46. *The World Bank Enhancing Carbon Stocks and Reducing CO2 Emissions in Agriculture and Natural Resource Management Projects*; World Bank: Washington, DC, USA, 2012.
47. Amoah-Antwi, C.; Kwiatkowska-Malina, J.; Thornton, S.F.; Fenton, O.; Malina, G.; Szara, E. Restoration of Soil Quality Using Biochar and Brown Coal Waste: A Review. *Sci. Total Environ.* **2020**, *722*, 137852.
48. Zhang, W.; Xu, M.; Wang, X.; Huang, Q.; Nie, J.; Li, Z.; Li, S.; Hwang, S.W.; Lee, K.B. Effects of Organic Amendments on Soil Carbon Sequestration in Paddy Fields of Subtropical China. *J. Soils Sediments* **2012**, *12*, 457–470. [CrossRef]
49. Singh, B.P.; Setia, R.; Wiesmeier, M.; Kunhikrishnan, A. *Agricultural Management Practices and Soil Organic Carbon Storage*; Elsevier Inc.: Amsterdam, The Netherlands, 2018; ISBN 9780128127667.
50. Rutkowska, A.; Piku, D. Effect of Crop Rotation and Nitrogen Fertilization on the Quality and Quantity of Soil Organic Matter. *Soil Process. Curr. Trends Qual. Assess.* **2013**, *249*, 267. [CrossRef]
51. Pradhan, S.; Hedberg, J.; Rosenqvist, J.; Jonsson, C.M.; Wold, S.; Blomberg, E.; Wallinder, I.O. Influence of Humic Acid and Dihydroxy Benzoic Acid on the Agglomeration, Adsorption, Sedimentation and Dissolution of Copper, Manganese, Aluminum and Silica Nanoparticles-A Tentative Exposure Scenario. *PLoS ONE* **2018**, *13*, e0192553. [CrossRef] [PubMed]

52. Volkov, D.S.; Rogova, O.B.; Proskurnin, M.A. Organic Matter and Mineral Composition of Silicate Soils: FTIR Comparison Study by Photoacoustic, Diffuse Reflectance, and Attenuated Total Reflection Modalities. *Agronomy* **2021**, *11*, 1879. [CrossRef]
53. Volkov, D.S.; Krivoshein, P.K.; Proskurnin, M.A. Detonation Nanodiamonds: A Comparison Study by Photoacoustic, Diffuse Reflectance, and Attenuated Total Reflection Ftir Spectroscopies. *Nanomaterials* **2020**, *10*, 2501. [CrossRef]
54. Dudek, M.; Kabała, C.; Łabaz, B.; Mituła, P.; Bednik, M.; Medyńska-Juraszek, A. Mid-Infrared Spectroscopy Supports Identification of the Origin of Organic Matter in Soils. *Land* **2021**, *10*, 215. [CrossRef]
55. Mylotte, R.; Verheyen, V.; Reynolds, A.; Dalton, C.; Patti, A.F.; Chang, R.R.; Burdon, J.; Hayes, M.H.B. Isolation and Characterisation of Recalcitrant Organic Components from an Estuarine Sediment Core. *J. Soils Sediments* **2015**, *15*, 211–224. [CrossRef]
56. Müller, C.M.; Pejcic, B.; Esteban, L.; Piane, C.D.; Raven, M.; Mizaikoff, B. Infrared Attenuated Total Reflectance Spectroscopy: An Innovative Strategy for Analyzing Mineral Components in Energy Relevant Systems. *Sci. Rep.* **2014**, *4*, 6764. [CrossRef]
57. Mierzwa-Hersztek, M.; Gondek, K.; Nawrocka, A.; Pińkowska, H.; Bajda, T.; Stanek-Tarkowska, J.; Szostek, M. FT-IR Analysis and the Content of Phenolic Compounds in Exogenous Organic Matter Produced from Plant Biomass. *J. Elem.* **2019**, *24*, 879–896. [CrossRef]
58. Ziolkowska, A.; Debska, B.; Banach-Szott, M. Transformations of Phenolic Compounds in Meadow Soils. *Sci. Rep.* **2020**, *10*, 19330. [CrossRef]
59. Min, K.; Freeman, C.; Kang, H.; Choi, S.U. The Regulation by Phenolic Compounds of Soil Organic Matter Dynamics under a Changing Environment. *BioMed Res. Int.* **2015**, *2015*, 825098. [CrossRef]
60. Misra, D.; Dutta, W.; Jha, G.; Ray, P. Interactions and Regulatory Functions of Phenolics in Soil-Plant-Climate Nexus. *Agronomy* **2023**, *13*, 280. [CrossRef]
61. Liptzin, D.; Norris, C.E.; Cappellazzi, S.B.; Mac Bean, G.; Cope, M.; Greub, K.L.H.; Rieke, E.L.; Tracy, P.W.; Aberle, E.; Ashworth, A.; et al. An Evaluation of Carbon Indicators of Soil Health in Long-Term Agricultural Experiments. *Soil Biol. Biochem.* **2022**, *172*, 108708. [CrossRef]
62. Rubin, R.; Oldfield, E.; Lavallee, J.; Griffin, T.; Mayers, B.; Sanderman, J. Climate Mitigation through Soil Amendments: Quantification, Evidence, and Uncertainty. *Carbon Manag.* **2023**, *14*, 2217785. [CrossRef]

agriculture

MDPI

Article

Photosynthetic Performance and Urea Metabolism After Foliar Fertilization with Nickel and Urea in Cotton Plants

Jailson Vieira Aguilar [1], Allan de Marcos Lapaz [1], Nayane Cristina Pires Bomfim [1], Thalita Fischer Santini Mendes [1], Lucas Anjos Souza [2], Enes Furlani Júnior [1] and Liliane Santos Camargos [1,*]

[1] School of Engineering, São Paulo State University (UNESP), Ilha Solteira 15385-000, SP, Brazil; aguilarsbio@gmail.com (J.V.A.); allanlapaz60@gmail.com (A.d.M.L.); nayanecristinapires@gmail.com (N.C.P.B.); thalita.fischer@unesp.br (T.F.S.M.); enes.furlani@unesp.br (E.F.J.)

[2] Instituto Federal Goiano, Campus Rio Verde, Rio Verde 75906-750, GO, Brazil; lucas.anjos@ifgoiano.edu.br

* Correspondence: liliane.camargos@unesp.br

Abstract: The use of nickel (Ni) as a fertilizer remains a topic of debate, particularly in non-legume species, as Ni is required only in trace amounts for optimal plant function. Urea application in plants, whether foliar or root-based, relies on the urease enzyme to convert urea into NH_{4^+} and CO_2, with Ni serving as an essential cofactor. In this study, we conducted an experiment using a 2×2 factorial design, combining two urea concentrations [4% and 8% (w/v)] with the absence or presence of Ni (0.3 g L^{-1} supplied as $NiSO_4{\cdot}6H_2O$). Gas exchange parameters were measured two days after fertilization. We quantified urease enzyme activity, urea content, photosynthetic pigments, carbohydrates, and other nitrogenous metabolites. The presence of Ni during foliar urea fertilization significantly increased the photosynthetic rate and photosynthetic pigments, which we attributed to improved urea assimilation. The combination of urea and Ni enhanced urease activity, leading to higher levels of various nitrogenous metabolites. Ni positively influenced foliar urea assimilation, promoting its conversion into organic compounds, such as proteins, while mitigating the toxic effects associated with urea accumulation.

Keywords: urease; nitrogen; amino acids; chlorophylls

1. Introduction

Production costs and increased food demand are relevant for fertilizer use; this is inducing continuous pressure on the agricultural sector and industries to meet the growing food needs [1]. Nutrients such as N are of significant importance as their scarcity leads to rapid response and plants become more susceptible to abiotic and biotic stresses [2]; in this situation, a greater demand for inputs is common, requiring efficient use. Urea is considered an efficient source of N when it is applied to the leaves due to its high N content (46%), which is advantageous for cotton crops since cotton tolerates concentrated N solutions. Urea concentrations between 3 to 15% have been reported in cotton [3–6]. The application of urea on the leaves of cotton plants aims to provide N in a high yield, supplying N demand when the root extraction capacity is reduced during the flowering peak or late fruiting [7].

Dixon et al. [8] verified the presence of two Ni ions in the constitution of the metalloenzyme urease (urea amidohydrolase; EC 3.5.1.5.), which catalyzes the urea hydrolysis into ammonia (NH_{4^+}) and carbon dioxide (CO_2). Since then, the scientific community has investigated the role of Ni in urea metabolism [9–13]. Studies have also deepened the knowledge of its role in other metabolic aspects, such as in the production of phytoalexins [14–17],

 https://doi.org/10.3390/agriculture15070699

in the hydrogenase enzyme [18], and in the glyoxalase I enzyme [19], which performs an important role in the degradation of a potent cytotoxic compound—methylglyoxal. This fact suggests that Ni may play an important role in the metabolism of antioxidants in plants, especially under stress [20].

Urease enzyme activity has been detected in several plants [21,22]. The absence of Ni compromises the biosynthesis of this enzyme, leading to urea accumulation and, consequently, necrotic lesions on leaf tips [10,14,23]. In situations where nitrogen (N) is supplied as urea in a hydroponic nutrient solution or via the leaves, the role of Ni becomes essential since it stimulates the primary metabolism and increases plants' yield traits [24–26]. Ni and N application faces challenges in soils due to the low prevalence of soluble and readily absorbable urea [27].

The herbaceous cotton (*Gossypium hirsutum* L. r. *latifolium* Hutch.) is widely used for fiber production and the process to produce fibers demands high technology; therefore, there is a high monetary value associated with the fiber production chain due to the wide range of products produced [28,29]. Nitrogen is one of the most expensive inputs in the cotton production chain and needs 48 to 85 kg ha^{-1} of N to produce 1 ton of seed cotton, depending on the variety used and edaphoclimatic conditions [30,31]. Nitrogen plays an important role in the plant [32] since it is a constituent or activator of enzymes, amino acids, nucleic acids, and chlorophyll [33], as well as stimulating the formation and development of flowering and fruiting buds [34].

Previous studies have reported that the application of Ni-associated N to the soil has no significant effect on the chlorophyll index (SPAD) [35] and on plant growth [36] in cotton plants. Other studies have found the beneficial effect of Ni in combination with the application of foliar urea in soybean plants, such as in decreasing the toxic effects of urea [37] and alleviating the harmful effects of glyphosate [38]. Nevertheless, few studies have investigated the physiological responses and N metabolism associated with Ni after foliar application in agronomic crops. This paper hypothesizes that, similar to leguminous plants, the presence of Ni during foliar urea fertilization will enhance its assimilation into organic nitrogen compounds, thereby improving the photosynthetic performance of the plant. Here, the behavior of gas exchange, carbohydrates, pigments, and N metabolism indicators were studied in cotton plants at the early adult reproductive stage newly fertilized with urea and Ni to demonstrate short-term changes in photosynthetic performance and metabolites related to urea metabolization.

2. Materials and Methods

2.1. Experimental Site

The experiment was carried out in an arch-type greenhouse, covered with a plastic film that diffuses transparent light, with a thickness of 1000 microns, in the municipality of Ilha Solteira, São Paulo State, Brazil. The experimental site belongs to the Agronomy Campus of the São Paulo State University Júlio de Mesquita Filho (UNESP) (20°25′06.0″ S and 51°20′29.7″ W).

2.2. Experimental Design and Treatments

The experimental design was completely randomized and arranged in a 2 × 2 factorial scheme comprising two N concentrations as CH_4N_2O [4% and 8% (*w*/*v*)] in the presence or absence of Ni as $NiSO_4{\cdot}6H_2O$ (0.3 g L^{-1}) applied on cotton leaves. The concentration applied was based on a dose screening performed by the author (U. Each plot consisted of a pot containing a cotton plant, with four plants per treatment.

2.3. Growing Conditions

The soil used was a Dystrophic Ultisol soil [39], collected in the experimental site of the Fazenda de Ensino e Pesquisa of UNESP in the municipality of Selvíria, Mato Grosso do Sul, Brazil (20°20′24.9″ S 51°24′19.7″ W) in areas adjacent to the plots for cultivation of agronomic crops. The particle size analysis of the composite sample of soil collected from the 0.00–0.40 cm layer contained the following fractions: 38% clay, 56% sand, and 6% silt. The chemical traits of the Ultisol used were as follows: pH = 5.2 ($CaCl_2$ 0.01M), organic matter = 15.0 g kg^{-1} (colorimetric method), P = 2 mg kg^{-1} (resin), K = 0.5 $mmol_c$ kg^{-1} (resin), Ca = 12 $mmol_c$ kg^{-1} (resin), Mg = 10 $mmol_c$ kg^{-1} (resin), potential acidity 18.0 $mmol_c$ kg^{-1} (SMP buffer), Al = 0.0 $mmol_c$ kg^{-1}, Cu = 0.8 mg dm^{-3} (DTPA), Fe = 11 mg dm^{-3} (DTPA), Mn = 7.2 mg dm^{-3} (DTPA), Zn = 0.3 mg dm^{-3} (DTPA), and Ni < 2 mg dm^{-3} (ICP-OES, HNO_3 EPA. 3051A).

Before the experiment, the soil field capacity was determined [100% of the water mass (g) that the soil withstood] as described by [40]. During the experimental conduction, the replenishment of evapotranspired water for the plots (pots + plants) was achieved using suspended micro-sprinklers, which were activated twice a day (morning and afternoon), providing the water amount necessary for soil moisture to remain at 80% of field capacity.

The soil was placed in pots with a capacity of 8 dm^{-3} and we added $CaCO_3$ and $MgCO_3$ at a 3:1 ratio, increasing the base saturation to 70% [41]. The soil remained incubated for 50 days, keeping moisture at 80% of field capacity to allow for the reaction. Seven days before sowing, the soil was dried, chipped, sieved, and incorporated with 40 mg dm^{-3} N as urea, 200 mg dm^{-3} P as P_2O_5, 80 mg dm^{-3} K as KCl, 4 mg dm^{-3} Zn as $ZnSO_4{\cdot}7H_2O$, and 0.8 mg dm^{-3} B as H_3BO_3. Eight cotton seeds cv. 'TMG 81 WS' were placed 2.5 cm below the soil surface of each pot. The plants were thinned to one representative plant per plot in the second week. The experiment was conducted during the 2019 year. Plants were grown until 42 days after planting; subsequently, treatments were applied as reported below.

2.4. Foliar Fertilization with Urea and Nickel

At 42 days after planting, automated irrigation was suspended to apply Ni ($NiSO_4{\cdot}6H_2O$) and urea (CH_4N_2O) on the adaxial surfaces of the cotton leaves, according to the treatments, using an individual brush. At this time, cotton plants were at the growth stage 51 [42], with the first buds detectable (pin-head square), but still without the visible petals.

2.5. Gas Exchange Measurement

At 44 days after planting (2 days after fertilization), coinciding with the emission of the first flower buds detectable, the photosynthetic performance of the cotton plant was evaluated through gas exchange with the portable CIRAS-3 analyzer (Portable Photosynthesis System-PP Systems). Ambient light under photosynthetically active radiation (PAR) was used with an average of 1189 μmol m^{-2} s^{-1} with 390 μmol^{-1} mol^1 of CO_2 as a reference to carry out the analyses.

The readings were taken from the apex on the first fully expanded leaves, between 9h30 and 12h00 in the morning, with the frequency of three analyses per plot on the same leaf, following the stability of measurements recorded in real time on the equipment screen for manual recording of each value. The equipment had been configured for automatic calibration. The photosynthetic rate values (A, μmol CO_2 m^{-2} s^{-1}), stomatal conductance (gs, mol H_2O m^{-2} s^{-1}), transpiration (E, mmol H_2O m^{-2} s^{-1}), and internal carbon (Ci, μmol CO_2 m^{-2} ar^{-1}) were measured. Subsequently, the instantaneous efficiency of carboxylation (EiC, mol ar^{-1}) was estimated (A/Ci).

2.6. Collection of Plant Material for Laboratory Analysis

Three days after fertilization with urea and Ni, the first and second leaves fully expanded were excised from the apex for analysis of N metabolites and soluble carbohydrates. The leaves were washed with deionized water and dried in paper towels. Then, a fraction of the fresh tissue was removed in vivo to evaluate the urease enzyme activity and content of urea and photosynthetic pigments. The remaining fresh tissue was placed in a −20 °C freezer.

2.7. Photosynthetic Pigments

Chlorophyll *a* (Chl *a*), chlorophyll *b* (Chl *b*), and carotenoid (CAR) contents were determined using the extracting agent dimethyl sulfoxide (DMSO). Leaf tissue (50 mg) was cut into 1 mm fragments and incubated in 7 mL of DMSO in the dark in a water bath at 65 °C for 30 min [43]. After readings in the spectrophotometer, the contents of the photosynthetic pigments were expressed in mg g^{-1} FW.

2.8. Extraction and Measurement of Urease Enzyme Activity

The urease activity was determined through N-NH_{4^+} production [44]. The extract was prepared with adaptations of the method described by Hogan [21]. Fresh leaf samples (200 mg) were immersed in 8 mL of 0.1 M sodium phosphate buffer (pH 7.4) containing 0.21 M urea and N-propanol (pH 7.4). This extract was kept for 3 h in a water bath with manual shaking of the tube rack every 15 min. The reaction consisted of 500 µL of extract, 2.5 mL of reagent I—0.1 M phenol and 170 µM sodium nitroprusside (SNP)—and 2.5 mL of reagent II: 0.125 M sodium hydroxide, 0.15 M disodium phosphate, and 3% sodium hypochlorite (*w*/*v*). Then, the samples were incubated in a water bath at 37 °C for 35 min and the determination was carried out in a spectrophotometer at λ = 625 nm with the results expressed in µmol N-NH_{4^+} g^{-1} FW h^{-1}.

2.9. Quantification of Urea Content

The urea content in cotton leaves was determined based on the reaction described by Kyllingsbæk [45] and adapted by Kojima et al. [46]. Fresh tissue (500 mg) was cut (1 mm) and then added to 2 mL of ice-cold formic acid. The material was centrifuged at 12,000 rpm at 4 °C for 15 min. A quantity of 60 µL of the supernatant was incubated in 2 mL of color-developing reagent [4.6 mM diacetylmonoxime, 1.28 mM thiosemicarbazide, 6.6% H_2SO_4 (*v*/*v*), 14.6 µM ferric chloride hexahydrate, and 0.006% orthophosphoric acid (*v*/*v*)] at 99 °C for 15 min followed by cooling in ice for 5 min. The determination was performed in a spectrophotometer at λ = 540 nm and the results were expressed in µmol g^{-1} FW.

2.10. Extraction of Soluble Compounds

The soluble compounds were extracted according to Bieleski and Turner [47], using the MCW extractor solution (60% methanol, 25% chloroform, and 15% water) for water-soluble compounds and 0.1 M NaOH for protein extraction, as described by Lapaz et al. [48].

2.11. Quantification of Ammonia Content

The ammonia content was determined using the method described by McCullough [44]. The reaction consisted of 100 µL of water-soluble extract, 500 µL of reagent I (0.1 M phenol, 170 µM sodium nitroprusside (SNP)), and 500 µL of reagent II (0.125 M sodium hydroxide, 0.15 M disodium phosphate, and (3%) sodium hypochlorite (*w*/*v*)), followed by incubation in a water bath at 37 °C for 60 min. Ammonia content determination was performed in a spectrophotometer at λ = 630 nm with the results expressed in µmol g^{-1} FW.

2.12. Quantification of Ureide Content

Total ureides were measured according to Van Der Drift et al. [49]. The test was divided into two phases. The first phase comprised the addition of 250 μL of the water-soluble extract in 500 μL distilled H_2O, one drop of 0.33% phenylhydrazine (*w*/*v*), and 250 μL of 0.5 M NaOH. Then, the samples were heated in a water bath (100 °C) for 8 min and cooled to room temperature. Quantities of 250 μL of 0.4 M phosphate buffer (pH 7) and 250 μL of 0.33% phenylhydrazine solution were added. After 5 min at room temperature, the assay was incubated in an ice bath for another 5 min, then 1.25 mL HCl p.a (−8 °C) and 250 μL of 1.65% P-ferrocyanide (*w*/*v*) were added. After 15 min at room temperature, the assay was removed from the ice bath and the determination was carried out in a spectrophotometer at λ = 535 nm. The results were expressed in μmol g^{-1} FW. The ureide content was measured using the standard curve of allantoin solution.

2.13. Quantification of Amino Acid Content

Total amino acids were quantified with the acid ninhydrin method [50]. We used 100 μL of the water-soluble extract in 900 μL of H_2O, 500 μL of citrate buffer (pH 5.5), 200 μL of 5% ninhydrin (*w*/*v*) in methyl glycol, and 1 mL of 0.0002 M potassium cyanide solution. The samples were heated in a water bath at 100 °C for 20 min and placed to cool at room temperature for subsequent inclusion of 1 mL of 60% ethanol (*v*/*v*). The determination was carried out in a spectrophotometer at λ = 570 nm and the results were expressed in μmol g^{-1} FW. Total amino acid content was measured using the standard curve of leucine solution.

2.14. Quantification of Protein Content

Total proteins were measured using the Coomassie blue reaction [51]. For the test, we used 50 μL of alkaline extract (sodium hydroxide fraction) in 2.5 mL of Bradford solution. After resting for 5 min, total protein content was determined in a spectrophotometer at λ = 595 nm. Protein content was measured using the standard curve of bovine serum albumin solution and the results were expressed in μmol g^{-1} FW.

2.15. Quantification of Carbohydrate Content

Total carbohydrates were determined according to the anthrone method described by Yemm and Willis [52]. A quantity of 1 mL of the water-soluble extract was added to a test tube with 1 mL of anthrone 0.2% in sulfuric acid (*w*/*v*), followed by stirring and subsequent incubation in a water bath for 10 min and cooling at room temperature. Then, the determination was carried out in a spectrophotometer at λ = 620 nm. The concentration of total carbohydrates was measured using the standard curve of glucose solution and the results were expressed in μmol g^{-1} FW.

2.16. Data Analysis

The data were first assessed for normality using the Shapiro–Wilk test ($p < 0.05$). Subsequently, analysis of variance (ANOVA) was conducted using the F test ($p \leq 0.05$), with treatments treated as quantitative factors. Mean comparisons were performed using the Tukey test ($p < 0.05$). All statistical analyses were carried out using custom protocols implemented in the R software.

3. Results

The results were presented in two ways. For simple effects, the behavior of the variables was displayed in a table (Table 1), highlighting the differences between the isolated factors. When the interaction was significant, bar graphs were plotted (Figures 1–3). We will begin by presenting the gas exchange data, followed by pigment analysis and concluding with N-metabolism data. The simple effects, along with any significant interactions, will be presented in three distinct moments within the same section. The corresponding table will be referenced to facilitate a clear and structured description of the results.

Table 1. Significance of the F-test and results of the Tukey test for variables with simple effect for substomatal concentration of carbon dioxide (Ci, µmol CO_2 m^{-2} ar^{-1}), total soluble carbohydrates (TSC, µmol g^{-1} FW), carotenoids (CAR, mg g^{-1} FW), urease activity (µmol N–NH_4^+ g^{-1} FW h^{-1}), ammonia (µmol g^{-1} FW), ureides (µmol g^{-1} FW), and amino acids (µmol g^{-1} FW) in cotton plants subjected to foliar fertilization with urea (4% and 8%) and Ni (0 and 0.3 g L^{-1}).

Factor	Ci	TSC	CAR	Urease	Ammonia	Ureides	Amino Acids
Urea	ns	**	ns	ns	**	*	***
Ni	**	ns	ns	***	ns	***	***
Urea × Ni	ns	ns	ns	ns	ns	ns	ns
Urea (%)							
4	300.5	1.95 b	0.75	5.65	0.51 b	13.17 b	11.70 b
8	322.0	2.38 a	0.72	6.20	0.79 a	13.95 a	23.15 a
Ni (g L^{-1})							
0	346,0 a	2.15	0.77	2.74 b	0.57	11.81 b	15.38 b
0.3	276.5 b	2.18	0.71	9.10 a	0.73	15.31 a	19.47 a

Significance by factorial analysis: ns, not significant; * $p < 0.05$; ** $p < 0.01$; *** $p < 0.001$. Different letters in the same column indicate significant differences according to the Tukey test ($p < 0.05$).

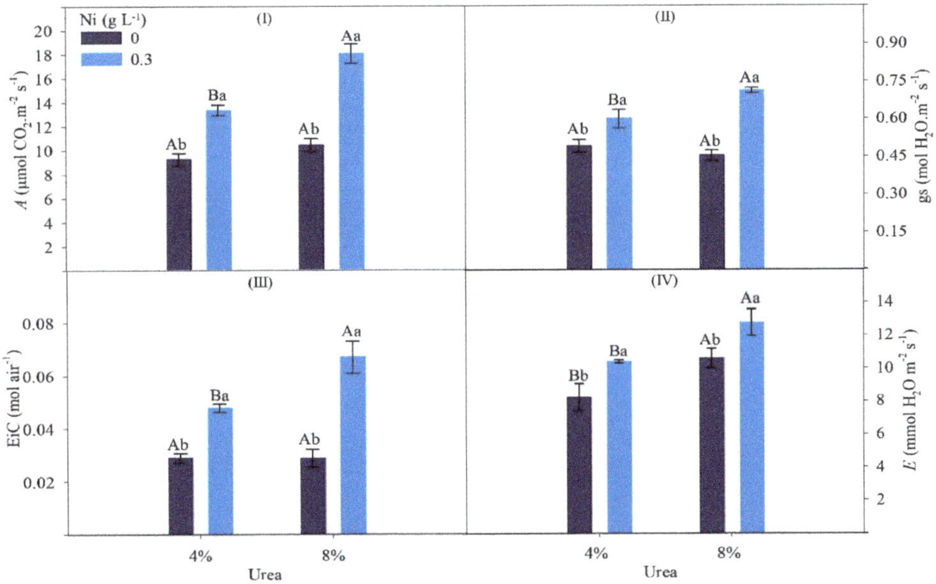

Figure 1. Photosynthetic rate ((**I**); *A*), stomatal conductance ((**II**); *gs*), instantaneous carboxylation efficiency ((**III**); *EiC*), and transpiration rate ((**IV**); *E*) in cotton plants subjected to foliar fertilization with urea (4% and 8%) and Ni (0 and 0.3 g L^{-1}). Different letters indicate significant differences according to the Tukey test ($p < 0.05$). Uppercase letters compare the concentrations of urea in the same Ni concentration while lowercase letters compare the concentrations of Ni at the same concentration of urea. Bars represent the standard error ($n = 4$ plants).

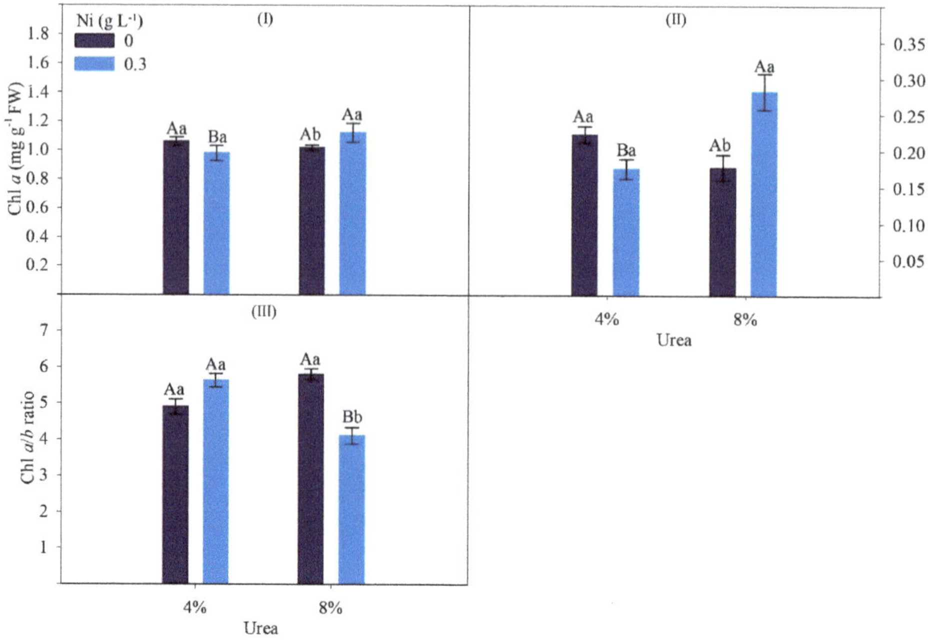

Figure 2. Chlorophyll *a* ((**I**); Chl *a*), chlorophyll *b* ((**II**); Chl *b*), and chlorophyll ratio ((**III**); Chl *a*/*b* ratio) contents in cotton plants subjected to foliar fertilization with urea (4% and 8%) and Ni (0 and 0.3 g L^{-1}). Different letters indicate significant differences according to the Tukey test ($p < 0.05$). Uppercase letters compare the concentrations of urea in the same Ni concentration while lowercase letters compare the concentrations of Ni at the same concentration of urea. Bars represent the standard error ($n = 4$ plants).

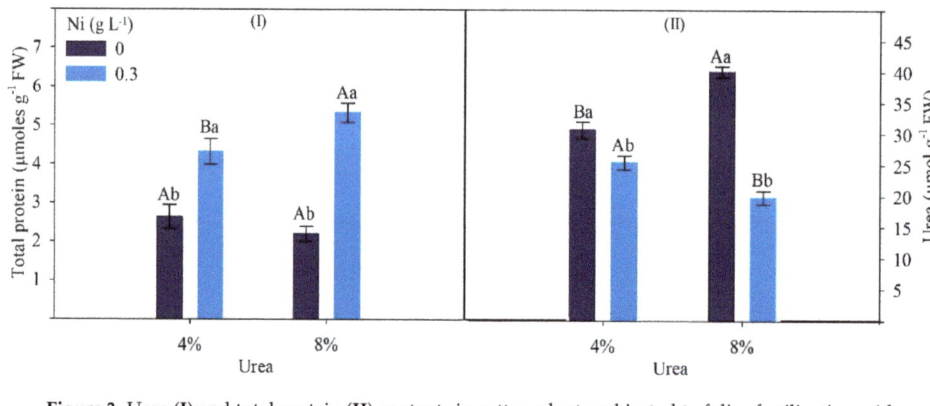

Figure 3. Urea (**I**) and total protein (**II**) contents in cotton plants subjected to foliar fertilization with urea (4% and 8%) and Ni (0 and 0.3 g L^{-1}). Different letters indicate significant differences according to the Tukey test ($p < 0.05$). Uppercase letters compare the concentrations of urea in the same Ni concentration while lowercase letters compare the concentrations of Ni at the same concentration of urea. Bars represent the standard error ($n = 4$ plants).

The traits TSC and ammonia showed an isolated effect only for the urea factor while Ci and urease showed an isolated effect only for the Ni factor on the ANOVA (Table 1). The concentrations of ureides and amino acids showed isolated effects for both factors on the ANOVA (Table 1). The average CAR concentration was 0.743 mg g^{-1} FW and it was not significant for both factors on the ANOVA (Table 1). The remaining traits (*A*, gs, EiC, *E*, Chl *a*, Chl *b*, Chl *a*/*b* ratio, urea, and total protein) showed a significant effect of double interaction on the ANOVA (Figures 1–3).

At 2 days after fertilization, *A*, gs, and EiC increased in the presence of Ni in both urea concentrations. This increase was more pronounced in plants that had received the 8% urea concentration, with increases of 94, 62, 45, 21, and 131% compared to plants fertilized with 8% urea in the absence of Ni (Figure 1I–III). The same behavior was also observed in E, although to a lesser extent compared to the previously mentioned variables (Figure 1IV). All these variables showed significance for the Ni × urea interaction. Our findings indicate that the combination of Ni and urea enhances gas exchange and that this effect is responsive to increasing concentrations of nitrogen fertilizer.

The variable Ci was not significant in the interaction between the factors. However, it was demonstrated that plants treated with Ni exhibited a 20% reduction in substomatal carbon compared to those without Ni application (Table 1). Soluble carbohydrates were also evaluated through simple effects based on the significance observed for the urea factor. The increase in urea concentration led to a 22% rise in this metabolite (Table 1).

In summary, we demonstrated that increasing the concentration of urea applied through foliar fertilization enhanced gas exchange. The presence of Ni reduced the internal CO_2 concentration, suggesting that carboxylation may have occurred more rapidly in plants treated with Ni. This hypothesis was supported by the higher instantaneous carboxylation efficiency observed in plants receiving Ni at both urea concentrations, with an additional increase of 8%. Furthermore, soluble carbohydrate data reinforced that higher urea levels also led to greater sugar availability.

In the presence of Ni in plants fertilized with 8% urea, Chl *a* and Chl *b* contents increased by 10.1% and 59.3%, respectively, compared to plants that did not receive Ni (Figure 2I,II). Thus, the Chl *a*/*b* ratio was also altered, decreasing from 5.8 to 4.1 when fertilized with 8% urea associated with Ni (Figure 2III). The CAR content was not significantly different for any fertilization condition (Table 1).

Chlorophyll *a* exhibited minimal variation across all treatments. However, a peak in chlorophyll *b* content was observed in the 8% urea treatment with Ni, which significantly altered the Chl *a*/*b* ratio. This suggests that, under these conditions, the plant may have enhanced its light-harvesting capacity by increasing the pool of chlorophyll *b* in the antenna complex, likely in response to the high nitrogen availability.

The protein content increased in the presence of Ni at both urea concentrations, and the highest values were observed in plants fertilized with 8% urea concentration, with increases of 141%. In plants fertilized with 8% urea in the absence of Ni, at 4%, following the same trend, the percentage increase was 63,7% (Figure 3I). This significant increase in the soluble protein pool suggests that a substantial portion of the N from the applied urea was assimilated into proteins, and Ni likely played a crucial role in facilitating this process.

The highest urea content was observed in plants fertilized with 8% urea in the absence of Ni. On the other hand, in the presence of Ni, the urea content decreased by 17 and 50.7% in plants fertilized with urea at 4 and 8%, respectively, compared to plants that had not received Ni (Figure 3II). The increase in foliar urea concentration, in contrast to the low protein content observed in treatments without Ni, reinforces the importance of Ni in the metabolization of the applied urea.

The activity of the urease enzyme was not significant for the urea factor or for the urea × Ni interaction. However, it showed high significance for the Ni factor. Plants treated with Ni exhibited urease activity more than three times higher than those without Ni application. The ammonia content showed significance only for the urea factor, with a 59.4% increase observed in plants treated with 8% urea (Table 1).

Significant variations were observed for both factors tested in relation to ureides and amino acids. For the urea factor, increases of 5.9% in ureides and 97.8% in amino acids were observed between the two concentrations tested. For the Ni factor, plants treated with Ni showed a 29.6% increase in ureide content and a 26.5% increase in amino acid content compared to those without Ni (Table 1). N metabolites responded to both N application and Ni alone, reinforcing our findings that Ni plays a key role in enhancing the conversion of urea into organic N compounds in this case.

Figure 4 showed that the Pearson correlation analysis highlighted significant relationships between physiological and metabolic variables in plants, providing a comprehensive way to uncover connections in the dataset that had not been explored in previous analyses. Transpiration showed a strong positive correlation with stomatal conductance and total protein content, indicating that higher stomatal opening and protein synthesis are associated with increased transpiration rates.

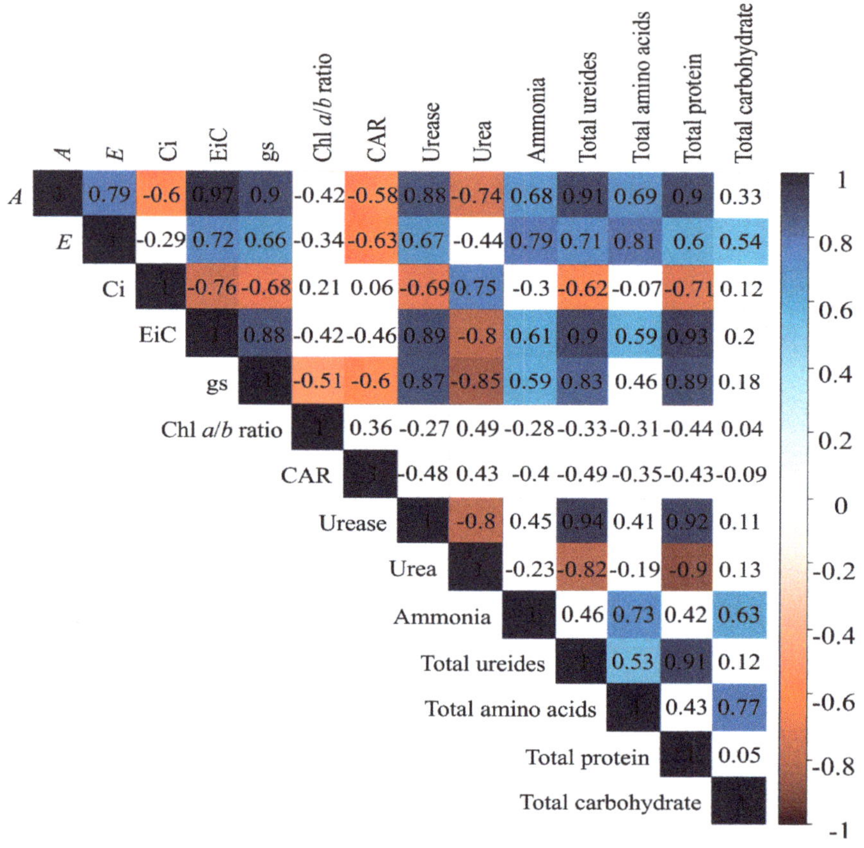

Figure 4. The Pearson correlations for the traits evaluated. The squares that have received the white color belong to the category of non-significant values. The figure shows photosynthetic rate (*A*), stomatal conductance (*gs*), substomatal concentration of carbon dioxide (*Ci*), transpiration rate (*E*), instantaneous carboxylation efficiency (*EiC*), total carbohydrate content, chlorophyll ratio (Chl *a*/*b* ratio), carotenoid content (CAR), urease activity and urea, ammonia, total ureide, total protein, and total amino acid contents in cotton plants subjected to foliar fertilization with urea (4% and 8%) and Ni (0 and 0.3 g L^{-1}).

Urease exhibited a strong positive correlation with total ureide and total protein contents, reinforcing its role in converting urea into organic nitrogen compounds and revealing an intriguing, unexplained behavior: the accumulation of ureides in non-nodulated plants. Total ureide contents were also strongly correlated with total protein contents, raising the possibility of their importance in protein synthesis under high urea-N conditions.

The photosynthetic rate showed a strong negative correlation with Ci, indicating efficient CO_2 utilization, and a strong positive correlation with gs, highlighting the importance of stomatal opening for photosynthesis. Additionally, the positive correlation with total proteins suggests that higher photosynthetic rates enhance protein synthesis. Finally, the negative correlation with urea implies that plants with high photosynthetic activity may metabolize urea more rapidly.

Carotenoids, however, showed weak correlations, except for a moderate negative correlation with A, E and gs, indicating limited direct involvement in nitrogen metabolism. Finally, ammonia was moderately correlated with total amino acid and gas exchange traits, reinforcing its role as an intermediate in nitrogen assimilation. This analysis underscores the importance of urease and stomatal conductance in N metabolism and plant growth while revealing unexpected relationships that warrant further investigation.

4. Discussion

N fertilization with urea on cotton leaves favored gas exchange, increasing the *A* with split applications of urea in the field during the reproductive phase of cotton, where the need for photosynthates is high [53]. The premature photosynthetic decline of the cotton canopy as a consequence of the apparent remobilization of ribulose 1,5 biphosphate carboxylase-oxygenase (Rubisco) [54] suggests that extra and timely N fertilization could maintain the canopy photosynthetic capacity longer in the reproductive stage, which is likely to provide even greater yield increases [53]. The increase in *A* (Figure 1I) in response to urea and Ni fertilization was associated with increased CO_2 influx mediated by higher *gs* (Figure 1II), and also by higher *EiC* (Figure 1III), corroborated by their significant correlation (Figure 4).

In our study, Chl *a* and *b* increased with 8% urea and Ni fertilization (Figure 2I,II), changing the Chl *a*/*b* ratio (Figure 2III) due to a substantial increase in Chl *b* content, improving light absorption capacity in the antenna complex. Chl *a* is involved in the transformation of light energy into electrical energy and Chl *b* contributes to improving the light absorption capacity, acting in the absorption of blue–violet light [55]. However, Barcelos [35] did not observe significant effects of N and Ni fertilization in the soil on chlorophylls (SPAD) in cotton. Importantly, growth promotion is related to the improvement of urea-N assimilation and chlorophyll contents in urea-fed wheat plants when Ni is supplemented (0.01 and 0.05 mg Ni L^{-1} in nutrient solution); the increase in chlorophyll content demonstrated the improvement in urea-N assimilation by the Ni supplement because chlorophyll is an N-containing compound [56].

Ni is a constituent of the urease enzyme responsible for breaking down urea into ammonia and carbon dioxide [8], participating in N recycling in the plant [20]. Therefore, some studies have been related to whether or not Ni is associated with urea applications to assess its Ni role, especially in the activity of the urease enzyme. Studies with cereals [11] pecan [15], and colza [57] reported increases in the urease enzyme activity in the presence of Ni and inefficiency in urea metabolism in Ni's absence. Ni enrichment in soybean seeds and the application in nutrient solution reduced the toxic effects of urea accumulation after foliar fertilization, increasing the content of photosynthetic pigments and improving the nutritional statuses of plants. Ni-deficient plants took longer to show positive responses to foliar fertilization with 2% urea, indicating significantly quicker assimilation of urea N by Ni-fed plants [37].

Three days after urea and Ni application, the urease enzyme activity increased in cotton plants in the early adult reproductive stages (Table 1) and showed a strong correlation between urease activity, urea, ureides, and proteins (Figure 4). The foliar application of urea of 31 and 16 mg of N plant^{-1}, combined or not with the supply of 0.05 mg of Ni plant^{-1}, showed that Ni impaired N uptake in coffee seedlings because when Ni was present, the leaf always absorbed less N than when Ni was not present [33]. In our study, N absorption did not seem to be impaired in Ni's presence since the urea content in the leaves decreased in both experimental urea concentrations (Figure 3II).

Once urea enters the plant cell or is generated as a metabolic by-product, its rapid metabolism or temporary accumulation is possible [58]. In senescent tissue, urea accumulation has already been verified as there is a need for the remobilization of reserves to sustain plant growth [59]. In non-senescent tissues, urea as the intact molecule is practically not accumulated [60]. Urea accumulation in cotton plants that have not received Ni can be a positive indicator of the Ni effect during urea metabolization after foliar fertilization, as demonstrated by the negative correlation of this metabolite with urease enzyme activity and protein content (Figure 4).

The ammonia content is expected to increase after urea fertilization as urease is dependent on Ni and urea for its activation [58]. In our study, the urease enzyme activity was unchanged at both experimental concentrations of urea (Table 1); however, the ammonia content was higher in plants that had received a greater amount of urea (Table 1). The increase in ammonia content was also observed after the application of 2% urea to potatoes; nevertheless, the urease enzyme activity did not appear to be induced by urea concentration [13]. Ammonia produced is assimilated mainly by the combined action of the enzymes glutamine synthetase (GS) and glutamate oxoglutarate aminotransferase (GOGAT) [61], which suggests the use of CO_2 resulting from the process of organic acid synthesis via tricarboxylic acid cycle [62]. This internal situation might be coordinated between C and N metabolism because glycolysis and the TCA cycle must produce energy and metabolic precursors, such as amino acids, for the storage of protein synthesis.

As expected, the amino acid contents also increased in the presence of urea and Ni (Table 1). The accumulation of amino acids in urea-fed plants is mainly related to the pool of the amino acids glutamine and asparagine used in the N translocation. High levels of these molecules decrease ammonia assimilation due to the feedback inhibition of assimilation products that are not efficiently translocated [57,61]. Ureides are found in several plants and their formation usually occurs after the oxidative degradation of nucleic acids and their purines [63,64]. Some plants contain high ureide contents in their tissues, such as legumes [32]. In this case, these ureides are substances associated with N storage and transport from biological fixation [65].

Ureides are economical molecules in situations of low *A*, such as in C3 plants in which the use of ureides for storage could be an alternative to save metabolic energy. The advantage of having ureides as N storage and transport molecules is their N: C ratio (1:1) while the amides glutamine and asparagine contain (1:2.5) and (1.2), respectively [66]. The positive correlation between urease enzyme activity, proteins, and ureides (Figure 4) indicates a possibility of using ureides as nitrogenous molecules for temporary storage in this situation, even under high *A* conditions (Figure 1I).

Many studies have addressed this topic, primarily in legumes, attributing productivity responses to Ni application [67–70]. However, in cotton, there is still limited information on the best practices for Ni application as a nutrient, with recent studies focusing mainly on root exposure [71,72]. We believe that the findings of this manuscript provide a comparative basis for more detailed studies on the use of foliar Ni, either combined or not with urea, in cotton.

5. Conclusions

The presence of Ni during foliar urea fertilization increased the photosynthetic rate and photosynthetic pigments, which we attributed to the enhanced urea assimilation process. Simultaneously, the combination of urea and Ni boosted urease enzyme activity, leading to higher levels of various N metabolites. The detection of ureides in cotton leaves following foliar fertilization with urea and Ni suggests their potential roles as storage forms of N. Furthermore, Ni can alleviate the stress induced by urea application, mitigating potential toxic effects and improving overall plant performance.

The increases in protein content and its positive correlation with gas exchange parameters suggest that short-term improvements in photosynthesis, coupled with elevated N metabolite levels in plants treated with urea and Ni, may have long-term benefits. These changes could support late-cycle cotton traits, such as flowering, by providing an abundant supply of N compounds to meet increased metabolic demands during reproductive stage.

Author Contributions: Conceptualization, L.S.C., J.V.A. and E.F.J.; methodology, J.V.A. and N.C.P.B.; formal analysis, J.V.A. and T.F.S.M.; writing—original draft preparation, J.V.A., A.d.M.L. and N.C.P.B.; writing—review and editing, L.S.C., E.F.J. and L.A.S.; supervision, L.S.C. All authors have read and agreed to the published version of the manuscript.

Funding: This research was funded by Fundação de Amparo à Pesquisa do Estado de São Paulo, (FAPESP—Brazil, grant number 2020/12421-4 to LSC), Coordenação de Aperfeiçoamento de Pessoal de Nível Superior (CAPES–Brazil, Finance Code 001 to J.V.A.), and Conselho Nacional de Desenvolvimento Científico e Tecnológico (CNPq-Brazil, grant number 302499/2021-0 to L.S.C.).

Institutional Review Board Statement: Not applicable.

Data Availability Statement: Data are contained within the article.

Acknowledgments: This work was carried out with the support of the Coordenação de Aperfeiçoamento de Pessoal de Nível Superior (CAPES-Brazil), which sponsored the PhD studentships to the first author.

Conflicts of Interest: The authors declare no conflicts of interest.

References

1. Hossain, A.; Krupnik, T.J.; Timsina, J.; Mahboob, M.G.; Chaki, A.K.; Farooq, M.; Bhatt, R.; Fahad, S.; Hasanuzzaman, M. Agricultural Land Degradation: Processes and Problems Undermining Future Food Security. In *Environment, Climate, Plant and Vegetation Growth*; Fahad, S., Hasanuzzaman, M., Alam, M., Ullah, H., Saeed, M., Khan, I.A., Adnan, M., Eds.; Springer International Publishing: Cham, Switzerland, 2020; pp. 17–61.
2. Asghar, M.G.; Bashir, A. Protagonist of Mineral Nutrients in Drought Stress Tolerance of Field Crops. In *Abiotic Stress in Plants*; Fahad, S., Saud, S., Chen, Y., Wu, C., Wang, D., Eds.; IntechOpen: London, UK, 2020.
3. Ferraz, C.A.M.; Fuzatto, M.G.; Gridi-Papp, I.L. Dados preliminares sôbre o emprêgo de adubos minerais nitrogenados em pulverização foliar no algodoeiro. *Bragantia* **1969**, *28*, 33–38. [CrossRef]
4. Rosolem, C.A.; Barretto, A.E.; Trivelin, P.C.O.; Victória, R.L. Absorção de uréia via foliar pelo algodoeiro em função do pH da solução. *Pesqui. Agropecu. Bras.* **1989**, *25*, 491–497.
5. Carvalho, M.A.C.D.; Paulino, H.B.; Furlani-Júnior, E.; Buzetti, S.; Sá, M.E.D.; Athayde, M.L.F.D. Uso da adubação foliar nitrogenada e potássica no algodoeiro. *Bragantia* **2001**, *60*, 239–244. [CrossRef]
6. Buriro, M.; Soomro, S.; Buriro, G.; Jogi, Q.; Muhaamad, N.; Kandhro Rais, N. Effect of foliar applied boron, zinc, and urea on growth and yield of cotton. *Sci. Int.* **2016**, *28*, 4113–4117.
7. Ali, N. Review: Nitrogen utilization features in cotton crop. *Am. J. Plant Sci.* **2015**, *6*, 987–1002.
8. Dixon, N.E.; Gazzola, C.; Blakeley, R.L.; Zerner, B. Jack bean urease (EC 3.5.1.5). metalloenzyme simple biological role for nickel. *J. Am. Chem. Soc.* **1975**, *97*, 4131–4133.

9. Eskew, D.L.; Welch, R.M.; Cary, E.E. Nickel: An essential micronutrient for legumes and possibly all higher plants. *Science* **1983**, *222*, 621–623.
10. Eskew, D.L.; Welch, R.M.; Norvell, W.A. Nickel in higher plants: Further evidence for an essential role. *Plant Physiol.* **1984**, *76*, 691–693.
11. Brown, P.H.; Welch, R.M.; Cary, E.E. Nickel: A micronutrient essential for higher plants. *Plant Physiol.* **1987**, *85*, 801–803.
12. Fageria, N.K.; Baligar, V.; Clark, R.B. Micronutrients in crop production. *Adv. Agron.* **2002**, *77*, 185–268.
13. Witte, C.-P.; Tiller, S.A.; Taylor, M.A.; Davies, H.V. Leaf urea metabolism in potato urease activity profile and patterns of recovery and distribution of 15N after foliar urea application in wild-type and urease-antisense transgenics. *Plant Physiol.* **2002**, *128*, 1129. [PubMed]
14. Walker, C.D.; Graham, R.D.; Madison, J.T.; Cary, E.E.; Welch, R.M. Effects of Ni deficiency on some nitrogen metabolites in cowpeas (*Vigna unguiculata* L. Walp). *Plant Physiol.* **1985**, *79*, 474–479.
15. Bai, C.; Reilly, C.C.; Wood, B.W. Nickel deficiency disrupts metabolism of ureides, amino acids, and organic acids of young pecan foliage. *Plant Physiol.* **2006**, *140*, 433–443.
16. Harasim, P.; Filipek, T. Nickel in the environment. *J. Elem.* **2015**, *20*, 525–534.
17. Macedo, F.G.; Santos, E.F.; Lavres, J. Agricultural crop influences availability of nickel in the rhizosphere; a study on base cation saturations, Ni dosages and crop succession. *Rhizosphere* **2020**, *13*, 100182.
18. Klucas, R.V.; Hanus, F.J.; Russell, S.A.; Evans, H.J. Nickel: A micronutrient element for hydrogen-dependent growth of rhizobium japonicum and for expression of urease activity in soybean leaves. *Proc. Natl. Acad. Sci. USA* **1983**, *80*, 2253–2257. [PubMed]
19. Mustafiz, A.; Ghosh, A.; Tripathi, A.K.; Kaur, C.; Ganguly, A.K.; Bhavesh, N.S.; Tripathi, J.K.; Pareek, A.; Sopory, S.K.; Singla-Pareek, S.L. A unique Ni^{2+}-dependent and methylglyoxal-inducible rice glyoxalase I possess a single active site and functions in abiotic stress response. *Plant J.* **2014**, *78*, 951–963. [CrossRef]
20. Fabiano, C.C.; Tezotto, T.; Favarin, J.L.; Polacco, J.C.; Mazzafera, P. Essentiality of nickel in plants: A role in plant stresses. *Front. Plant Sci.* **2015**, *6*, 754.
21. Hogan, M.E.; Swift, I.E.; Done, J. Urease assay and ammonia release from leaf tissues. *Phytochemistry* **1983**, *22*, 663–667. [CrossRef]
22. Witte, C.-P.; Medina-Escobar, N. In-gel detection of urease with nitroblue tetrazolium and quantification of the enzyme from different crop plants using the indophenol reaction. *Anal. Biochem.* **2001**, *290*, 102–107.
23. Khoshgoftarmanesh, A.H.; Hosseini, F.; Afyuni, M. Nickel supplementation effect on the growth, urease activity and urea and nitrate concentrations in lettuce supplied with different nitrogen sources. *Sci. Hortic.* **2011**, *130*, 381–385.
24. Follmer, C. Insights into the role and structure of plant ureases. *Phytochemistry* **2008**, *69*, 18–28. [PubMed]
25. Lopes, J.F.; Coelho, F.C.; Rabello, W.S.; Rangel, O.J.P.; Gravina, G.d.A.; Vieira, H.D. Produtividade e composição mineral do feijão em resposta às adubações com molibdênio e níquel. *Rev. Ceres* **2016**, *63*, 419–426.
26. Macedo, L.O.; Favarin, J.L.; Tezotto, T.; Neto, A.P.; Andrade, S.A.L.; Mazzafera, P. Soil and foliar nickel application in coffee seedlings alters leaf nutrient balance. *Agrochim. Int. J. Plant Chem. Soil Sci. Plant Nutr. Univ. Pisa* **2020**, *64*, 167–180.
27. Tsadilas, C.D.; Rinklebe, J.; Selim, H.M. *Nickel in Soils and Plants*; CRC Press: New York, NY, USA, 2019.
28. Américo, G.H.P.; Américo-Pinheiro, J.H.P.; Furlani, E., Jr. Hormesis effect of dichlorophenoxy acetic acid sub-doses and mepiquat chloride on cotton plant. *Planta Daninha* **2017**, *35*, 1–9.
29. Shareef, M.; Gui, D.; Zeng, F.; Waqas, M.; Ahmed, Z.; Zhang, B.; Iqbal, H.; Xue, J. Nitrogen leaching, recovery efficiency, and cotton productivity assessments on desert-sandy soil under various application methods. *Agric. Water Manag.* **2019**, *223*, 105716.
30. Rosolem, C.A.; van Mellis, V. Monitoring nitrogen nutrition in cotton. *Rev. Bras. Ciência Solo* **2010**, *34*, 1601–1607.
31. Rosolem, C.A.; Echer, F.R.; Lisboa, I.P.; Barbosa, T.S. Acúmulo de nitrogênio, fósforo e potássio pelo algodoeiro sob irrigação cultivado em sistemas convencional e adensado. *Rev. Bras. Ciência Solo* **2012**, *36*, 457–466.
32. Camargos, L.; Sodek, L. Nodule growth and nitrogen fixation of Calopogonium mucunoides L. show low sensitivity to nitrate. *Symbiosis* **2010**, *51*, 167–174.
33. Pereira Bruno, I.; Moraes, M.F.; Damin, V.; Dourado-Neto, D.; Reichardt, K. Does nickel influence leaf nitrogen uptake in coffee seedlings? *Braz. J. Agric.* **2019**, *94*, 259–269.
34. Malavolta, E.; Vitti, G.C.; de Oliveira, S.A. *Avaliação do Estado Nutricional das Plantas: Princípios e Aplicações*, 2nd. ed; Potafos: Piracicaba, Brazil, 1997.
35. Barcelos, J.; Furlani Junior, E.; Reis, H.; Putti, F.; Reis, A. Diagnóstico da exigência do algodoeiro em nitrogênio e níquel pela utilização do medidor portátil de clorofila [Diagnosis of nitrogen and nickel requirements for cotton plants using a portable chlorophyll meter]. *Rev. Bras. Eng. Biossistemas* **2016**, *10*, 97–106.
36. Rigon, J.P.G.; Neto, J.F.B.; Capuani, S.; Beltrão, N.E.M.; Silva, F.V.F. Utilização de nitrogênio e níquel durante o crescimento do Algodão. *Enciclopedia Biosf.* **2011**, *7*, 1019–1026.
37. Kutman, B.Y.; Kutman, U.B.; Cakmak, I. Nickel-enriched seed and externally supplied nickel improve growth and alleviate foliar urea damage in soybean. *Plant Soil* **2013**, *363*, 61–75. [CrossRef]

38. Einhardt, A.M.; Ferreira, S.; Oliveira, L.M.; Ribeiro, D.M.; Rodrigues, F.Á. Glyphosate and nickel differently affect photosynthesis and ethylene in glyphosate-resistant soybean plants infected by Phakopsora pachyrhizi. *Physiol. Plant.* **2020**, *170*, 592–606. [CrossRef]
39. Santos, H.G.; Jacomine, P.K.T.; Anjos, L.H.C.; Oliveira, V.A.; Lumbrearas, J.F.; Coelho, M.R.; Almeida, J.A.; Filho, J.C.A.; Oliveira, J.B.; Cunha, T.J.F. *Sistema Brasileiro de Classificação de Solos*; Embrapa: Rio de Janeiro, Brazil, 2018; Volume 5.
40. Ibañez, T.B.; Santos, L.F.M.; Lapaz, A.M.; Ribeiro, I.V.; Ribeiro, F.V.; Reis, A.R.; Moreira, A.; Heinrichs, R. Sulfur modulates yield and storage proteins in soybean grains. *Sci. Agric.* **2021**, *78*, e20190020. [CrossRef]
41. Quaggio, J.A.; van Raij, B.; Malavolta, E. Alternative use of the SMP-buffer solution to determine lime requirement of soils. *Commun. Soil Sci. Plant Anal.* **1985**, *16*, 245–260. [CrossRef]
42. Munger, P.; Bleiholder, H.; Hack, H.; Hess, M.; Stauß, R.; van den Boom, T.; Weber, E. Phenological Growth Stages of the Cotton Plant (*Gossypium hirsutum* L.): Codification and Description according to the BBCH Scale1. *J. Agron. Crop Sci.* **1998**, *180*, 143–149. [CrossRef]
43. Hiscox, J.; Israelstam, G.F. A method for the extraction of chlorophyll from leaf tissue without maceration. *Can. J. Bot.* **1979**, *57*, 1332–1334. [CrossRef]
44. McCullough, H. The determination of ammonia in whole blood by a direct colorimetric method. *Clin. Chim. Acta* **1967**, *17*, 297–304. [CrossRef]
45. Kyllingsbæk, A. Extraction and colorimetric determination of urea in plants. *Acta Agric. Scand.* **1975**, *25*, 109–112. [CrossRef]
46. Kojima, S.; Bohner, A.; Gassert, B.; Yuan, L.; Wirén, N. AtDUR3 represents the major transporter for high-affinity urea transport across the plasma membrane of nitrogen-deficient Arabidopsis roots. *Plant J.* **2007**, *52*, 30–40. [PubMed]
47. Bieleski, R.L.; Turner, N.A. Separation and estimation of amino acids in crude plant extracts by thin-layer electrophoresis and chromatography. *Anal. Biochem.* **1966**, *17*, 278–293. [CrossRef]
48. Lapaz, A.M.; Camargos, L.S.; Yoshida, C.H.P.; Firmino, A.C.; de Figueiredo, P.A.M.; Aguilar, J.V.; Nicolai, A.B.; Silva de Paiva, W.D.; Cruz, V.H.; Tomaz, R.S. Response of soybean to soil waterlogging associated with iron excess in the reproductive stage. *Physiol. Mol. Biol. Plants* **2020**, *26*, 1635–1648. [PubMed]
49. Van Der Drift, C.; De Windt, F.E.; Vogels, G.D. Allantoate hydrolysis by allantoate amidohydrolase. *Arch. Biochem. Biophys.* **1970**, *136*, 273–279. [CrossRef] [PubMed]
50. Yemm, E.W.; Cocking, E.C.; Ricketts, R.E. The determination of amino-acids with ninhydrin. *Analyst* **1955**, *80*, 209–214.
51. Bradford, M.M. A rapid and sensitive method for the quantitation of microgram quantities of protein utilizing the principle of protein-dye binding. *Anal. Biochem.* **1976**, *72*, 248–254.
52. Yemm, E.W.; Willis, A.J. The estimation of carbohydrates in plant extracts by anthrone. *Biochem. J.* **1954**, *57*, 508–514.
53. Giri, M.; Dhonde, M.B.; Tumbare, A.D. Effect of split and foliar application of nitrogen on leaf nitrogen concentration, SPAD index and photosynthesis in Bt. Cotton (*Gossypium hirsutum* L.). *SAARC J. Agric.* **2017**, *14*, 1–11.
54. Pettigrew, W.T.; McCarty, J.C.; Vaughn, K.C. Leaf senescence-like characteristics contribute to cotton's premature photosynthetic decline. *Photosynth. Res.* **2000**, *65*, 187–195.
55. Wientjes, E.; Philippi, J.; Borst, J.W.; van Amerongen, H. Imaging the photosystem i/photosystem ii chlorophyll ratio inside the leaf. *Biochim. Biophys. Acta.* **2017**, *1858*, 259–265.
56. Gheibi, M.N.; Malakouti, M.J.; Kholdebarin, B.; Ghanati, F.; Teimouri, S.; Sayadi, R. Significance of nickel supply for growth and chlorophyll content of wheat supplied with urea or ammonium nitrate. *J. Plant Nutr.* **2009**, *32*, 1440–1450. [CrossRef]
57. Gerendas, J.; Zhu, Z.; Sattelmacher, B. Influence of N and Ni supply on nitrogen metabolism and urease activity in rice (*Oryza sativa* L.). *J. Exp. Bot.* **1998**, *83*, 65–71. [CrossRef]
58. Polacco, J.C.; Mazzafera, P.; Tezotto, T. Opinion: Nickel and urease in plants: Still many knowledge gaps. *Plant Sci.* **2013**, *199–200*, 79–90. [CrossRef]
59. Polacco, J.C.; Holland, M.A. Roles of Urease in Plant Cells. In *International Review of Cytology*; Jeon, K.W., Jarvik, J., Eds.; Academic Press: New York, NY, USA, 1993; pp. 65–103.
60. Winkler, R.G.; Polacco, J.C.; Eskew, D.L.; Welch, R.M. Nickel is not required for apourease synthesis in soybean seeds. *Plant Physiol.* **1983**, *72*, 262–263. [PubMed]
61. Garnica, M.; Houdusse, F.; Zamarreño, A.M.; Garcia-Mina, J.M. Nitrate modifies the assimilation pattern of ammonium and urea in wheat seedlings. *J. Sci. Food Agric.* **2010**, *90*, 357–369.
62. Matiz, A.; Mioto, P.T.; Aidar, M.P.M.; Mercier, H. Utilization of urea by leaves of bromeliad Vriesea gigantea under water deficit: Much more than a nitrogen source. *Biol. Plant.* **2017**, *61*, 751–762.
63. Thomas, R.; Schrader, L. Ureide metabolism higher plants. *Phytochemistry* **1981**, *20*, 361–371.
64. Werner, A.K.; Witte, C.-P. The biochemistry of nitrogen mobilization: Purine ring catabolism. *Trends Plant Sci.* **2011**, *16*, 381–387.
65. Mothes, K. The metabolism of urea and ureides. *Can. J. Bot.* **2011**, *39*, 1785–1807.
66. Raso, M.J.; Muñoz, A.; Pineda, M.; Piedras, P. Biochemical characterization of an allantoate-degrading enzyme from French bean (*Phaseolus vulgaris*): The requirement of phenylhydrazine. *Planta* **2007**, *226*, 1333–1342.

67. Delfim, J.; Dameto, L.S.; Moraes, L.A.C.; Moreira, A. Nitrogen and Nickel Foliar Application on Grain yield, Yield Components, and Quality of Soybean. *Commun. Soil Sci. Plant Anal.* **2022**, *53*, 1226–1234. [CrossRef]
68. Zhran, M.; Moursy, A.; Lynn, T.M.; Fahmy, A. Effect of urea fertilization on growth of broad bean (*Vicia faba* L.) under various nickel (Ni) levels with or without acetic acid addition, using 15N-labeled fertilizer. *Environ. Geochem. Health* **2021**, *43*, 2423–2431. [CrossRef] [PubMed]
69. Kochenborger, A.C.; Orioli Júnior, V.; Silva, G.A.; Sargentim, M.M.; Torres, J.L.R. Aplicação foliar de ureia, níquel e sacarose em estádio reprodutivo da soja. *Nativa* **2024**, *11*, 82–89. [CrossRef]
70. Mendes, N.A.C.; Cunha, M.L.O.; Bosse, M.A.; Silva, V.M.; Moro, A.L.; Agathokleous, E.; Vicente, E.F.; Reis, A.R.D. Physiological and biochemical role of nickel in nodulation and biological nitrogen fixation in *Vigna unguiculata* L. Walp. *Plant Physiol. Biochem.* **2023**, *201*, 107869. [CrossRef] [PubMed]
71. Aguilar, J.V.; Lapaz, A.M.; Sanches, C.V.; Yoshida, C.H.P.; Camargos, L.S.; Furlani-Júnior, E. Application of 2,4-D hormetic dose associated with the supply of nitrogen and nickel on cotton plants. *J. Environ. Sci. Health Part B* **2021**, *56*, 852–859. [CrossRef]
72. Aguilar, J.V.; Ferreira, T.C.; Bomfim, N.C.P.; Mendes, T.F.S.; Lapaz, A.M.; Brambilla, M.R.; Coscione, A.R.; Souza, L.A.; Furlani-Júnior, E.; Camargos, L.S. Different responses to phenological stages: A role for nickel in growth and physiology of herbaceous cotton. *Plant Growth Regul.* **2023**, *101*, 663–678. [CrossRef]

agriculture

MDPI

Article

Characterization of Cupuaçu (*Theobroma grandiflorum*) Waste for Substrate in Seedling Production

Isaac Manoel Rocha de Sousa Filho [1], Clodoaldo Alcino Andrade dos Santos [1,*], Geomarcos da Silva Paulino [2], Anselmo Júnior Corrêa Araújo [1,2], Wandicleia Lopes de Sousa [3,4], Helionora da Silva Alves [1,4], Thiago Almeida Vieira [1,2,4] and Denise Castro Lustosa [1]

1 Institute of Biodiversity and Forests, Federal University of Western Pará, Santarém 68040255, Brazil; isaacrochafilho99@gmail.com (I.M.R.d.S.F.); anselmo.araujo@ufopa.edu.br (A.J.C.A.); helionora.alves@ufopa.edu.br (H.d.S.A.); thiago.vieira@ufopa.edu.br (T.A.V.); denise.lustosa@ufopa.edu.br (D.C.L.)
2 Doctor Program Society, Nature and Development, Federal University of Western Pará, Santarém 68040255, Brazil; geomarcospaulino19@gmail.com
3 Campus of Alenquer, Federal University of Western Pará, Alenquer 68200000, Brazil; wandicleia@hotmail.com
4 Master Program Society, Environment and Quality of Life, Federal University of Western Pará, Santarém 68040255, Brazil
* Correspondence: clodoaldo.santos@ufopa.edu.br

Abstract: Cupuaçu (*Theobroma grandiflorum* (Willd. ex Spreng.) K. Schum.), a fruit native to the Amazon, is widely used in the food industry. However, it generates large amounts of shell and seed residues after processing, which are generally discarded inappropriately. We characterized cupuaçu residues to determine their potential as substrates for seedling production. Shells and fruit seeds were collected from a rural community in the municipality of Belterra, Brazil. The residues were weighed, dried, and crushed for chemical (macro and micronutrients, elemental analysis, and ash content) and physical (density and porosity) analyses. Different proportions of each residue (9:1, 3:1, 2:1, 1:1, and 0:1) were used, and soil alone was used as the control. The chemical analysis of the plant material showed that cupuaçu seed residues had higher concentrations of macro- and micronutrients among the analyzed elements when compared to the fruit shell residues. The macronutrient concentrations (P, K, Ca, and Mg) were, respectively, 5, 1.3, 2.3, and 5.6 times higher than those in the shell residues. Compared with the soil sample, the concentrations of the macronutrients (P, K, Ca, and Mg) in the shell residues were, respectively, 600, 162, 1283, and 12.35 times higher. Analysis of variance and comparison of treatment means were performed using Tukey's test ($p \leq 0.05$). Chemical analysis showed that cupuaçu seed and shell residues had higher concentrations of macro- and micronutrients than soil. All proportions tested with residues had lower densities and greater porosities than soil. The cupuaçu residues showed desirable chemical and physical characteristics for their use as substrate in seedling production.

Keywords: fruit shell; seeds; alternative substrate; organic waste

1. Introduction

The production of seedlings is an activity of great importance, especially when planning reforestation programs and the recovery of degraded areas. However, in addition to propagation material, the choice of substrate is crucial for the production of high-quality seedlings.

An alternative for use as a substrate in seedling production is plant residues, which can help reduce environmental impacts while lowering costs [1]. In the Amazon, the tropical fruit production chain generates a lot of waste, mainly from shells and seeds, and its disposal is almost always neglected, as in the case of cupuaçu (*Theobroma grandiflorum* (Willd. ex Spreng.) K. Schum.).

Cupuaçu is native to the Amazon and is widely used in the food industry. It is one of the most popular fruits in Northern Brazil and is used to produce juice, sweets, and ice cream. The growing commercial interest in the fruit has led to the development of new industries and research on this food species [2].

In 2017, 21,240 tons of fresh cupuaçu were recorded in Brazil [3]. The physical composition of the fruit, on average, consists of 42% shells, 14% seeds, and 39% pulp [4]. Although the shell represents a considerable portion of the total fruit weight, it is not commercially used and is frequently discarded inappropriately [5], generating a large volume of agro-industrial waste after pulping.

Despite the various potential uses of cupuaçu residues, such as substrates for seedling production, energy generation, gasification technology to replace diesel, and human consumption in the form of flour with excellent nutritional value [6–8], these residues have not received the same level of attention from the scientific community as açaí (*Euterpe oleracea* Mart.) seeds and coconut (*Cocos nucifera* L.) shells in the Amazon biome [9–11].

The use of plant residues as substrates for seedling production requires prior physical and chemical evaluation of the material [12,13], as the quantity and availability of mineral nutrients influence seedling growth. Moreover, identifying critical levels of elements and species requirements can impose limitations on methods used to diagnose nutritional deficiencies [14,15].

Thus, the physical characteristics of materials are essential for evaluating their volumetric density, porosity, and water retention capacity. Based on these properties, it is possible to indicate the quality and suggest the use and limitations of substrates [16]. In this context, we evaluated the potential of cupuaçu fruit shells and seed waste as a substrate for use in seedling production.

2. Materials and Methods

Cupuaçu fruit shells and seeds were collected from a property in the Vila Mensalista community, located in the municipality of Belterra, Pará, in the Brazilian Amazon, between January and April 2023. The plant material was weighed and dried in an oven with forced air circulation for three days at a temperature of 75 °C for the shells and 105 °C for the seeds, due to their resistance to drying, until their weight stabilized. After drying, the shells and seeds were ground in a knife mill, and the seeds were passed through a 2-mm sieve for laboratory analysis.

For chemical characterization, soil samples of Umbric Ferralsols (Latossolo Amarelo com A húmico) under forest vegetation and cupuaçu residues were sent to the Soil Laboratory of the Federal University of Viçosa (UFV). A complete fertility analysis was performed on the soil samples. Soil samples were analyzed for pH (1:2.5 soil:water); organic matter (OM), determined by organic carbon analysis by wet digestion with potassium dichromate and multiplication by the factor 1.724; potassium (K^+), phosphorus (P), sodium (Na^+), copper (Cu^{2+}), iron (Fe^{2+}), manganese (Mn^{2+}), aluminum (Al^{+3}), and zinc (Zn^{2+}) extracted by Mehlich^{-1} (H_2 SO_4 0.0125 mol L^{-1} and HCl 0.05 mol L^{-1}), with K^+ determined using a flame photometer, P by colorimetry, and cationic micronutrients using an atomic absorption spectrophotometer; calcium (Ca^{2+}) and magnesium (Mg^{2+}) were extracted with 1 mol L^{-1} KCl and determined using atomic absorption spectrophotometry; potential acidity (H + Al) was extracted with 0.5 mol L^{-1} calcium acetate and quantified by titration; sulfur (S) was extracted using barium chloride, with subsequent reading using a spectrophotometer; and boron (B) was determined by hot water [17]. At last, the sum of bases (SB = K^+ + Ca^{2+} + Mg^{2+}), effective cation exchange capacity (ECEC = SB + Al^{+3}), total cation exchange capacity (TCEC = SB + H + Al), base saturation (BS = SB/TCEC × 100), and aluminum saturation (AS = (Al^{+3}/ECEC × 100) were calculated.

The total levels of macro- and micronutrients in the cupuaçu residue samples were determined. Elemental analysis of carbon (C), nitrogen (N), hydrogen (H), and sulfur (S) was performed at the Laboratory of Wood Technology and Bioproducts of the Federal University of Western Pará (UFOPA) using a Vario Macro Cube elemental analyzer with 99.9% accuracy. Ash content was analyzed according to a methodology adapted from the Manual of Physical–Chemical Methods for Food Analysis [18].

The physical analysis of the soil and cupuaçu residues was performed under nursery conditions. Soil, fruit shell, and seed residues were placed in 400-mL plastic containers in proportions (soil:residues) of 9:1, 3:1, 2:1, 1:1, and 0:1 (Table 1). The control treatment consisted of soil without the addition of any cupuaçu residue. The experimental design was completely randomized (CRD) in a factorial scheme (6 × 3), with three replications. The containers with the soil and residues were kept in a greenhouse covered with transparent plastic for 30 days, during which they were maintained at field capacity by adding distilled water daily.

Table 1. Proportion of soil (So) and residues of fruit shells (FS) and seeds (Sd) of cupuaçu in the evaluated treatments.

Treatment	Proportion (Soil:Residues)	Soil (g)	Waste of Cupuaçu (g)	
			Fruit Shells (FS)	Seed (Sd)
So + FS	9:1	180	20	0
So + FS	3:1	150	50	0
So + FS	2:1	130	70	0
So + FS	1:1	100	100	0
FS	0:1	0	100	0
So + Sd	9:1	180	0	20
So + Sd	3:1	150	0	50
So + Sd	2:1	130	0	70
So + Sd	1:1	100	0	100
Sd	0:1	0	0	100
So + FS + Sd	18:1:1	180	10	10
So + FS + Sd	6:1:1	150	25	25
So + FS + Sd	4:1:1	130	35	35
So + FS + Sd	2:1:1	100	50	50
FS + Sd	0:1:1	0	100	100
So (control)	1:0:0	200	0	0

After 30 days, the physical attributes of the total porosity (TP) and substrate density (SD) were evaluated using the methodology recommended by the soil analysis method [17]. The treatment samples were collected using volumetric rings and placed in air forced circulation greenhouse at 105 °C for 24 and 48 h. After these periods, the samples were placed in a desiccator and weighed on an analytical balance. The data obtained were subjected to analysis of variance and Tukey's test ($p \leq 0.05$) to compare the means of the treatments and Dunnett's test to compare the control and the treatments using R software version 4.1.2 [19].

3. Results and Discussion

The soil had high acidity (pH 4.7), and the fruit and seed shell residues had medium acidity, with pH values of 5.3 and 5.2, respectively (Table 2). Acidic soils are common in Amazonian ecosystems [20], and soils can be considered good for agricultural crops when their pH is between 5.5 and 6.0 [21]. The pH of organic substrates should be in the range of 5.2–5.5 [20]. The residues evaluated in this study were within the recommended pH levels for the use of organic substrates.

Table 2. Chemical attributes of soil (So) and residues of cupuaçu shells (FS) and seeds (Se) and optimal substrate levels for seedling production.

Properties	Unit	FS	Se	So	Optimal Levels [21,22]
pH H_2O		5.23	5.20	4.67	5.2–6.3
C/N		28.26	26.39	14.27	20–40
P	$g.kg^{-1}$	0.63	3.26	0.01	0.006–0.01
S		0.86	1.59	N.A.	-
K		6.79	9.07	0.07	>7
Ca		0.45	1.04	0.23	0.15–0.25
Mg		0.68	3.83	0.06	>0.2
N		2.90	23.30	4.45	0.0–0.2
Al^{3+}	$cmol_c.dm^{-3}$	N.A.	N.A	1.79	-
H + Al		N.A.	N.A	15.00	-
O.C.	%	8.16	61.49	6.44	-
O.M.		4.01	2.30	11.07	>80
Ashes		6.60	10.00	N.A.	-
Cu	$mg.kg^{-1}$	2.91	16.91	N.A.	0.001–0.5
Fe		N.A.	3.35	N.A.	>70
Zn		6.77	43.14	N.A.	0.3–3.0
Mn		7.36	8.54	N.A.	0.3–3.0
B		2.78	11.23	N.A.	0.005–0.5

O.C.: organic carbon; O.M.: organic matter; N.A. = not available.

The soil sample had high exchangeable acidity, with an exchangeable aluminum (Al^{3+}) content of 1.79 $cmol_c.dm^{-3}$ and low base sum (Table 2). High exchangeable acidity can limit root growth, affect the availability of other nutrients, and impair the mineralization of organic matter [21], thereby impairing plant growth in nurseries. The potential acidity (H + Al) of the soil sample was 15 $cmol_c.dm^{-3}$, which is classified as very high, which is considered beneficial for soil fertility purposes. However, because the base saturation (BS = 7.7%) was remarkably low and aluminum saturation was high (AS = 58.7%) (Table 2), the soil sample was characterized as having low fertility. Considering the evaluated acidities (active, exchangeable, and potential acidity), it is evident that the residues present better physicochemical conditions for the production of seedlings, mainly in the development of the root system.

The chemical analysis of the plant material showed that cupuaçu seed residues had higher concentrations of macro- and micronutrients among the elements analyzed compared to the fruit shell residues (Table 2). The macronutrient concentrations (P, K, Ca, and Mg) in the seeds were 5.0, 1.3, 2.3, and 5.6 times higher, respectively, than those in the shell residues. Compared with the soil sample, the mineral richness of the shell residues in terms of macronutrients was 600, 162, 1283, and 12.35, respectively. Plants require the presence of calcium, magnesium, phosphorus, potassium, and other elements in considerable quantities, as they perform vital functions throughout their life cycle [23]. The cupuaçu shell and seed residues contained high concentrations of nutrients, indicating their potential for use in agriculture as organic substrates that can provide essential elements to plants.

The greatest differences in carbon, nitrogen, and phosphorus concentrations between cupuaçu seed and shell residues were observed, with levels 7.5, 8.0, and 5.5 times higher, respectively, in seeds than in fruit shell residues. The P concentration in the shell was 0.63 $g.kg^{-1}$ (Table 2), which is above the recommended level for substrates for seedling production, which is typically between 0.006 and 0.01 $g.kg^{-1}$ [22]. The P concentration of the Brazilian nutshell residue was 18.3 $g.kg^{-1}$ [24], which was much higher than that found in our research.

The highest P content was observed in seed residues (3.26 $g.kg^{-1}$), a value well above that of soil (0.012 $g.kg^{-1}$) (Table 2). These results were higher than the optimal levels (0.006–0.01 $g.kg^{-1}$) recommended by Lopes et al. [22]. The presence of adequate concentrations of phosphorus is essential in the initial stages of plant growth, since its deficiency results in stunted growth in the juvenile phase of the plant [25]. Phosphorus (P) is a key element in photosynthesis, and the availability of this element in cupuaçu residue can favor better seedling growth, both in terms of speed (reduction in time) and quality.

The K levels found in the residues of cupuaçu seeds and fruit shells and in soil were 9.07 $g.kg^{-1}$, 6.79 $g.kg^{-1}$, and 0.07 $g.kg^{-1}$, respectively (Table 2), and were considered adequate only in seed residues, according to the recommendations of Ribeiro et al. [21] in their studies with soils in the state of Minas Gerais, Brazil. Potassium plays several vital roles in plants and is mainly important for activating various enzyme systems, many of which are involved in the processes of photosynthesis and respiration [26].

The Mg content in the residues was 0.68 $g.kg^{-1}$ for the shells and 3.83 $g.kg^{-1}$ for the seeds (Table 2), which was above the optimum level recommended by Lopes et al. [22]. The soil had a Mg content (0.06 $g.kg^{-1}$) below the recommended optimum level. These results demonstrate the mineral richness and nutritional potential of cupuaçu residues as a substrate for seedling production when compared to the soil usually used for this purpose.

The greater concentration of essential elements (calcium, magnesium, and potassium) in the residues, when compared to the soil, demonstrates that the use of cupuaçu residues can reduce the use of mineral fertilizers, shorten production time, and obtain more vigorous seedlings.

The soil presented a higher organic matter (OM) content compared to the cupuaçu shell and seed residues, with values of 11.07%, 4.01%, and 2.3%, respectively (Table 2). The ideal OM content in substrates for plant seedlings may vary depending on the species and cultivation conditions. However, the substrate for seedlings should have an OM content >80% to promote healthy plant development [22]. In addition to contributing to the supply of nutrients, OM has a significant influence on the physical properties of growing media.

Elemental analysis indicated that fruit and seed shell residues have C/N ratios of 26 and 28, respectively. These values are in accordance with the ideal standards for substrates used in plant cultivation (Table 2), which range from 20 to 40 [22,24]. The results obtained for the residues contrast those of the soil sample with a C/N ratio of 14.27. The C/N ratio of cupuaçu residues shows that, during its decomposition, mineralization will predominate, favoring the release of essential elements for plant seedlings. This fact corroborates the tendency of cupuaçu residues to reduce the use of mineral fertilizers and produce healthier seedlings in a shorter time.

The cupuaçu seed residues had an ash content of 10%, which was higher than that of the shell residues (6.6%), indicating that the seeds had higher levels of inorganic and organic material than the shells. The ash contents obtained for the seeds are compatible with the greater mineral richness observed in these residues (Table 2), being higher than those observed in cupuaçu seed powder, which presented an ash content of 2.12% [27], and in the nibs of cupuaçu almond (2.18%) when evaluating the physical and chemical changes of the seeds during the fermentation process [28]. The ash content of the cupuaçu shell observed in this study was higher than that found in shell residues (1.99%) of this fruit, which were evaluated for flour production [29].

Cupuaçu shells and seed residues are rich in nutrients (mineral and organic) and moderately acidic compared to soil. Therefore, they can be an alternative substrate for seedling production, replacing the conventional substrate (anthropogenic horizon A), which is generally used in this process. The use of these residues for seedling production, in addition to the high levels of essential nutrients for plants, will play a relevant environmental and economic role in agriculture. In particular, agroecological and organic production systems have gained prominence as a sustainable alternative to conventional cultivation methods, as they are based on principles and dynamics that seek to reduce dependence on chemical fertilizers [30,31].

In the evaluation of total porosity (TP) and substrates density (SD), a significant difference was found between the treatments ($p \leq 0.01$). All proportions of the treatments containing cupuaçu residues had lower SD and higher TP values than the control treatment (Table 3). These results demonstrate that cupuaçu residues improve the physical properties of soil substrates.

Table 3. Substrate density (SD) and total porosity (TP) of the soil substrate (So) and the residues of shells (FS) and seeds (Sd) of cupuaçu fruits used in different proportions.

Treatments	Proportion (Soil:Residues)	SD (kg.dm^{-3})	TP (m^3.m^{-3})
So + FS	9:1	0.57 **	0.78 **
So + FS	3:1	0.44 **	0.83 **
So + FS	2:1	0.43 **	0.83 **
So + FS	1:1	0.38 **	0.85 **
FS	0:1	0.26 **	0.90 **
So + Sd	9:1	0.49 **	0.81 **
So + Sd	3:1	0.41 **	0.84 **
So + Sd	2:1	0.37 **	0.85 **
So + Sd	1:1	0.32 **	0.87 **
Sd	0:1	0.22 **	0.91 **
So + FS + Sd	18:1:1	0.51 **	0.80 **
So + FS + Sd	6:1:1	0.48 **	0.81 **
So + FS + Sd	4:1:1	0.48 **	0.81 **
So + FS + Sd	2:1:1	0.44 **	0.83 **
FS + Sd	0:1:1	0.36 **	0.86 **
So (control)	1:0:0	0.73 **	0.72 **
CV (%)		7.54	1.48

** Significant by Dunnett's test ($p < 0.01$). CV: coefficient of variation.

The substrates density (SD) of the residues ranged from 0.26 to 0.57 kg.dm^{-3} for the shells, from 0.22 to 0.49 kg.dm^{-3} for the seeds, and from 0.36 to 0.51 kg.dm^{-3} for the shells + seeds combination, values much lower than those observed for the soil (0.73 kg.dm^{-3}) (Table 3). The results indicate that cupuaçu residues improve the density of the substrates, which favors gas exchange, rooting, infiltration, and the availability of water and nutrients for plants. The optimum level of density for plant cultivation substrates can range from 0.45 to 0.55 kg cm^{-3} [22], and the values observed for cupuaçu residues in our research are within the recommended levels, indicating that they can be used as a plant substrate, regarding this physical attribute.

The low values found for substrate density in the proportion of 100% of the residues of shells (0.26 kg.dm^{-3}), seeds (0.22 kg.dm^{-3}), and for the combination of shells + seeds (0.36 kg.dm^{-3}) (Table 3) indicate that these residues present optimal levels for this variable, better than those considered by Lopes et al. [22]. The substrate composed of Brazil nut shell + acerola seed has a low density (0.29 kg.dm^{-3}), being recommended for the production of single açaí (*Euterpe precatoria* Mart.) seedlings, because it promotes greater growth and dry biomass, resulting in better quality seedlings [32].

The total porosity (TP) of the residues ranged from 0.78 to 0.90 m^3.m^{-3} for the shell, from 0.81 to 0.91 m^3.m^{-3} for the seeds, and from 0.80 to 0.86 m^3.m^{-3} for the shell + seed combination (Table 3), values within the optimum level recommended for substrates used in seedling production, which is above 0.85 m^3.m^{-3} [22]. TP is closely linked to the space available for plant development, which varies according to how the particles are arranged [33].

The TP values found in the proportion of 10% of the residues were 0.78 m^3.m^{-3} for the shells, 0.81 m^3.m^{-3} for seeds, and 0.80 m^3.m^{-3} for shell + seeds (Table 3), which are lower than the optimum level recommended by Lopes et al. [22]. However, the use of this proportion of residues can still be recommended, because even below the level considered optimum, they present higher TP than agricultural soils, which vary between 20% and 40%.

Substrates of processed coconut fiber (coconut powder) and composted pine bark, used alone and in combination, aiming to promote the growth of umbu seedlings (*Spondias tuberosa* Arruda), presented a porosity of 83.5% (0.835 m^3.m^{-3}) in the proportion of 75% coconut powder + 25% composted pine bark, which is below the recommended level; however, its use was still indicated, as it provides the formation of seedlings with greater vegetative quality [34].

The higher the proportion of cupuaçu fruit shells, seeds, and their mixtures in the substrate, the lower the density value and the higher the TP value (Table 3). This can be explained by the fact that the porosity is inversely proportional to the density of the substrate [35]. These results show the potential of cupuaçu residues as substrates for plants, which can replace the use of soil due to their greater nutritional richness (chemical properties) and desirable physical characteristics (SD and TP), in addition to bringing environmental benefits by reducing the inappropriate disposal of these residues into the environment and enabling economic gains for the producer by reducing the costs of purchasing substrates.

These aspects are in line with what is based on agroecological and organic production, which encourages the use of plant residues to produce organic compounds to improve soil fertility, promote biodiversity enrichment, provide suitable habitats for beneficial organisms, and seek to maximize resource efficiency and minimize waste within agricultural systems by transforming plant residues into fertilizer, which can contribute to the circular economy. In addition, the presence of residues can help create an environment less conducive to pests [30,36,37].

The interaction of factors (residue versus concentration) demonstrated that the mixture of shell + seed residues, in the proportion of 100%, resulted in the best results for SD and TP (Tables 4 and 5), with a lower density value (0.22 kg.dm^{-3}) and greater total porosity (0.91 m^3.m^{-3}). Although the highest concentrations of cupuaçu residues provided the best physical attributes, treatments with lower proportions of cupuaçu residues also increased in SD and TP compared to 100% soil. The results show that small amounts of cupuaçu residues are sufficient to promote improvements in the density and porosity of the substrate.

Table 4. Total density of soil (So) and residues of shells (FS) and seeds (Sd) of cupuaçu fruits.

Residues	Total Density (kg.dm^{-3})				
	Proportion (Soil:Residues)				
	9:1	3:1	2:1	1:1	0:1
FS	0.57 (±0.02) bB	0.44 (±0.04) bC	0.43 (±0.04) bC	0.38 (±0.04) bcC	0.26 (±0.04) bcD
Sd	0.51 (±0.01) bB	0.48 (±0.05) bBC	0.48 (±0.03) bC	0.44 (±0.02) bCD	0.36 (±0.02) bD
FS + Sd	0.49 (±0.01) bB	0.41 (±0.02) bB	0.37 (±0.03) bB	0.32 (±0.02) cBC	0.22 (±0.01) cC
So (control)	0.74 (±0.07) aA	0.74 (±0.07) aA	0.74 (±0.07) aA	0.74 (±0.07) aA	0.74 (±0.07) aA
CV (%) = 9.11					

Means followed by the same lowercase letters in the columns and the same uppercase letters in the rows do not differ from each other by Tukey's test ($p < 0.05$). Values in parentheses represent the standard deviation. CV = coefficient of variation.

Table 5. Total porosity of soil (So) and residues of shells (FS) and seeds (Sd) of cupuaçu fruits.

Residues	Total Porosity (kg.dm^{-3})				
	Proportion (Soil:Residues)				
	9:1	3:1	2:1	1:1	0:1
FS	0.78 (±0.01) aC	0.83 (±0.02) aB	0.83 (±0.02) aB	0.85 (±0.01) abB	0.90 (±0.01) abA
Sd	0.80 (±0.01) aB	0.82 (±0.02) aB	0.82 (±0.01) aB	0.83 (±0.01) bAB	0.86 (±0.01) bA
FS + Sd	0.81 (±0.00) aC	0.84 (±0.01) aBC	0.85 (±0.01) aB	0.87 (±0.01) aAB	0.91 (±0.00) aA
So (control)	0.72 (±0.03) bA	0.72 (±0.03) bA	0.72 (±0.03) bA	0.72 (±0.03) cA	0.72 (±0.03) cA
CV (%) = 3.21					

Means followed by the same lowercase letters in the columns and the same uppercase letters in the rows do not differ from each other by Tukey's test ($p < 0.05$). Values in parentheses represent the standard deviation. CV = coefficient of variation.

The characteristics of the evaluated cupuaçu residues are promising for use as substrates in seedling production. The suitability of recycled residues as raw materials for substrate production is assessed based on their nutrient content, water retention capacity, pH, salinity, porosity, and stability. Therefore, understanding the specific attributes of these waste materials is essential to explore their potential applications and broader implications in the production of recycled waste substrates (RWS) [38].

4. Conclusions

The residues from cupuaçu fruits demonstrated desirable chemical and physical characteristics for their use as substrates in seedling production, as they are rich in macro- and micronutrients essential for plant growth and development, in addition to having low density and high porosity, which make them favorable for rooting and plant support.

The proportion of 100% residues (shell + seed) achieved the highest density and porosity. However, regardless of the proportion used, cupuaçu residues promoted improvements in these characteristics, making their use viable in small quantities or in situations of low availability of residues.

The results obtained in this study are very promising and of great importance, as they indicate the use of cupuaçu shells and seeds as substrate for seedling production, in addition to showing a possible destination for the large amount of residue generated during fruit processing, thus avoiding improper disposal and accumulation in the environment. They also show that cupuaçu fruit residues can be a valuable tool for agroecological and organic agriculture, as they contribute to the sustainability of agricultural systems.

Future research can investigate the costs and economic gains that the use of residues from this species can provide when incorporated into forest or agricultural seedling production programs. There is a need to further research the release of nutrients from cupuaçu residues to plants, as well as to evaluate the growth and development of seedlings using these residues as substrates.

Author Contributions: Conceptualization, I.M.R.d.S.F., C.A.A.d.S., T.A.V. and D.C.L.; methodology, I.M.R.d.S.F., C.A.A.d.S., G.d.S.P., A.J.C.A., W.L.d.S., H.d.S.A., T.A.V. and D.C.L.; software, I.M.R.d.S.F., C.A.A.d.S., A.J.C.A., T.A.V. and D.C.L.; validation, I.M.R.d.S.F., C.A.A.d.S., T.A.V. and D.C.L.; formal analysis, I.M.R.d.S.F., C.A.A.d.S., G.d.S.P., A.J.C.A., W.L.d.S., H.d.S.A., T.A.V. and D.C.L.; investigation, I.M.R.d.S.F. and G.d.S.P.; resources, I.M.R.d.S.F., C.A.A.d.S., G.d.S.P., A.J.C.A., W.L.d.S., H.d.S.A., T.A.V. and D.C.L.; data curation, H.d.S.A., T.A.V. and D.C.L.; writing—original draft preparation, I.M.R.d.S.F., C.A.A.d.S., T.A.V. and D.C.L.; writing—review and editing, I.M.R.d.S.F., C.A.A.d.S., G.d.S.P., A.J.C.A., W.L.d.S., H.d.S.A., T.A.V. and D.C.L.; visualization, I.M.R.d.S.F., C.A.A.d.S., G.d.S.P., A.J.C.A., W.L.d.S., H.d.S.A., T.A.V. and D.C.L.; supervision, C.A.A.d.S., W.L.d.S., H.d.S.A., T.A.V. and D.C.L.; project administration, T.A.V. and D.C.L.; funding acquisition, H.d.S.A., T.A.V. and D.C.L. All authors have read and agreed to the published version of the manuscript.

Funding: This research was funded by the Fundação Coordenação de Aperfeiçoamento de Pessoal de Nível Superior (Capes), Programa: 13179-PDPG-Pós-Doutorado Estratégico, grant number 88881.692754/2022-01, and the Federal University of Western Pará (Pibic/Ufopa 2022), and the APC was funded by Capes, Programa: 13179-PDPG-Pós-Doutorado Estratégico, grant number 88881.692754/2022-01.

Institutional Review Board Statement: Not applicable.

Data Availability Statement: The data presented in this study are available on request from the corresponding author.

Conflicts of Interest: The authors declare no conflicts of interest.

Abbreviations

The following abbreviations are used in this manuscript:

CEC	Total cation exchange capacity (CEC)
So	Soil
Se	Seed
FS	Fruit shells
OM	Organic matter
RWS	Recycled waste substrates

References

1. Bezerra, F.C.; Ferreira, F.V.M.; Silva, T.C. Produção de mudas de berinjela em substratos à base de resíduos orgânicos e irrigadas com água ou solução nutritiva. *Hortic. Bras.* **2009**, *27*, S1348–S1352.
2. Pereira, A.L.F.; Abreu, V.K.G.; Rodrigues, S. Cupuassu—*Theobroma grandiflorum*. *Exot. Fruits* **2018**, 159–162. [CrossRef]
3. IBGE. Censo Agropecuário: Valor da Produção, Quantidade Produzida, Área Colhida, Maior Produtor, Estabelecimentos, Número de pés. Produção de Cupuaçu. 2017. Available online: https://www.ibge.gov.br/explica/producao-agropecuaria/cupuacu/br (accessed on 31 January 2025).
4. Lima, M.C.F. Caracterização de Substâncias Fenólicas e Alcaloides dos Resíduos do Cupuaçu (*Theobroma grandiflorum* (Willd. ex Spreng.) Schum). Master's Thesis, Universidade Federal do Amazonas, Manaus, Brazil, 18 December 2013. Available online: https://tede.ufam.edu.br/bitstream/tede/4405/2/Disserta%C3%A7%C3%A3o%20-%20Milena%20Freitas%20de%20Lima.pdf (accessed on 31 March 2024).
5. Gondim, T.M.S.; Thomazini, M.J.; Cavalcante, M.J.B.; Souza, J.M.L. *Aspectos da Produção de Cupuaçu*; Embrapa: Rio Branco, Brazil, 2001.
6. Silva, L.S.; Pierre, F.C. Aplicabilidade do cupuaçu (*Theobroma grandiflorum* (Willd. ex Spreng.) Schum.) em produtos e subprodutos processados. *Tekhne e Logos* **2021**, *12*, 19–33.
7. Mendes, R.F.; Araújo, J.C.; Andrade, R.C.N.; Araújo, J.M.; Guilherme, J.P.M. Crescimento de mudas de maracujazeiro em substrato alternativo com fertilizante de liberação controlada. *Rev. Bras. Agropecuária Sustentável* **2019**, *9*, 34–40. [CrossRef]

8. Silva, A.C.L.; Santos, E.C.S. Estimativa de Geração de Energia Elétrica Utilizando o Carvão da Casca do Fruto do Cupuaçuzeiro no estado do Amazonas. Available online: https://www.osti.gov/etdeweb/servlets/purl/21379855 (accessed on 31 January 2025).
9. Padilha, N.A.; Brigido, T.S.; Cardozo, C.F.; Fernandes, L.F.; Oliveira, A.S.; Mattos, J.C.P.; Venâncio, N.M. Investigação preliminar das propriedades químicas dos resíduos do cupuaçú (*Theobroma grandiflorum*) através da técnica de espectroscopia de fluorescência de raio-X por energia dispersiva. *Sci. Nat.* **2021**, *3*, 114–123. [CrossRef]
10. Maranho, A.S.; Paiva, A.V. Produção de mudas de *Physocalymma scaberrimum* em substratos compostos por diferentes porcentagens de resíduo orgânico de açaí. *Floresta* **2012**, *42*, 399–408. [CrossRef]
11. Gomes, A.D.V.; Freire, A.L.O. Crescimento e qualidade de mudas de cedro (*Cedrela fissilis* L.) em função do substrato e sombreamento. *Sci. Plena* **2019**, *15*, 11. [CrossRef]
12. Fersi, M.; Louati, I.; Hadrich, B.; Smaoui, Y.; Jerbi, B.; Jedidi, N.; Hassen, A.; Hachicha, R. Positive influences of the application of inoculated lignocellulosic waste compost on the agronomic potential of a sandy loam soil. *Clean Technol. Environ. Policy* **2024**, *27*, 1521–1533. [CrossRef]
13. Xie, Y.; Zhou, L.; Dai, J.; Chen, J.; Yang, X.; Wang, X.; Wang, Z.; Feng, L. Effects of the C/N ratio on the microbial community and lignocellulose degradation, during branch waste composting. *Bioprocess Biosyst. Eng.* **2022**, *45*, 1163–1174. [CrossRef] [PubMed]
14. Mula, H.C.A. Avaliação de Diferentes Substratos na Produção de Mudas de *Sebastiania commersoniana* (Baillon) L. B. Smith & R. J. Downs. Master's Thesis, Universidade Federal do Paraná, Curitiba, Brazil, 2011.
15. Gomes, J.M.; Paiva, H.N. *Viveiros florestais (Propagação sexuada)*; Editora UFV: Viçosa, Brazil, 2011.
16. Kämpf, A.N. *Produção Comercial de Plantas Ornamentais*; Agropecuária: Guaíba, Brazil, 2000.
17. Teixeira, P.C.; Donagemma, G.K.; Fontana, A.; Teixeira, W.G. *(Orgs.) Manual de Métodos de Análise de Solos*; Embrapa Solos: Rio de Janeiro, Brazil, 2017.
18. Instituto Adolfo Lutz. *Normas Analíticas: Métodos Químicos e Físicos de Composição de Alimentos*, 4th ed.; IAL: São Paulo, Brazil, 2005.
19. R Core Team. *R: A Language and Environment for Statistical Computing*; R Foundation for Statistical Computing: Vienna, Austria, 2021. Available online: https://www.R-project.org/ (accessed on 14 October 2024).
20. Souza, J.J.L.L.; Fontes, M.P.F.; Gilkes, R.; Costa, L.M.; Oliveira, T.S. Geochemical signature of Amazon tropical rainforest soils. *Rev. Bras. De Ciência Do Solo* **2018**, *42*, e0170192. [CrossRef]
21. Ribeiro, A.C.; Guimarães, P.T.G.; Alvarez, V.V.H. *Recomendações Para o Uso de Corretivos e Fertilizantes em Minas Gerais*, 5th ed.; CFSEMG: Viçosa, Brazil, 1999.
22. Lopes, J.L.W.; Guerrini, I.A.; Saad, J.C.C.; Silva, M.R. Atributos químicos e físicos de dois substratos para produção de mudas de eucalipto. *Cerne* **2008**, *14*, 358–367.
23. Prado, R.M. *Nutrição de Plantas*, 2nd ed.; Editora Unesp: São Paulo, Brazil, 2021.
24. Gomes, I.B.; Ferreira, M.S.; Queiroz, G.M.; Santos, L.A.C. Uso de resíduos de castanha-do-Brasil para o cultivo de cactos globulares na região Amazônica. *Rev. Verde De Agroecol. E Desenvolv. Sustentável* **2022**, *17*, 221–226. [CrossRef]
25. Taiz, L.; Zeiger, E. *Plant Physiology*, 5th ed.; Sinauer Associates: Sunderland, Brazil, 2010; 623p.
26. Ernani, P.R.; Almeida, J.A.; Santos, F.C.; Potássio. *Fertilidade do Solo*; Novais, R.F., Alvarez, V.V.U., Barros, N.F., Fontes, R.L.F., Cantarutti, R.B., Neves, J.C.L., Eds.; UFV: Viçosa, Brazil, 2007.
27. Marchese, D.A. Estudo do Processo de Obtenção do pó de Cupuaçu (*Theobroma grandiflorum* Schum) Alcalinizado. Master's Thesis, Universidade Estadual de Campinas, Campinas, Brazil, 2002. Available online: https://repositorio.unicamp.br/Busca/Download?codigoArquivo=461941&tipoMidia=0 (accessed on 31 January 2025).
28. Aragão, C.G. Mudanças Físicas e Químicas da Semente de Cupuaçu (*Theobroma grandiflorum* Schum) Durante o Processo Fermentativo. Master's Thesis, Instituto Nacional de Pesquisa da Amazônia Fundação Universidade do Amazonas, Manaus, Brazil, 1992.
29. Silva, G.M.; Moreira, R.J.A.; Cândido, J.E.; Silva, I.R.; Morais, B.A.; Gouveia, M.J.; Leite, T.C.C.; Sena, A.R. Obtenção de farinhas das cascas de cupuaçu e pitomba: Análise bromatológica e fitoquímica. In *Extensão Rural: Práticas & Pesquisas Para o Fortalecimento da Agricultura Familiar*; Oliveira, R.S., Ed.; Editora Científica: Guarujá, Brazil, 2021.
30. Gliessman, S.R.; Friedmann, H.; Howard, P.H. Agroecology and Food Sovereignty. *IDSBulletin* **2019**, *50*, 91–109. [CrossRef]
31. Gliessman, S.R. 'Transforming food systems with Agroecology'. *Agroecol. Sustain. Food Syst.* **2016**, *40*, 187–189. [CrossRef]
32. Araújo, C.S.; Lunz, A.M.P.; Santos, V.B.; Andrade, R.C.N.; Nogueira, S.R.; Santos, S.R. Use of agro-industry residues as substrate for the production of *Euterpe precatoria* seedlings. *Pesqui. Agropecuária Trop.* **2020**, *50*, e58709. [CrossRef]
33. Jorge, M.H.A.; Melo, R.A.C.; Resende, F.V.; Costa, E.; Silva, J.; Guedes, I.M.R. *Informações Técnicas Sobre Substratos Utilizados na Produção de Mudas de Hortaliças*; Embrapa Hortaliças: Brasília, Brazil, 2020. Available online: https://www.infoteca.cnptia.embrapa.br/infoteca/bitstream/doc/1125796/1/DOC-180-18-set-2020.pdf (accessed on 30 October 2024).
34. Neves, O.S.C.; Horlle, J.C.A.; Avrella, E.D.; Paim, L.P.; Fior, C.S. Potencial dos substratos pó-de-coco e casca de pinus compostada na promoção de crescimento de mudas de umbuzeiro. *Rev. Agrária Acadêmica* **2021**, *4*, 5–12. [CrossRef]

35. Santos, P.L.F.; Castilho, R.M.M. Atributos físicos de diferentes substratos para fins de desenvolvimento de plantas. In *Mobilizar o Conhecimento Para Alimentar o Brasil*; Magnoni Júnior, L., Silva Junior, E.C., Tondato, C., Colombo, A.S., Silva, A.P., Tonin, G.A., Branco Júnior, G.A., Magnoni, M.G.M., Figueiredo, W.S., Eds.; Centro Paula Souza: São Paulo, Brazil, 2018.
36. Machado, L.C.P.; Machado Filho, L.C.P. *Dialética da Agroecologia*; Expressão Popular: São Paulo, Brazil, 2014; 360p.
37. Altieri, M. *Agroecologia: Bases Científicas Para Uma Agricultura Sustentável*, 3rd ed.; Expressão Popular: São Paulo, Brazil, 2012; 400p.
38. Kader, S.; Gratchev, I.; Michael, R.N. Recycled waste substrates: A systematic review. *Sci. Total Environ.* **2024**, *953*, 176029. [CrossRef] [PubMed]

agriculture

MDPI

Article

Replacing Nitrogen Fertilizers with Incorporation of Rice Straw and Chinese Milk Vetch Maintained Rice Productivity

Peng Li, Linlin Zhao, Donghui Li, Qiaoli Leng, Mingjian Geng and Qiang Zhu *

Key Laboratory of Arable Land Conservation (Middle and Lower Reaches of Yangtze River), Ministry of Agriculture and Rural Affairs, Microelement Research Center, Huazhong Agricultural University, Wuhan 430070, China; pli@whrsm.ac.cn (P.L.); 19937514663@163.com (L.Z.); lidh@webmail.hzau.edu.cn (D.L.); lengqiaoli@webmail.hzau.edu.cn (Q.L.); mjgeng@mail.hzau.edu.cn (M.G.)

* Correspondence: zhuqiang@mail.hzau.edu.cn

Abstract: The cultivation of Chinese milk vetch (CMV) during the winter fallow season and the return of rice straw are important practices for increasing the soil fertility of paddy fields in southern China. In order to provide data-based evidence for the scientific strategy of nitrogen (N) fertilizer reduction through the incorporation of rice straw and CMV, a three-year field trial was conducted. The treatments included the three N application rates of 0%, 60%, and 100% of the local conventional rate (165 kg ha^{-1}), with the incorporation of CMV alone (MN0, MN60, and MN100) or with both CMV and rice straw (SMN60 and SMN100). The rice grain yield, N uptake, and dynamic changes in inorganic N in the soil and surface water were determined for the period from 2019 to 2021. The results show that both the rice grain yield and plant N uptake of the MN60 and SMN60 treatments were not significantly different from those of the treatment with only conventional N application (N100). Although the SMN100 treatment significantly increased the uptakes of N in the aboveground part in the tillering and shooting stages compared with SMN60, no significant differences were found between the grain yields in 2021. Meanwhile, the SMN60 treatment significantly increased the soil microbial biomass N and NH_4^+-N contents during the maturity stage in 2020 and 2021, respectively, compared with MN60. Furthermore, the SMN100 treatment resulted in higher NO_3^--N concentrations in the surface water at days 3 and 6 after transplantation in 2020 than those under SMN60. In conclusion, the incorporation of CMV and rice straw with an application rate of 60% of conventional N fertilizer is an essential approach to reducing the risk of N loss while maintaining rice grain yields in the Jianghan Plain of China.

Keywords: green manure; rice grain yield; plant nitrogen uptake; soil inorganic nitrogen; microbial biomass nitrogen

1. Introduction

Planting leguminous green manure in summer or winter in rotation with the main crop is a way of making full use of fallow fields. Also, it benefits the growth and development of leguminous green manure itself, as well as the subsequent main crop [1]. Previous studies have shown that leguminous green manure can increase N supply, improve N-use efficiency, and maintain the N balance in a crop rotation system through its biological N fixation, thus increasing the yield of the main crop and reducing the application rate of N chemical fertilizers [2–4]. As a high-quality source of organic fertilizer, Chinese milk vetch (CMV, *Astragalus sinicus* L.) can be planted to reduce soil erosion, optimize the soil's physical structure, and increase the soil's nutrient content to a certain extent [5,6].

When chemical fertilizers are applied to a field, the nutrients are rapidly released. After incorporation, the CMV decomposes at a fast rate during the early stage and at a slow rate during the later stage, so CMV and chemical fertilizers can be applied together to ensure both a high demand for nutrients in the early growth stage of a crop, and the supply of nutrients in the later stage [7]. Studies have shown that the biological N fixation by CMV ranged from 60 to 115 kg ha^{-1} [8]. In southern China, the incorporation of CMV can replace 20–40% of N fertilizer without reducing the rice (*Oryza sativa* L.) yield [3].

It has been pointed out that returning straw to the soil may result in adverse effects [9,10]. When the C/N ratio of straw exceeds the demand of soil microorganisms, the addition of organic carbon will promote microbial reproduction, which causes N competition between microorganisms and crops [11–13]. Therefore, the incorporation of crop residues with an appropriate C/N ratio is essential for soil fertility and crop growth. Recently, the incorporation of green manure and rice straw has become a hot spot in research. The C/N ratio of rice straw (generally 50~70:1) is much higher than that of CMV (10~20:1); thus, the combination of them is helpful for the optimization of the C/N ratio [14]. The incorporation of CMV and rice straw promotes the activities of β-glucosidase and cellulose hydrolase, thus enhancing the decomposition of rice straw and nutrient release [15].

The N released through mineralization after the incorporation of CMV and rice straw into a field can replenish the soil's N pool [16]. Combined with that, an appropriate amount of N fertilizer will meet the N demand of rice for the whole growth period. Previous studies have mostly focused on the nutrient release of CMV and rice straw when they are incorporated into a field alone. There is a lack of data on the N cycle during each growing stage of rice after incorporating both of them. The objective of the present study was to compare the N-supplying capacity of soils after the incorporation of CMV alone with that after the incorporation of CMV and rice straw at both 60% and 100% of the conventional N fertilizer application rate.

2. Materials and Methods

2.1. Overview of the Field Trial

The present study was conducted at Taihu farm, Jianghan Plain, Hubei Province (N 30°22′1″, E 112°2′57″; altitude: 44.6 m), where the local climate was subtropical monsoon, with a monthly average temperature of 17.6 °C and an annual average precipitation of 1069 mm, during the 2019–2021 growing seasons (Figure 1). The paddy soil was developed from alluvial deposits with the following properties (0–20 cm): pH of 7.6, organic matter of 22.4 g kg^{-1}, total N of 2.0 g kg^{-1}, available phosphorus of 10.6 mg kg^{-1}, and available potassium of 156.0 mg kg^{-1}. The local traditional crop rotation pattern was mid-season rice, grown from May to September of every year, followed by a winter fallow.

The field trial was established in the fall of 2018. There were seven treatments, and these included two treatments with N fertilizer application rates of 0 (N0) and 165 kg ha^{-1} (100% of local conventional rate, N100), without the addition of crop residues. With the other three treatments, only CMV was incorporated, and N fertilizer was applied at 0 (MN0), 99 (60% of conventional rate, MN60), and 165 kg ha^{-1} (MN100). The remaining two treatments involved the incorporation of CMV and rice straw, with N fertilizer application rates of 99 (SMN60) and 165 kg ha^{-1} (SMN100). The total N input of each treatment is shown in Table 1. Phosphorus and potassium were applied at 60 kg ha^{-1} of P_2O_5 and 90 kg ha^{-1} of K_2O for all the treatments.

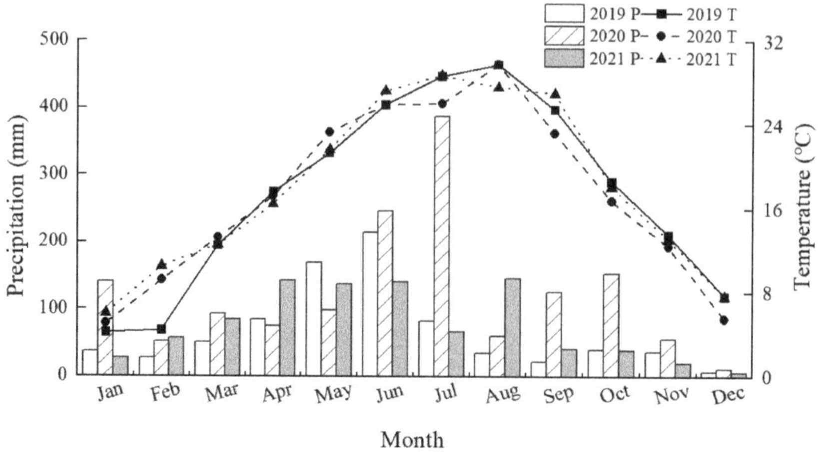

Figure 1. The monthly mean temperature and accumulative precipitation of the experimental site over three years (2019–2021).

Table 1. The nitrogen (N) input (kg ha^{-1}) of different treatments in 2020 and 2021.

Treatment	Urea	Chinese Milk Vetch (M)		Rice Straw (S)		Total	
		2020	2021	2020	2021	2020	2021
N0							
MN0		75.4	41.6			75.4	41.6
MN60	99.0	65.8	47.5			164.8	146.5
SMN60	99.0	62.3	47.4	35.2	43.4	196.5	189.9
N100	165.0					165.0	165.0
MN100	165.0	59.9	38.0			224.9	203.0
SMN100	165.0	57.6	62.2	40.0	50.0	262.6	277.2

The N input of Chinese milk vetch is biological N fixation with a N fixation rate of 66%, which was determined by the authors through the ^{15}N Natural Abundance Method. The dry-weight biomass and N concentrations of Chinese milk vetch and rice straw are shown in Table S1.

Treatments were arranged in a randomized complete block design with three replications and were applied to the same plot every year. Each plot was 4 m wide and 5 m long and was spaced with 30 cm-wide ridges. For the rice straw treatments, all the rice straws were incorporated into the soil after rice grain harvest; otherwise they were removed from the plots. The CMV seeds were sown in the first week of October at a rate of 30 kg ha^{-1}, with biomass incorporated on the spot during the full blooming stage (around 15 April in the following year). After CMV incorporation (with a depth of 10–15 cm), the field was subject to flooding for 4–5 weeks. The rice variety was "Fengliangyou No. 2" and 400 rice hills were transplanted for each plot. No fertilizers were added in the CMV growing season, with urea, calcium superphosphate, and potassium chloride applied in the rice season. On the day of rice transplantation, 70% of the total N fertilizers and all phosphorus and potassium fertilizers were spread on the surface of each plot. The remaining N fertilizers were applied at 13 and 70 days after transplantation (DAT). The pest control and irrigation management followed local practices.

2.2. Plant Sampling and Analysis

Before incorporation into soil, the aboveground fresh weight of CMV in the whole plot was measured; subsamples were randomly taken to measure the dry weight and N concentration. In the rice growing season, the aboveground biomasses of five hills per plot were collected in the tillering, shooting, grain filling, and maturity stages (i.e., approximately 30, 60, 90, and 120 DAT). In the latter two stages, grain and shoot were separated to measure dry weight individually. During the maturity stage, rice grains in the whole plot were harvested and the yield was determined after air-drying. In this study, rice yield data were from 2019 to 2021, and the rest of the data were from 2020 to 2021.

All plant samples were oven-dried at 105 °C for 30 min and then at 65 °C to a constant weight. After drying, they were ground to pass a 0.84 mm sieve and digested by H_2SO_4-H_2O_2. The N concentration was measured using the semi-micro Kjeldahl method [17]. Plant N uptake was calculated by multiplying N concentration by dry weight.

2.3. Soil Sampling and Analysis

Soil samples were taken one day before rice transplantation (18 May 2020 and 20 May 2021) and in the rice tillering, shooting, and maturity stages. Five subsamples were collected at a depth of 0 to 20 cm from each plot. The soil samples were air-dried, ground, and passed through a 2 mm sieve for the measurement of microbial biomass N (MBN), ammonium-N (NH_4^+-N), and nitrate-N (NO_3^--N) contents. The MBN content was measured by the chloroform fumigation–UV spectrophotometer method [18]. Soil NH_4^+-N and NO_3^--N were extracted by 1 mol L^{-1} KCl and determined using an AA3 Continuous Flow Injection Analyzer (SEAL Analytical, Norderstedt, Germany).

2.4. Water Sampling and Analysis

Field surface water samples were taken by syringe before fertilizer application, and at 3, 6, 13, 16, and 28 DAT in 2020 and 2, 8, 13, 15, and 29 DAT in 2021. Five subsamples were randomly collected in each plot and mixed as one sample. These samples were sent to the laboratory in ice boxes, filtered through a 0.45 μm membrane, and stored in a 4 °C refrigerator before analysis. The NH_4^+-N and NO_3^--N concentrations in water samples were determined by an AA3 Continuous Flow Injection Analyzer.

2.5. Statistical Analysis

The data from each year were analyzed separately because of the different weather conditions. One-way analysis of variance was performed through SAS (Version 9.2). Duncan's Multiple Range Test was used to compare the significance of differences at the $p < 0.05$ level.

3. Results

3.1. Rice Grain Yield

The rice grain yields of all the treatments in 2021 were lower than those in 2019 and 2020, which might be due to the higher rainfall accumulation in the grain filling stage of 2021, but a similar trend was shown among the treatments over the three growing seasons (Figure 2). The yield of the MN0 treatment was significantly higher than that of N0 over three years, but significantly lower than that of MN60 in 2019 and 2020. No significant differences were found among the MN60, SMN60, and N100 treatments, indicating that CMV alone or in combination with rice straw incorporated into the field could replace 40% of conventional N fertilizer without reducing rice yield. There were no significant differences in rice yields between the MN60 and MN100 treatments in three years. Additionally, no significant differences were found between the SMN60 and SMN100 treatments except in 2020, indicating that increasing the N application rate from 60% to 100% of conventional rate could not improve rice yield any more when CMV was incorporated with or without rice straw.

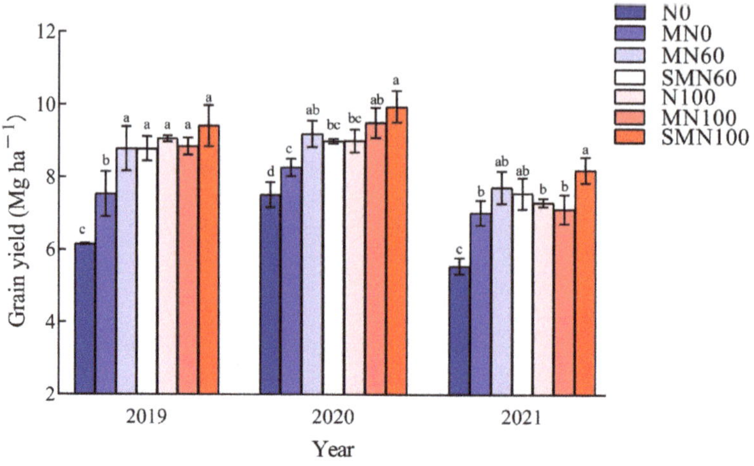

Figure 2. Response of rice grain yield to the incorporation of Chinese milk vetch (M) and rice straw (S) at different nitrogen (N) application rates. Note: Different letters above the bars within the same year indicate significant difference at the $p < 0.05$ level.

3.2. Nitrogen Uptake

The MN60 and SMN60 treatments did not significantly decrease the N concentrations in rice plant compared with the N100 treatment (except rice straw in the grain filling stage in 2020) (Table 2). The N concentrations in the rice straw under the MN100 and SMN100 treatments were significantly higher than those under N100 in the grain filling stage of both growing seasons and the maturity stage in 2021. When CMV was incorporated alone, there were no significant differences in the N concentrations of rice plant between the MN60 and MN100 treatments in rice tillering and shooting stages. However, the N concentration in rice straw under the MN100 treatment was significantly higher than that under MN60 in the grain filling and maturity stages. When CMV and rice straw were incorporated, the N concentrations in both the whole aboveground part in the tillering and shooting stages and the rice straw in the grain filling stage under the SMN100 treatment were significantly higher than those under SMN60 in 2020.

The variation in N uptake by rice was similar to that in N concentration, with the lowest N uptake seen in the N0 treatment in both years (Table 3). Plant N uptakes among the MN60, SMN60, and N100 treatments did not differ significantly during all stages, indicating that N fertilizer reduction by 40% with crop residues incorporation did not significantly affect N accumulation in the rice plant. In the maturity stage, N uptakes in rice grain and straw under MN100 and SMN100 increased by 10.4–24.1% and 6.4–61.1%, respectively, compared with those under N100, indicating that CMV incorporation alone or in combination with rice straw at the conventional N application rate could improve N accumulation. Nonetheless, the absorbed extra N tended to accumulate in rice straw, especially when the grain yield was low (in 2021). There were no significant differences in the grain N uptakes between the MN60 and MN100 treatments in the grain filling and maturity stages. However, the MN100 treatment resulted in a significantly higher N uptake in rice straw compared with MN60, showing that the higher N application rate only promoted N accumulation in rice straw when CMV was incorporated alone. The SMN100 treatment significantly increased N uptake in the aboveground part during the tillering and shooting stages and in the rice grain in the maturity stage compared with SMN60, indicating that more N fertilizers promoted N uptake in the whole rice plant, including grain, when CMV and rice straw were incorporated together.

Table 2. Plant nitrogen (N) concentrations (%) during different rice growing stages in response to the incorporation of Chinese milk vetch (M) and rice straw (S) at different N application rates.

Year	Treatments	Tillering Stage	Shooting Stage	Grain Filling Stage		Maturity Stage	
		Aboveground	Aboveground	Panicle	Straw	Grain	Straw
2020	N0	2.28 d	1.50 c	0.90 b	0.68 d	0.98 b	0.40 d
	MN0	2.53 cd	1.72 bc	0.94 b	0.72 d	1.05 b	0.48 cd
	MN60	3.21 b	1.94 ab	1.09 ab	1.00 c	1.20 a	0.59 bc
	SMN60	2.98 b	1.67 bc	0.99 ab	0.99 c	1.17 a	0.62 bc
	N100	2.63 c	1.95 ab	1.18 a	1.20 b	1.17 a	0.67 ab
	MN100	2.98 b	2.13 a	1.18 a	1.55 a	1.22 a	0.77 a
	SMN100	3.52 a	2.05 a	1.19 a	1.39 a	1.26 a	0.64 ab
2021	N0	2.80 b	1.31 d	0.97 c	0.46 e	0.81 c	0.57 b
	MN0	2.90 b	1.43 cd	1.04 c	0.61 de	1.01 b	0.61 b
	MN60	3.30 ab	1.85 abc	1.04 c	0.91 bcd	1.11 ab	0.62 b
	SMN60	3.04 ab	1.65 bcd	1.13 bc	0.95 bc	1.11 ab	0.68 b
	N100	3.25 ab	1.68 bcd	1.08 c	0.83 cd	1.09 ab	0.69 b
	MN100	3.56 a	2.15 a	1.45 a	1.48 a	1.25 a	1.04 a
	SMN100	3.51 a	1.95 ab	1.35 ab	1.22 ab	1.20 ab	1.07 a

Data in the table are the means of three replications and different lowercase letters in each column indicate significant difference at the $p < 0.05$ level.

Table 3. Plant nitrogen (N) uptake (kg ha^{-1}) during different rice growing stages in response to the incorporation of Chinese milk vetch (M) and rice straw (S) at different N application rates.

Year	Treatments	Tillering Stage	Shooting Stage	Grain Filling Stage		Maturity Stage	
		Aboveground	Aboveground	Panicle	Straw	Grain	Straw
2020	N0	24.5 ± 1.5 e	43.6 ± 2.7 c	29.4 ± 2.9 a	21.2 ± 2.0 c	66.5 ± 5.0 d	21.9 ± 3.2 d
	MN0	37.0 ± 0.8 d	70.7 ± 8.4 b	33.6 ± 1.9 a	24.7 ± 2.1 c	78.3 ± 3.0 c	31.0 ± 4.2 c
	MN60	45.3 ± 1.3 c	78.7 ± 13.1 ab	48.0 ± 6.9 a	45.4 ± 6.0 b	99.0 ± 3.4 b	40.8 ± 0.7 b
	SMN60	50.1 ± 4.5 c	68.2 ± 6.6 b	41.7 ± 7.3 a	43.3 ± 3.7 b	94.3 ± 0.3 b	43.4 ± 2.1 b
	N100	47.2 ± 1.7 c	91.2 ± 2.6 ab	44.4 ± 3.0 a	55.6 ± 3.2 b	94.6 ± 3.1 b	47.0 ± 5.8 b
	MN100	62.5 ± 3.0 b	80.9 ± 11.9 ab	43.6 ± 7.1 a	80.2 ± 3.1 a	104.4 ± 8.1 ab	59.1 ± 0.5 a
	SMN100	70.3 ± 3.7 a	100.8 ± 8.3 a	50.0 ± 2.8 a	71.2 ± 7.3 a	112.3 ± 2.3 a	50.0 ± 2.6 b
2021	N0	27.3 ± 5.5 d	53.2 ± 3.0 d	51.9 ± 1.9 a	24.2 ± 2.3 c	42.2 ± 3.3 d	26.2 ± 1.9 c
	MN0	39.3 ± 5.2 cd	77.0 ± 8.3 cd	65.2 ± 5.7 a	39.7 ± 3.5 bc	66.7 ± 5.0 c	33.0 ± 0.6 bc
	MN60	50.8 ± 3.7 bc	120.5 ± 21.1 ab	59.4 ± 6.5 a	64.3 ± 9.9 b	80.0 ± 1.4 b	42.7 ± 1.4 bc
	SMN60	57.0 ± 3.6 b	105.1 ± 3.9 bc	71.6 ± 7.9 a	75.8 ± 10.2 ab	78.1 ± 4.4 bc	47.7 ± 5.2 b
	N100	58.2 ± 1.4 b	102.3 ± 24.0 bc	71.0 ± 11.5 a	61.1 ± 0.1 bc	74.3 ± 3.1 bc	45.0 ± 4.3 b
	MN100	53.2 ± 6.4 b	123.5 ± 18.7 ab	61.3 ± 4.9 a	103.9 ± 21.6 a	82.4 ± 4.2 ab	64.0 ± 5.5 a
	SMN100	80.4 ± 9.7 a	154.4 ± 9.9 a	73.4 ± 8.4 a	105.4 ± 11.7 a	92.2 ± 3.8 a	72.5 ± 10.0 a

Data in the table are mean ± standard error and different lowercase letters in each column indicate significant difference at the $p < 0.05$ level.

3.3. Soil Nitrogen Content

In 2020, there were no significant differences in soil MBN contents among all the treatments before the maturity stage (Table 4). Compared with the N0 treatment, MN0 significantly increased the MBN by 59.8% in the maturity stage. No significant differences were found in the MBN for the MN60 and N100 treatments, but the SMN60 treatment significantly increased the MBN by 60.2% and 24.0% compared with N100 and MN60. There were no significant differences in MBN for the three treatments with 100% of the conventional N application rate. In 2021, the MBN values under MN0 were significantly higher than those under N0 in both tillering and shooting stages. Throughout the whole growing season, the MBN values in soil with SMN60 or SMN100 were not significantly different from those with N100. These results indicate that the incorporation of CMV and rice straw maintained a high soil N supply capacity by improving MBN content even when using 60% of the conventional N application rate, but when the conventional rate was applied, the extra N fertilizers were not reserved in the soil in the form of MBN.

Table 4. Soil microbial biomass nitrogen content (mg kg^{-1}) during different stages in response to the incorporation of Chinese milk vetch (M) and rice straw (S) at different N application rates.

Year	Treatment	Before Transplantation	Tillering Stage	Shooting Stage	Maturity Stage
2020	N0	19.4 ± 4.1 a	19.3 ± 0.8 a	20.7 ± 4.2 a	12.7 ± 0.1 d
	MN0	18.5 ± 4.5 a	39.9 ± 10.1 a	26.3 ± 1.9 a	20.3 ± 0.7 c
	MN60	33.8 ± 4.5 a	40.9 ± 7.6 a	40.8 ± 2.7 a	27.9 ± 4.4 b
	SMN60	35.3 ± 4.7 a	26.8 ± 3.2 a	43.5 ± 7.8 a	34.6 ± 1.1 a
	N100	22.8 ± 4.1 a	20.9 ± 0.5 a	28.0 ± 1.1 a	21.6 ± 2.4 bc
	MN100	34.8 ± 5.1 a	32.8 ± 2.3 a	33.1 ± 8.0 a	26.5 ± 1.8 bc
	SMN100	28.3 ± 3.9 a	32.4 ± 2.3 a	33.0 ± 4.7 a	20.6 ± 1.3 c
2021	N0	44.7 ± 0.8 a	36.9 ± 11.4 c	41.7 ± 5.2 b	14.0 ± 3.1 c
	MN0	47.4 ± 4.4 a	65.6 ± 11.7 ab	61.2 ± 4.5 a	24.9 ± 5.1 bc
	MN60	40.0 ± 1.7 a	44.6 ± 3.2 bc	35.3 ± 0.7 b	32.2 ± 6.2 ab
	SMN60	46.5 ± 4.9 a	34.8 ± 3.2 c	52.9 ± 7.0 ab	35.3 ± 1.2 ab
	N100	40.0 ± 5.1 a	48.9 ± 3.7 abc	61.0 ± 2.0 a	32.0 ± 7.5 ab
	MN100	45.7 ± 7.6 a	61.5 ± 3.5 ab	65.0 ± 1.9 a	46.5 ± 3.8 a
	SMN100	53.5 ± 10.9 a	68.7 ± 4.3 a	68.8 ± 11.6 a	30.2 ± 1.1 b

Data in the table are mean ± standard error and different lowercase letters in each column indicate significant difference at the $p < 0.05$ level.

There were no significant differences in soil NH_4^+-N contents among all the treatments before rice transplantation and in the tillering stages of the two growing seasons (Table 5). In 2020, the NH_4^+-N content under the MN60 treatment was significantly higher than that under MN0 and N100 in the shooting and maturity stages, but not significantly different from that under SMN60 and MN100. In the maturity stage, the SMN100 treatment significantly increased soil NH_4^+-N content by 93.9% and 50.6% compared with the N100 and MN100 treatments, respectively. No significant differences were found in the NH_4^+-N contents between the two treatments with the incorporation of CMV and rice straw in the shooting stage. However, in the maturity stage, the NH_4^+-N content under SMN100 was significantly higher than that under SMN60. In the maturity stage in 2021, the NH_4^+-N content in soil with SMN60 was not significantly different from that with N100 and SMN100, but was significantly higher than that with MN60 and N0. These results indicate that the SMN60 and SMN100 treatments could maintain higher soil NH_4^+-N contents, especially in the season with a higher rainfall accumulation (in 2020, Figure 1). However, the soil NH_4^+-N content under the SMN100 treatment tended to be surplus.

There were no significant differences in soil NO_3^--N contents among all the treatments throughout the rice growing season in 2020 (Table 6). In the shooting stage in 2021, the NO_3^--N contents under MN60, SMN60, and N100 treatments were not significantly different. Compared with the N100 treatment, SMN100 significantly increased the NO_3^--N content, but not compared to the MN100 treatment. This indicates that the conventional N application rate with the incorporation of CMV and rice straw increased the risk of NO_3^--N loss.

Table 5. Soil ammonium–nitrogen content (mg kg^{-1}) in different stages in response to the incorporation of Chinese milk vetch (M) and rice straw (S) at different N application rates.

Year	Treatment	Before Transplantation	Tillering Stage	Shooting Stage	Maturity Stage
2020	N0	4.4 ± 0.9 a	5.1 ± 0.8 a	4.4 ± 0.9 c	3.8 ± 0.6 d
	MN0	5.1 ± 0.6 a	6.7 ± 1.1 a	7.7 ± 0.5 bc	6.2 ± 0.9 cd
	MN60	3.1 ± 0.5 a	4.7 ± 0.7 a	12.0 ± 1.1 a	10.2 ± 0.6 ab
	SMN60	3.7 ± 0.2 a	5.8 ± 0.3 a	10.3 ± 1.6 ab	8.0 ± 1.1 bc
	N100	3.3 ± 0.1 a	4.6 ± 0.8 a	8.2 ± 1.4 b	6.6 ± 0.9 cd
	MN100	4.3 ± 0.7 a	5.1 ± 0.4 a	10.9 ± 1.5 ab	8.5 ± 1.8 bc
	SMN100	3.5 ± 0.1 a	5.5 ± 0.1 a	10.5 ± 0.4 ab	12.8 ± 0.4 a
2021	N0	3.8 ± 0.6 a	9.2 ± 2.5 a	9.0 ± 0.3 a	4.4 ± 0.7 bc
	MN0	2.8 ± 0.4 a	11.4 ± 1.2 a	10.6 ± 0.4 a	4.9 ± 0.5 abc
	MN60	3.2 ± 0.9 a	8.9 ± 0.7 a	12.9 ± 0.4 a	4.1 ± 0.2 c
	SMN60	2.8 ± 0.5 a	9.2 ± 1.8 a	11.6 ± 1.9 a	6.1 ± 0.4 a
	N100	2.9 ± 0.4 a	9.7 ± 1.8 a	6.9 ± 2.1 a	5.7 ± 0.5 ab
	MN100	2.4 ± 0.9 a	10.8 ± 1.5 a	9.1 ± 1.4 a	5.6 ± 0.0 abc
	SMN100	2.4 ± 0.2 a	13.2 ± 1.1 a	10.3 ± 1.2 a	6.3 ± 0.8 a

Data in the table are mean ± standard error and different lowercase letters in each column indicate significant difference at the $p < 0.05$ level.

Table 6. Soil nitrate–nitrogen content (mg kg^{-1}) during different stages in response to the incorporation of Chinese milk vetch (M) and rice straw (S) at different N application rates.

Year	Treatment	Before Transplantation	Tillering Stage	Shooting Stage	Maturity Stage
2020	N0	4.5 ± 0.4 a	5.7 ± 0.9 a	9.0 ± 2.3 a	2.4 ± 0.2 a
	MN0	4.2 ± 0.7 a	6.6 ± 0.5 a	7.5 ± 1.1 a	2.5 ± 0.1 a
	MN60	4.4 ± 0.6 a	6.5 ± 0.9 a	6.4 ± 1.2 a	2.7 ± 0.1 a
	SMN60	5.0 ± 0.4 a	5.1 ± 0.4 a	11.8 ± 0.3 a	3.1 ± 0.1 a
	N100	4.1 ± 0.7 a	4.3 ± 1.0 a	6.0 ± 0.3 a	3.1 ± 0.1 a
	MN100	4.5 ± 0.5 a	5.6 ± 0.8 a	6.5 ± 1.7 a	3.0 ± 0.5 a
	SMN100	4.0 ± 0.1 a	5.5 ± 1.1 a	6.4 ± 1.4 a	3.0 ± 0.2 a
2021	N0	2.3 ± 0.0 b	3.9 ± 0.2 a	5.9 ± 1.3 c	4.8 ± 1.9 a
	MN0	7.3 ± 0.5 a	3.9 ± 1.8 a	5.9 ± 1.9 c	4.5 ± 0.6 a
	MN60	3.3 ± 0.5 b	5.6 ± 1.6 a	8.1 ± 0.9 bc	6.6 ± 1.8 a
	SMN60	3.8 ± 0.4 b	4.7 ± 0.2 a	11.9 ± 1.6 ab	5.8 ± 1.0 a
	N100	3.5 ± 0.7 b	4.0 ± 0.5 a	8.8 ± 2.4 bc	4.5 ± 0.3 a
	MN100	3.7 ± 0.2 b	5.0 ± 0.7 a	4.0 ± 1.2 c	7.2 ± 1.5 a
	SMN100	7.6 ± 0.5 a	3.7 ± 1.0 a	14.9 ± 1.0 a	5.6 ± 1.2 a

Data in the table are mean ± standard error and different lowercase letters in each column indicate significant difference at the $p < 0.05$ level.

3.4. NH_4^+-N and NO_3^--N in Surface Water

The changes in NH_4^+-N concentrations in surface water were consistent over the two seasons (Figure 3). Compared with N100, the treatments with 60% of conventional N application rate significantly reduced the NH_4^+-N concentrations by 24.4–47.3% and 66.3–85.1% at 2–3 and 15–16 DAT, respectively. The MN100 and SMN100 treatments significantly increased the NH_4^+-N concentrations by 26.3% and 72.4%, respectively, compared with the N100 treatment at 2 DAT in 2021.

In 2020, the NO_3^--N concentrations in surface water were significantly higher under the SMN100 treatment than those under N100 and SMN60 at 3 and 6 DAT. At 16 DAT, the SMN60 treatment significantly decreased the NO_3^--N concentration by 59.2% compared with N100. In 2021, the SMN100 treatment also resulted in a significantly higher NO_3^--N concentration compared with N100 and SMN60 at 8 DAT. However, at 2 and 15 DAT, there were no significant differences in the NO_3^--N concentrations among the treatments with 0% and 100% of the conventional N application rate.

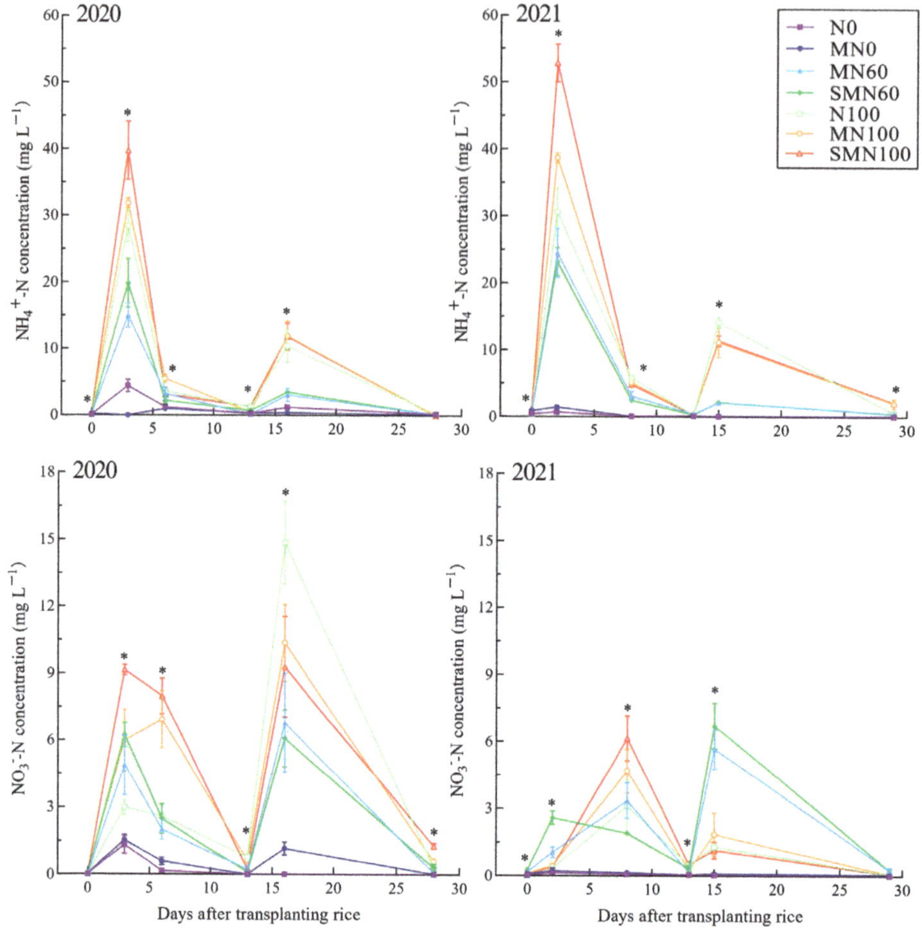

Figure 3. Dynamic changes in the ammonium– (NH_4^+-N) and nitrate–nitrogen (NO_3^--N) concentrations in the surface water after transplanting rice. Note: M, Chinese milk vetch. S, rice straw. * indicates significant difference among the treatments at the $p < 0.05$ level.

4. Discussion

The present study found that the treatments incorporating CMV alone or combined with rice straw and with 60% of N fertilizer did not reduce rice grain yield compared with N100. This is consistent with the previous results, which showed that planting CMV could replace 20–40% of N fertilizer in the paddy fields in southern China [3,19]. The CMV, as a type of leguminous green manure, released a large amount of N after incorporation, which met the N demand of rice in the early growth stage and thus replaced the partial N fertilizer [20,21]. In addition, the C sources provided by the fresh organic residues increased the biomass and activity of soil microorganisms, leading to a positive priming effect and organic N mineralization [22]. When organic materials were incorporated, the treatments with the conventional N application rate did not significantly increase rice grain yield compared with the N-reduced treatments, but significantly increased the N uptake in the rice plant.

The results show that the plant N uptake increased with increasing N application rate. However, the N uptakes under the treatments with a 40% reduction in N fertilizer (MN60 and SMN60) were not significantly different from that under N100. This confirms that the incorporation of CMV alone or in combination with rice straw could replace 40% of N fertilizers. Comparing with CMV alone, the incorporation of both CMV and rice straw with chemical fertilizers could harmonize the supply and demand of N nutrient, promoting rice growth [20,23]. The incorporation of CMV and rice straw at the conventional N application rate promoted the N uptake in rice straw and grain, but did not significantly improve rice grain yield. Thus, such a high amount of N accumulation in rice plants could be considered as surplus uptake. Abe et al. [24] found that excessive N fertilizer application even reduced rice grain yield. Therefore, a proper application rate of N fertilizer is crucial for improving rice grain yield and minimizing the waste of resources and environmental risk.

In the tillering and shooting stages, the incorporation of CMV and rice straw with a 40% reduction in N fertilizer did not significantly reduce the soil inorganic N content compared with the N100 treatment. The input of nutrients from CMV and rice straw increased the soil organic N pool and promoted the mineralization of organic matter, which increased N release and thus the inorganic N concentration [25]. In addition, the incorporation of CMV and rice straw optimized the C/N ratio of the inputs [26], increasing the activity of soil microorganisms and the MBN content [27].

Inorganic N in the surface water can be directly absorbed by rice and represent the primary source of N loss from paddy field [28]. When N was applied at 60% of the conventional rate, the NH_4^+-N concentrations in the surface water with CMV alone and in combination with rice straw were lower than that under N100. This indicates that the 40% N fertilizer replacement reduced the risk of N loss, especially at 2–3 days after N fertilizer application. The changes in NO_3^--N concentration in the surface water generally lagged behind changes in NH_4^+-N [29]. The first peak of NO_3^--N concentration occurred at 2 and 8 DAT in 2020 and 2021, respectively, which might be related to the sampling time. The first sampling date after initial fertilizer application in 2021 was one day earlier than that in 2020, probably resulting in the missed peak of NO_3^--N concentration. Chu et al. [30] and Ai et al. [31] showed that an increased N application rate could enhance nitrification and soil NO_3^--N content. Therefore, the high N input is one of the key factors causing N loss from the paddy field. In the present study, especially in 2020, a 40% N fertilizer replacement reduced the NO_3^--N concentrations in the surface water, thus reducing the potential for N loss through leaching and runoff.

5. Conclusions

Based on the rice grain yield and N uptake in rice plant, both the incorporation of CMV alone and its combination with rice straw could be used to replace 40% of conventional N fertilizer, but the latter maintained a higher level of soil N supply capacity. Moreover, the conventional N application rate with the incorporation of crop residues could result in a luxury N uptake and a high risk of N loss. Consequently, the incorporation of both CMV and rice straw with a N application rate at 60% of the conventional rate would be sufficient for use in the mid-season rice–green manure rotation system.

Supplementary Materials: The following supporting information can be downloaded at: https://www.mdpi.com/article/10.3390/agriculture15060623/s1, Table S1: The dry-weight biomass and nitrogen (N) concentrations of Chinese milk vetch (M) and rice straw (S) for different treatments.

Author Contributions: Conceptualization, M.G.; methodology, Q.Z. and L.Z.; software, P.L.; validation, D.L.; formal analysis, L.Z.; investigation, L.Z., D.L. and Q.L.; resources, M.G.; data curation, P.L. and D.L.; writing—original draft preparation, P.L.; writing–review and editing, Q.Z.; visualization, D.L.; supervision, Q.Z.; project administration, M.G.; funding acquisition, M.G. All authors have read and agreed to the published version of the manuscript.

Funding: The present study was financially supported by the National Key Research and Development Program of China (2021YFD1700200) and the Key Research and Development Program of Hubei Province (2023BBB049).

Institutional Review Board Statement: Not applicable.

Data Availability Statement: The data presented in this study are available on request from the corresponding author. The data are not publicly available due to the requirements of the funding agency.

Conflicts of Interest: The authors declare no conflicts of interest.

Abbreviations

The following abbreviations are used in this manuscript:

CMV	Chinese milk vetch
DAT	Days after transplantation
MBN	Microbial biomass nitrogen

References

1. Zhou, X.; Liao, Y.L.; Lu, Y.H.; Rees, R.M.; Cao, W.D.; Nie, J.; Li, M. Management of rice straw with relay cropping of Chinese milk vetch improved double-rice cropping system production in southern China. *J. Integr. Agr.* **2020**, *19*, 2103–2115. [CrossRef]
2. Gollner, G.; Starz, W.; Friedel, J.K. Crop performance, biological N fixation and pre-crop effect of pea ideotypes in an organic farming system. *Nutr. Cycl. Agroecosys.* **2019**, *115*, 391–405. [CrossRef]
3. Xie, Z.J.; Shah, F.; Tu, S.X.; Xu, C.X.; Cao, W.D. Chinese milk vetch as green manure mitigates nitrous oxide emission from monocropped rice system in south China. *PLoS ONE* **2016**, *11*, e0168134. [CrossRef]
4. Zhang, D.B.; Yao, P.W.; Zhao, N.; Yu, C.W.; Cao, W.D.; Gao, Y.J. Contribution of green manure legumes to nitrogen dynamics in traditional winter wheat cropping system in the Loess Plateau of China. *Eur. J. Agron.* **2016**, *72*, 47–55.
5. Chen, J.R.; Qin, W.J.; Chen, X.F.; Cao, W.D.; Qian, G.M.; Liu, J.; Xu, C.X. Application of Chinese milk vetch affects rice yield and soil productivity in a subtropical double-rice cropping system. *J. Integr. Agr.* **2020**, *19*, 2116–2126. [CrossRef]
6. Yang, Z.P.; Xu, M.G.; Zheng, S.X.; Nie, J.; Gao, J.S.; Liao, Y.L.; Xie, J. Effects of long-term winter planted green manure on physical properties of reddish paddy soil under a double-rice cropping system. *J. Integr. Agr.* **2012**, *11*, 655–664. [CrossRef]
7. Zhu, B.; Yi, L.X.; Hu, Y.G.; Zeng, Z.H.; Lin, C.W.; Tang, H.M.; Yang, G.L.; Xiao, X.P. Nitrogen release from incorporated 15N-labelled Chinese milk vetch (*Astragalus sinicus* L.) residue and its dynamics in a double rice cropping system. *Plant Soil* **2014**, *374*, 331–344. [CrossRef]
8. Papastylianou, I.; Danso, S.K.A. Nitrogen fixation and transfer in vetch and vetch-oats mixtures. *Soil Biol. Biochem.* **1991**, *23*, 447–452. [CrossRef]
9. Eagle, A.J.; Bird, J.A.; Horwath, W.R.; Linquist, B.A.; Brouder, S.M.; Hill, J.E.; van Kessel, C. Rice yield and nitrogen utilization efficiency under alternative straw management practices. *Agron. J.* **2000**, *92*, 1096–1103. [CrossRef]
10. Huang, S.; Zeng, Y.J.; Wu, J.F.; Shi, Q.H.; Pan, X.H. Effect of crop residue retention on rice yield in China: A meta-analysis. *Field Crops Res.* **2013**, *154*, 188–194. [CrossRef]
11. Gentile, R.; Vanlauwe, B.; van Kessel, C.; Six, J. Managing N availability and losses by combining fertilizer-N with different quality residues in Kenya. *Agr. Ecosyst. Environ.* **2009**, *131*, 308–314. [CrossRef]
12. Wang, J.G.; Bakken, L.R. Competition for nitrogen during mineralization of plant residues in soil: Microbial response to C and N availability. *Soil Biol. Biochem.* **1997**, *29*, 163–170.
13. Yang, L.J.; Zhang, L.L.; Yu, C.X.; Li, D.P.; Gong, P.; Xue, Y.; Song, Y.C.; Cui, Y.L.; Doane, T.A.; Wu, Z.J. Nitrogen fertilizer and straw applications affect uptake of ^{13}C, ^{15}N-glycine by soil microorganisms in wheat growth stages. *PLoS ONE* **2017**, *12*, e0169016. [CrossRef] [PubMed]
14. Restovich, S.B.; Andriulo, A.E.; Armas-Herrera, C.M.; Beribe, M.J.; Portela, S.I. Combining cover crops and low nitrogen fertilization improves soil supporting functions. *Plant Soil* **2019**, *442*, 401–417. [CrossRef]

15. Zhou, G.P.; Gao, S.J.; Chang, D.N.; Rees, R.M.; Cao, W.D. Using milk vetch (*Astragalus sinicus* L.) to promote rice straw decomposition by regulating enzyme activity and bacterial community. *Bioresour. Technol.* **2021**, *319*, 124215. [CrossRef]
16. Yao, Z.Y.; Xu, Q.; Chen, Y.P.; Liu, N.; Li, Y.Y.; Zhang, S.Q.; Cao, W.D.; Zhai, B.N.; Wang, Z.H.; Zhang, D.B.; et al. Leguminous green manure enhances the soil organic nitrogen pool of cropland via disproportionate increase of nitrogen in particulate organic matter fractions. *Catena* **2021**, *207*, 105574. [CrossRef]
17. Bao, S.D. *Soil and Agricultural Chemistry Analysis*; China Agricultural Press: Beijing, China, 2008; pp. 265–267.
18. Lu, R. *Soil and Agricultural Chemistry Analysis*; China Agricultural Science and Technology Press: Beijing, China, 2000; pp. 128–129.
19. Lu, Y.H.; Liao, Y.L.; Nie, J.; Zhou, X.; Xie, J.; Yang, Z.P. Effect of different incorporation of Chinese milk vetch coupled with urea or controlled release urea on yield and nitrogen and potassium nutrient use efficiency in double-cropping rice system. *J. Plant Nutr. Fert.* **2017**, *23*, 360–368. (In Chinese with English Abstract)
20. Ma, Y.Q.; Qian, C.C.; Deng, L.P.; Huang, G.Q. Effects of combining Chinese milk vetch with nitrogen fertilizer on grain and dry matter yield, nitrogen absorption and utilization of double-cropping rice. *J. Nucl. Agr. Sci.* **2017**, *31*, 2399–2407. (In Chinese with English Abstract)
21. Meng, X.T.; Li, Y.Y.; Zhang, Y.; Yao, H.Y. Green manure application improves rice growth and urea nitrogen use efficiency assessed using 15N labeling. *Soil Sci. Plant Nutr.* **2019**, *65*, 511–518. [CrossRef]
22. Fontaine, S.; Mariotti, A.; Abbadie, L. The priming effect of organic matter: A question of microbial competition? *Soil Biol. Biochem.* **2003**, *35*, 837–843. [CrossRef]
23. Li, F.C.; Wang, Z.H.; Dai, J.; Li, Q.; Wang, X.; Xue, C.; Liu, H.; He, G. Fate of nitrogen from green manure, straw, and fertilizer applied to wheat under different summer fallow management strategies in dryland. *Biol. Fert. Soils* **2015**, *51*, 769–780. [CrossRef]
24. Abe, S.S.; Hashimoto, S.; Umezane, T.; Yamaguchi, T.; Yamamoto, S.; Yamada, S.; Endo, T.; Nakata, N. Excessive application of farmyard manure reduces rice yield and enhances environmental pollution risk in paddy fields. *Arch. Agron. Soil Sci.* **2016**, *62*, 1208–1221. [CrossRef]
25. Zhou, G.P.; Cao, W.D.; Bai, J.S.; Xu, C.X.; Zeng, N.H.; Gao, S.J.; Rees, R.M.; Dou, F.G. Co-incorporation of rice straw and leguminous green manure can increase soil available nitrogen (N) and reduce carbon and N losses: An incubation study. *Pedosphere* **2020**, *30*, 661–670. [CrossRef]
26. Aljerib, Y.M.; Geng, M.J.; Xu, P.D.; Li, D.H.; Rana, M.S.; Zhu, Q. Equivalent incorporation of Chinese milk vetch and rice straw enhanced nutrient mineralization and reduced greenhouse gas emissions. *Soil Sci. Plant Nutr.* **2022**, *68*, 167–174. [CrossRef]
27. Devêvre, O.C.; Horwáth, W.R. Decomposition of rice straw and microbial carbon use efficiency under different soil temperatures and moistures. *Soil Biol. Biochem.* **2000**, *32*, 1773–1785. [CrossRef]
28. Bashir, M.A.; Zhai, L.M.; Wang, H.Y.; Liu, J.; Raza, Q.U.; Geng, Y.C.; Rehim, A.; Liu, H.B. Apparent variations in nitrogen runoff and its uptake in paddy rice under straw incorporation. *J. Integr. Agr.* **2022**, *21*, 3356–3367. [CrossRef]
29. Zhang, S.J.; Zhang, G.; Wang, D.J.; Liu, Q.; Wang, S.W. Effects of straw returning coupled with application of nitrogen fertilizer on rice yield and dynamics of nitrogen in surface water of paddy field. *Acta Pedol. Sin.* **2020**, *57*, 435–445. (In Chinese with English Abstract)
30. Chu, H.Y.; Fujii, T.; Morimoto, S.; Lin, X.G.; Yagi, K.; Hu, J.L.; Zhang, J.B. Community structure of ammonia-oxidizing bacteria under long-term application of mineral fertilizer and organic manure in a sandy loam soil. *Appl. Environ. Microb.* **2007**, *73*, 485–491. [CrossRef] [PubMed]
31. Ai, C.; Liang, G.Q.; Sun, J.W.; Wang, X.B.; He, P.; Zhou, W. Different roles of rhizosphere effect and long-term fertilization in the activity and community structure of ammonia oxidizers in a calcareous fluvo-aquic soil. *Soil Biol. Biochem.* **2013**, *57*, 30–42. [CrossRef]

agriculture

MDPI

Article

Cheese Whey Characterization for Co-Composting with Solid Organic Wastes and the Agronomic Value of the Compost Obtained

Steven Ramos-Romero [1], Irene Gavilanes-Terán [1], Julio Idrovo-Novillo [1], Alessandro Idrovo-Gavilanes [2], Víctor Valverde-Orozco [3] and Concepción Paredes [4,*]

1 Faculty of Science, Higher Polytechnic School of Chimborazo, Riobamba 060155, Chimborazo, Ecuador; stalin.ramos@espoch.edu.ec (S.R.-R.); irene.gavilanes@espoch.edu.ec (I.G.-T.); julio.idrovo@espoch.edu.ec (J.I.-N.)

2 Faculty of Medicine and Health Sciences, Ghent University, Corneel Heymanslaan 10, 9000 Ghent, Belgium; alessandro.idrovo@ugent.be

3 Faculty of Engineering, National University of Chimborazo, Riobamba 060108, Chimborazo, Ecuador; victor.valverde@unach.edu.ec

4 Agri-Food and Agro-Environmental Research and Innovation Institute (CIAGRO-UMH), Miguel Hernandez University, EPS-Orihuela, ctra. Beniel km 3.2, 03312 Orihuela, Alicante, Spain

* Correspondence: c.paredes@umh.es

Abstract: Cheese production generates a large amount of liquid waste called cheese whey (CW). The management of CW is not optimized in Ecuador since a large proportion of it is discharged into the soil or effluents, causing significant environmental impacts. For this reason, the co-composting of whey with solid organic wastes can be a suitable method for its treatment for small companies generating this liquid waste due to its effectiveness and low cost. In this study, we analyzed 10 CW samples from different small companies in the Mocha canton (Tungurahua, Ecuador) to determine specific physicochemical and chemical parameters. Subsequently, a waste pile was formed with crop residues (corn and beans) and cow manure, which was composted using the turned pile composting system. Throughout the composting process, the temperature of the pile was controlled, its moisture was maintained between 40 and 60% by adding whey, and several physicochemical, chemical, and biological properties were determined. The results showed that the CW presented a high organic load, notable macronutrient content, and low heavy metal concentrations, all of which are beneficial for its co-composting with other organic solid wastes. The only limiting factors involved in using large amounts of whey in the composting process were the low pH values of the acid CW and the high concentrations of salts. It was also observed that co-composting CW with agro-livestock wastes was a viable strategy to treat these wastes and produce compost with stabilized and humified organic matter and remarkable agricultural value.

Keywords: cheese industry wastewater; lactose; protein; fat; mineral salts; agro-livestock waste; composting; phytotoxicity; humification organic matter

1. Introduction

In South America, the average cheese production was 1.14 million tons per year in the period 2011–2021. In 2021, Ecuador was the fourth-largest producer, with 94,422 tons of cheese [1]. The main by-product/waste produced during cheesemaking is cheese whey (CW), with around 9 L of CW produced for each kilogram of cheese [2]. Therefore, in 2021, the CW production in Ecuador was approximately 850,000 m^3 [1,2]. This liquid residue comprises mainly milk water, lactose, proteins, and mineral salts, and the concentration of these components depends on the origin of the milk and the process used to make the cheese. There are two types of CW: sweet CW (pH = 6.0–7.0; produced from casein coagulation by the enzymatic action of rennet) and acid CW (pH = 4.5–5.8; obtained from milk coagulation by the acid action of lactobacilli or inorganic acids) [3,4].

CW has different uses in the food and non-food industries, such as the production of fermented or non-fermented dairy beverages [5] and the development of a wide range of food products (whey cheese, ice cream, confectionery products, soups, processed meats, salad dressings, etc.) [4]. Bioactive peptides, sugars, oligosaccharides, and sugar alcohols derived from CW have physiological properties beneficial to human health and are used as a functional ingredient in foods, dietary supplements, food for special medical purposes, and pharmaceutical products, as well as in toiletries and cosmetics [6–8]. Moreover, CW has been used for animal feed [9], as a biofertilizer [10], and to obtain biopolymers through the biological transformation of lactose, enabling the production of biodegradable plastic components [11]. CW is considered a promising resource for biofuel production, such as biomethane, bioethanol, biohydrogen, and biodiesel derived from microbial lipids obtained by fermenting the sugars contained in CW by highly oleaginous microorganisms [12].

However, in Ecuador, only 1.5% of CW is incorporated in the production process of dairy products (unfermented and fermented beverages, yogurt, fresh cheese, spreadable cheese, etc.) [13]. This low percentage of recovered CW is mainly because the cheese industry in most Latin American countries, including Ecuador, is made up of small family companies, which cannot make the necessary investments to transform CW into value-added products [4,14]. Therefore, most of the CW is discharged into effluents or the soil [15]. These practices have adverse environmental impacts since they damage aquatic life by reducing the dissolved oxygen caused by the degradation of the high organic load of this waste [4]. They also provoke eutrophication processes [16] and the acidification, salinization, and reduction of the redox potential of the soil, as well as altering its physical structure [17].

Hence, co-composting CW with solid organic waste can be an appropriate way to treat it within the context of small companies due to its effectiveness and low technology and cost [18]. Several investigations have focused on the reuse of liquid wastes through composting, such as the co-composting of liquid manure with vegetable wastes [19] or municipal solid wastes [20], where this liquid waste has been used as a source of moisture or easily degradable soluble organic matter, achieving final composts with notable nutrient content and stabilized organic matter in less time. The liquid residue from olive oil extraction has been co-composted with olive mill sludge from evaporation ponds, grape marc, and green wastes [21] or with olive mill pomace, olive mill solid husk, poultry manure, and green wastes [22]. Both studies confirmed the success of using the co-composting process to valorize this liquid waste and stabilize and transform the other wastes into mature and hygienic composts that can be used as biofertilizer. Moreover, Ghasemzadeh et al. [23] co-composted rice straw with poultry manure or sugarcane vinasse as cheap and accessible C/N reducers to accelerate the composting process of a highly lignocellulosic waste such as rice straw. However, these authors did not find that applying vinasse had positive effects on accelerating the composting process or on the agricultural quality of the compost obtained. Studies have also been performed on co-composting liquid wastes from wine-processing with other wine-processing solid wastes and urban wastes [24] or with wastes from olive milling, grape marc, and green wastes [21]. In these studies, it was observed that adding these liquid wastes did not negatively affect the composting process, and composts rich in nutrients, without phytotoxicity, and with stabilized and humified organic matter were obtained.

However, there is little information related to treating CW through composting. Alfonzo et al. [25] evaluated the effects of adding CW to the composting of grape pomace mixed with green herbaceous crop and pruning residues, using this liquid waste as a source of moisture and nutrients. These authors observed that adding CW resulted in a more substantial reduction of total and dissolved organic carbon content and more accelerated microbial carbon mineralization than the pile where the moisture content was maintained by water irrigation, due to the stimulation of microbial activity by the nitrogen compounds in this liquid waste. CW has been co-composted with woodchips and biochar as a water-soluble carbon source to extract the heat generated during composting through a hydraulic circuit [26]. Considering that the efficiency and effectiveness of the composting process for the treatment of CW have been little explored, the objective of this work was to determine the composition of CW generated in different small cheese companies in Canton Mocha (Tungurahua, Ecuador) to evaluate its potential reuse through co-composting with agro-livestock residues. The results obtained could be useful for the joint management of organic wastes generated by cheese, agricultural and livestock farms in centralized composting facilities close to the production areas.

2. Materials and Methods

2.1. Surveys on the Dairy and Agro-Livestock Sectors of the Study Area

The study was conducted in the Mocha canton, in the southwestern part of the province of Tungurahua, Ecuador. This canton is located at altitudes ranging from 2500 to 4965 m above sea level (a.s.l.), with a total area of 85.8 km^2. The climate varies widely, but a high mountain equatorial climate covers the majority of this area. Average annual temperatures vary between 2–10 °C, and the average annual rainfall ranges between 1000–500 mm. The main economic activities in the Mocha canton are agriculture and cattle raising. Livestock farming in this canton focuses on milk production, which is mostly used to make cheese [27].

In this study, ten small dairy companies were surveyed to obtain information on the type of dairy products produced, the daily milk processing volume, and the use or disposal of CW in the Mocha canton. Twenty-five farmers and eleven randomly selected livestock breeders were surveyed to determine the types of crops and cattle species produced in the canton, along with the type of management carried out on the waste generated by these activities.

2.2. Sampling of the Cheese Whey

Ten samples of CW were collected from the cheese companies participating in the survey. These samples were collected at the same time of year (January 2022) to reduce climatic effects on sample composition. Approximately 1000 mL of CW were taken directly from the refrigeration tanks, buckets, or collection containers of each participating company. The samples were transported in containers with coolants, maintaining a temperature of 2–6 °C throughout the transportation process until they arrived at the laboratory, where they were stored at 6 °C for later analysis. All determinations were made in triplicate.

2.3. Design of Co-Composting Experiment

To evaluate the feasibility of treating the CW through composting, a waste pile weighing approximately 1000 kg was formed, with dimensions of 2 × 3 m at the base and a height of 1.5 m. The pile composition consisted of a mixture of wastes from corn and bean crops, cow manure, and CW as a source of moisture. For the preparation of this pile, plant residues were shredded to a particle size of 1–5 cm. These residues were mixed with cow dung manually, preparing layers of residues in the following proportions (on a fresh weight basis): 60% corn residues + 5% bean residues + 35% cow manure. The selection of these waste ratios was made to achieve an initial ratio of total organic carbon (Corg)/total nitrogen (Nt) close to the 25–35 range, suitable for an active composting process. The main characteristics of these wastes were as follows: for corn residues, pH 4.84 ± 0.02, 3.42 ± 0.35 dS/m electrical conductivity (EC), 21.2 ± 0.01% moisture, 91.3 ± 0.1% organic matter (OM), 43.7 ± 0.2% Corg, 0.74 ± 0.04% Nt and 59.1 ± 3.1 Corg/Nt ratio; for bean residues, pH 4.96 ± 0.05, 5.52 ± 0.13 dS/m EC, 14.8 ± 0.01% moisture, 88.6 ± 0.1% OM, 41.1 ± 1.4% Corg, 3.32 ± 0.18% Nt and 12.4 ± 0.3 Corg/Nt ratio; and for cow manure, pH 7.79 ± 0.06, 6.06 ± 0.10 dS/m EC, 61.5 ± 0.01% moisture, 77.7 ± 0.9% OM, 41.0 ± 0.2% Corg, 2.16 ± 0.16% Nt and 19.0 ± 1.4 Corg/Nt ratio. Throughout the bio-oxidative phase, the moisture content was controlled by adding CW (total volume = 0.35 L CW/kg fresh weight of the initial waste mixture), ensuring appropriate values in the 40–60% range. This liquid residue was added homogeneously throughout the pile by means of a hose with a sprinkler, which transferred the CW from a tank with the help of a submersible pump. The characteristics of this CW are those shown for sample CW-2, indicated below.

During the composting process, daily average temperatures were recorded at five different points, at a depth of 30 cm, using a portable probe. Daily ambient temperatures were also measured. A total of two turnings were conducted on the pile to homogenize the mixture and provide the oxygen necessary for the aerobic degradation of the organic matter. These turnings were carried out when temperatures were below the minimum value considered for the thermophilic stage (<40 °C). The bio-oxidative phase lasted for 80 days, followed by a 30-day maturation period. The mixture was sampled four times during the bio-oxidative phase and at the end of maturation by collecting sub-samples from seven randomly selected points covering the entire profile of the pile. These subsamples were then mixed to obtain a representative sample of 2 kg. These samples were dried at 60 °C, then ground and sieved to 0.5 mm for subsequent analysis. All determinations were performed in triplicate.

2.4. Analytical Methods

In the evaluation of the CW, physical-chemical and chemical analyses were conducted, addressing parameters such as pH, electrical conductivity (EC), total and suspended solids (TS and SS, respectively), biochemical oxygen demand (BOD_5), chemical oxygen demand (COD), macro and micronutrients, and heavy metals determined according to the Standard Methods for the Examination of Water and Wastewater [28]. Fat, protein, and lactose contents of the CW were also analyzed using the methods described in Standard Methods for the Examination of Dairy Products [29].

In the analysis of the initial materials and samples from the composting pile, pH, EC, OM, Corg, water-soluble anions and polyphenols, cation exchange capacity (CEC), germination index (GI), macro- and micronutrients, and heavy metals were measured following the analytical techniques used by Idrovo-Novillo et al. [30].

2.5. Statistical Methods

Regarding the results of the physicochemical and chemical analyses of the CW, the mean value, range of values, and coefficient of variation (CV) of each parameter were calculated. A principal component analysis (PCA) was applied to the different parameters using the Varimax rotational method with normalization to identify groups of interrelated variables.

For the initial materials and samples from the composting pile, the standard deviation was determined for all the mean values of the parameters analyzed. The LSD (least significant difference; $p < 0.05$) method was used to determine the significant differences in the evolution of each parameter in the pile during the composting process. The IBM SPSS 27 statistical program was used to perform the statistical processing of all the results.

3. Results and Discussion

3.1. Situation of the Dairy and Agro-Livestock Sectors Studied and Destinations of the Generated Wastes

The results derived from the surveys conducted in small and medium-sized cheese, agricultural, and livestock farms are shown in Table 1. In the dairy industry of the Mocha canton, the dairy products with the highest production were cheese and yogurt, with all the companies producing cheese and 20% of them also producing yogurt. Seventy percent of the companies processed between 600 and 1000 L of milk daily, compared to the other companies processing quantities of 300 to 500 L (20%) and 50 to 200 L (10%). Milk processing for cheese production produced CW as a by-product, which was generally considered waste by most of the companies. It was not incorporated into the production process of dairy products mainly due to the lack of knowledge about the nutritional benefits of this by-product and the difficulties small family companies have acquiring the appropriate technologies for its handling and processing in the dairy industry. Regarding the destination of the CW, 50% of the producers used it as feed for cattle and pigs, 20% allocated it for pasture irrigation, 10% sold it at a very low price (0.01 US dollar/L), and the remaining 20% did not make any use of this liquid waste, dumping it into the environment (ravines, surface water bodies, etc.). It is important to highlight that pouring this liquid residue into the soil can affect the physical structure and chemical composition of the soil, as well as negatively influencing crop yields. The suspended solids contained in CW can clog soil pores, reducing its infiltration rate, and discharging it into the soil can also produce acidification and salinization and reduce the soil's redox potential [17]. Continuously dumping large amounts of CW into the soil or surface water bodies can lead to groundwater and effluent contamination [4,16].

The surveys conducted in the agricultural sector indicated that in the Mocha canton, potatoes represented 24.3% of the predominant crops, followed by corn at 20.3% (Table 1). Other significant crops included red and white onions, at 13.5% and 8.1%, respectively. Percentages of 8.1, 6.8, and 5.4 were obtained for peas, beans, and carrots, respectively. Vegetables such as quinoa, cilantro, and tomatoes, among others, were also grown, each accounting for less than 5% of the total crops produced. Crops such as corn, potatoes, peas, beans, quinoa, and tomatoes produce higher amounts of waste because only the edible part of the plant is marketed. However, crops such as carrots, cilantro, and white and red onions create less waste since most of their plant structure is marketed. Regarding the destination of harvest residues, 51% of producers used them as feed for livestock, whereas 14% incinerated them, emitting aerosols and greenhouse gases [31,32], as well as decreasing carbon storage in the soil and reducing nutrients through runoff during rainfall [33]. The remaining 35% of farmers applied these residues directly to the soil without prior stabilization. Applying fresh crop residues to the soil can modify its redox potential due to oxygen consumption in the mineralization of unstabilized organic matter, increase soil microbial activity, which can mineralize endogenous soil organic carbon, modify soil pH [34], and reduce crop germination and development due to the presence of phytotoxic compounds such as polyphenols [35].

Table 1. Results derived from the surveys conducted in small and medium-sized cheese, agricultural, and livestock farms. Data expressed in %.

	Dairy industry							
Dairy product produced	Cheese only	Cheese and yogurt						
	80	20						
Milk processing volume (L/day)	600–100	500–300	50–200					
	70	20	10					
Use or disposal of CW	Animal feed	Irrigation for pastures	Sale for other uses	Dumped into the environment				
	50	20	10	20				
	Agricultural farms							
Type of crop produced	Potato	Corn	Red onion	White onion	Pea	Bean	Carrot	Other crops
	24.3	20.3	13.5	8.1	8.1	6.8	5.4	13.5
Destination of harvest residues	Animal feed	Incineration	Direct application to soil					
	51	14	35					
	Livestock farms							
Type of cattle species raised	Native-breed cattle	Improved breed cattle						
	56.8	43.2						
Destination of cattle manure	Direct application to soil	Applied to soil after drying	Left where the cattle excrete					
	58	35	7					

Regarding the livestock sector, the data indicated that 56.8% of participants bred native-breed cattle, compared to 43.2% who opted for improved breeds (Table 1). Native-breed cattle are able to adapt to extreme conditions, including low-quality forage, low temperatures, and reduced humidity, which are characteristics of the study area. Furthermore, they are very versatile, as they are used for milk and meat production and work tasks [36]. Among the surveyed farmers, 71.6% identified manure as a waste; of this group, 58.0% dumped it directly on the soil, while 35.0% used it as organic amendment after drying. The rest of the respondents preferred to leave it where the cattle excrete. These practices cause significant environmental impacts, such as soil and water contamination by pathogens, excess nutrients and emerging contaminants [37,38], and greenhouse gas emissions [39].

This scenario highlights the need to seek environmentally friendly alternatives, such as composting, to treat dairy, agricultural, and livestock wastes. This approach will not only favor the sustainability of the dairy and agricultural sectors in this region but will also address existing environmental issues, thereby contributing to improving the profitability of the activities of these sectors.

3.2. Cheese Whey Characterization

The physicochemical characteristics of the CW samples are shown in Table 2. The acidity levels recorded in these samples ranged from 4.88 to 6.77, with sweet whey (pH around 6.0–7.0) represented in CW-2, CW-6, CW-7, CW-8, and CW-9, and the rest were acid whey (pH around 4.5–5.8) [4]. The variability in pH values is due to the process used for coagulating milk casein by adding rennet, by the action of lactobacillus, or by adding lactic acid or mineral acids [4]. Adding large volumes of CW, especially acidic CW, during solid organic waste co-composting can be a limiting factor for adequate composting since the optimal pH range for microbiota growth during this process is 5.5–9.0 [40]. Considerable salt concentrations were evident in all the samples evaluated through EC, especially in the case of CW-6 (EC = 8.19 dS/m). This high salt content was possibly due to the solubilization of minerals from the milk and salts added to the cheese during its production, such as $CaCl_2$ to promote rennet coagulation, HCl for milk coagulation through acidification, NaCl for salting, etc. [3]. The variability of salts that can be solubilized in CW could also explain the variation in the EC values of the whey samples studied (CV = 20). The elevated concentrations of mineral salts may limit the addition of this liquid residue to small quantities during the composting of other organic waste because there is an increase in salts as a consequence of OM mineralization, and because of the concentration effect caused by the lost mass during composting [40]. This could result in an excessively saline final compost, the agricultural use of which could adversely impact soil quality and crops.

Table 2. Physicochemical characterization of the cheese whey (CW) samples.

Samples	pH	EC (dS/m)	BOD_5 (mg O_2/L)	COD (mg O_2/L)	TS (mg/L)	SS (mg/L)
CW-1	4.88	6.23	33,312	64,450	78,200	3719
CW-2	6.77	5.03	32,244	66,000	73,400	3518
CW-3	5.11	5.76	36,521	73,150	78,700	3733
CW-4	5.40	6.46	35,065	70,200	69,167	3175
CW-5	5.19	6.72	30,365	62,200	68,000	3306
CW-6	6.19	8.19	31,491	66,000	73,167	3526
CW-7	6.55	4.58	34,379	71,000	68,067	3198
CW-8	6.57	4.58	37,359	76,500	69,533	3401
CW-9	6.16	7.46	36,209	65,150	71,233	3431
CW-10	5.78	7.01	30,529	64,200	59,155	2851
Mean	5.86	6.20	33,747	67,885	70,862	3386
Range	4.88–6.77	4.58–8.19	30,365–37,359	62,200–76,500	59,155–78,700	2851–3733
CV (%)	12	20	8	7	8	8

EC: electrical conductivity; BOD_5: biochemical oxygen demand; COD: chemical oxygen demand; TS: total solids; SS: suspended solids; CV: coefficient of variation.

The elevated levels of BOD_5 and COD reflected the high organic load in the CW samples (Table 2), indicating the environmental pollution problem that its disposal creates. The BOD_5 and COD values were within the range found by Carvalho et al. [41] in a review study of the characteristics of cheese effluents (27,000–60,000 mg/L and 50,000–102,000 mg/L for BOD_5 and COD, respectively). According to these authors, lactose is the main whey component responsible for its high organic load, representing 70–75% of the ST. In the present study, these solids were in the range of 59,155–78,700 mg/L, whereas the SS ranged between 2851 and 3733 mg/L, both parameters having values within the intervals reported by other studies [41].

The contents of the main organic components and macronutrients of the CW samples are shown in Table 3. Lactose, proteins, and fats are the most abundant organic constituents of whey, considered the primary solids that are not incorporated into the cheese and remain in the CW (about 90% of the lactose, 20% of the protein, and 10% of the fat of the milk are present in the whey) [3]. In the CW samples analyzed, the lactose content ranged from 34.5 to 42.1 g/L, below the range observed by other authors (42.6–60.0 g/L) [3,4,41]. Regarding the protein content, CW-2 and CW-9 had the lowest and highest concentrations of this organic compound, respectively, and this range of values was within that reported by Carvalho et al. [41] in a review study of the characteristics of cheese effluents (1.42–8.00 g/L). Most of the CW samples presented fat concentrations within the values found by previous authors (0.99–9.44 g/L). However, substantial variability was found in the values of this parameter among the samples analyzed (CV = 108), probably because the fat content of the CW depends on factors such as the fat concentration of the milk used to make the cheese. In addition, the extent of some operations carried out during the process (milk pasteurization and pumping and cutting and cooking the curd) can favor free fat formation (non-globular) that is lost in the whey [3].

Table 3. Main organic compounds and macronutrients in the cheese whey (CW) samples.

Samples	Lactose (g/L)	Protein (g/L)	Fat (g/L)	N (g/L)	P (g/L)	Ca (g/L)	K (mg/L)
CW-1	42.1	4.62	1.77	0.72	0.11	0.16	2.81
CW-2	40.1	3.93	0.74	0.62	0.10	0.13	2.94
CW-3	38.5	4.53	8.47	0.71	0.11	0.15	2.03
CW-4	38.1	4.40	0.94	0.69	0.06	0.20	2.15
CW-5	37.1	4.72	0.79	0.74	0.10	0.27	2.92
CW-6	39.5	4.02	1.40	0.63	0.11	0.24	2.44
CW-7	36.7	4.49	1.35	0.70	0.11	0.15	2.90
CW-8	38.3	4.36	1.47	0.68	0.11	0.11	2.18
CW-9	37.2	4.75	3.80	0.75	0.11	0.13	2.69
CW-10	34.5	4.47	1.28	0.70	0.11	0.14	2.07
Mean	38.2	4.43	2.20	0.69	0.10	0.17	2.51
Range	34.5–42.1	3.93–4.75	0.74–8.47	0.62–0.75	0.06–0.11	0.11–0.27	2.03–2.94
CV (%)	5	6	108	6	14	31	15

CV: coefficient of variation.

The values of the macronutrient contents were in the ranges 0.62–0.75 g/L, 0.06–0.11 g/L, 0.11–0.27 g/L, and 2.03–2.94 mg/L for N, P, Ca, and K, respectively (Table 3). These data were within or below the scale of values found by other authors (0.20–1.76 g/L for N and 0.12–0.5 g/L for P [41] and 0.4–1.6 g/L for Ca [3,42]). Ca was the macronutrient with the greatest variability among the samples analyzed (CV = 31), probably because this element can be transferred from milk to whey and added in cheese-making as $CaCl_2$ to promote rennet coagulation [3].

According to the chemical analyses of the micronutrients and heavy metals in the CW samples, most of these elements had values below the detection limits of the equipment used to measure them, with only very low concentrations of Fe and Zn found (Table 4). The CW samples presented notable variability in Fe content, probably due to factors such as the breed of cattle, dietary intake, and, especially, the process used for milk casein coagulation (by adding rennet or by acidification) and the type of cheese that is produced [3]. From an agronomic perspective, the presence of macro- and micronutrients in CW suggests that its co-composting with other organic solid wastes can improve the fertilizing value of the final compost. Furthermore, its low heavy metal content guarantees the safe agricultural use of this compost.

Table 4. Micronutrients and heavy metals in the cheese whey (CW) samples.

Samples	Fe (mg/L)	Cu (mg/L)	Zn (mg/L)	Ni (mg/L)	Cr (mg/L)	Cd (mg/L)	Pb (mg/L)
CW-1	0.46	<0.10	0.75	<0.10	<0.10	<0.10	<0.10
CW-2	0.36	<0.10	0.71	<0.10	<0.10	<0.10	<0.10
CW-3	0.55	<0.10	0.67	<0.10	<0.10	<0.10	<0.10
CW-4	0.57	<0.10	0.67	<0.10	<0.10	<0.10	<0.10
CW-5	0.70	<0.10	0.68	<0.10	<0.10	<0.10	<0.10
CW-6	0.50	<0.10	0.57	<0.10	<0.10	<0.10	<0.10
CW-7	0.50	<0.10	0.65	<0.10	<0.10	<0.10	<0.10
CW-8	0.53	<0.10	0.55	<0.10	<0.10	<0.10	<0.10
CW-9	0.26	<0.10	0.61	<0.10	<0.10	<0.10	<0.10
CW-10	0.24	<0.10	0.60	<0.10	<0.10	<0.10	<0.10
Mean	0.47	<0.10	0.65	<0.10	<0.10	<0.10	<0.10
Range	0.24–0.70	<0.10	0.55–0.75	<0.10	<0.10	<0.10	<0.10
CV (%)	31	-	10	-	-	-	-

CV: coefficient of variation.

3.3. Relationships Among the Cheese Whey Characteristics

PCA was applied to the following parameters analyzed in the CWs: pH, EC, TS, SS, COD, BOD_5, lactose, protein, fat, N, P, Ca, K, Fe, and Zn (n = 15). The model obtained through PCA explains > 70.0% of the variability by establishing three principal components, with the following contribution of each principal component: component 1, 26.4%; component 2, 24.0%; and component 3, 20.6%. In the proposed model, the KMO (Kaiser-Meyer–Olkin's measure of adequacy) value was >0.5 (KMO = 0.539), and the p-value was 0.000 for Bartlett's test of sphericity. None of the variables presented an extraction value < 0.5. All of this indicates that the model is adequate.

In principal component 1, the variables TS, SS, lactose, and fat were grouped and directly correlated (Table 5). This indicated that the TS and SS were mainly composed of the lactose and fats contained in the CW samples, agreeing with Fox et al. [3] that most of the lactose and part of the milk fat are present in the whey.

Table 5. Principal component loadings for the parameters determined in the cheese whey samples.

Paremeter	Principal Component 1	2	3
TS	**0.948**	0.172	−0.160
SS	**0.916**	0.152	−0.180
Lactose	**0.757**	0.247	−0.502
Fat	**0.584**	−0.158	0.553
COD	0.407	**−0.764**	0.167
Ca	−0.189	**0.753**	−0.116
pH	−0.263	**−0.687**	−0.510
Zn	0.432	**0.596**	−0.127
BOD_5	0.571	**−0.595**	0.404
EC	−0.337	**0.507**	0.113
K	0.061	**0.448**	−0.429
Fe	0.324	**0.338**	0.024
P	0.101	**−0.182**	0.043
Protein	0.019	0.334	**0.863**
N	0.010	0.340	**0.862**

For the abbreviations, see Table 1. The different loadings for the parameters in each principal component are shown in bold type in their highest absolute value.

Principal component 2 was associated with COD, Ca, pH, Zn, BOD_5, EC, K, Fe, and P, with COD, BOD_5, pH, and P negatively correlated with the rest of the variables. However, the factorial loading of P was low, indicating the least association of this variable with this principal component. The direct relationship between the variables COD, BOD_5, and pH indicated that in this study, the reduction of CW acidity favored the content of the organic load of this liquid waste. The negative correlation between the pH and the determined macro- and micronutrients has also been observed by other authors, where the acidic CW samples had a higher mineral content [3,4,41]. EC was positively correlated with the determined mineral elements, corroborating the relationship between CW salinity and its macro- and micronutrient content.

Significant variables for principal component 3 included protein and N, which correlated positively due to their chemical relationship and their determination (total protein = N $\times$ 6.38).

3.4. Co-Composting Cheese Whey with Agro-Livestock Wastes

The temperature in the pile rose rapidly at the onset of the process, reaching thermophilic values (>40 °C) within the first 72 h. These values were maintained for approximately 37 days until the first turning (Figure 1). The maximum average temperature was recorded on day 15, at 59.2 °C. Other researchers also observed a rapid increase in temperature early in the composting process when co-composting plant residues with CW as a source of moisture and nutrients [25,26]. The supply of oxygen and non-degraded materials was controlled by turnings, which were carried out when the temperature dropped below 40 °C. After the first turning, the temperature increased to above 40 °C for 16 days, while no thermophilic temperature values were observed after the second turning.

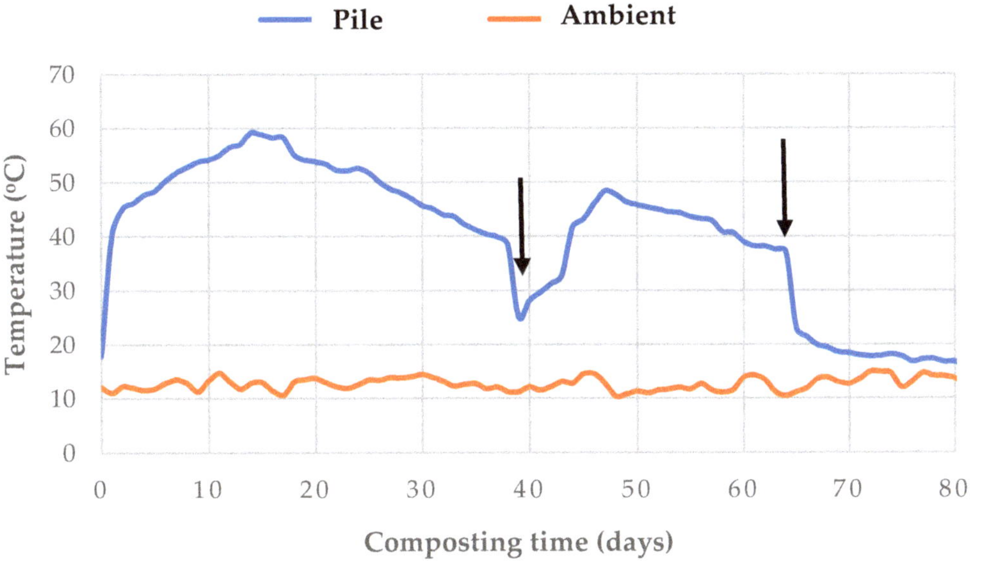

Figure 1. Temperature evolution during the composting process. The black arrows indicate the days on which turnings were performed.

The high temperatures recorded during the thermophilic stage, together with their duration, were not those recommended by the United States Environmental Protection Agency [43] to ensure effective sanitization of the composted material in piles with aeration by turning (temperature $\geq$ 55 °C for 15 consecutive days). These requirements were not met in the present composting experiment, as temperatures were ≥55 °C for only seven consecutive days. However, the maintenance of high temperatures for a given period of time is not the only factor acting on pathogen reduction during the composting process. Pathogen inactivation is also favored when the thermophilic process is carried out under alkaline conditions, since the presence of certain compounds, such as ammonia, hydroxide anions, etc., contribute to the reduction of this type of microorganism [44]. In this composting experiment, high pH values (>9) were recorded during the thermophilic stage (Table 6). Also, the reduction of pathogens during the composting process may be due to antagonism or competition between the bacterial species present in the pile [45]. In addition, lactic acid bacteria can inhibit different pathogenic microorganisms through the formation of antibacterial metabolites [44], and this type of bacteria was present in the CW [3]. Therefore, sanitization of the studied waste pile was probably adequate although temperatures ≥ 55 °C were not maintained for at least 15 days.

Table 6. Evolution of the principal physicochemical parameters during the composting process.

Composting Time (Days)	pH	EC (dS/m)	OM (%)	Nt (%)	Corg/Nt	Polyphenols (g/kg)	GI (%)
0	7.85 ± 0.10	4.39 ± 0.01	80.9 ± 0.4	1.20 ± 0.02	35.7 ± 0.8	7.62 ± 0.07	46.8 ± 2.5
39	9.16 ± 0	4.98 ± 0.01	76.2 ± 0.8	1.38 ± 0.05	28.9 ± 0.2	6.71 ± 0.06	78.7 ± 7.2
65	9.08 ± 0	6.68 ± 0	67.9 ± 0.2	1.73 ± 0.07	22.4 ± 0.6	3.01 ± 0.04	88.4 ± 5.0
80	8.82 ± 0	7.50 ± 0	64.9 ± 0.8	2.26 ± 0.13	16.1 ± 0.2	4.29 ± 0.06	83.4 ± 5.4
mature	8.93 ± 0	8.35 ± 0	67.6 ± 1.1	3.07 ± 0.01	10.3 ± 0.5	3.69 ± 0.05	90.9 ± 4.6
LSD	0.16	0.02	3.9	0.35	3.9	0.21	10.2

EC: electrical conductivity; OM: organic matter; Nt: total nitrogen; Corg: total organic carbon; GI: germination index; and LSD: least significant difference at $p < 0.05$. Values reported as mean ± standard error.

The bio-oxidative phase of composting was considered complete when the pile temperature stabilized and remained close to ambient [46]. After day 80, the composting process entered the maturation phase for one month, evidenced by a stabilization of the pile temperature.

The initial pH value in the pile (7.85; Table 6) was within the range of 5.5–8.0, suggesting it is suitable for composting [46]. This parameter increased significantly during the first 65 days, which could be attributed to acid-type compound degradation and the mineralization of organic nitrogen to ammonia [46]. However, a reduction in pH was observed at the end of the bio-oxidative and maturity stages, probably due to the release of H^+ from nitrification processes [40], as explained below.

The EC values increased continuously throughout the composting process (Table 6), possibly due to soluble salts provided by adding CW during composting and OM mineralization, as well as the concentration effect of ions caused by the mass loss of the pile [40]. This progressive increase in salt content throughout the composting process has also been observed in other experiments where liquid wastes have been used as a moisture source, such as winery wastewater [21] and olive mill wastewater [22].

OM degradation during the process led to its content being reduced from 80.9% to 67.6% (Table 6). This degradation produced a loss of mass in the pile that contributed to a significant increase in Nt throughout the process [46]. However, an accumulation of nitrogen due to the contribution of this nutrient with the CW applied during the composting process cannot be excluded. The degradation of Corg and increase of Nt caused the Corg/Nt ratio to decrease from an initial value of 35.7 to a final value of 10.3, indicating the evolution of OM maturity [46].

Soluble polyphenol content is one of the main parameters used to monitor phytotoxicity reduction in vegetable residues during composting. Plants synthesize these compounds to defend themselves from attack by phytopathogens and the damaging effects of solar UV rays [47]. At the same time, it has been observed that these compounds have phytotoxic effects on seed germination and seedling development [48]. Therefore, their degradation throughout the process is closely related to the detoxification of the material to be composted, as can be observed in an increase in GI as polyphenolic compounds were degraded in the composting process, showing an absence of phytotoxicity from day 39 of composting (GI > 50% [49]) (Table 6).

Soluble anions contributed to the salinity of the composted materials. Most of these ions increased throughout the process as a consequence of their production when the OM degraded and of the ion concentration effect of pile weight loss [40], as well as through the CW applied during the process (Table 7). Only in the case of sulfates was no significant increase in their concentration observed during the composting process. However, the rest of the anions studied increased their content from initial values of 0.53 g/kg, 0.86 g/kg, and 4.94 g/kg to final values of 0.89 g/kg, 15.99 g/kg, and 6.43 g/kg for Cl^-, NO_3^-, and PO_4^{3-}, respectively. Notably, the highest nitrate concentrations were observed at the end of the bio-oxidative and maturation stages, when temperatures were <40 °C and NH_3 concentrations were probably low. This could be because the nitrification process is inhibited at thermophilic temperatures and when there is an excessive amount of NH_3 from organic N degradation in the initial stages of the process, as reported by Cáceres et al. [50] in a review study on nitrification in composting.

Table 7. Evolution of the main soluble anions during the composting process.

Composting Time (Days)	Cl^- (g kg^{-1})	SO_4^{2-} (g kg^{-1})	NO_3^- (g kg^{-1})	PO_4^{3-} (g kg^{-1})
0	0.53 ± 0	5.13 ± 0.11	0.86 ± 0.02	4.94 ± 0.10
39	0.59 ± 0	3.86 ± 0.12	1.42 ± 0.06	5.12 ± 0.06
65	0.61 ± 0	3.06 ± 0.11	6.61 ± 0.13	5.20 ± 0.17
80	0.78 ± 0	3.85 ± 0.23	8.84 ± 0.30	5.74 ± 0.12
mature	0.89 ± 0	4.66 ± 0.23	15.99 ± 0.28	6.43 ± 0.16
LSD	0.01	0.62	0.72	0.46

For abbreviations, see Table 5.

Table 8 shows the values of the parameters determined in the final compost. In this compost, the range of pH and EC values established as suitable in compost for different agricultural applications and field conditions by the American guidelines of the US Composting Council [51] was exceeded. Despite this, this compost did not show phytotoxic effects on germination and root elongation of *Lepidium sativum* L. seeds, as corroborated by the GI value (GI > 50%, absence of phytotoxicity [49]). The compost presented an OM content within the range indicated by the American guidelines, while the Nt and P contents were above the recommended values (OM = 50–60% and Nt and P ≥ 10 g/kg [51]), showing the remarkable fertilizing capacity of the compost obtained. In addition, the compost presented an adequate degree of maturity, as can be observed from the values of the Corg/Nt ratio and CEC (Corg/Nt < 20 and CEC > 67 meq/100 g OM [52]). Finally, the contents of potentially toxic elements were below the limit values recommended by the US Composting Council [49] for safe agricultural compost use.

Table 8. The main characteristics of the final compost.

Parameter	Value	US Composting Council [49]
pH	8.93 ± 0	6.0–7.5
CE (dS/m)	8.35 ± 0	<5
OM (%)	67.6 ± 1.1	50–60
Corg/Nt	10.3 ± 0.5	-
CEC (meq/100 g OM)	97.0 ± 1.0	-
GI (%)	90.9 ± 4.6	-
Nt (g/kg)	30.7 ± 0.1	≥10
P (g/kg)	10.1 ± 0.1	≥10
K (g/kg)	32.0 ± 0.7	-
Fe (mg/kg)	2550 ± 32	-
Cu (mg/kg)	36 ± 5	1500
Mn (mg/kg)	148 ± 1	-
Zn (mg/kg)	73 ± 2	2800
Ni (mg/kg)	12 ± 0	420
Cr (mg/kg)	100 ± 2	-
Cd (mg/kg)	0.54 ± 0.08	39
Pb (mg/kg)	2.51 ± 0.11	300
Se (mg/kg)	1.74 ± 0.11	100
Hg (mg/kg)	<0.05	17
As (mg/kg)	1.31 ± 0.06	41

CEC: cation exchange capacity; for other abbreviations, see Table 5.

4. Conclusions

According to the results obtained from the characterization of the CW produced in the Mocha canton (Tungurahua, Ecuador), this liquid waste presented a high organic load, consisting mainly of lactose, proteins, and fats, and a notable content of macronutrients, as well as low concentrations of potentially toxic elements. All these characteristics are beneficial for co-composting CW with other organic solid wastes. However, the low pH values of whey, especially in the case of CW with milk casein coagulation by acid action (pH = 4.88–5.78), and its high salt content can be limiting factors for developing the composting process and obtaining a compost of agricultural quality.

Based on the results of treating CW by composting, we can infer that co-composting this liquid waste with plant residues and cow manure can be an effective way to valorize these residues in remote areas with a low-income population and limited access to expensive technologies. The quantities of residues used (60% corn residues + 5% bean residues + 35% cow manure—on a fresh weight basis—and 0.35 L CW/kg fresh weight of initial waste mixture) were adequate to achieve the thermophilic stage of the composting process, which is necessary to reduce phytotoxic compounds and pathogenic microorganisms. However, it was observed that irrigation with CW increased salinity levels, so caution should be taken with the amounts supplied to the composting process to avoid compost with excess salts. Nevertheless, the overall results were compatible with plant growth due to the significant OM, Nt, and P contents of the final compost and the final values of Corg/Nt ratio (<20), CEC (>67 meq/100 g OM) and GI (>50%), indicating OM stabilization and humification and phytotoxicity reduction during the composting process.

Therefore, co-composting agro-livestock wastes with CW can reduce the environmental impact of whey disposal, particularly in regions where small-scale dairy operations lack access to efficient processing technologies, thus providing an alternative valorization method. It can also contribute to synergies between small and medium-sized cheese, agricultural, and livestock farms, which could benefit from centralized composting facilities to manage their wastes and mitigate environmental impacts. However, further studies are needed to optimize the amount or dilution of CW to reduce the salt content of the final compost, as this was the main limiting factor of using this liquid waste as a moisture source during the co-composting of solid organic wastes.

Author Contributions: Conceptualization, C.P. and I.G.-T.; Methodology, S.R.-R., I.G.-T., J.I.-N., A.I.-G., V.V.-O. and C.P.; Data curation, C.P., J.I.-N. and S.R.-R.; Validation, C.P. and I.G.-T.; Formal analysis, S.R.-R., I.G.-T., J.I.-N. and C.P.; Investigation, C.P. and I.G.-T.; Writing—original draft preparation, S.R.-R., C.P. and I.G.-T.; Writing—review and editing, C.P.; Visualization, C.P., I.G.-T. and J.I.-N.; Supervision, C.P. and I.G.-T.; Project administration, I.G.-T.; Funding acquisition, I.G.-T. and C.P. All authors have read and agreed to the published version of the manuscript.

Funding: This work was supported by the Higher Polytechnic School of Chimborazo (Ecuador), in the framework of the 1824.IDI.ESPOCH.2021 project.

Institutional Review Board Statement: Not applicable.

Data Availability Statement: The data and original contributions presented in this study are included in the article. Any additional information beyond that presented in the article is available upon request from the corresponding author.

Acknowledgments: This study is part of the research project entitled: "Estudio de biorremediación de suelos contaminados con la ceniza volcánica empleando compost proveniente de residuos agroindustriales en la provincia de Chimborazo", carried out between the Associated Research Group in Biotechnology, Environment and Chemistry (GAIBAQ) of the Higher Polytechnic School of Chimborazo and the Environmental Research Group of Agrochemistry and Environment (GIAAMA) of the Miguel Hernández University of Elche, for which the authors appreciate their financial support and scientific contribution.

Conflicts of Interest: The authors declare no conflicts of interest.

References

1. FAOSTAT. Available online: http://www.fao.org/faostat/en/#data (accessed on 8 May 2024).
2. Mazorra-Manzano, M.A.; Moreno-Hernández, J.M. Propiedades y opciones para valorizar el lactosuero de la quesería artesanal. *Cienc. UAT* **2019**, *14*, 133–144. [CrossRef]
3. Fox, P.F.; Guinee, T.P.; Cogan, T.M.; McSweeney, P.L.H. *Fundamentals of Cheese Science*, 2nd ed.; Springer: New York, NY, USA, 2017.
4. Pires, A.F.; Marnotes, N.G.; Rubio, O.D.; Garcia, A.C.; Pereira, C.D. Dairy by-products: A review on the valorization of whey and second cheese whey. *Foods* **2021**, *10*, 1067. [CrossRef] [PubMed]
5. Panghal, A.; Patidar, R.; Jaglan, S.; Chhikara, N.; Khatkar, S.K.; Gat, Y.; Sindhu, N. Whey valorization: Current options and future scenario—A critical review. *Nutr. Food Sci.* **2018**, *48*, 520–535. [CrossRef]
6. Ostertag, F.; Schmidt, C.M.; Berensmeier, S.; Hinrichs, J. Development and validation of an RP-HPLC DAD method for the simultaneous quantification of minor and major whey proteins. *Food Chem.* **2021**, *342*, 128176. [CrossRef]
7. de Almeida, M.P.G.; Mockaitis, G.; Weissbrodt, D.G. Got whey? Sustainability endpoints for the dairy industry through resource biorecovery. *Fermentation* **2023**, *9*, 897. [CrossRef]
8. Zhang, W.; Chen, J.; Chen, Q.; Wu, H.; Mu, W. Sugar alcohols derived from lactose: Lactitol, galactitol, and sorbitol. *Appl. Microbiol. Biotechnol.* **2020**, *104*, 9487–9495. [CrossRef]
9. Arshad, U.-E.; Hassan, A.; Ahmad, T.; Naeem, M.; Chaudhary, M.T.; Abbas, S.Q.; Randhawa, M.A.; Pimentel, T.C.; da Cruz, A.G.; Aadil, R.M. A recent glance on the valorisation of cheese whey for industrial prerogative: High-value-added products development and integrated reutilising strategies. *Int. J. Food Sci. Technol.* **2023**, *58*, 2001–2013. [CrossRef]
10. Ketterings, Q.; Czymmek, K.; Gami, S.; Godwin, G.; Ganoe, K. *Guidelines for Land Application of Acid Whey*; Department of Animal Science, College of Agriculture & Life Sciences, Cornell University: Ithaca, NY, USA, 2017.
11. Chalermthai, B.; Giwa, A.; Schmidt, J.E.; Taher, H. Life cycle assessment of bioplastic production from whey protein obtained from dairy residues. *Bioresour. Technol. Rep.* **2021**, *15*, 100695. [CrossRef]
12. Osorio-González, C.S.; Gómez-Falcon, N.; Brar, S.K.; Ramírez, A.A. Cheese whey as a potential feedstock for producing renewable biofuels: A review. *Energies* **2022**, *15*, 6828. [CrossRef]
13. Estudio de Mercado. Sector Lácteo. 2021. Available online: https://www.sce.gob.ec/sitio/wp-content/uploads/2021/04/estudio_de_mercado_sector_lacteo_SCPM-IGT-INAC-002-2019.pdf (accessed on 23 July 2024).

14. Acosta, A.; Galetto, A.; Valdés, A.; Londinsky, A. *Más Allá de la Finca Lechera—Enmarcando el Diálogo de Política Lechera en América Latina*; FAO and FEPALE: Rome, Italy, 2022.
15. Poveda, E. Suero lácteo, generalidades y potencial uso como fuente de calcio de alta biodisponibilidad. *Rev. Chil. Nutr.* **2013**, *40*, 397–403. [CrossRef]
16. Prazeres, A.R.; Carvalho, F.; Rivas, J. Cheese whey management: A review. *J. Environ. Manag.* **2012**, *110*, 48–68. [CrossRef] [PubMed]
17. Caballero, P.; Rodríguez-Morgado, B.; Sandra, M.; Manuel, T.; Juan, P. Obtaining plant and soil biostimulants bywaste whey fermentation. *Waste Biomass Valoriz.* **2020**, *11*, 3281–3292. [CrossRef]
18. Ayilara, M.; Olanrewaju, O.; Babalola, O.; Odeyemi, O. Waste management through composting: Challenges and potentials. *Sustainability* **2020**, *12*, 4456. [CrossRef]
19. Vázquez, M.A.; de la Varga, D.; Plana, R.; Soto, M. Integrating liquid fraction of pig manure in the composting process for nutrient recovery and water re-use. *J. Clean. Prod.* **2015**, *104*, 80–89. [CrossRef]
20. Rastogi, M.; Nandal, M.; Nain, L. Additive effect of cow dung slurry and cellulolytic bacterial inoculation on humic fractions during composting of municipal solid waste. *Int. J. Recycl. Org. Waste Agricult.* **2019**, *8*, 325–332. [CrossRef]
21. Majbar, Z.; Lahlou, K.; Ben Abbou, M.; Ammar, E.; Triki, A.; Abid, W.; Nawdali, M.; Bouka, H.; Taleb, M.; El Haji, M.; et al. Co-composting of olive mill waste and wine-processing waste: An application of compost as soil amendment. *J. Chem.* **2018**, *2018*, 7918583. [CrossRef]
22. Bargougui, L.; Guergueb, Z.; Chaieb, M.; Mekki, A. Co-composting of olive industry wastes with poultry manure and evaluation of the obtained compost maturity. *Waste Biomass Valor.* **2020**, *11*, 6235–6247. [CrossRef]
23. Ghasemzadeh, S.; Sharafi, R.; Salehi Jouzani, G.; Karimi, E.; Ardakani, M.R.; Vazan, S. Efficient lignocellulose degradation during rice straw composting with native effective microorganisms and chicken manure. *Org. Agric.* **2022**, *12*, 397–409. [CrossRef]
24. Bustamante, M.A.; Paredes, C.; Moral, R.; Moreno-Caselles, J.; Pérez-Murcia, M.D.; Pérez-Espinosa, A.; Bernal, M.P. Co-composting of distillery and winery wastes with sewage sludge. *Water Sci. Technol.* **2007**, *56*, 187–192. [CrossRef]
25. Alfonzo, A.; Laudicina, V.A.; Muscarella, S.M.; Badalucco, L.; Moschetti, G.; Spanò, G.M.; Francesca, N. Cellulolytic bacteria joined with deproteinized whey decrease carbon to nitrogen ratio and improve stability of compost from wine production chain by-products. *J. Environ. Manag.* **2022**, *304*, 114194. [CrossRef]
26. Pivato, A.; Malesani, R.; Bocchi, S.; Rafieenia, R.; Schievano, A. Biochar addition to compost heat recovery systems improves heat conversion yields. *Front. Energy Res.* **2024**, *11*, 1327136. [CrossRef]
27. Memoria Técnica Cantón Mocha/Bloque 1.1. Available online: http://metadatos.sigtierras.gob.ec/pdf/Memoria_tecnica_Coberturas_MOCHA_20150306.pdf (accessed on 29 July 2024).
28. Rice, E.W.; Baird, R.B.; Eaton, A.D.; Clesceri, L.S. *Standard Methods for the Examination of Water and Wastewater*; American Public Health Association: Washington, DC, USA, 2012; Volume 10.
29. Wehr, H.M.; Frank, J.F. *Standard Methods for the Examination of Dairy Products*, 17th ed.; American Public Health Association: Washington, DC, USA, 2004.
30. Idrovo-Novillo, J.; Gavilanes-Terán, I.; Bustamante, M.A.; Paredes, C. Composting as a method to recycle renewable plant resources back to the ornamental plant industry: Agronomic and economic assessment of composts. *Process Saf. Environ. Protect.* **2018**, *116*, 388–395. [CrossRef]
31. Rahman, M.H.; Singh, N.; Kundu, S.; Datta, A. Potential areas of crop residue burning contributing to hazardous air pollution in Delhi during the post-monsoon season. *J. Environ. Qual.* **2022**, *51*, 181–192. [CrossRef]
32. Mgalula, M.E.; Wasonga, O.V.; Hülsebusch, C.; Richter, U.; Hensel, O. Greenhouse gas emissions and carbon sink potential inEastern Africa rangeland ecosystems: A review. *Pastoralism* **2021**, *11*, 19. [CrossRef]
33. Snyman, H.A. Short-term responses of Southern African semi-arid rangelands to fire: A review of impact on soils. *Arid Land Res. Manag.* **2015**, *29*, 222–236. [CrossRef]
34. Medina, J.; Monreal, C.; Barea, J.M.; Arriagada, C.; Borie, F.; Cornejo, P. Crop residue stabilization and application to agricultural and degraded soils: A review. *Waste Manag.* **2015**, *42*, 41–54. [CrossRef]
35. Fu, B.; Chen, L.; Huang, H.; Qu, P.; Wei, Z. Impacts of crop residues on soil health: A review. *Environ. Pollut. Bioavailab.* **2021**, *33*, 164–173. [CrossRef]
36. Mendoza, D.; Marini, P.; Zambrano, J. Los bovinos criollos un recurso zoogenético de seguridad alimentaria para Ecuador y Latinoamérica. *Rev. Cient. Arbitr. Multidiscip. PENTACIENCIAS* **2022**, *4*, 175–185.
37. Urra, J.; Alkorta, I.; Garbisu, C. Potential benefits and risks for soil health derived from the use of organic amendments in agriculture. *Agronomy* **2019**, *9*, 542. [CrossRef]
38. Ghirardini, A.; Grillini, V.; Verlicchi, P. A review of the occurrence of selected micropollutants and microorganisms in different raw and treated manure-Environmental risk due to antibiotics after application to soil. *Sci. Total Environ.* **2020**, *707*, 136118. [CrossRef]

39. van derWeerden, T.J.; Noble, A.; de Klein, C.A.M.; Hutchings, N.; Thorman, R.E.; Alfaro, M.A.; Amon, B.; Beltran, I.; Grace, P.; Hassouna, M.; et al. Ammonia and nitrous oxide emission factors for excreta deposited by livestock and land-applied manure. *J. Environ. Qual.* **2021**, *50*, 1005–1023. [CrossRef] [PubMed]
40. Onwosi, C.O.; Igbokwe, V.C.; Odimba, J.N.; Eke, I.E.; Nwankwoala, M.; Iroh, I.N.; Ezeogu, L.I. Composting technology in waste stabilization: On the methods, challenges and future prospects. *J. Environ. Manag.* **2017**, *190*, 140–157. [CrossRef] [PubMed]
41. Carvalho, F.; Prazeres, A.R.; Rivas, J. Cheese whey wastewater: Characterization and treatment. *Sci. Total Environ.* **2013**, *445*, 385–396. [CrossRef] [PubMed]
42. Sebastián-Nicolás, J.L.; González-Olivares, L.G.; Vázquez-Rodríguez, G.A.; Lucho-Constatino, C.; Castañeda-Ovando, A.; Cruz-Guerrero, A.E. Valorization of whey using a biorefinery. *Biofuels Bioprod. Bioref.* **2020**, *14*, 1010–1027. [CrossRef]
43. EPA, United States Environment Protection Agency. *Environmental Regulations and Technology Control of Pathogens and Vector Attraction in Sewage Sludge*; EPA625-/R-92/-103; EPA: Cincinnati, OH, USA, 2003.
44. Arthurson, V. Proper sanitization of sewage sludge: A critical issue for a sustainable society. *Appl. Environ. Microbiol.* **2008**, *74*, 5267–5275. [CrossRef]
45. Hanajima, D.; Aoyagi, T.; Hori, T. Dead bacterial biomass-assimilating bacterial populations in compost revealed by high-sensitivity stable isotope probing. *Environ. Int.* **2019**, *133*, 105235. [CrossRef]
46. Bernal, M.P.; Sommer, S.G.; Chadwick, D.; Qing, C.; Guoxue, L.; Michel, F.C., Jr. Current approaches and future trends in compost quality criteria for agronomic, environmental, and human health benefits. *Adv. Agron.* **2017**, *144*, 143–233.
47. Fereidoon, S.; Vamadevan, V.; Won Young, O.; Han, P. Phenolic compounds in agri-food by-products, their bioavailability and health effects. *J. Food Bioact.* **2019**, *5*, 57–119.
48. Hegab, M.M.; Abdelgawad, H.; Abdelhamed, M.S.; Hammouda, O.; Pandey, R.; Kumar, V.; Zinta, G. Effects of tricin isolated from jungle rice (*Echinochloa colona* L.) on amylase activity and oxidative stress in wild oat (*Avena fatua* L.). *Allelopath. J.* **2013**, *31*, 345–354.
49. Zucconi, F.; Pera, A.; Forte, M.; de Bertoldi, M. Evaluating toxicity of immature compost. *Biocycle* **1981**, *22*, 54–57.
50. Cáceres, R.; Malińska, K.; Marfà, O. Nitrification within composting: A review. *Waste Manag.* **2018**, *72*, 119–137. [CrossRef] [PubMed]
51. US Composting Council. Field Guide to Compost Use. 2001. Available online: http://www.mncompostingcouncil.org/uploads/1/5/6/0/15602762/fgcu.pdf (accessed on 16 December 2024).
52. Bernal, M.P.; Alburquerque, J.A.; Moral, R. Composting of animal manures and chemical criteria for compost maturity assessment: A review. *Bioresour. Technol.* **2009**, *100*, 5444–5453. [CrossRef] [PubMed]

agriculture

MDPI

Article

Exploring the Role of Biostimulants in Sweet Cherry (*Prunus avium* L.) Fruit Quality Traits

Sílvia Afonso [1,*], Ivo Oliveira [1], Carlos Ribeiro [2], Alice Vilela [3], Anne S. Meyer [4] and Berta Gonçalves [1]

[1] Centre for the Research and Technology for Agro-Environmental and Biological Sciences (CITAB), Institute for Innovation, Capacity Building and Sustainability of Agri-food Production, Inov4Agro, Department of Biology, University of Trás-os-Montes e Alto Douro (UTAD), Quinta de Prados, 5000-801 Vila Real, Portugal; ivo.vaz.oliveira@utad.pt (I.O.); bertag@utad.pt (B.G.)

[2] Centre for the Research and Technology for Agro-Environmental and Biological Sciences (CITAB), Institute for Innovation, Capacity Building and Sustainability of Agri-food Production, Inov4Agro, Department of Agronomy, University of Trás-os-Montes e Alto Douro (UTAD), Quinta de Prados, 5000-801 Vila Real, Portugal; cribeiro@utad.pt

[3] Chemistry Research Centre, CQ-VR, Department of Agronomy, University of Trás-os-Montes e Alto Douro (UTAD), Quinta de Prados, 5000-801 Vila Real, Portugal; avimoura@utad.pt

[4] Department of Biotechnology and Biomedicine, Technical University of Denmark, DTU Building 221, 2800 Kongens Lyngby, Denmark; asme@dtu.dk

* Correspondence: safonso@utad.pt

Abstract: Sweet cherries are among consumers' preferred fresh fruits, known for their attractive organoleptic properties and high nutritional value. Agronomical practices, which are now shifting to more environmentally sustainable options, can influence several key quality traits of sweet cherries. In this context, reducing conventional agrochemicals and increasing the application of preharvest biostimulants has emerged as an innovative strategy. This approach can not only enhance cherry production and quality but also ensure the economic and environmental sustainability of the cherry supply chain. Hence, this work is aimed at studying the effect of the application of two concentrations of glycine betaine (GB) and Ecklonia maxima-based (EM) biostimulants, and their combination, in two cultivars of sweet cherry: the early-maturing 'Early Bigi' and the late-maturing 'Lapins', both grafted onto SL-64 rootstock. Evaluated parameters included fruit weight and dimensions, color, firmness, total soluble solids (TSS), titratable acidity (TA), phenolic and anthocyanin contents, and sensory profile. Key findings highlight that, with a few exceptions, biostimulant treatments had a positive impact on the studied parameters, although the responses varied between cultivars. For instance, fruit size increased by 13.41% in 'Early Bigi' and 47.20% in 'Lapins'. Additionally, reduced color values, coupled with higher TSS/TA ratios, indicate advanced fruit maturation, which could allow for an earlier harvest. The total phenolic content rose by 56.88% in 'Early Bigi' and 30.24% in 'Lapins', while anthocyanin levels surged by 88.28% and 36.10%, respectively. Fruit firmness also improved following biostimulant application. Sensory analysis further revealed enhancements in key descriptors such as "overall aspect", "firmness", and "cherry flavor", underscoring the beneficial effects of these treatments. These combined results indicate that the preharvest application of glycine betaine or *Ecklonia maxima*-based (EM) biostimulants significantly improves key quality traits of sweet cherries. This approach offers benefits not only from a commercial perspective but also for the sweet cherry supply chain sustainability by reducing the application of chemical-based products and replacing them with ecofriendly substances while enhancing the quality of the fruit.

Keywords: chemical attributes; glycine betaine; fruit quality; fruit skin color; seaweed *Ecklonia maxima*-based extract; sensory evaluation; sustainable agriculture; texture

1. Introduction

Sweet cherries are among the preferred fresh fruits for consumers owing to their appealing organoleptic properties, high nutritional value, and recognized health benefits, which are attributed to their rich antioxidant content [1–4]. These quality features, associated with fruit size, uniform skin color, firmness, sweetness, absence of defects, and a green stem, give this fruit substantial economic viability [5].

According to FAOSTAT (2024), the world cherry production reached 2,765,827 tons in 2022, with a substantial increase of 137% in Portugal over the last decade [6]. This growth highlights sweet cherry cultivation's significance in the country's agricultural landscape. In particular, the Entre Douro e Minho region plays a crucial role in cherry cultivation, particularly in the municipality of Resende, due to its unique edaphoclimatic conditions, resulting in the early market availability of sweet cherries with superior quality and excellent flavor [7]. However, this crop faces some challenges and constraints related to its sensitivity to climate conditions [8]. Climatic fluctuations throughout its vegetative cycle can affect fruit formation, flowering, and fertilization, resulting in production losses. Additionally, physiological disorders can compromise its quality and, consequently, its market value [9,10]. As such, several agronomic strategies have been employed to enhance cherry production and quality, with recent emphasis placed on the use of novel biostimulants [3,11,12]. These are considered environmentally friendly and promising alternatives, and their application has proven to be highly effective in mitigating both abiotic and biotic stresses, thereby improving tree productivity and performance while enhancing the overall quality [3,13]. Among the entire categories of biostimulants, glycine betaine (GB) and seaweed extracts have garnered attention from the scientific and agronomic community [3,14] and according to Colla et al. [15,16], they represent the two most important categories of biostimulants.

Despite the limited studies on the effects of the foliar application of the osmoregulatory GB, it emerges as a promising alternative to enhance cherry tree performance [17,18]. Moreover, it shows potential to improve fruit quality, particularly in increasing fruit size, total soluble solids, pH, and reducing acidity [3,19,20]. Recently, foliar applications of seaweed extracts have demonstrated potential effects on fruit, including increases in organic acid concentration and resistance to fruit cracking [20–22]. However, little literature is available regarding the foliar application of these two biostimulants in sweet cherry fruit quality. Therefore, this study aimed to assess the preharvest application effects of GB and seaweed-based *Ecklonia maxima* extract on the cherries' quality, and the sensory attributes of the early-maturing 'Early Bigi' and the late-maturing 'Lapins'.

2. Materials and Methods

2.1. Site Description and Weather Conditions

Experiments were performed in a commercial sweet cherry orchard located at Quinta da Alufinha, Municipality of Resende (latitude 41°06′ N and longitude 7°54′ W) in three consecutive years: 2019, 2020, and 2021. The orchard was seven years old and located at a low elevation (140 m above sea level). The spacing was 2.5 m between trees in a row and 3.0 m between rows. Irrigation was applied through a drip irrigation system between each tree in the row. All trees were treated with standard fertilizer, herbicides, and pesticides, ensuring similar agronomic practices throughout the trial years.

An automatic weather station was installed near the orchard, and the meteorological data from the 2019, 2020, and 2021 growing seasons were recorded.

2.2. Plant Material, Treatments, and Sampling

This research specifically concentrated on the two most representative cultivars of the region: the early-maturing 'Early Bigi' and the late-maturing 'Lapins', both grafted onto SL-64 rootstock.

Eight trees from each cultivar were selected for each treatment. Six different formulations were applied: two concentrations of glycine betaine (GB 0.25% and 0.40%), two concentrations of *Ecklonia maxima* seaweed-based biostimulants (EM 0.15% and 0.30%), a combination of the lowest concentrations of both biostimulants (Mix-GB 0.25% and EM 0.15%), and the control (C), which was performed using tap water.

The treatments were applied using a backpack sprayer and repeated at three different stages of the sweet cherry tree growth cycle, according to the BBCH scale (Biologische Bundesanstalt, Bundessortenamt, and Chemical industry scale) [23]: stage 77 (representing 70% of final fruit size), stage 81 (the beginning of fruit coloring), and stage 86 (indicating advanced coloring, three days before fruit harvesting). Table 1 shows the dates of the treatment applications over the three years. However, due to the COVID-19 pandemic, in 2020 it was not possible to harvest the fruit and thus, results from the present study are focused only on data collected for 2019 and 2021.

Table 1. Biostimulant treatments applied to sweet cherry cultivars 'Early Bigi' and 'Lapins'.

Cultivars	Biostimulants	Concentrations	Sweet Cherries Developmental Stages	Date of Spraying		
				2019	2020	2021
'Early Bigi'	Glycine betaine (GB) *Ecklonia maxima* extract (EM)	GB 0.25% and GB 0.40% EM 0.15% and EM 0.30%	BBCH 77	11 April	1 April	6 April
			BBCH 81	19 April	11 April	15 April
			BBCH 86	30 April	16 April	24 April
'Lapins'	Glycine betaine (GB) *Ecklonia maxima* extract (EM)	GB 0.25% and GB 0.40% EM 0.15% and EM 0.30%	BBCH 77	11 April	1 April	6 April
			BBCH 81	16 May	4 May	11 May
			BBCH 86	24 May	12 May	31 May

In 2019, cv. 'Early Bigi' was harvested on 3 May and cv. 'Lapins' on 27 May, whereas in 2021, cv. 'Early Bigi' was harvested on 27 April and cv. 'Lapins' on 3 June.

For quality measurements, 100 fruits from each tree, treatment, and cultivar were randomly collected at commercial maturity, transported in a portable freezer under refrigeration to the laboratory, and then carefully divided into three groups. The first group was designated for biometric measurements, epidermis rupture force (ERF), flesh firmness (FF), skin color, total soluble solids (TSS), and titratable acidity (TA). The second group was used to determine total phenolics and anthocyanins. The third group was allocated for sensory analysis. Fruits from the second group were frozen in liquid nitrogen, lyophilized (SCANVAC 55-4 Pro, LaboGene, Lynge, Denmark) for 120 h at −55 °C, and then ground into a dried powder using a commercial blender (Model BL41, Waring Commercial, Torrington, WY, USA).

2.3. Physical Characteristics of Sweet Cherries

Weight, Dimensions, Texture, and Skin Color

Fruit weight (g), length (mm), width (mm), and diameter (mm) were measured in 30 fruits per treatment.

The epidermis rupture force (ERF, N) and flesh firmness (FF, N mm^{-1}) were determined using a TA.XTPlus texture analyzer (Stable Micro Systems, Godalming, Surrey, UK), employing a 5 kg loading cell and a 2 cm diameter cylindrical probe, with a total displacement of 5 mm at a speed of 1 mm s^{-1}.

The fruit color was measured on opposite sides of the fruit using a colorimeter (Model CR-300, Minolta, Osaka, Japan) and expressed in CIE Lab* values. Before measurement, the fruit surfaces were gently cleaned with a dry cloth to remove any residues that could affect the accuracy of the readings. In this color space, L* represents lightness (0 = black, 100 = white), a* represents the green-red spectrum (negative values indicate greenness, positive values indicate redness), and b* represents the blue-yellow spectrum (negative values indicate blueness, positive values indicate yellowness). Subsequently, the hue angle (Hue°), which indicates color intensity (purity), was determined using the formula Hue° = arctan(b*/a*). Additionally, the chroma (C*), representing color saturation, was calculated using the formula $(a^{*2} + b^{*2})^{1/2}$ (calculations performed using Microsoft Excel 365, version 2407).

2.4. Chemical Properties of Sweet Cherries

2.4.1. Total Soluble Solids, pH, and Titratable Acidity

Thirty fruits from the first group were divided into three sub-groups of ten fruits each. The juice was extracted from each group using an electrical extractor (ZN350C70, Tefal Elea, China) for 1 min.

The juice's total soluble solids content (TSS, °Brix) was measured using a digital refractometer (PR-101, Atago, Japan) at ±20 °C. Titratable acidity (TA, % malic acid) was determined by automatic titration (Schott Easy Titroline automatic titrator, Germany) with 0.1 N NaOH to pH 8.2 after diluting 10 mL of juice with 10 mL of distilled water.

The maturity index, calculated as the ratio of TSS and TA, was expressed as the average of three replicates with standard deviation (SD).

2.4.2. Bioactive Compounds

For total phenolic content, the methods of Singleton and Rossi [24] and Dewanto et al. [25] were applied with minor modifications. Forty milligrams of the freeze-dried sample were dissolved in 1 mL of 70% (*v*/*v*) aqueous methanol, heated to 70 °C, and vortexed every 5 min for 30 min. The extracts were centrifuged at 13,000 rpm (Centrifuge 5804R, Eppendorf, Hamburg, Germany) at 1 °C for 15 min. The supernatants were filtered through a 0.2 µm, 13 mm diameter PTFE filter (Teknokroma, Spain). Subsequently, 20 µL of each extract was added to 100 µL of Folin–Ciocalteu phenol reagent (1:10 in bidistilled H_2O) and 80 µL of 7.5% Na_2CO_3 in a 96-well microplate (Multiskan™ FC Microplate Photometer, Waltham, MA, USA). The microplate was incubated for 15 min at 45 °C in the dark. Absorbance values against a blank were recorded immediately at 765 nm using a microplate reader (Multiskan GO Microplate Spectrophotometer, Thermo Scientific, Vantaa, Finland). A standard curve using gallic acid at various concentrations was created, and the total phenolic content was expressed as mg gallic acid equivalent (GAE) per gram of dry weight (DW), with results presented as the mean ± standard error (SE) of three replicates.

Total anthocyanins were quantified according to Nicoué et al. [26]. Methanolic extracts were prepared by dissolving 0.5 g of the freeze-dried sample in 5 mL of acidified methanol (1% HCl, *v*/*v*) and keeping the solution at 4 °C for 1 h. The extracts were then centrifuged at 4000 rpm at 4 °C for 15 min (Centric 250, UniEquip, Munich, Germany) and filtered using Whatman™ No. 1 90 mm filter paper. The supernatants were diluted with 0.025 M of KCl buffer (pH 1) and 0.4 M of sodium acetate (CH_3COONa) buffer (pH 4.5) at a ratio of 1:6 and allowed to stand for 10 min. Absorbance readings at 541 nm and 700 nm were then taken spectrophotometrically for both pH 1 and pH 4.5 buffers (U-2000, serial 121-0120, Hitachi Ltd., Tokyo, Japan). Net anthocyanin absorbance was calculated as $(A_{541}-A_{700})$ pH 1.0—$(A_{541}-A_{700})$ pH 4.5, and anthocyanin content was determined using the molar absorptivity (ε= 26,900) and molecular weight (MW = 449.2) of cyanidin-3-*O*-rutinoside. Results are expressed as cyanidin-3-*O*-rutinoside equivalents (mg cy-3-rut g^{-1} DW), reported as the mean ± standard error (SE) of three replicates.

2.5. Sensory Evaluation of Sweet Cherries

Fifteen cherry attributes (Chauvin et al. [27]) (Table S2) were evaluated by a trained panel of twelve experts. Sensory evaluation sessions were conducted annually in a laboratory for sensory analysis maintained at approximately 20 °C (ISO 6658 [28]), specifically once a year in 2019 and 2021. The attributes were rated on a five-point scale, with 1 indicating that the attribute was not present and 5 that it was present at the highest intensity (ISO 4121 [29]).

For each session, three randomly coded cherries per treatment were presented to each panelist on white Pyrex plates. To ensure consistent conditions, cherries were equilibrated at room temperature for 2 h before the sessions. Panelists cleansed their palates with a sip of water or a bite of a low-salt cracker between samples. To maintain the integrity of the sensory evaluation, panelists refrained from wearing perfume and avoided consuming food, drink, or smoking for one hour before the tasting.

2.6. Statistical Analysis

The statistical analysis was performed using SPSS version 27 (SPSS-IBM Corp., Armonk, NY, USA). Before conducting the study, the assumptions for an analysis of variance (ANOVA) were verified, including the homogeneity of variances using Levene's mean test and normality using the Shapiro–Wilk test. Statistical differences among treatments within each variety and year were assessed using a one-way ANOVA, followed by Tukey's post hoc test for multiple comparisons. To evaluate the main effects of treatment, year, and their interactions, a multivariate analysis of variance (MANOVA) was performed, utilizing Pillai's trace as the test statistic. Differences were considered statistically significant at $p < 0.05$. Sensory data were analyzed using a one-way ANOVA and Duncan's multiple range test for post hoc comparisons ($p < 0.05$).

3. Results

3.1. Weather Conditions

Weather conditions are summarized in Figure 1. Based on data from 2019 to 2021, there was a clear trend of hot and dry summers followed by wetter winters. The year 2020 was the warmest, with maximum temperatures reaching 32.1 °C in July and 23.3 °C in May, surpassing the values recorded in 2019 and 2021. In opposition, 2019 stood out as the rainiest year, with significant precipitation levels that exceeded those observed in 2020 (1091 mm) and 2021 (1076 mm), making 2019 the year with the highest rainfall (1163 mm).

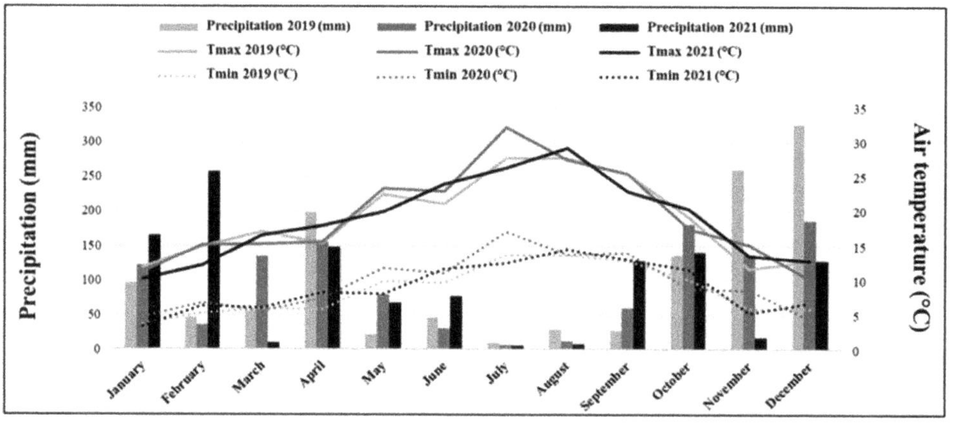

Figure 1. Monthly maximum (Tmax) and minimum (Tmin) air temperature (°C) and precipitation (mm) in 2019, 2020, and 2021.

3.2. Biometric and Physical Characteristics of Sweet Cherries

For both cultivars, treatment and year significantly influenced the biometric parameters of sweet cherries, as well as the interaction between these two factors (Table S1). The application of GB and EM biostimulants significantly influenced fruit weight and dimensions for both cultivars (Table 2), with a low concentration of biostimulants and their combination being particularly effective for the cv. 'Early Bigi'. For cv. 'Lapins', the best overall results were observed with the higher concentrations of both biostimulants.

Table 2. Physical parameters of 'Early Bigi' and 'Lapins' cherries after different preharvest treatments, with results for 2019 and 2021.

Cultivar	Parameter	Year	C	EM 0.15%	EM 0.30%	GB 0.25%	GB 0.40%	Mix	Treatment Effect
'Early Bigi'	Weight (g)	2019	9.44 ± 0.74 ab	7.67 ± 0.74 c	7.46 ± 0.62 c	9.57 ± 0.83 a	8.84 ± 0.89 b	8.00 ± 0.70 c	***
		2021	8.65 ± 0.69 c	9.54 ± 0.70 ab	9.06 ± 0.71 bc	9.49 ± 0.87 ab	8.83 ± 0.87 c	9.81 ± 0.86 a	***
	Length (mm)	2019	24.84 ± 0.58 a	23.46 ± 0.85 cd	22.83 ± 0.61 d	24.47 ± 0.66 ab	24.12 ± 0.71 bc	23.67 ± 0.89 c	***
		2021	22.40 ± 1.46 c	23.48 ± 0.78 b	22.82 ± 0.94 bc	23.35 ± 1.07 b	23.26 ± 0.95 b	24.33 ± 0.77 a	***
	Diameter (mm)	2019	22.87 ± 0.76 ab	21.40 ± 0.70 c	20.60 ± 0.59 d	23.18 ± 1.01 a	22.14 ± 0.97 bc	21.84 ± 0.96 c	***
		2021	22.11 ± 0.87 bc	22.66 ± 1.08 ab	21.69 ± 1.06 c	22.40 ± 1.02 abc	22.63 ± 0.84 ab	23.02 ± 1.08 a	***
	Width (mm)	2019	28.48 ± 0.98 ab	26.07 ± 0.92 c	26.48 ± 1.01 c	28.79 ± 0.88 a	27.66 ± 1.34 b	26.47 ± 1.01 c	***
		2021	28.09 ± 1.21	28.13 ± 1.14	27.68 ± 1.12	28.57 ± 1.67	27.66 ± 0.98	28.09 ± 1.16	n.s.
'Lapins'	Weight (g)	2019	8.41 ± 1.12 d	11.12 ± 1.23 c	12.00 ± 0.75 ab	11.62 ± 1.12 bc	12.76 ± 1.08 a	11.82 ± 1.10b c	***
		2021	7.50 ± 1.30 d	10.37 ± 0.76 b	11.04 ± 0.83 a	10.35 ± 0.83 b	9.25 ± 0.82 c	8.89 ± 0.58 c	***
	Length (mm)	2019	26.65 ± 0.99 ab	23.41 ± 0.97 d	26.03 ± 0.75 bc	26.08 ± 1.33 bc	26.84 ± 0.98 a	25.81 ± 0.99 c	***
		2021	23.09 ± 1.30 d	25.01 ± 0.81 ab	25.31 ± 0.89 a	25.36 ± 0.94 a	24.26 ± 0.61 c	24.34 ± 0.89b c	***
	Diameter (mm)	2019	25.03 ± 1.16 a	22.24 ± 1.06 d	24.88 ± 0.67 ab	24.18 ± 0.99 bc	24.88 ± 0.91 ab	24.00 ± 0.97 c	***
		2021	21.42 ± 1.22 d	24.02 ± 1.06 a	23.88 ± 0.86 a	23.74 ± 0.64 ab	23.12 ± 0.75 bc	22.58 ± 0.70 c	***
	Width (mm)	2019	29.00 ± 1.29 ab	25.91 ± 1.45 c	29.06 ± 1.03 ab	28.68 ± 0.98 b	29.71 ± 0.89 a	29.10 ± 1.06 ab	***
		2021	24.64 ± 1.72 c	27.61 ± 1.00 a	28.25 ± 1.07 a	27.89 ± 0.89 a	26.63 ± 0.99 b	26.53 ± 0.87 b	***

C—Control treatment; EM 0.15%—Ecklonia maxima seaweed-based biostimulant applied at a concentration of 0.15%; EM 0.30%—Ecklonia maxima seaweed-based biostimulant applied at a concentration of 0.30%; GB 0.25%—Glycine betaine applied at a concentration of 0.25%; GB 0.40%—Glycine betaine applied at a concentration of 0.40%; Mix—EM 0.15% plus GB 0.25%. Data are the mean ± SD. For each row, different letters indicate significant differences between treatments (*** $p < 0.001$). n.s.—Not significant.

This positive effect on fruit weight and dimension from biostimulants is well-documented in several species, including sweet cherry [3]. For instance, increases in weight and dimension have been documented in cvs. 'Skeena', 'Sweetheart', 'Simone', 'Ziraat', 'Kordia', and 'Regina', following the application of various biostimulants [17,20,21,30–32]. Glycine betaine has been shown to increase these parameters in fruits of cv. 'Skeena' [21], while Ecklonia maxima-based biostimulants produced a similar effect in the fruits of cv. 'Bing' [22].

Cherry fruit growth follows a double-sigmoid growth curve consisting of three distinct stages: stage I, with mesocarp growth consisting of both cell division and cell enlargement; stage II, a lag period coinciding with endocarp hardening and embryo development; and stage III, a second period of exponential fruit growth with rapid cell expansion, classically referred to as cell division, endocarp lignification, and cell expansion, respectively [33]. The first application of biostimulants, in the present work, was performed when cherries were in the BBHC stage 77, with fruits about 70% of final size [23], and within the first phase of growth (stage I), which includes cellular division. Considering this timing of application, our results suggest that these biostimulants might have induced cytokinin-like effects, an activity previously described for other biostimulants [34,35].

The brown seaweed Ecklonia maxima extracts are rich in organic and mineral compounds. They contain unique and complex polysaccharides found exclusively in seaweeds, as well as a variety of plant hormones, including cytokinins, auxins, abscisic acid, gibberellins, and hormone-like compounds such as sterols and polyamines [3]. To date, there are limited studies on the effects of Ecklonia maxima extract on cherry fruits. However, Ureta Ovalle [22] conducted a study over three consecutive years on 'Bing' cherry trees and observed notable improvements in fruit-related biometric parameters following the application of this seaweed extract. The increase in cherry fruit size can be attributed to both the regulation of growth hormones [36] and endogenous hormonal homeostasis, which is influenced by the presence of polysaccharides [3,37]. These polysaccharides can activate biochemical pathways responsible for synthesizing secondary metabolites, thus contributing to the increase in fruit size and the overall improvement in plant quality. However, the exact mechanisms involved are not fully understood [37].

More complex interaction with fruit growth can be linked to the effect of glycine betaine on the expression of auxin-responsive IAA gene levels [38], which is linked to fruit dimensions and weight in sweet cherries [39–41].

Regarding color characteristics, results indicate that, for cv. 'Early Bigi', both treatment and year significantly influence L* ($p < 0.01$ and $p < 0.05$, respectively). At the same time, C* and Hue° were mainly affected by the year ($p < 0.05$), with no significant interaction between treatment and year. For cv. 'Lapins', treatment had a strong impact on L* and C* ($p < 0.001$), and the significant interaction between treatment and year suggests that treatment effects on color parameters vary by year, particularly for L* and C* (Table S1). Overall results show that the application of biostimulants decreased the L*, C*, and Hue° values compared to control samples (Table 3). Usually, lower values of these parameters indicate darker, redder, and mature fruit with lower water content and increased concentration of anthocyanins [42–45]. Furthermore, other studies point out the link between chroma reduction and increases in polyphenols, higher TSS, and reduced acidity of the fruits [46]. Considering that C* values usually decrease as ripening progresses, the present results may indicate earlier ripening due to the application of biostimulants [20]. Similar effects on color parameters have been recorded elsewhere when using biostimulants in sweet cherries. For instance, Ecklonia maxima seaweed-based biostimulants change the color of sweet cherries cv. 'Bing' [22], while glycine betaine and Ascophyllum nodosum-based extract reduced the values of L*, C*, and Hue° values in cv. 'Stacatto' [20].

Table 3. Color parameters (L*, C*, and Hue°) of 'Early Bigi' and 'Lapins' cherries after applying different preharvest treatments.

Cultivar	Parameter	Year	C	EM 0.15%	EM 0.30%	GB 0.25%	GB 0.40%	Mix	Treatment Effect
'Early Bigi'	*L**	2019	51.02 ± 2.89 a	37.37 ± 1.81 c	40.43 ± 4.42 bc	43.16 ± 4.17 b	43.25 ± 4.17 b	40.33 ± 3.81 bc	***
		2021	48.69 ± 2.27 a	45.40 ± 2.97 c	47.90 ± 2.85 cb	46.52 ± 2.94 abc	45.83 ± 4.36 bc	42.84 ± 3.31 d	***
	*C**	2019	41.60 ± 1.65 ab	34.92 ± 3.69 d	37.20 ± 5.31 cd	42.23 ± 3.09 a	40.56 ± 4.32 abc	37.98 ± 4.72 bcd	***
		2021	42.06 ± 2.19 b	44.01 ± 1.95 a	42.50 ± 2.01 ab	42.02 ± 2.26 b	41.86 ± 3.09 b	41.77 ± 2.77 b	***
	Hue°	2019	35.09 ± 1.14 a	27.14 ± 2.61 c	30.03 ± 2.13 b	31.40 ± 2.13 b	30.88 ± 3.06 b	30.12 ± 2.32 b	***
		2021	33.97 ± 2.07 a	33.39 ± 1.42 a	33.83 ± 1.88 a	33.29 ± 1.82 a	32.98 ± 2.27 a	31.45 ± 2.28 b	***
'Lapins'	*L**	2019	41.58 ± 3.52 a	36.89 ± 3.40 b	34.99 ± 2.21bc	33.47 ± 2.40 c	34.17 ± 3.09 c	33.85 ± 1.81 c	***
		2021	35.46 ± 1.84 b	38.40 ± 1.98 a	38.18 ± 2.59 a	37.61 ± 1.89 a	37.78 ± 2.54 a	36.95 ± 2.28 ab	***
	*C**	2019	42.22 ± 3.57 a	35.37 ± 4.96 b	35.01 ± 3.52 b	31.33 ± 4.47 c	34.83 ± 4.95 b	30.29 ± 4.05 c	***
		2021	38.58 ± 2.39	38.55 ± 2.31	39.69 ± 2.56	38.98 ± 2.48	38.78 ± 2.28	37.97 ± 2.73	n.s.
	Hue°	2019	29.89 ± 2.08 a	25.87 ± 3.82 b	25.61 ± 1.96 b	23.12 ± 2.76 c	26.36 ± 2.01 b	22.85 ± 2.34 c	***
		2021	26.79 ± 2.59	27.62 ± 2.03	28.09 ± 2.64	26.33 ± 2.58	26.81 ± 2.76	26.88 ± 2.62	n.s.

C—Control treatment; EM 0.15%—*Ecklonia maxima* seaweed-based biostimulant applied at a concentration of 0.15%; EM 0.30%—*Ecklonia maxima* seaweed-based biostimulant applied at a concentration of 0.30%; GB 0.25%—Glycine betaine applied at a concentration of 0.25%; GB 0.40%—Glycine betaine applied at a concentration of 0.40%; Mix—EM 0.15% plus GB 0.25%. Data are the mean ± SD. For each row, different letters indicate significant differences between treatments (*** $p < 0.001$). n.s.—Not significant.

A possible explanation for variations in fruit color might be linked to gene expression changes caused by hormones or hormone-like compounds present in the biostimulants that can upregulate, for instance, anthocyanin regulatory and biosynthetic genes [40]. Otherwise, exogenous applications of ABA are known to improve fruit color [47] but also indirectly affect color through the influence of auxin and cytokinin biosynthesis [48]. Even so, differences in color development occur naturally between cultivars with different ripening times [49] but also when using exogenous compounds, depending on the timing of application [50].

Observing cv. 'Early Bigi', the year had a significant effect on both epidermis rupture force (ERF) and flesh firmness (FF) ($p < 0.01$). However, the treatment and the interaction between treatment and year were not significant for ERF ($p > 0.05$). On the other hand, a significant interaction effect was observed for FF ($p < 0.05$), suggesting that the impact of treatments on firmness might vary across different years. For cv. 'Lapins', the year influenced both ERF and FF ($p < 0.001$), but neither the treatment nor the interaction between the treatment and year was significant for either parameter ($p > 0.05$). (Table S1).

Epidermis rupture force (RF) and flesh firmness (FF) were, in both cultivars, significantly affected by the treatments (Figure 2). Regarding RF, all treatments resulted in a higher force needed to cause a rupture of the epidermis of the fruits, with particular relevance to GB sprays, in cv. 'Early Bigi' and EM applications for cv. 'Lapins'. Positive effects of using biostimulants were also observed for FF, where almost all treatments resulted in firmer fruit compared to the control. Positive effects of foliar-applied compounds have been reported in sweet cherry [51], including the use of glycine betaine [19] and seaweed-based biostimulants [52]. Even so, other works did not find any influence of treatments on this specific parameter of sweet cherries [20]. The positive effects of such treatments in firmness appear to be linked to calcium metabolism. Indeed, it is known that calcium pectin cross-links have a crucial part in the physical and structural properties of fruit [53,54], including firmness, and several studies point out an increase in calcium concentration in cherry fruits after the application of exogenous compounds, like glycine betaine or plant extracts [1,52]. These increases in fruit firmness are of significant importance, as this parameter is strongly correlated to consumers' acceptance of cherries and with improved suitability for handling during postharvest operations [55,56], and to a reduced susceptibility to fruit cracking by a higher concentration of calcium within the fruit [57].

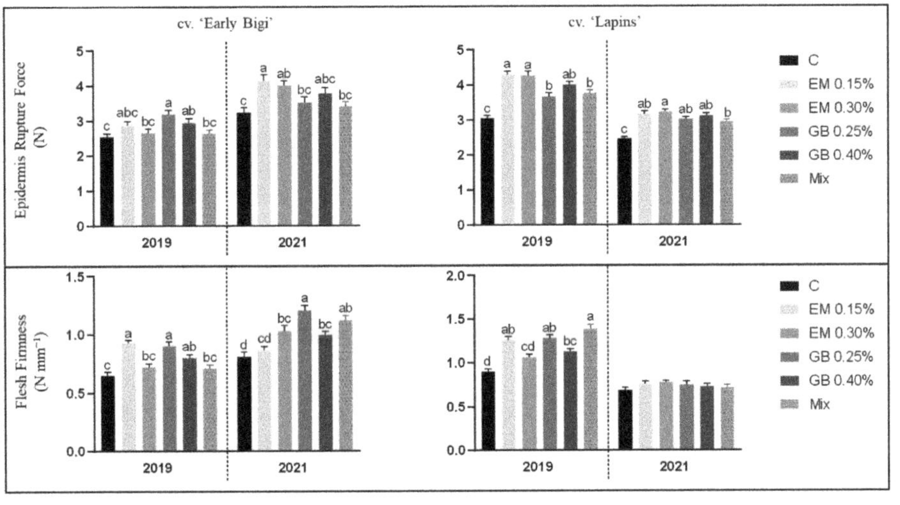

Figure 2. Epidermis rupture force (N) and flesh firmness (N mm^{-1}) of 'Early Bigi' and 'Lapins' cherries after applying different preharvest treatments. Values presented are expressed as mean ± standard error (SE). Different letters indicate significant differences between treatments within the sampling year. The absence of letters indicates that no significant differences were observed. C—Control treatment; EM 0.15%—*Ecklonia maxima* seaweed-based biostimulant applied at a concentration of 0.15%; EM 0.30%—*Ecklonia maxima* seaweed-based biostimulant applied at a concentration of 0.30%; GB 0.25%—Glycine betaine applied at a concentration of 0.25%; GB 0.40%—Glycine betaine applied at a concentration of 0.40%; Mix—EM 0.15% plus GB 0.25%.

3.3. Chemical Properties of Sweet Cherries

For both 'Early Bigi' and 'Lapins' cultivars, the treatments, year, and their interaction significantly affected the total soluble solids (TSS), titratable acidity (TA), and the maturity index (MI) ($p < 0.001$) (Table S1), highlighting the impact of biostimulants on the chemical characteristics of the fruits (Table 4), despite these values remaining within the expected range for sweet cherries [58–61].

Table 4. Chemical parameters (TSS—total soluble solids, TA—titratable acidity, and MI—maturity Index) of 'Early Bigi' and 'Lapins' cherries after the application of different preharvest treatments.

Cultivar	Parameters	Year	C	EM 0.15%	EM 0.30%	GB 0.25%	GB 0.40%	Mix	Treatment Effect
'Early Bigi'	TSS (°Brix)	2019	10.38 ± 0.13 c	12.50 ± 0.28 a	11.60 ± 0.23 b	12.38 ± 0.34 a	11.65 ± 0.19 b	12.03 ± 0.40 ab	***
		2021	8.58 ± 0.03 c	9.93 ± 0.12 ab	9.53 ± 0.29 b	9.50 ± 0.56 b	10.17 ± 0.23 ab	10.47 ± 0.32 a	*
	TA (% malic acid)	2019	0.58 ± 0.01 ab	0.54 ± 0.06 ab	0.49 ± 0.02	0.62 ± 0.03 a	0.53 ± 0.06 b	0.53 ± 0.01 b	***
		2021	0.36 ± 0.01	0.39 ± 0.01	0.39 ± 1.78	0.37 ± 0.01	0.48 ± 0.05	0.39 ± 0.03	n.s.
	MI (TSS/TA)	2019	17.97 ± 0.09 b	23.92 ± 1.34 a	23.35 ± 0.92 a	19.96 ± 0.99 b	23.59 ± 0.81 a	22.64 ± 0.47 a	***
		2021	23.89 ± 0.32 ab	25.03 ± 0.50 a	26.05 ± 0.59 a	25.51 ± 1.53 a	21.17 ± 2.32 b	27.07 ± 1.36 a	*
'Lapins'	TSS (°Brix)	2019	15.13 ± 0.25 d	19.37 ± 0.07 b	18.18 ± 0.16 c	19.32 ± 0.79 b	19.42 ± 0.41 b	21.20 ± 0.23 a	***
		2021	14.28 ± 0.19 c	16.13 ± 0.50 ab	16.30 ± 0.36 a	15.93 ± 0.42 ab	15.23 ± 0.2 bc	16.33 ± 0.38 a	***
	TA (% malic acid)	2019	0.63 ± 0.02 bc	0.70 ± 0.02 ab	0.59 ± 0.04 c	0.57 ± 0.04 c	0.62 ± 0.02 c	0.76 ± 0.02 a	***
		2021	0.49 ± 0.01 a	0.46 ± 0.01 b	0.45 ± 0.01 b	0.50 ± 0.02 a	0.46 ± 0.01 b	0.45 ± 0.01 b	***
	MI (TSS/TA)	2019	24.03 ± 0.56 b	27.70 ± 0.68 bc	31.00 ± 1.96 ab	34.26 ± 2.34 a	31.39 ± 1.39 ab	27.80 ± 0.89 bc	***
		2021	28.59 ± 0.61 d	34.91 ± 0.75 ab	36.27 ± 1.18 a	31.65 ± 0.31 c	33.43 ± 1.14 bc	36.25 ± 0.93 a	***

C—Control treatment; EM 0.15%—*Ecklonia maxima* seaweed-based biostimulant applied at a concentration of 0.15%; EM 0.30%—*Ecklonia maxima* seaweed-based biostimulant applied at a concentration of 0.30%; GB 0.25%—Glycine betaine applied at a concentration of 0.25%; GB 0.40%—Glycine betaine applied at a concentration of 0.40%; Mix—EM 0.15% plus GB 0.25%. Data are the mean ± SD. For each row, different letters indicate significant differences between treatments (*** $p < 0.001$ and * $p < 0.05$). n.s.—Not significant.

For cv. 'Early Bigi', values of TSS between 10% and 18% have been reported [62–66], while for cv. 'Lapins', values between 13 and 20% [62,65,67–69] are common. In the present work, the use of biostimulants increased TSS when compared to control samples in either cultivar, with overall results pointing out the use of the Mix treatment. Several previous works have shown the effect of biostimulants in TSS, namely its increased content. TSS increase has been found with the use of gibberellins [33], auxins [39,40], a tropical plant extract biostimulant [1], salicylic acid, Ascophyllum nodosum seaweed extract [20], but, more importantly, the biostimulants studied in the present work, Ecklonia maxima seaweed product [22] and glycine betaine [20]. This effect on sugars is probably linked to changes caused in the metabolism through the alteration of the expression of some genes [70–73], by increasing their translocation to the fruit [74], or by increments in net photosynthesis and stomatal conductance [75]. Different effects of the use of biostimulants, depending on the cultivars, have already been described, linked to the different sensitivities of the cultivars to the application of biostimulants or a differential response on fruit set, determining source–sink relations that were less favorable for carbon partitioning to fruits [1].

Titratable acidity (TA) varied from 0.36 to 0.62 in cv. 'Early Bigi', and between 0.45 and 0.76, in cv. 'Lapins' (Table 4) values well within the usual ranges for these cultivars [66, 68,69]. The effect of the use of biostimulants was more evident in cv. 'Lapins', than in cv. 'Early Bigi', where no significant effect of spraying was observed in 2021. Overall results show that biostimulant application reduces the TA of sweet cherries, even though higher values were found for some treatments, such as the lower concentrations or the mixed treatment. Variations in the TA of sweet cherries have been reported, either increasing or decreasing when using different compounds [39,76–79]. Previous results using glycine betaine and an algae-based biostimulant also showed a decrease in TA (even though not significant) in sweet cherries of cv. 'Stacatto' [20]. The authors suggested that this reduction might have been caused by the influence of exogenous compounds on normal metabolism and gene expression related to acid [77,80].

The maturity index (TSS/TA) varied from 17.97 to 27.07 in cv. 'Early Bigi', and for cv. 'Lapins', from 24.03 to 36.25, usual values for sweet cherry [59], with an increase in this parameter recorded in all treatments for both cultivars and years. Previous work regarding the use of exogenous compounds, including glycine betaine and algae-based products, in sweet cherries has shown an increase in the maturation index [19–21,40,78], which has also been observed in the present. Some of these compounds can modulate gene expression related to the anthocyanin synthesis pathway and promote ethylene synthesis and fruit ripening [81]. This increased maturation index can be significant if aiming at anticipation of harvest to allow an earlier availability of sweet cherries to consumers.

3.4. Bioactive Compounds

As with the previous parameters, the total phenolic content was significantly affected by the exogenous application of biostimulants for both cultivars, 'Early Bigi' and 'Lapins' (Tables S1 and 5), even though the overall values are similar to those in previous works [4,52,61,82,83]. Significant increases in total phenolic content were found in fruits of cv. 'Early Bigi' treated with EM 0.30% and GB 0.40% in 2019, and in 2021, this increase was noticeable for treatments with GB 0.25% and EM 0.15%. The latter treatment also increased the total phenolic content in cv. 'Lapins' in 2019, along with GB 0.40% treatment, while in 2021, GB 0.40% and seaweed-based biostimulant sprays presented a higher content of these compounds. In contrast, the control fruits consistently had the lowest phenolic content for both sweet cherry cultivars. Furthermore, the application of biostimulants, either algae-based products, glycine betaine, or others [20,52], has proven to increase the presence of phenolic compounds in sweet cherries. This increase might be due to the reported rise in carbon metabolism and to the stimulation of the biosynthesis of secondary metabolites [84] or the activated expression of two pathways, the flavone biosynthesis pathway, and ascorbate-glutathione [85,86], hence enhancing antioxidant content via synthesis and decreasing the depletion of these compounds.

Table 5. Total phenolic (mg GAE g^{-1} DW) and total anthocyanin (mg cy-3-rut g^{-1} DW) contents of 'Early Bigi' and 'Lapins' cherries after applying different preharvest treatments.

Cultivar	Parameters	Year	C	EM 0.15%	EM 0.30%	GB 0.25%	GB 0.40%	Mix	Treatment Effect
'Early Bigi'	Total phenolics (mg GAE g^{-1} DW)	2019	7.19 ± 0.26 c	8.22 ± 0.63 bc	9.54 ± 0.71 a	8.08 ± 0.38 bc	9.09 ± 0.76 ab	7.45 ± 0.84 c	**
		2021	5.82 ± 0.35 d	9.13 ± 0.27 a	8.31 ± 0.28 b	8.58 ± 0.31 ab	7.95 ± 0.26 b	6.89 ± 0.11 c	***
	Total anthocyanin (mg cy-3-rut g^{-1} DW)	2019	14.18 ± 0.63 e	37.77 ± 1.39 a	38.52 ± 1.89 a	25.04 ± 0.32 c	31.32 ± 1.28 b	18.88 ± 0.63 d	***
		2021	31.11 ± 2.93 c	54.13 ± 2.29 ab	58.57 ± 3.72 a	47.26 ± 3.32 b	58.42 ± 0.95 a	33.35 ± 2.14 c	***
'Lapins'	Total phenolics (mg GAE g^{-1} DW)	2019	5.96 ± 0.51 b	7.77 ± 0.50 a	6.78 ± 0.52 b	6.28 ± 0.57 b	8.07 ± 0.42 a	6.45 ± 0.30 b	***
		2021	8.20 ± 0.38 c	10.04 ± 0.51 ab	10.68 ± 0.89 a	8.96 ± 0.15 bc	9.68 ± 0.52 ab	8.93 ± 0.91 bc	**
	Total anthocyanin (mg cy-3-rut g^{-1} DW)	2019	64.03 ± 1.04 b	80.15 ± 2.02 a	84.35 ± 3.53 a	81.95 ± 6.61 a	82.92 ± 1.31 a	76.03 ± 1.95 a	***
		2021	82.68 ± 2.59 c	102.54 ± 4.60 b	112.53 ± 3.64 a	88.64 ± 1.74 c	107.61 ± 2.89 b	91.63 ± 2.22 c	***

C—Control treatment; EM 0.15%—*Ecklonia maxima* seaweed-based biostimulant applied at a concentration of 0.15%; EM 0.30%—*Ecklonia maxima* seaweed-based biostimulant applied at a concentration of 0.30%; GB 0.25%—Glycine betaine applied at a concentration of 0.25%; GB 0.40%—Glycine betaine applied at a concentration of 0.40%; Mix—EM 0.15% plus GB 0.25%. Data are the mean ± SD. For each row, different letters indicate significant differences between treatments (*** $p < 0.001$ and ** $p < 0.01$).

The anthocyanin content was also significantly affected by the exogenous application of biostimulants (Tables S1 and 5). Overall, anthocyanin content increased with the application of EM by 0.30% for both cultivars and years. In contrast, fruits from the control treatment presented the lowest contents. Seaweed Ecklonia maxima extracts contain several phytohormones, including high levels of auxins [36]. Auxin application has already been reported to increase anthocyanin amounts in sweet cherries by modifying gene expression [40]. The positive effects of GB in anthocyanin content have already been reported elsewhere [20]. These effects can be linked to changes in enzyme activity, specifically phenylalanine ammonia-lyase (PAL) [87] and gene expression [88].

3.5. Sensory Evaluation

The different treatments affected both cultivars' sensory profiles in distinct manners (Figure 3). The cv. 'Early Bigi' was less prone to changes due to the use of biostimulants. Indeed, in this cultivar, only the attributes "color intensity", "sweet taste", and "cherry flavor" in 2019 and "color intensity" and "overall aspect" in 2021 showed significant differences across different treatments. For "color intensity", the control fruit exhibited significantly lower intensity, which correlates with the reduction in measured color parameters (Table 3) and decrease in anthocyanins (Table 5), compounds responsible for the color in sweet cherries [45], and for the delay in the maturation process [81] caused by the use of biostimulants. Additional attributes of cv. 'Early Bigi' fruits were enhanced by the exogenous application of biostimulants, namely "sweet taste" and "cherry flavor" (with all treatments increasing the values compared to control), and "overall aspect" (only in 2021), with C and GB 0.40% displaying similar and lower values compared to the remaining treatments. The evaluation of "sweet taste" in sensory tests, as well as the "acidic taste", aligns with the trend observed in the chemical analysis (Table 4), as their chemical composition largely affects the sensory quality of fruits [89].

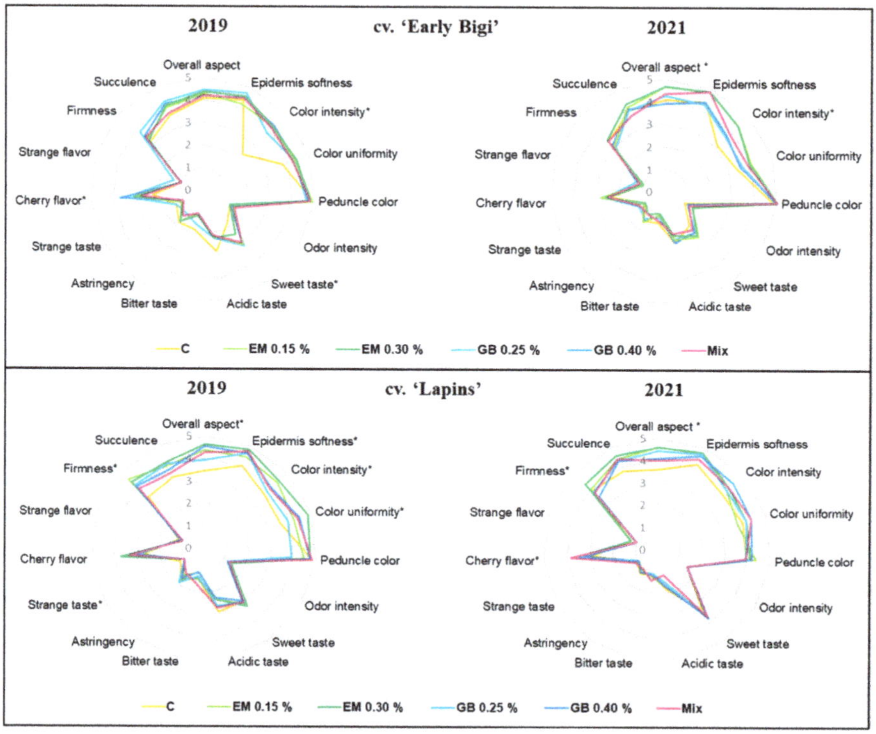

Figure 3. Spider plot of the sensory profile of the 'Early Bigi' and 'Lapins' cherries after spray treatment application in 2019 and 2021. * $p < 0.05$ represents significant differences between treatments by Duncan's test. The absence of superscripts indicates no significant differences. C—Control treatment; EM 0.15%—*Ecklonia maxima* seaweed-based biostimulant applied at a concentration of 0.15%; EM 0.30%—*Ecklonia maxima* seaweed-based biostimulant applied at a concentration of 0.30%; GB 0.25%—Glycine betaine applied at a concentration of 0.25%; GB 0.40%—Glycine betaine applied at a concentration of 0.40%; Mix—EM 0.15% plus GB 0.25%.

The effect of biostimulants on the sensory traits of sweet cherries was more pronounced for cv. 'Lapins' than for cv. 'Early Bigi'. In 2019, six attributes were significantly different when comparing treatments, with fruits from C treatment obtaining lower scores for "overall aspect", "epidermis softness", "color intensity", "color uniformity", and "firmness", and higher scores for "strange taste". In contrast, data from the 2021 analysis showed significant differences only for three attributes, specifically "overall aspect", "cherry flavor", and "firmness", with sweet cherries from the control treatment presenting the lowest scores. Similar variations in these quality parameters have been documented when using hydrogen cyanamide or gibberellic acid, with sweeter fruit, increased cherry flavor observed in 'Bing' cherries [90], or enhancement of several traits in fruits of cvs. 'Skeena' and 'Sweetheart', with foliar application of GB or gibberellic acid [21]. Significant correlations were observed between physical and chemical parameters and sensory traits. For cv. 'Early Bigi', color parameters (L*, C*, and Hue°) showed positive correlations with the "Overall aspect" rating, with correlation coefficients of 0.233 for L*, 0.256 for C*, and 0.233 for Hue°. In cv. 'Lapins', fruit weight and anthocyanin content were positively correlated with "Overall aspect", with coefficients of 0.227 and 0.543, respectively. Additionally, "Acidic taste" was positively correlated with total acidity (TA) values (0.476) and negatively correlated with the maturity index (MI) (TSS/TA) (−0.344). "Firmness" showed a positive correlation with RF (0.240), and "Color intensity" was associated with L* values (0.219).

The improvement of fruit sensory quality through the application of specific plant biostimulants can be achieved and has also been observed for several other fruits, with implications in firmness, coloration, carotenoid, and soluble solids content [91]. Hence, biostimulants have the potential to improve the sensory characteristics of sweet cherries, aiming at higher consumer acceptability [92,93].

4. Conclusions

The present results highlight the positive effects of the preharvest application of exogenous compounds, such as glycine betaine and *Ecklonia maxima*-based biostimulants, on enhancing quality-related physical, chemical, and sensory traits of sweet cherry fruits in cultivars 'Early Bigi' and 'Lapins'. *Ecklonia maxima* seaweed-based biostimulant, at a concentration of 0.30%, proved to be the treatment that, overall, yielded the most favorable results in the assessed parameters. Seaweed-based biostimulants may offer a more sustainable and environmentally friendly approach to sweet cherry production, with additional benefits expected in sweet cherry trees' physiology and abiotic stress tolerance.

Given the increasing interest in the preharvest application of biostimulants, this research fills a notable gap in the scientific literature regarding their impact on cherry culture. By elucidating the effects of these biostimulants on fruit quality, this study contributes valuable insights and enhances the current understanding of sustainable practices in cherry production.

Supplementary Materials: The following supporting information can be downloaded at https://www.mdpi.com/article/10.3390/agriculture14091521/s1. Table S1: Statistical analysis of the effects of Treatment (T), Year (Y), and their interaction on sweet cherry quality parameters. Table S2: Adapted vocabulary from Chauvin et al. [27] and reference standards for the descriptive sensory analysis of cherries.

Author Contributions: Conceptualization, S.A., A.S.M. and B.G.; methodology, S.A., I.O., C.R. and A.V.; software, S.A.; formal analysis, S.A.; investigation, S.A.; data curation, S.A.; writing—original draft preparation, S.A.; writing—review and editing, S.A., I.O., C.R., A.V., A.S.M. and B.G.; supervision, A.S.M. and B.G. All authors have read and agreed to the published version of the manuscript.

Funding: This research was funded by "Fundo Europeu Agrícola de Desenvolvimento Rural (FEADER)" and by "Estado Português" in the context of "Ação 1.1 «Grupos Operacionais»", integrated in "Medida 1. «Inovação» do PDR 2020—Programa de Desenvolvimento Rural do Continente—Grupo Operacional para a valorização da produção da Cereja de Resende e posicionamento da subfileira nos mercados (iniciativa no. 362)". (https://doi.org/10.54499/UIDB/04033/2020 (accessed on 11 July 2014)) and LA/P/0126/2020, Inov4Agro (https://doi.org/10.54499/LA/P/0126/2020 (accessed on 11 July 2014)).

Institutional Review Board Statement: Not applicable.

Data Availability Statement: The data presented in this study are available on request from the corresponding author.

Acknowledgments: Sílvia Afonso is grateful to FCT, MCTES, and FSE for the PhD Fellowship SFRH/BD/139922/2018.

Conflicts of Interest: The authors declare no conflicts of interest.

References

1. Basile, B.; Brown, N.; Valdes, J.M.; Cardarelli, M.; Scognamiglio, P.; Mataffo, A.; Rouphael, Y.; Bonini, P.; Colla, G. Plant-based biostimulant as sustainable alternative to synthetic growth regulators in two sweet cherry cultivars. *Plants* **2021**, *10*, 619. [CrossRef]
2. Gonçalves, B.; Alfredo, A.; Oliveira, I.; Afonso, S.; Morais, M.C.; Correia, S.; Martins, S.; Silva, A.P. Sweet Cherry. In *Temperate Fruits*; Apple Academic Press: Cambridge, MA, USA, 2021; pp. 333–415.
3. Afonso, S.; Oliveira, I.; Meyer, A.S.; Gonçalves, B. Biostimulants to improved tree physiology and fruit quality: A review with special focus on sweet cherry. *Agronomy* **2022**, *12*, 659. [CrossRef]
4. Afonso, S.; Oliveira, I.; Ribeiro, C.; Vilela, A.; Meyer, A.M.; Gonçalves, B. Innovative edible coatings for postharvest storage of sweet cherries. *Sci. Hortic.* **2023**, *310*, 111738. [CrossRef]
5. Romano, G.S.; Cittadini, E.D.; Pugh, B.; Schouten, R. Sweet cherry quality in the horticultural production chain. *Stewart Postharvest Rev.* **2006**, *6*, 1–9.
6. FAOSTAT. Data. Available online: https://www.fao.org/faostat/en/#home (accessed on 11 July 2024).
7. Gonçalves, B.; Silva, A.P.; Vilela, A.; Malheiro, A.; Ribeiro, C.; Bacelar, E.; Raimundo, F.; Guedes, F.; Cotez, I.; Sousa, J.R.; et al. Manual de Boas Práticas da Cultura da Cerejeira | Resende. Dolmen-Desenvolvimento Local e Regional, Crl. 2022. Available online: https://gocerejaresende.pt/manual.pdf (accessed on 8 July 2024).
8. Wenden, B.; Campoy, J.; Lecourt, J. A collection of European sweet cherry phenology data for assessing climate change. *Sci. Data* **2016**, *3*, 160108. [CrossRef] [PubMed]
9. Engin, H.; Sen, F.; Pamuk, G.; Gokbayrak, Z. Investigation of physiological disorders and fruit quality of sweet cherry. *Eur. J. Hortic. Sci.* **2009**, *74*, 118.
10. Medda, S.; Fadda, A.; Mulas, M. Influence of Climate Change on Metabolism and Biological Characteristics in Perennial Woody Fruit Crops in the Mediterranean Environment. *Horticulturae* **2022**, *8*, 273. [CrossRef]
11. Rouphael, Y.; Colla, G. Biostimulants in agriculture. *Front. Plant Sci.* **2020**, *11*, 40. [CrossRef]
12. Verma, N.; Sehrawat, K.D.; Mundlia, P.; Sehrawat, A.R.; Choudhary, R.; Rajput, V.D.; Minkina, T.; van Hullebusch, E.D.; Siddiqui, M.H.; Alamri, S. Potential use of ascophyllum nodosum as a biostimulant for improving the growth performance of *Vigna aconitifolia* (Jacq.) marechal. *Plants* **2021**, *10*, 2361. [CrossRef]
13. Rojas, G.; Fernandez, E.; Whitney, C.; Luedeling, E.; Cuneo, I.F. Adapting sweet cherry orchards to extreme weather events–Decision Analysis in support of farmers' investments in Central Chile. *Agric. Syst.* **2021**, *187*, 103031. [CrossRef]
14. Du Jardin, P. Plant biostimulants: Definition, concept, main categories and regulation. *Sci. Hortic.* **2015**, *196*, 3–14.
15. Colla, G.; Hoagland, L.; Ruzzi, M.; Cardarelli, M.; Bonini, P.; Canaguier, R.; Rouphael, Y. Biostimulant action of protein hydrolysates: Unraveling their effects on plant physiology and microbiome. *Front. Plant Sci.* **2017**, *8*, 2202. [CrossRef]
16. Colla, G.; Cardarelli, M.; Bonini, P.; Rouphael, Y. Foliar applications of protein hydrolysate, plant and seaweed extracts increase yield but differentially modulate fruit quality of greenhouse tomato. *HortScience* **2017**, *52*, 1214–1220. [CrossRef]
17. Correia, S.; Oliveira, I.; Queirós, F.; Ribeiro, C.; Ferreira, L.; Luzio, A.; Silva, A.P.; Gonçalves, B. Preharvest application of seaweed based biostimulant reduced cherry (*Prunus avium* L.) cracking. *Procedia Environ. Sci.* **2015**, *29*, 251–252.
18. Serapicos, M.; Afonso, S.; Gonçalves, B.; Silva, A.P. Exogenous application of glycine betaine on sweet cherry tree (*Prunus avium* L.): Effects on tree physiology and leaf properties. *Plants* **2022**, *11*, 3470. [CrossRef]
19. Li, M.; Zhi, H.; Dong, Y. Influence of Preharvest and Postharvest Applications of Glycine Betaine on Fruit Quality Attributes and Storage Disorders of 'Lapins' and 'Regina' Cherries. *HortScience Horts* **2019**, *54*, 1540–1545. [CrossRef]
20. Gonçalves, B.; Morais, M.C.; Sequeira, A.; Ribeiro, C.; Guedes, F.; Silva, A.P.; Aires, A. Quality preservation of sweet cherry cv.'staccato' by using glycine-betaine or Ascophyllum nodosum. *Food Chem.* **2020**, *322*, 126713.
21. Correia, S.; Queirós, F.; Ribeiro, C.; Vilela, A.; Aires, A.; Barros, A.I.; Schouten, R.; Silva, A.P.; Gonçalves, B. Effects of calcium and growth regulators on sweet cherry (*Prunus avium* L.) quality and sensory attributes at harvest. *Sci. Hortic.* **2019**, *248*, 231–240.
22. Ureta Ovalle, A.; Atenas, C.; Larraín, P. Application of an Ecklonia maxima seaweed product at two different timings can improve the fruit set and yield in 'Bing' sweet cherry trees. In *VIII International Cherry Symposium*; ISHS: Leuven, Belgium, 2017; pp. 319–326.
23. Fadón, E.; Herrero, M.; Rodrigo, J. Flower development in sweet cherry framed in the BBCH scale. *Sci. Hortic.* **2015**, *192*, 141–147.
24. Singleton, V.; Rossi, J. Colorimetry of total phenolics with phosphomolybdic-phosphotungstic acid reagents. *Am. J. Enol. Vitic.* **1965**, *16*, 144–158.
25. Dewanto, V.; Wu, X.; Adom, K.; Liu, R. Thermal processing enhances the nutritional value of tomatoes by increasing total antioxidant activity. *J. Agric. Food Chem.* **2002**, *50*, 3010–3014.

26. Nicoue, E.E.; Savard, S.; Belkacemi, K. Anthocyanins in wild blueberries of Quebec: Extraction and identification. *J. Agric. Food Chem.* **2007**, *55*, 5626–5635. [CrossRef] [PubMed]
27. Chauvin, M.A.; Whiting, M.; Ross, C.F. The influence of harvest time on sensory properties and consumer acceptance of sweet cherries. *Horttechnology* **2009**, *19*, 748–754. [CrossRef]
28. *ISO 6658*; Sensory Analysis-Methodology—General Guidance. 3rd ed. International Organization for Standardization: Geneva, Switzerland, 2017; pp. 1–26.
29. *ISO 4121*; Sensory Analysis-Guidelines for the Use of Quantitative Response Scales. 2nd ed. International Organization for Standardization: Geneva, Switzerland, 2003; pp. 1–9.
30. Bund, S.; Norre, J. Seaweed extract improve cherry fruit quality. In Proceedings of the Australian Palnt Health Conference. Australian Society of Horticultural Science. New Zealand Institute of Agriculture and Horticuleture Science, Joint Conference, Lorne, Australia, 18–22 September 2011; pp. 18–22.
31. Sajid, M.; Basit, A.; Ullah, Z.; Shah, S.T.; Ullah, I.; Mohamed, H.I.; Ullah, I. Chitosan-based foliar application modulated the yield and biochemical attributes of peach (*Prunus persica* L.) cv. Early Grand. *Bull. Natl. Res. Cent.* **2020**, *44*, 150.
32. Esitken, A.; Pirlak, L.; Turan, M.; Sahin, F. Effects of floral and foliar application of plant growth promoting rhizobacteria (PGPR) on yield, growth and nutrition of sweet cherry. *Sci. Hortic.* **2006**, *110*, 324–327.
33. Zhang, C.; Whiting, M. Plant growth regulators improve sweet cherry fruit quality without reducing endocarp growth. *Sci. Hortic.* **2013**, *150*, 73–79. [CrossRef]
34. Stirk, W.A.; Van Staden, J. Comparison of cytokinin-and auxin-like activity in some commercially used seaweed extracts. *J. Appl. Phycol.* **1996**, *8*, 503–508.
35. Pizzeghello, D.; Francioso, O.; Ertani, A.; Muscolo, A.; Nardi, S. Isopentenyladenosine and cytokinin-like activity of different humic substances. *J. Geochem. Explor.* **2013**, *129*, 70–75.
36. Kocira, S.; Szparaga, A.; Kuboń, M.; Czerwińska, E.; Piskier, T. Morphological and biochemical responses of *Glycine max* (L.) Merr. to the use of seaweed extract. *Agronomy* **2019**, *9*, 93. [CrossRef]
37. Baltazar, M.; Correia, S.; Guinan, K.J.; Sujeeth, N.; Bragança, R.; Gonçalves, B. Recent advances in the molecular effects of biostimulants in plants: An overview. *Biomolecules* **2021**, *11*, 1096. [CrossRef]
38. Valenzuela-Soto, E.M.; Figueroa-Soto, C.G. Biosynthesis and Degradation of Glycine Betaine and Its Potential to Control Plant Growth and Development. In *Osmoprotectant-Mediated Abiotic Stress Tolerance in Plants*; Hossain, M., Kumar, V., Burritt, D., Fujita, M., Mäkelä, P., Eds.; Springer: Cham, Switzerland, 2019. [CrossRef]
39. Stern, R.A.; Flaishman, M.; Applebaum, S.; Ben-Arie, R. Effect of synthetic auxins on fruit development of 'Bing'cherry (*Prunus avium* L.). *Sci. Hortic.* **2007**, *114*, 275–280. [CrossRef]
40. Clayton-Cuch, D.; Yu, L.; Shirley, N.; Bradley, D.; Bulone, V.; Böttcher, C. Auxin treatment enhances anthocyanin production in the non-climacteric sweet cherry (*Prunus avium* L.). *Int. J. Mol. Sci.* **2021**, *22*, 10760. [CrossRef]
41. Flaishman, M.; Ben-Arie, R.; Stern, R.A.; Applebaum, S. Auxins increase fruit size of 'bing'(*Prunus avium* L.) cherry in a warm climate. In *XXVII International Horticultural Congress-IHC2006: International Symposium on Endogenous and Exogenous Plant Bioregulators*; ISHS: Leuven, Belgium, 2006; pp. 243–250.
42. Champa, W.H.; Gill MI, S.; Mahajan BV, C.; Arora, N.K. Preharvest salicylic acid treatments to improve quality and postharvest life of table grapes (*Vitis vinifera* L.) cv. Flame Seedless. *J. Food Sci. Technol.* **2015**, *52*, 3607–3616.
43. Serradilla, M.J.; Martín, A.; Ruiz-Moyano, S.; Hernández, A.; López-Corrales, M.; de Guía Córdoba, M. Physicochemical and sensorial characterisation of four sweet cherry cultivars grown in Jerte Valley (Spain). *Food Chem.* **2012**, *133*, 1551–1559.
44. Korley Kortei, N.; Tawia Odamtten, G.; Obodai, M.; Appiah, V.; Toah Akonor, P. Determination of color parameters of gamma irradiated fresh and dried mushrooms during storage. *Croat. J. Food Technol. Biotechnol. Nutr.* **2015**, *10*, 66–71.
45. Giménez, M.J.; Valverde, J.M.; Valero, D.; Guillén, F.; Martínez-Romero, D.; Serrano, M.; Castillo, S. Quality and antioxidant properties on sweet cherries as affected by preharvest salicylic and acetylsalicylic acids treatments. *Food Chem.* **2014**, *160*, 226–232.
46. Gonçalves, B.; Silva, A.P.; Moutinho-Pereira, J.; Bacelar, E.; Rosa, E.; Meyer, A.S. Effect of ripeness and postharvest storage on the evolution of colour and anthocyanins in cherries (*Prunus avium* L.). *Food Chem.* **2007**, *103*, 976–984.
47. Kondo, S.; Inoue, K. Abscisic acid (ABA) and 1-aminocyclopropane-1-carboxylic acid (ACC) content during growth of 'Satonishiki' cherry fruit, and the effect of ABA and ethephon application on fruit quality. *J. Hort. Sci.* **1997**, *72*, 221–227.
48. Kuhn, N.; Arellano, M.; Ponce, C.; Hodar, C.; Correa, F.; Multari, S.; Martens, S.; Carrera, E.; Donoso, J.M.; Meisel, L.A. RNA-Seq and WGBS Analyses During Fruit Ripening and in Response to ABA in Sweet Cherry (*Prunus avium*) Reveal Genetic and Epigenetic Modulation of Auxin and Cytokinin Genes. *J. Plant Growth Regul.* **2024**, 1–23. [CrossRef]
49. Ponce, C.; Kuhn, N.; Arellano, M.; Time, A.; Multari, S.; Martens, S.; Carreras, E.; Sagredo, B.; Donoso, J.; Meisel, L.A. Differential phenolic compounds and hormone accumulation patterns between early-and mid-maturing sweet cherry (*Prunus avium* L.) cultivars during fruit development and ripening. *J. Agric. Food Chem.* **2021**, *69*, 8850–8860.
50. Zhang, C.; Whiting, M.D. Improving 'Bing' sweet cherry fruit quality with plant growth regulators. *Sci. Hortic.* **2011**, *127*, 341–346.
51. Sabir, I.A.; Liu, X.; Jiu, S.; Whiting, M.; Zhang, C. Plant growth regulators modify fruit set, fruit quality, and return bloom in sweet cherry. *HortScience* **2021**, *56*, 922–931.
52. Zhi, H.; Dong, Y. Seaweed-based biostimulants improves quality traits, postharvest disorders, and antioxidant properties of sweet cherry fruit and in response to gibberellic acid treatment. *Sci. Hortic.* **2024**, *336*, 113454.
53. Hocking, B.; Tyerman, S.D.; Burton, R.A.; Gilliham, M. Fruit calcium: Transport and physiology. *Front. Plant Sci.* **2016**, *7*, 569.

54. Dong, Y.; Zhi, H.; Wang, Y. Cooperative effects of pre-harvest calcium and gibberellic acid on tissue calcium content, quality attributes, and in relation to postharvest disorders of late-maturing sweet cherry. *Sci. Hortic.* **2019**, *246*, 123–128.
55. Ricardo-Rodrigues, S.; Laranjo, M.; Agulheiro-Santos, A.C. Methods for quality evaluation of sweet cherry. *J. Sci. Food Agric.* **2023**, *103*, 463–478.
56. Hampson, C.R.; Stanich, K.; McKenzie, D.L.; Herbert, L.; Lu, R.; Li, J.; Cliff, M.A. Determining the Optimum Firmness for Sweet Cherries Using Just-About-Right Sensory Methodology. *Postharvest Biol. Technol.* **2014**, *91*, 104–111.
57. Wang, Y.; Long, L.E. Physiological and Biochemical Changes Relating to Postharvest Splitting of Sweet Cherries Affected by Calcium Application in Hydrocooling Water. *Food Chem.* **2015**, *181*, 241–247. [PubMed]
58. Ballistreri, G.; Continella, A.; Gentile, A.; Amenta, M.; Fabroni, S.; Rapisarda, P. Fruit quality and bioactive compounds relevant to human health of sweet cherry (*Prunus avium* L.) cultivars grown in Italy. *Food Chem.* **2013**, *140*, 630–638.
59. Hayaloglu, A.A.; Demir, N. Physicochemical characteristics, antioxidant activity, organic acid and sugar contents of 12 sweet cherry (*Prunus avium* L.) cultivars grown in Turkey. *J. Food Sci.* **2015**, *80*, C564–C570.
60. Guyer, D.E.; Sinha, N.K.; Chang, T.-S.; Cash, J.N. Physicochemical and sensory characteristics of selected Michigan sweet cherry (*Prunus avium* L.) cultivars. *J. Food Qual.* **1993**, *16*, 355–370.
61. Faniadis, D.; Drogoudi, P.D.; Vasilakakis, M. Effects of cultivar, orchard elevation, and storage on fruit quality characters of sweet cherry (*Prunus avium* L.). *Sci. Hortic.* **2010**, *125*, 301–304. [CrossRef]
62. Pereira, S.; Silva, V.; Bacelar, E.; Guedes, F.; Silva, A.P.; Ribeiro, C.; Gonçalves, B. Cracking in Sweet Cherry Cultivars 'Early Bigi' and 'Lapins': Correlation with Quality Attributes. *Plants* **2020**, *9*, 1557. [CrossRef]
63. López, L.; Larrigaudière, C.; Giné-Bordonaba, J.; Echeverria, G. Defining key parameters and predictive markers of 'Early Bigi'cherry consumer satisfaction by means of differential storage scenarios. *Postharvest Biol. Technol.* **2023**, *195*, 112117.
64. Martins, V.; Silva, V.; Pereira, S.; Afonso, S.; Oliveira, I.; Santos, M.; Ribeiro, C.; Vilela, A.; Bacelar, E.; Silva, A.P.; et al. Rootstock Affects the Fruit Quality of 'Early Bigi' Sweet Cherries. *Foods* **2021**, *10*, 2317. [CrossRef]
65. García-Montiel, F.; López, D.; Guirao, P.; García, F.; Frutos, D.; López-Ortega, G.; Carrillo, A.; López, D.; Cos, J. Preliminary results of sweet cherry (*Prunus avium* L.) collection in Jumilla, Murcia, Spain. *Acta Hortic.* **2017**, *1161*, 281–286.
66. Radičević, S.; Marić, S.; Milošević, N.; Glišić, I.; Đorđević, M. Phenological characteristics and fruit quality of introduced sweet cherry (*Prunus avium* L.) cultivars in agroecological conditions of Čačak. *Voćarstvo* **2022**, *56*, 93–99.
67. Drake, S.R.; Fellman, J.K. Indicators of Maturity and Storage Quality of 'Rainier' Sweet Cherry. *HortScience* **1987**, *22*, 283–285. [CrossRef]
68. Velardo-Micharet, B.; Peñas Díaz, L.; Tapia García, I.M.; Nieto Serrano, E.; Campillo Torres, C. Effect of irrigation on postharvest quality of two sweet cherry cultivars (*Prunus avium* L.). *Acta Hortic.* **2017**, *1161*, 667–672. [CrossRef]
69. Usenik, V.; Fajt, N.; Mikulic-Petkovsek, M.; Slatnar, A.; Stampar, F.; Veberic, R. Sweet cherry pomological and biochemical characteristics influenced by rootstock. *J. Agric. Food Chem.* **2010**, *58*, 4928–4933.
70. Yao, L.; Liang, D.; Xia, H.; Pang, Y.; Xiao, Q.; Huang, Y.; Zhang, W.; Pu, C.; Wang, J.; Lv, X. Biostimulants promote the accumulation of carbohydrates and biosynthesis of anthocyanins in 'Yinhongli'plum. *Front. Plant Sci.* **2023**, *13*, 1074965.
71. Gupta, N.; Thind, S.K. Foliar Application of Glycine Betaine Alters Sugar Metabolism of Wheat Leaves Under Prolonged Field Drought Stress. *Proc. Natl. Acad. Sci. India Sect. B Biol. Sci.* **2018**, *89*, 877–884. [CrossRef]
72. Wang, L.I.; Shan, T.; Xie, B.; Ling, C.; Shao, S.; Jin, P.; Zheng, Y. Glycine betaine reduces chilling injury in peach fruit by enhancing phenolic and sugar metabolisms. *Food Chem.* **2019**, *272*, 530–538. [PubMed]
73. Shukla, P.S.; Prithiviraj, B. Ascophyllum nodosum biostimulant improves the growth of Zea mays grown under phosphorus impoverished conditions. *Front. Plant Sci.* **2021**, *11*, 601843.
74. Arteca, R.N. *Plant Growth Substances: Principles and Applications*; Springer Science & Business Media: Berlin/Heidelberg, Germany, 1996.
75. Abou-Aly, H.E.; Mady, M.A. Complemented effect of glycine betaine and biofertilizers on growth and produtivity of sweet pepper (*Capsicum annuum* L.) plant under high temperature condition. *J. Plant Prod.* **2014**, *5*, 505–526.
76. Ruiz-Aracil, M.C.; Valverde, J.M.; Lorente-Mento, J.M.; Carrión-Antolí, A.; Castillo, S.; Martínez-Romero, D.; Guillén, F. Sweet Cherry (*Prunus avium* L.) Cracking during Development on the Tree and at Harvest: The Impact of Methyl Jasmonate on Four Different Growing Seasons. *Agriculture* **2023**, *13*, 1244. [CrossRef]
77. Ruiz-Aracil, M.C.; Valverde, J.M.; Beltrà, A.; Carrión-Antolí, A.; Lorente-Mento, J.M.; Nicolás-Almansa, M.; Guillén, F. Putrescine Increases Frost Tolerance and Effectively Mitigates Sweet Cherry (*Prunus avium* L.) Cracking: A Study of Four Different Growing Cycles. *Agronomy* **2024**, *14*, 23. [CrossRef]
78. Palacios-Peralta, C.; Ruiz, A.; Ercoli, S.; Reyes-Díaz, M.; Bustamante, M.; Muñoz, A.; Osorio, P.; Ribera-Fonseca, A. Plastic Covers and Potassium Pre-Harvest Sprays and Their Influence on Antioxidant Properties, Phenolic Profile, and Organic Acids Composition of Sweet Cherry Fruits Cultivated in Southern Chile. *Plants* **2023**, *12*, 50. [CrossRef]
79. Ozturk, B.; Akkaya, H.; Aglar, E.; Saracoglu, O. Effect of preharvest biofilm application regimes on cracking and fruit quality traits in '0900 Ziraat' sweet cherry cultivar. *BMC Plant Biol.* **2024**, *24*, 574. [CrossRef]
80. Souza, M.L.D.; Morgado, C.M.A.; Marques, K.M.; Mattiuz, C.F.M.; Mattiuz, B.H. Pós-colheita de mangas' Tommy Atkins' recobertas com quitosana. *Rev. Bras. De Frutic.* **2011**, *33*, 337–343.
81. Luo, H.; Dai, S.; Ren, J.; Zhang, C.; Ding, Y.; Li, Z.; Sun, Y.; Ji, K.; Wang, Y.; Li, Q.; et al. The Role of ABA in the Maturation and Postharvest Life of a Nonclimacteric Sweet Cherry Fruit. *J. Plant Growth Regul.* **2014**, *33*, 373–383. [CrossRef]

82. Ockun, M.A.; Gercek, Y.C.; Demirsoy, H.; Demirsoy, L.; Macit, I.; Oz, G.C. Comparative evaluation of phenolic profile and antioxidant activity of new sweet cherry (*Prunus avium* L.) genotypes in Turkey. *Phytochem. Anal.* **2022**, *33*, 564–576. [CrossRef] [PubMed]
83. Kelebek, H.; Selli, S. Evaluation of chemical constituents and antioxidant activity of sweet cherry (*Prunus avium* L.) cultivars. *Int. J. Food Sci. Technol.* **2011**, *46*, 2530–2537. [CrossRef]
84. Ertani, A.; Pizzeghello, D.; Francioso, O.; Sambo, P.; Sanchez-Cortes, S.; Nardi, S. *Capsicum chinensis* L. growth and nutraceutical properties are enhanced by biostimulants in a long-term period: Chemical and metabolomic approaches. *Front. Plant Sci.* **2014**, *5*, 375. [CrossRef]
85. De Saeger, J.; Van Praet, S.; Vereecke, D.; Park, J.; Jacques, S.; Han, T.; Depuydt, S. Toward the molecular understanding of the action mechanism of Ascophyllum nodosum extracts on plants. *J. Appl. Phycol.* **2020**, *32*, 573–597. [CrossRef]
86. Ali, O.; Ramsubhag, A.; Jayaraman, J. Biostimulant properties of seaweed extracts in plants: Implications towards sustainable crop production. *Plants* **2021**, *10*, 531. [CrossRef]
87. Habibi, F.; Valero, D.; Serrano, M.; Guillén, F. Exogenous application of glycine betaine maintains bioactive compounds, antioxidant activity, and physicochemical attributes of blood orange fruit during prolonged cold storage. *Front. Nutr.* **2022**, *9*, 873915. [CrossRef] [PubMed]
88. Monteiro, E.; De Lorenzis, G.; Ricciardi, V.; Baltazar, M.; Pereira, S.; Correia, S.; Ferreira, H.; Alves, F.; Cortez, I.; Gonçalves, B.; et al. Exploring Seaweed and Glycine Betaine Biostimulants for Enhanced Phenolic Content, Antioxidant Properties, and Gene Expression of Vitis vinifera cv. "Touriga Franca" Berries. *Int. J. Mol. Sci.* **2024**, *25*, 5335. [CrossRef]
89. Fazzari, M.; Fukumoto, L.; Mazza, G.; Livrea, M.A.; Tesoriere, L.; Marco, L.D. In vitro bioavailability of phenolic compounds from five cultivars of frozen sweet cherries (*Prunus avium* L.). *J. Agric. Food Chem.* **2008**, *56*, 3561–3568. [CrossRef]
90. Clayton, M.; Biasi, W.V.; Agar, I.T.; Southwick, S.M.; Mitcham, E.J. Sensory quality of 'Bing' sweet cherries following preharvest treatment with hydrogen cyanamide, calcium ammonium nitrate, or gibberellic acid. *HortScience* **2006**, *41*, 745–748. [CrossRef]
91. Soteriou, G.A.; Rouphael, Y.; Emmanouilidou, M.G.; Antoniou, C.; Kyratzis, A.C.; Kyriacou, M.C. Biostimulatory Action of Vegetal Protein Hydrolysate and the Configuration of Fruit Physicochemical Characteristics in Grafted Watermelon. *Horticulturae* **2021**, *7*, 313. [CrossRef]
92. Rodrigues, M.; Baptistella, J.L.C.; Horz, D.C.; Bortolato, L.M.; Mazzafera, P. Organic Plant Biostimulants and Fruit Quality—A Review. *Agronomy* **2020**, *10*, 988. [CrossRef]
93. Drobek, M.; Frąc, M.; Cybulska, J. Plant Biostimulants: Importance of the Quality and Yield of Horticultural Crops and the Improvement of Plant Tolerance to Abiotic Stress—A Review. *Agronomy* **2019**, *9*, 335. [CrossRef]

agriculture

Article

Growth, Photosynthesis and Yield Responses of Common Wheat to Foliar Application of *Methylobacterium symbioticum* under Decreasing Chemical Nitrogen Fertilization

Francesco Valente [1,*], Anna Panozzo [1], Francesco Bozzolin [1], Giuseppe Barion [1], Pranay Kumar Bolla [1], Vittorio Bertin [1], Silvia Potestio [2], Giovanna Visioli [2], Yu Wang [3] and Teofilo Vamerali [1]

1 Department of Agronomy, Food, Natural Resources, Animals and the Environment, University of Padua, 35020 Legnaro, Italy; anna.panozzo@unipd.it (A.P.); francesco.bozzolin@gmail.com (F.B.); giuseppe.barion@unipd.it (G.B.); pranaykumar.bolla@unipd.it (P.K.B.); vittorio.bertin@unipd.it (V.B.); teofilo.vamerali@unipd.it (T.V.)

2 Department of Chemistry, Life Sciences and Environmental Sustainability, University of Parma, 43124 Parma, Italy; silvia.potestio@unipr.it (S.P.); giovanna.visioli@unipr.it (G.V.)

3 Key Laboratory for Water Quality and Conservation of the Pearl River Delta, Institute of Environmental Research at Greater Bay Area, Ministry of Education, Guangzhou University, Guangzhou 510006, China; wangyu@gzhu.edu.cn

* Correspondence: francesco.valente@unipd.it; Tel.: +39-049-8272856

Abstract: Current agriculture intensifies crop cultivation to meet food demand, leading to unsustainable use of chemical fertilizers. This study investigates a few physiological and agronomic responses of common wheat following the inoculation with plant growth-promoting bacteria to reduce nitrogen inputs. A field trial was conducted in 2022–2023, in Legnago (Verona, Italy) on *Triticum aestivum* var. LG-Auriga comparing full (180 kg ha^{-1}) and reduced (130 kg ha^{-1}) N doses, both with and without foliar application at end tillering of the N-fixing bacterium *Methylobacterium symbioticum*. Biofertilization did not improve shoot growth, while it seldom increased the root length density in the arable layer. It delayed leaf senescence, prolonged photosynthetic activity, and amplified stomatal conductance and PSII efficiency under the reduced N dose. Appreciable ACC-deaminase activity of such bacterium disclosed augmented nitrogen retrieval and reduced ethylene production, explaining the ameliorated stay-green. Yield and test weight were unaffected by biofertilization, while both glutenin-to-gliadin and HMW-to-LMW ratios increased together with dough tenacity. It is concluded that *Methylobacterium symbioticum* can amplify nitrogen metabolism at a reduced nitrogen dose, offering a viable approach to reduce chemical fertilization under suboptimal growing conditions for achieving a more sustainable agriculture. Further research over multiple growing seasons and soil types is necessary to corroborate these preliminary observations.

Keywords: biofertilizer; biostimulant; plant growth promoting bacteria; leaf senescence; stomatal conductance; sustainable agriculture; nitrogen-fixing bacteria

1. Introduction

Nitrogen (N) is an essential nutrient for living organisms as it is a component of a vast range of fundamental biomolecules such as proteins and nucleic acids [1]. In plants, N serves to build amino acids, proteins, enzymes, chlorophyll, and other related important chemicals. Plants can assimilate inorganic N from the soil in the form of nitrate, and seldom ammonium. On the other hand, atmospheric N is abundant but inert and needs to be converted into usable forms through biological fixation. Some major cereals like maize, wheat, and rice rely substantially on N for growth and productivity. Unlike legume plants, cereals lack a stable symbiotic relationship with the N-fixing bacteria, although they can establish associations with various N-fixing microorganisms [2–5].

In recent decades, there has been a decreasing trend in soil organic matter, leading to a decline in nitrogen content in the soil. This decline is predominantly due to intensive cultivation cycles since the 1960s, which has promoted the development of nitrogen fertilizers to remedy the deficits of this plant nutrient.

The Haber–Bosch process, developed in 1909 by Fritz Haber and Carl Bosch, devised an artificial nitrogen-fixing process, which enabled the large-scale industrial production of ammonia [6]. Projections indicate that the world's ability to produce ammonia is around 224.6 million tons, with an actual production output of about 182.6 million tons [7]. However, this is a highly energy-demanding process, it uses fossil fuels, and it is currently considered one of the largest greenhouse gas emitters, responsible for 1.2% of CO_2 emissions globally [8].

Excessive nitrogen fertilizer usage leads to significant environmental losses, such as nitrate leaching into water bodies, and greenhouse gas emissions [9]. In the European Union, N losses are estimated to cost at least EUR 70 billion per year [10].

In response to these challenges, the European Commission presented the European Green Deal in December 2019, with the goal of making Europe the first continent to achieve a zero-climate footprint by 2050. From this standpoint, the EU aims to reduce greenhouse gas emissions by at least 55% by 2030, compared to 1990 levels [11]. Worldwide, there is a clear need to find alternatives to the use of synthetic fertilizers, due to their negative impact on the environment.

Recent advances have highlighted the potential of plant growth-promoting bacteria (PGPB) as a sustainable alternative to chemical fertilizers. Inoculation with the associative N-fixing bacteria *Azospirillum* has shown some promise in promoting growth and N fixation in cereals. Under microaerobic conditions, *Azospirillum* allows the conversion of atmospheric N into ammonium by the action of a nitrogenase enzyme complex, although this bacterium fixes N at a lower rate than *Rhizobium* bacteria in legume plants. It is imperative to understand the nitrogen metabolism molecular mechanisms in plants, and the link with N-fixing bacteria should be exploited for the progress of sustainable means to enhance nitrogen use efficiency in agriculture [12–15].

Microbial inoculants containing isolated beneficial microorganisms from natural environments are an emerging approach to both increasing food production and minimizing harm to human health and the environment; therefore, they are a possible alternative to major synthetic chemicals in modern agriculture [16,17]. The application of specific PGPB strains has demonstrated improved plant growth and stress tolerance, with examples provided by recent studies showcasing their effectiveness in enhancing crop productivity and resilience. These bacteria not only boost nutrient uptake but also help plants tolerate abiotic stresses such as drought and salinity, improve soil health, and reduce dependency on chemical pesticides [18–20].

PGPB have been proven to enhance soil health by sustaining organic matter decomposition and nutrient cycling, which can lead to long-term increases in soil fertility and structure. Certain PGPB strains, such as *Azospirillum*, can also release phytohormones that stimulate root growth and boost nutrient uptake efficiency, and develop systemic resistance against plant diseases, which improves plant defense systems and reduces dependency on chemical pesticides [21,22].

As a recent definition from the European regulation no. 1009/2019 provides, biostimulants are products containing substances and/or micro-organisms which, when applied to the plant or rhizosphere, stimulate natural processes that improve nutrient uptake and assimilation efficiency, tolerance to abiotic stresses and/or product quality, independently of their nutrient content. In recent years, the market for biostimulant products has been continuously expanding, with the European market leading the way, showing the largest volumes, i.e., ~50% of the world market. The growth potential of this market has attracted significant interest from the major companies in the sector, leading to a surge of biostimulant product launches on the market and an exponential increase in scientific publications [23].

Among PGPBs, the strains of *Methylobacterium* have been researched extensively for their plant growth-promoting activities, thereby positively impacting the growth of numerous plant species [24–27]. These bacteria are Gram-negative, rod-shaped, strictly aerobic and are a part of the *Alphaproteobacteria* class. They are characterized by pinkish pigmentation due to the synthesis of carotenoids [28]. For growth, *Methylobacterium* can utilize organic compounds containing only one carbon (C1), such as methanol or methylamine [29]. It can be found in various habitats such as soil, leaf surfaces and air [30]. These pink-pigmented facultative methylotrophic bacteria (PPFMs) have been observed to form relationships with over 70 plant species [31]. The epiphytic *Methylobacterium* species often cross the leaf surface and penetrate leaf stomatal openings, establishing endophytic bacterial communities [32]. *Methylobacterium* spp. can fix atmospheric nitrogen in plant leaves [33], representing a significant advantage for plants hosting these bacteria, that can convert atmospheric nitrogen N_2 into ammonium (NH_4^+) through the nitrogenase enzyme [29].

Methylobacterium symbioticum is a novel species recently isolated from spores of the symbiotic fungus *Glomus iranicum* var. *tenuihypharum* [34], demonstrating the potential to reduce nitrogen chemical fertilization in maize, rice, and wine grapes following its application, without affecting their growth, but seldom obtaining an even higher yield than controls.

Given this background, in this study, we aimed to evaluate the effect of a biostimulant containing *Methylobacterium symbioticum* on the growth and yield of a common wheat variety under decreasing doses of synthetic nitrogen fertilization. Specifically, the effects of the application of a commercial product BlueN® containing the strain SB23 of such bacterium at late tillering stage were investigated, to potentially reduce the mineral nitrogen fertilization, while trying to maintain or enhance the performance of the crop. This study investigated the effects of *Methylobacterium symbioticum* application on (i) shoot and root growth of wheat; (ii) dynamics of leaf chlorophyll content and canopy greenness; (iii) photosynthesis response; and (iv) grain yield and quality and consequences on flour rheological properties.

2. Materials and Methods

2.1. Field Trial Set-Up

The field trial was conducted at the AGRITAB farm, located in the municipality of Legnago, Verona, Italy, 15 m above sea level, during the 2022–2023 growing season. The trial was conducted over a single growing season. The soil was sandy loam with a pH of 8.4, 1.56% organic matter, a CEC of 11.6 cmol (+) kg^{-1}, and a total N content before sowing of 0.97 g kg^{-1}. Climatic data were obtained from a meteorological station in Legnago-Vangadizza and managed by the ARPAV regional weather service center.

The preceding crop was grain maize, fertilized with approximately 240 kg N ha^{-1}. The trial was conducted in a 3.80-ha field area, on the soft wheat *Triticum aestivum* var. "LG Auriga" (LG seeds, Parma—Italy). The field was divided into four sections/treatments, each comprising three plots/replicates (n = 3), following a randomized block design. Each plot, extending 4 m in length and thirty-three wheat rows in breadth, was positioned centrally within the treated zone and at least 10 m apart. Treatments were 100% of N dose, as from local recommendation (i.e., 180 kg N ha^{-1}), with (named "bact") and without *Methylobacterium symbioticum* inoculation, and a reduced N dose, corresponding to 75% of N dose (130 kg N ha^{-1}), again with and without the bacterial inoculum (Table 1).

During October 2022, the soil was ploughed to a depth of 0.35 m and then fertilized with 250 kg ha^{-1} of an N-P fertilizer (9% N, 20% P_2O_5), equivalent to 22.5 kg N ha^{-1}. The seedbed was then prepared by refining the soil with a rotary harrow to facilitate sowing, which took place on 3 November 2022. Seeds treated with fungicides (a.i. Sedaxane + Fludioxonil + Difenconazole), were distributed with a sowing density of 210 kg ha^{-1}, (~530 seeds m^{-2}) with a 12 cm interrow.

Table 1. Fertilization plan and distribution of biostimulant in each treatment.

Treatment Name	N (kg ha^{-1})	Bacterial Inoculation
100%N	180	NO
100%N + bact	180	YES
75%N	130	NO
75%N + bact	130	YES

A first dressing of N fertilization was performed at the end of February before stem elongation on all the experimental area, regardless of the treatment, with 370 kg ha^{-1} of N-S fertilizer (30% N, 15% SO_3), corresponding to an additional 111 kg N ha^{-1} to the soil. A second fertilization was carried out at the end of April, approximately at ear emergence, by applying 175 kg ha^{-1} of ammonium nitrate (27% N), equivalent to 47.3 kg N ha^{-1}, only in the 100%N and 100%N + bact treatments.

The crop was protected against weeds, pests, and diseases following local recommendations and weather conditions. On 20 March, a chemical weed control was applied with a post-emergence treatment with a.i. Tritosulfuron + Florasulam, together with a.i. Pinoxaden + Clodinafop. Weeding was associated with a fungicide treatment with a.i. Mefentrifluconazole + Pyraclostrobin, against leaf rust, septoria leaf spot, and powdery mildew. The application of 333 g ha^{-1} of the biostimulant BlueN® containing *Methylobacterium symbioticum* strain SB23 at a concentration of 3 × 10^7 CFU g^{-1}, as a wettable powder, in the 100%N + bact and 75%N + bact treatments occurred on 25 March.

A second fungicide treatment containing the a.i. Metconazole was applied at the beginning of May 2023, against *Fusarium* along with the above-mentioned pathogens, in combination with an insecticide treatment based on a.i. Tau-Fluvalinate for controlling aphids.

All herbicide and fungicide treatments and the distribution of BlueN® were conducted with 350 L ha^{-1} of water. The harvesting took place on 28 June 2023.

2.2. *Environmental Scanning Electron Microscope (ESEM) Imaging of Bacteria–Leaf Interactions*

ESEM was used to assess the ability of the *Methylobacterium symbioticum* SB23-based inoculum to colonize wheat leaves. This method facilitates the examination of biological specimens without the need for histological processing of samples, enabling the direct visualization of bacterial colonization both externally and internally within plant cells and tissues, providing a true depiction of the interaction [35].

A suspension of *Methylobacterium symbioticum* SB23 was plated in Luria–Bertani (LB) solid medium at different dilutions and incubated at 28 °C for 72 h. After obtaining single isolates, single colonies were cultured in 100 mL of LB medium on a rotary shaker at 28 °C for 48 h. After centrifugation, bacterial cells were re-suspended in sterile water to attain a final inoculum density of 3 × 10^7 colony-forming units (CFU) per mL.

The bacterial suspension was spread on the first leaf of 7-day-old wheat plantlets grown in sterile conditions on ½ MS agar medium [36], 1% sucrose, and inoculated plants were allowed to grow for another 7 days. Uninoculated plants were used as controls.

At the end of the incubation period, 5 mm large fresh leaves sections were excised with a sterile lancet and morphological analysis was performed using ESEM instrument Quanta™ 250 FEG (FEI, Hillsboro, OR, USA) in wet mode as previously described by Visioli et al., 2014 [37].

2.3. *Shoot Growth Analysis*

Monitoring of vegetational indexes was carried out during the vegetative growth of wheat, from the beginning of stem elongation to almost maturity. The leaf chlorophyll content was indirectly assessed through soil and plant analysis development (SPAD) values. Measurements were taken on 12 plants for each plot/replicate of all the treatments by using a SPAD 502 chlorophyll meter (Konica-Minolta, Hong Kong): each plot was divided into 3 sub-plots of approximately 4 m × 1.30 m size and SPAD readings were taken on 4 representative plants randomly distributed in the sub-plot. For each plant, the SPAD value was derived from the average of two readings taken at about one-third and two-thirds of the leaf blade of the last completely developed leaf, and the flag leaf when emerged. SPAD measurements were taken weekly from 15 April to 11 June 2023.

Concurrently with the SPAD measurements and until 18 June 2023, the normalized difference vegetation index (NDVI) of the canopy of each plot was monitored with an active handheld Greenseeker spectrometer (NTech Industries, Ukiah, CA, USA). The sensor detects canopy reflectance at wavelengths of 590 nm (ref_{RED}) and 880 nm (ref_{NIR}) and provides a ratio value as follows:

$$\text{NDVI} = \frac{ref_{NIR} - ref_{RED}}{ref_{NIR} + ref_{RED}}$$

The measurement was made by advancing longitudinally along the plots, keeping the sensor approximately 30 cm above the crop.

On 13 May 2023, during milk development, 0.12 m^2 wheat samples (1 m long row) within each plot were collected to measure some morphological parameters and shoot biomass. Plant height and the uppermost internode length were measured in 3 plants randomly selected within each sample. Leaves, culms, and spikes of each 0.12 m^2 area were assessed for fresh and dry weights (DW) of each component. DW was measured following oven-drying at 65 °C for 48 h, allowing the leaf-to-culm DW ratio to be also calculated.

Before drying, the photosynthesizing/green surface of leaves and culms was assessed using the LI-3100C Area Meter (Li-Cor instruments, Lincoln, NE, USA), allowing the leaf area index (LAI), the culm area index (CAI), as well as the LAI-to-CAI ratio to be calculated.

On 29 May 2023, during dough development, the leaf photosynthetic activity was measured using an infrared gas analyzer LI-6800 (Li-COR Inc., Lincoln, NE, USA). PSII photosynthetic efficiency (Fv'/Fm'), stomatal conductance (gsw), transpiration rate (Emm), and CO_2 net assimilation (A) were determined following the methods described by Murchie and Lawson [38].

The Fv'/Fm' ratio was measured as an index of efficiency in energy harvesting by the oxidized (open) reaction centers of photosystem II (PSII) of the flag leaf, where Fv' and Fm' represent variable and maximal fluorescence, respectively. As Fm' includes minimal fluorescence (F_0') of a dark-adapted leaf, Fv' is calculated as the difference $Fm'–F_0'$. Dark adaptation of the measured leaf was obtained using the far-red light to excite photosystem I (PSI), thus forcing electrons to drain from PSII. Only a few seconds of far-red light are needed to obtain this effect. The fluorimeter provides a "dark pulse" routine used to determine F_0'. Eight Fv'/Fm' records were registered on one leaf for each replicate.

Methylobacterium symbioticum strain was assessed for 1-aminocyclopropane-1-carboxylate (ACC) deaminase activity through a qualitative test; the growth of the bacterial strain was compared on two types of substrates, one containing ACC as the sole nitrogen source and the other with no nitrogen source. The bacterium was grown overnight in 10 mL of Luria and Bertani's liquid medium by shaking (130 rpm) at 28 °C. After measuring the optical density (OD) at 600 nm with a spectrophotometer (Varian Cary 50 UV–Visible, Santa Clara, CA, USA), the bacterial suspension was diluted to 10^8 cells mL^{-1} with sterile double-distilled water and plated on Luria–Bertani agar medium. Following 24 h of incubation at 28 °C, a bacterial colony was streaked to a Petri plate containing modified DF [39], minimal salts medium (glucose, 4.0 g/L; citric acid, 2.0 g/L; KH_2PO_4, 4.0 g/L; Na_2HPO_4, 6.0 g/L; $MgSO_4 \cdot 7H_2O$, 0.2 g/L; micro-nutrient solution ($FeSO_4$, 100 mg/100 mL; H_3BO_3, 10 mg/100 mL; $ZnSO_4 \cdot 7H_2O$, 124.6 mg/100 mL; Na_2MoO_4, 10 mg/100 mL; $CuSO_4$, 78.22 mg/100 mL; $MnSO_4$, 11.19 mg/100 mL)) and 3 mM 1-aminocyclopropane-1-carboxylic acid (ACC) as the sole nitrogen source. For the negative control, DF minimal medium without ACC was used. Growth of isolates on DF medium supplemented with ACC was assessed after incubation for 72 h at 28 °C and compared to controls. Additionally, *Bosea robiniae* and *Pseudomonas umsongensis* were used as negative and positive ACC deaminase producers, respectively [40].

2.4. Root Growth Analysis

The root system was investigated down to a depth of 1 m on 11 June 2023, roughly at the end of dough development, by soil coring. A probe was placed along the seed row to extract a 1 m long soil core for each plot/replicate, which was then divided into 10 sub-samples, each 10 cm long. Samples were kept at a temperature of −18 °C until washing. To separate the roots from the soil, a hydraulic sieving–centrifugation device on a 500 μm mesh sieve was used, and roots were stored in a 15% v/v ethanol solution at 4 °C until digitalization.

An EPSON Expression 11000XL PRO scanner (EPSON, Suwa—Nagano, Japan) was used to digitize the root images. Each sample was taken from the preservative solution and washed with demineralized water, and the roots were placed on a plexiglass tray with the help of a thin film of water (~3 mm). The scanner resolution for image acquisition was set to 400 DPI, in binary (1 bit) format in order to save computer memory resources.

Images were processed using KS 300 Rel. 3.0 software (Carl Zeiss Vision GmbH, Munich, Germany), with a minimum area of 40 pixels and elongation index E.I. (perimeter^2/area) > 40 for thresholding background noise. Objects with a round shape (E.I. < 40) were considered extraneous objects and thus not detected by the software.

Root length and area were calculated using the skeletonization/thinning algorithm. The area of skeletons (1 pixel width root objects) was multiplied by a correction factor of 1.12 [41,42] to account for all the possible angles of the roots arranged on the tray. The mean root diameter was calculated as the area-to-length ratio of root objects for each 10 cm long root sub-sample. Successively, root area density (RAD) and root length density (RLD) were determined and expressed as cm of roots per cm^3 of soil and cm^2 of roots per cm^3 of soil, respectively. Lastly, root dry weight was calculated after oven-drying for 2 days at 105 °C.

2.5. Yield Parameters and Grain Quality

At maturity, 1 m^2 samples of wheat plants were taken from the central part of each plot/replicate. Threshing was conducted using a stationary combine harvester to determine yield and yield parameters. The harvest index HI (grain-to-total shoot DW ratio) was measured after oven drying (65 °C for 48 h), while the thousand seed weight (TSW) was measured by averaging the weights of three subsamples of 1000 seeds from each plot/replicate.

The nitrogen content in straw and grain was determined using the Kjeldahl method [43], following the determination of nitrogen content, the total grain protein content (GPC) was obtained by multiplying the percentage of N by the coefficient 5.7.

Gluten composition was assessed by quantifying gliadins, high-molecular-weight glutenins (HMW-GS), and low-molecular-weight glutenin subunits (LMW-GS) extracted from refined flour following the procedure of Visioli et al. [44].

Rheological properties of dough produced from milled grains of the different treatments were investigated with the Chopin Alveograph Alveo PC model (Buhler, Leobendorf—Austria) to determine dough strength (W), tenacity-to-extensibility ratio (P/L index), and protein degradation index (PDI). The latter is expressed as the percentage of strength (W) loss of the dough remaining for 180 min in the alveograph compared to the normally calculated W at 28 min from dough preparation. Additionally, a Brabender FarinoGraph was used for the determination of other parameters, such as dough stability. The grain samples of the 3 replicates were pooled together to have sufficient sample size for rheological analyses, so these only have one replicate.

The total content of macro and micronutrients (Ca, Fe, K, Mg, P, Zn, Na, and Mn) in the grain and straw of the harvested plants was determined after sample mineralization with a Milestone Ethos 900 microwave mineralizer (Milestone S.r.l., Sorisole—Bergamo, Italy) (EPA Method No. 3052, USEPA, 1995). Mineralized samples were analyzed by inductively coupled plasma—optical emission spectroscopy (ICP-OES) using a Spectro Ciros Vision EOP instrument (Spectro Italia S.r.l., Lainate, Milan, Italy).

2.6. Statistical Analysis

The data from all the assessed parameters were subjected to ANOVA (Table S1) and the separation of homogenous groups of means was completed with the Tukey test. Statistical analyses were performed in R ver. 3.4.4 (The R Foundation for Statistical Computing Platform, 2024) and RStudio ver. 2023.12.1 (Posit Software, PBC, 2009–2024) ($p \leq 0.05$).

3. Results

3.1. Climatic Conditions during the Trials

During the wheat growing cycle from November 2022 to June 2023, there were notable deviations in temperature and rainfall values from the historical climate averages (Figure S1). December and January recorded warmer temperatures with increases of +2.4 °C in December and +2.6 °C in January. Contrastingly, April was cooler with temperatures dropping by −1.4 °C vs. the historical average. Overall precipitation during the wheat growth cycle totaled 540 mm (historical average: 470 mm), including intense rainfall in December, January, and during grain filling in May. Early season warmth boosted growth, but dry February and chilly April weather slightly retarded tillering and stem elongation.

3.2. Shoot Growth and Physiology

The treatment with biostimulant demonstrated varying impacts on plant growth and physiology, enhancing photosynthetic factors while creating minimal morphological alterations, especially under low nitrogen fertilization.

3.2.1. Bacterial Colonization

In asepsis, environmental scanning electron microscopy (ESEM) micrographs confirmed a discrete colonization of *Methylobacterium symbioticum* in wheat leaves (Figure S2) of biostimulant-treated plantlets, with bacterial cells primarily found along the veins of the leaves. In contrast, no such presence was found on the leaves of non-inoculated plantlets.

3.2.2. Leaf Chlorophyll Content and Canopy Greenness (SPAD and NDVI)

Data obtained over the growing season indicate that the leaf chlorophyll content, which directly correlates to the SPAD index, was highest at 100% nitrogen fertilization compared to lower doses. The fixed effect of fertilization was highly significant ($p \leq 0.001$), and an interaction between fertilization and bacterial inoculation was significant on May 6 and 13 ($p \leq 0.05$) (Table S1). The "100%N" treatment had the highest SPAD value on May 6 (50.3 units), whereas the "75%N" treatment showed a substantial reduction (−6% vs. "100%N"; $p < 0.05$) (Figure 1). This pattern remained stable on 13 May during milk development. SPAD values began to diminish in late May as wheat reached maturity, with a slower reduction in fully fertilized plots. By the 11th of June, the decreases were −21% for "75%N" and −17% for "75%N + bact" as compared with the reference treatment. Although there were no significant changes between the two treatments, "75%N + bact" consistently had higher SPAD values than "75%N".

NDVI readings mirrored SPAD results. Full fertilization increased canopy greenness compared to "75%N." Although without significant differences, inoculation with bacteria resulted in greater NDVI, particularly around harvest (Figure 1). On June 11 and 18, the "100%N + bact" and "75%N + bact" treatments had greater NDVI values than the non-inoculated counterparts, highlighting delayed canopy senescence.

In addition, the in vitro investigation demonstrated that the *Methylobacterium symbioticum* strain has ACC deaminase activity (Figure S3), although it was not quantified.

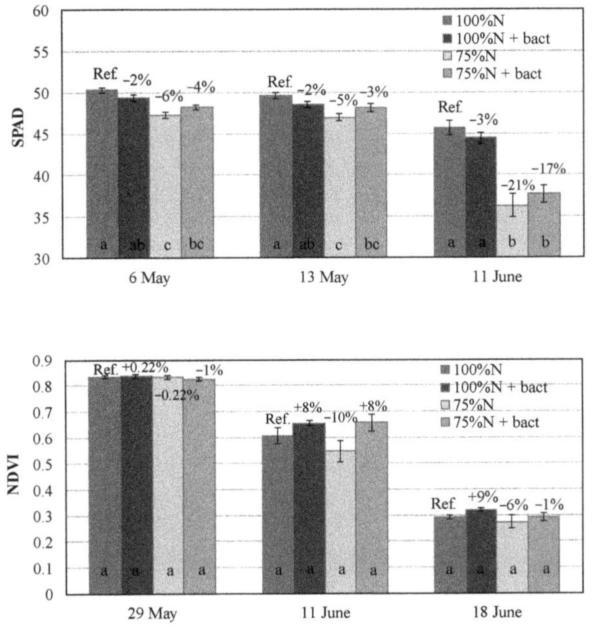

Figure 1. Dynamics of leaf chlorophyll content (as SPAD values) and canopy greenness (NDVI) (mean ± S.E.; n = 3) in wheat under "100%N", "100%N + bact", "75%N", and "75%N + bact" treatments during 2023. Percentages: variation vs. reference treatment 100%N (Ref.). Letters: statistical comparison among treatments (Tukey test, $p \leq 0.05$).

3.2.3. Plant Morphology

The analysis of wheat plant morphology during milk development yielded numerous noteworthy findings (Table 2). Plant height was greatly influenced by treatments. The "100%N" reference had the highest plants, whereas "100%N + bact" had the shortest, with a 15% drop compared to the reference. The biostimulant-treated plots "75%N + bact" were 6% lower in height than the "100%N" plots, whereas the "75%N" plots exhibited a 11% reduction. ANOVA showed significant effects of bacterial inoculation ($p \leq 0.05$) and fertilization × bacteria interaction ($p \leq 0.001$) on plant height (Table S1). However, there were no significant variations in the length of the uppermost internode among treatments.

Table 2. Morphological parameters: plant height, uppermost internode length, LAI, LAI/CAI, shoot dry weight (DW), and leaf-to-culm dry weight ratio (DW) (mean ± S.E.; n = 3) in wheat plants under "100%N", "100%N + bact", "75%N", and "75%N + bact" treatments. In brackets: % variation vs. reference treatment 100%N. Letters: statistical comparison between the 4 different treatments (Tukey test, $p \leq 0.05$).

Treatment	Plant Height		Uppermost Internode Length		LAI		LAI/CAI		Shoot DW		Leaf-to-Culm DW Ratio	
	cm	%var/C	cm	%var/C	$m^2\ m^{-2}$	%var/C		%var/C	$g\ m^{-2}$	%var/C		%var/C
100%N	92.2 ± 1.8	a	8.94 ± 0.65	a	8.48 ± 1.00	a	2.73 ± 0.10	a	1766.9 ± 194.4	a	0.329 ± 0.016	a
100%N + bact	78.8 ± 1.2	c (−15%)	9.44 ± 0.61	a (+6%)	7.39 ± 0.52	a (−13%)	2.52 ± 0.04	a (−8%)	1662.5 ± 124.1	a (−6%)	0.321 ± 0.004	ab (−3%)
75%N	82.1 ± 1.7	bc (−11%)	9.66 ± 0.60	a (+8%)	7.98 ± 0.17	a (−6%)	2.53 ± 0.06	a (−7%)	1770.2 ± 40.8	a (+0.2%)	0.289 ± 0.004	b (−12%)
75%N + bact	86.5 ± 2.1	ab (−6%)	9.83 ± 0.68	a (+10%)	7.62 ± 0.96	a (−10%)	2.55 ± 0.10	a (−6%)	1659.7 ± 176.1	a (−6%)	0.309 ± 0.005	ab (−6%)

LAI = leaf area index; CAI = culm area index.

The leaf area index (LAI) did not alter significantly across treatments; however, the biostimulant-treated plots had the lowest LAI values, with decreases of 13% and 10% for "100%N + bact" and "75%N + bact", respectively, compared to the reference. The LAI-to-CAI ratio indicated no significant variations, with the highest value in the reference treatment and ~7% decline in the other treatments.

The treatments did not have a significant effect on shoot dry weight, although the biostimulant-treated plots showed slightly lower DW values (−6% for both N dosages compared to the reference). Significant differences were seen in the leaf-to-culm DW ratio, with the "100%N" and "100%N + bact" plots exhibiting greater values. The "75%N + bact" treatment had an intermediate value, demonstrating that the biostimulant is effective at boosting leaf development in low N environments.

3.2.4. Photosynthetic Parameters

The application of the microbial biostimulant, especially at a low nitrogen dosage (75%N), improved various photosynthetic parameters (Figure 2 and Figure S4).

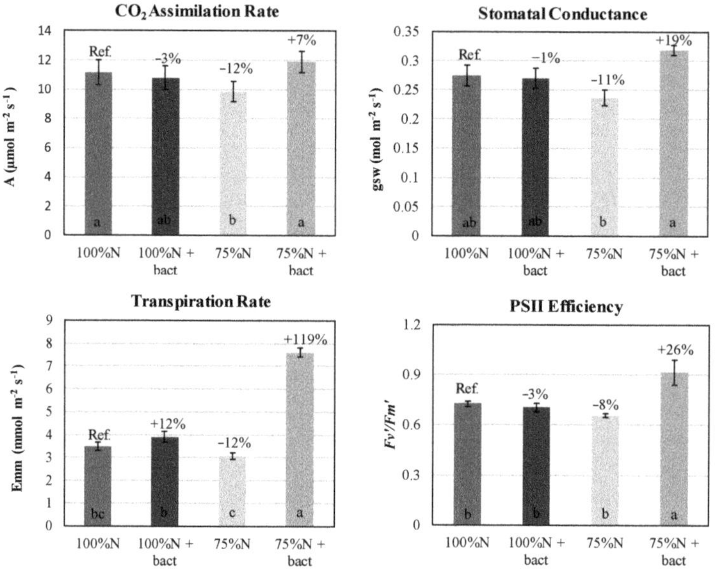

Figure 2. Photosynthetic parameters: CO_2 assimilation rate (A), stomatal conductance (gsw), transpiration rate (Emm), and PSII efficiency (Fv'/Fm') (mean ± S.E.; n = 3) in wheat plants under "100%N", "100%N + bact", "75%N", and "75%N + bact" treatments. Percentages: variation vs. reference treatment 100%N (Ref.). Letters: statistical comparison among treatments (Tukey test, $p \leq 0.05$).

The "75%N + bact" treatment enhanced stomatal conductance and CO_2 assimilation rates. However, only the increase in transpiration rate (119% compared to the reference, $p < 0.05$) was statistically significant. Furthermore, PSII efficiency was best in "75%N + bact," indicating improved photosynthesis.

ANOVA revealed significant impacts of bacterial inoculation and fertilization × bacteria interaction on stomatal conductance, transpiration rate, and PSII efficiency ($p < 0.01$) (see Table S1).

3.3. *Root Growth*

Statistical analysis of root length density (RLD) in the 0–100 cm soil layer revealed no significant variations between treatments (Figure 3). Both "100%N" treatments, with and without microbial biostimulant, showed a mean RLD of approximately 2.8 cm^{-3}.

The "75%N" and "75%N + bact" treatments increased RLDs by 13% and 3%, respectively, compared to "100%N," although these changes were not statistically significant (Table S2).

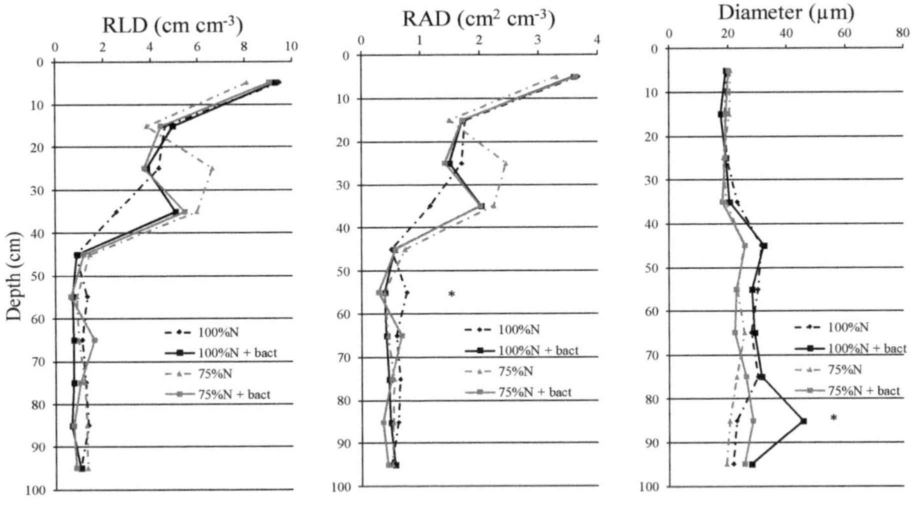

Figure 3. Root length density (RLD), root area density (RAD), and root diameter (mean; n = 3) in wheat plants under "100%N", "100%N + bact", "75%N", and "75%N + bact" treatments at various soil depths. Asterisk (*): statistically significant differences among treatments (Tukey test, $p \leq 0.05$).

The biostimulant impact was more evident in the 0–50 cm depth, where the "100%N + bact" and "75%N + bact" treatments increased RLD by 10% and 9%, respectively, compared to the reference; however, these differences were not statistically significant. The maximum RLD at this depth was observed for "75%N" without biostimulant. Contrastingly, at the 50–100 cm depth, treatments with the biostimulant resulted in a considerable drop in RLD (−36% and −20% for "100%N + bact" and "75%N + bact," respectively, compared to the reference.

Root area density (RAD) followed the RLD trends. There were no significant variations in RAD values across treatments in the 0–100 cm profile. Treatments with the biostimulant had more root area in the 0–50 cm profile (+7% and +6% for "100%N + bact" and "75%N + bact", respectively, compared to "100%N"). However, in the 50–100 cm profile, RAD was reduced by 26% and 28% for "100%N + bact" and "75%N + bact", respectively, vs. "100%N", although these changes were not statistically significant. A considerable RAD difference was notably seen at 50–60 cm depth interval (Figure 3).

3.3.1. Root Diameter

The treatment with the microbial biostimulant had no significant effect on overall root diameter (Table S2). Nevertheless, the biostimulant resulted in increased root diameter over the 0–100 cm soil profile when compared to plots fertilized with mineral nitrogen alone. Specifically, in the 0–50 cm depth, the "100%N + bact" and "75%N + bact" treatments resulted in smaller root diameters (−4% and −9%, respectively) vs. "100%N". In the 50–100 cm depth, "100%N + bact" and "75%N + bact" treatments showed larger root diameters (+22% and +10%, respectively) than their non-inoculated counterparts.

A statistically significant change was seen at the 80–90 cm deep interval (Figure 3). Fertilization and bacterial treatment had statistically significant effects on root diameter over the 0–100 cm profile ($p \leq 0.05$) and 50–100 cm depth ($p \leq 0.001$), although their interaction was not significant (Table S1).

3.3.2. Root Dry Weight

Root dry weight did not vary significantly over the whole soil profile (0–100 cm) or its sub-profiles (0–50 cm and 50–100 cm). Higher nitrogen dosages often reduced root biomass, although the microbial biostimulant slightly offset this impact. The "75%N" treatments, both with and without the biostimulant, had greater root dry weights than the "100%N" reference, but the differences were not statistically significant. Specifically, in the 0–100 cm profile, the "75%N" and "75%N + bact" treatments increased root dry weights by 17% and 26%, respectively, as compared to "100%N". In the 0–50 cm depth, both "75%N" and "75%N + bact" treatments again produced larger root dry weights than the reference. In the deeper 50–100 cm profile, there were little variations between treatments (Table S2).

3.4. Yield Parameters and Grain Quality

The yield metrics revealed no statistically significant changes across treatments, while certain trends were identified (Table 3).

Table 3. Productivity parameters of wheat: grain yield, harvest index (HI), thousand seed weight (TSW), and test weight (mean ± S.E.; n = 3) in wheat under "100%N", "100%N + bact", "75%N", and "75%N + bact" treatments. In brackets: % variation vs. reference treatment 100%N. Letters: statistical comparison among treatments (Tukey test, $p \leq 0.05$).

Treatments	Grain Yield		HI		TSW		Test Weight	
	g m^{-2}	%var/C	%	%var/C	g	%var/C	kg hL^{-1}	%var/C
100%N	1023 ± 41	a	42.6 ± 0.55	a	39.7 ± 0.44	a	81.0 ± 0.29	a
100%N + bact	956 ± 43	a (−7%)	42.0 ± 1.18	a (−2%)	39.6 ± 0.88	a (−0.2%)	80.7 ± 0.47	a (−0.4%)
75%N	947 ± 52	a (−8%)	41.1 ± 1.00	a (−4%)	39.6 ± 0.51	a (−0.4%)	82.2 ± 0.52	a (+1%)
75%N + bact	959 ± 87	a (−6%)	43.2 ± 0.81	a (+1%)	39.0 ± 0.07	a (−2%)	81.3 ± 0.41	a (+0.3%)

The "100%N" reference treatment had the highest grain yield, while the "100%N + bact" treatment had a 7% lower value. The "75%N" treatment had the lowest grain yield (−8% vs. reference); however, the microbial biostimulant in the "75%N + bact" treatment marginally reduced this gap to −6% vs. reference.

The harvest index (HI) varied from about 41% to 43% across all treatments, with "75%N + bact" having the highest value, albeit the difference among treatments was not statistically significant.

The average thousand seed weight (TSW) was around 39.5 g, and the test weight ranged from 80.7 to 82.1 kg hL^{-1}, with no significant variations or trends found among treatments.

In terms of grain quality, the application of the microbial biostimulant had no significant effects on total grain protein content (GPC), but the "100%N + bact" treatment showed a 3% increase over the reference, with protein content ranging from 12.7% to 13.0% across all treatments (Figure 4).

Further analysis of protein fractions revealed distinct impacts of treatments with the bacterium (Figure 4).

The use of biostimulants resulted in a decrease in gliadin content; however, it was not statistically significant, and the "100%N + bact" and "75%N + bact" treatments resulted in 6% and 4% decreases, respectively, compared to the "100%N" reference. On the other hand, the "100%N + bact" and "75%N + bact" treatments had the greatest glutenin content, increasing by 11% and 10%, respectively, compared to the reference. As a result, in the biostimulant-treated plots, the gliadin-to-glutenin ratio was lowered by 16% and 13% for "100%N + bact" and "75%N + bact," respectively, relative to the reference.

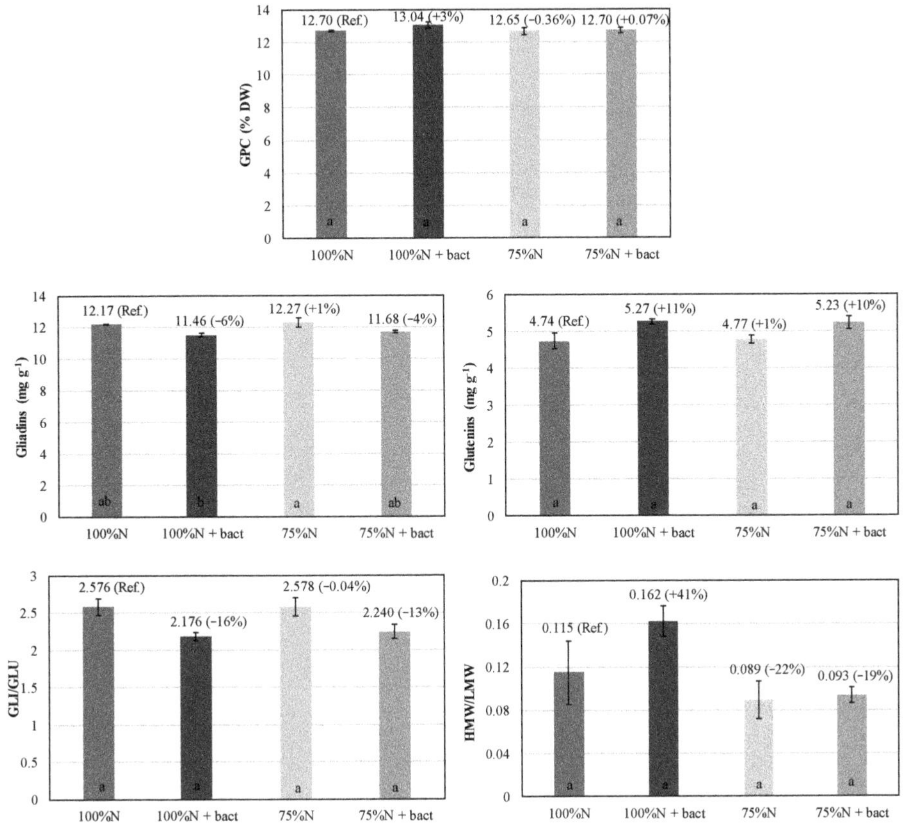

Figure 4. Qualitative characteristics of wheat grains: grain protein content (GPC), gliadins, glutenins, gliadin-to-glutenin ratio (GLI/GLU), and high-to-low molecular weight glutenin subunits ratio (HMW/LMW) (mean ± S.E.; n = 3) in wheat plants under "100%N", "100%N + bact", "75%N", and "75%N + bact" treatments. Percentages: variation vs. reference treatment 100%N (Ref.). Letters: statistical comparison among treatments (Tukey test, $p \leq 0.05$).

Biostimulant-treated plots also had a higher high-to-low molecular weight glutenin subunit ratio (HMW/LMW) than non-inoculated plots, with increases of 41% for "100%N + bact" and 3% for "75%N + bact" compared to the corresponding treatments not treated with the bacteria.

ANOVA revealed significant effects of bacterial treatment on gliadins ($p \leq 0.001$), glutenins ($p \leq 0.05$), and the gliadin-to-glutenin ratio ($p \leq 0.001$), while the effect of fertilization was significant on HMW/LMW ($p \leq 0.05$) (Table S1).

3.5. Dough Rheological Properties

Dough rheological characteristics indicated little variations across treatments (Table S3).

The "100%N" reference showed the maximum dough strength (W), and the "75%N + bact" management had a small lower value than the reference (−3%), whereas the "75%N" treatment showed a 18% decrease.

The "75%N + bact" treatment resulted in a 6% rise in the tenacity-to-extensibility ratio (P/L) compared to the reference (0.69 vs. 0.65), whereas the "100%N + bact" treatment had a significantly lower P/L value than the reference.

As regards the protein degradation index (PDI), the "100%N + bact" treatment had a 45% lower loss of dough strength than the reference, showing superior resistance to protein degradation. The "75%N + bact" treatment also showed a tendency towards higher dough strength preservation as compared to "75%N".

The Brabender FarinoGraph highlighted that treatments with the biostimulant generally showed higher stability, regardless of the N dose, with the highest stability (>8 min) observed in "75%N + bact" and "100%N + bact".

3.6. Nutrient Content in Wheat Organs

Following the analyses on the nitrogen content in grain using the "Kjeldahl" method, which showed no statistically significant differences among treatments, a study was carried out on the content of macro and micro-nutrients in grain and straw (Table S4).

3.6.1. Nutrient Content in Grain

As regards macronutrients, in less fertilized plots, the grain's content of Ca, K, Mg, and P tended to be lower, and the "75%N + bact" treatment showed the greatest drop in these four nutrients as compared to the "100%N" reference. When comparing the same nitrogen dosage with the bacteria removed, the macronutrient values obtained via microbial inoculation were marginally improved.

As regards micronutrients in wheat grains, the levels of Na and Zn did not significantly change across treatments. Every treatment showed reduced values for Fe when compared to the reference; however, the "75%N + bact" treatment had the most significant drop (−21% vs. reference). With the exception of the "75%N + bact" treatment, which exhibited a 33% drop, Mn content was typically consistent. Microbial inoculation led to slightly lower Fe and Mn values compared to the same N dose without the bacteria.

3.6.2. Nutrient Content in Straw

Analysis of macronutrients in wheat straw highlighted that the amounts of Ca, K, and Mg did not change significantly. On the other hand, P content was significantly lower in the "75%N + bact" treatment as compared to the reference "100%N".

The concentrations of Na, Fe, and Zn were comparable among treatments. Significantly less Mn was present in the "75%N + bact" treatment (compared to the reference by 27% and to the "75%N" treatment by 36%).

As regards N content, there were no appreciable variations in the straw among treatments. However, microbial inoculation produced somewhat greater values than the same N dosage without bacteria, just like it did with grain.

4. Discussion

The development of ecological practices to increase productivity and decrease dependence on chemical fertilizers in cereal production is a current challenge for agriculture. In recent years, an emerging approach has involved applying foliar biostimulant products, given their ease of application and low cost. Several studies have focused on the efficacy of beneficial microorganisms such as PGP bacteria [45,46]. However, the new strain *Methylobacterium symbioticum*, isolated by Pascual et al. in 2020 [34], lacks an extensive bibliography and the effects of an inoculum based on this bacterium in common wheat have not yet been tested.

Consistent with the findings by some authors in crops such as maize, strawberry, and apple [29,34], this trial confirmed the positive effects on some vegetational parameters (SPAD and NDVI) following the inoculation with *Methylobacterium symbioticum*. This inoculum allowed a more prolonged stay-green period of the crop, thereby extending the photosynthetic activity during the crop cycle. This may represent an agronomic advantage by allowing the plant to continue photosynthesizing for longer, thereby being able to assimilate more CO_2, and increasing total biomass at harvest. This study on common wheat showed that SPAD, near harvesting, as expected, was significantly lower in plots fertilized with the reduced dose of nitrogen ("75%N") with respect to the reference. However, the "75%N + bact" treatment reduced this difference, demonstrating that the bacterium is able to delay chlorophyll degradation close to the end of the crop cycle. Additionally, the tendency, albeit not significant, for the plots treated with the microbial biostimulant to show higher SPAD values, which is an index indirectly correlated to chlorophyll content, than untreated plots, especially at the lower N fertilization dose, highlights how in the case of low N fertilization level the bacterium is better able to express its effects. It is probable that a plant with a worse nitrogen nutritional status is more favorable to establishing and maintaining an endophytic relationship with these plant-aiding bacteria. Similar evidence emerged in an in vitro test in maize [29], in which, at a reduced dose of 50%N, the application of the bacterium increased the SPAD values when compared to the full fertilized control, thus confirming a better establishment of the association if the plants are under nutritional deficit/stress.

The analysis also revealed that canopy NDVI at the end of the cycle was statistically similar across all treatments, but with better values in the "75%N + bact" plots than in the "75%N" ones. Previous experiments have shown that various strains of the genus *Methylobacterium* exhibit ACC-deaminase activity [47]. Under this assumption, the preliminary in-vitro test revealed that the strain tested here also possesses this enzymatic activity capable of delaying plant senescence by the degradation of the ethylene precursor, thus resulting in an extension of the foliar stay-green period. This enzyme breaks down ACC, lowering ethylene synthesis while increasing stress tolerance [48]. This mechanism is consistent with the observations of longer canopy stay-green and increased photosynthetic efficiency. Indeed, while the bacterial treatment did not lead to substantial effects on morphological traits (e.g., plant height, uppermost internode length, LAI, and shoot biomass), there were notable effects on photosynthetic parameters. The "75%N + bact" treatment had a positive effect on the wheat plants, increasing stomatal conductance and transpiration rate, compared to both "100%N" reference and the corresponding N fertilization dose without the inoculum ("75%N"). A better CO_2 assimilation rate and PSII efficiency were also observed. This evidence aligns with the SPAD findings and demonstrates how plants seek a bond with microorganisms to compensate for a nutritional deficit. Conversely, at the higher dose of nitrogen fertilization, this does not seem to occur given the higher availability of nitrogen in the soil: it can be seen in vegetative parameters, in which there is a tendency for the "100%N + bact" treatment to be very similar to the reference. Therefore, it emerges that the bacterium has a physiological influence on the plant mostly in the case of reduced N fertilization.

Similar studies with other PGPB also indicated similar benefits. For example, Bhattacharyya and Jha [21] reported that the association with *Azospirillum brasilense* enhances the growth of various crops by fixing atmospheric N_2 and synthesizing growth-promoting substances. Egamberdieva and Lugtenberg [22] observed that *Pseudomonas fluorescens* enhanced the tolerance of plants to saline stress with the production of phytohormones and ACC deaminase, which ultimately decreases the level of ethylene and improved growth.

Methylobacterium symbioticum is claimed to be N-fixing in its endophytic life in plants [33], and it is expected to increase the fixation ability under reduced N availability, as occurring in many N-fixing symbiosis. This would explain the better success of inoculation under "75%N" vs. "100%N". Nitrogen fixation was not investigated in this trial, but better N content in wheat grains and straw and enhanced photosynthetic parameters, especially at a reduced N fertilizer dose, would corroborate the ability of *Methylobacterium symbioticum* to fix atmospheric N_2. Surely, under reduced N fertilization, this bacterium retrieves some N at metabolic level by degrading the plant ethylene precursor ACC thanks to the ACC deaminase enzyme, and in this way, also delaying plant senescence.

No studies can be currently found in the literature assessing the effect of the *Methylobacterium symbioticum* strain on root growth and the presence of this bacterium in plant roots. However, some PGPBs including the genus *Methylobacterium*, and particularly *Azospirillum*, are known to create associations with plants that lead to an improvement in the root system [49], probably due to the production and exchange of bioregulators such as auxins or auxin-like compounds. This study showed, in agreement with the literature, good RLD and RAD values in the 0–100 cm soil profile of wheat in all treatments, with an average RLD > 2.8 cm cm^{-3} and an RAD > 1.16 cm^2 cm^{-3} [50]. Here, the *Methylobacterium symbioticum* association with common wheat did not show significant effects on root growth, except for a certain increase in RLD and RAD in the arable layer. The "75%N" treatment, having reduced chemical N supply, exhibited higher root growth than "100%N", aligning with the well-known plant response to nitrogen deficiency by encouraging root growth for better soil exploration [51,52]. The "75%N + bact" treatment showed root growth levels very similar to the reference "100%N", probably due to nitrogen compensation from atmospheric N_2 fixation by the bacterium, thus decreasing the need for an extensive root system.

Yield and other production parameters remained stable across treatments, indicating that the bacterium does not have a significant influence on these parameters. In contrast, other studies show that the application of *Methylobacterium symbioticum* increased yields in maize, grapevine, rice and strawberry, particularly at lower doses of N supplied [29,34]. These authors suggest that the enhanced crop performances might be attributed to the capability of microorganisms to assimilate ammonium through N fixation. Despite this being an energy-consuming process, it ultimately leads to the conservation of energy for the plant during the assimilation of nitrogen. This is because the plant can bypass the energy-intensive step of converting nitrate, deriving from root uptake, to ammonia, as it directly receives ammonium from the microorganisms. This would make the utilization of nitrogen more efficient, leading to an improved plant growth [29].

Regarding grain quality, the microbial biostimulant did not increase significantly the protein content (GPC), consistent with the N content findings. However, it impacted protein quality, with an increase in glutenins over gliadins, and HMW over LMW. Therefore, it was associated to an increase in the high molecular weight protein component over the low molecular weight component, with the final effect of increasing dough tenacity and stability.

To assess the possible presence of proteases released by phytophagous bugs (e.g., *Eurygaster*, *Aelia*, and particularly *Halyomorpha halys*), which facilitate nutrient suction from the developing kernels by injecting protease-rich saliva [53], the PDI measurement of the flours from the various treatments highlighted interesting effects. The bacterial treatment reduced protein degradation at both levels of nitrogen fertilization, a positive effect that deserves to be further investigated. Previous studies have shown that the degradation of durum wheat glutenins and gliadins have different dynamics concerning the flour incubation temperature, and that glutenins are probably also degraded by other substances released by bugs into grains [54]. Our PDI results highlight lower protein degradation in bacteria-treated plots, probably due to the greater difficulty for protease enzymes to degrade larger protein subunits, and thus preserving the dough strength.

Earlier studies on dough rheological properties and gluten quality demonstrated a significant correlation between dough extensibility and the gliadin/glutenin (GLI/GLU) ratio. In addition, HMW subunits have twice the effect on maximum dough resistance than LMW subunits [55]. Based on this evidence, a higher P/L was expected in treatments with the microbial biostimulant, given lower GLI/GLU and higher HMW/LMW ratios, with consequently lower dough extensibility and increased tenacity. This behavior was only observed at the reduced N fertilization dose, with an increase in P/L in the "75%N + bact" treatment compared to the "75%N". According with this trend, the Brabender FarinoGraph analysis indicated an improved dough stability due to treatment with the bacterium, at both nitrogen levels.

From an economic perspective, the application of the microbial inoculum does not require additional field machinery costs, as it could be applied along with the post-emergence weeding or fungicide treatments, in this way saving time and fuel costs. Although any possible interaction between the bacterium and pesticides requires to be preliminary investigated, at this time the cost-effectiveness would lie in the compensation point between the cost of the applied biostimulant with that of saved fertilizer following the inoculum application. This depends mainly on fertilizer prices, and recent raw material shortages have increased fertilizer prices, consequently making biostimulants more cost-effective. However, regardless of existing prices, the environmental benefits of decreased synthetic mineral fertilizer use, such as reduced nitrogen leaching following heavy rainfall, must be noted. In conditions of drought and difficulties for soil nutrient uptake, foliar spraying of effective microbial biostimulants should be considered as an additional agronomic benefit.

5. Conclusions

This preliminary study on the foliar application of *Methylobacterium symbioticum* demonstrates that it had no significant effect with respect to the yield or grain protein content of soft wheat, while marked improvements in vegetational indexes such as SPAD and NDVI were observed, reflecting an improved nitrogen status and extended leaf greenness. These findings align with the plant growth-promoting characteristics of the bacterium probably attributable to its ACC-deaminase activity, with the major benefits achievable under reduced N fertilization.

Interesting changes in gluten protein composition were also noted, referred to increases in high molecular weight protein components, which influence the dough rheological properties, to be exploited for specific baking applications.

Considering that the plant–bacteria interactions are variable, the N fixation potential of this bacterium across reduced N regimes, as well as the effects of inoculation in other wheat varieties and under various environmental conditions are worth of investigation.

In the context of climate change and associated plant stress, improved root colonization in the arable layer due to microbial inoculation may offer an additional benefit from plant inoculation with *Methylobacterium symbioticum* in nitrogen-deficient and organic farming systems. However, to fully validate these benefits, more studies are needed across diverse agricultural contexts and multiple growing seasons.

In conclusion, *Methylobacterium symbioticum* demonstrates to be promising and a viable sustainable agricultural tool, but further research needs to be devoted to fully understand its mechanisms and optimize its usage. The potential of this bacterium to improve nitrogen availability, and therefore support plant growth under suboptimal conditions, underscores its potential in sustainable agriculture.

Supplementary Materials: The following supporting information can be downloaded at: https://www.mdpi.com/article/10.3390/agriculture14101670/s1, Figure S1: monthly mean temperatures and precipitations during the wheat growth cycle (2022–2023), compared with the historical average of the last decade 2012–2022 (Legnago-Vangadizza—Verona, ARPAV station); Figure S2: ESEM micrographs of soft wheat leaf surface in control (non-inoculated) (A–C) and inoculated (+BACT) (D–F) in 14-day-old wheat seedlings. Note bacteria colonization on the right side only; Figure S3: qualitative test for 1-aminocyclopropane-1-carboxylic acid (ACC) deaminase activity. (A) DF minimal medium without any nitrogen source; (B) DF minimal medium with ACC as a unique nitrogen source. 1. *Methylobacterium symbioticum* SB23; 2. negative control: *Bosea robiniae* [40]; 3. positive control: *Pseudomonas umsongensis* [40]. Note colonization of *Methylobacterium symbioticum SB23* and *Pseudomonas umsongensis* (colored bands) when only ACC was present in the growing medium; Figure S4: model lines of some photosynthetic parameters: stomatal conductance (gsw), transpiration rate (Emm), and photosynthetic efficiency (Fv'/Fm') (mean; n = 3) in wheat plants under "100%N", "100%N + bact", "75%N", and "75%N + bact" treatments, at different photosynthetic photon flux density (PPFD). Letters: statistical comparison among treatments (Tukey test, $p \leq 0.05$); Table S1: ANOVA results for SPAD, NDVI, shoot characteristics, photosynthetic parameters, root attributes, yield components, and nutrient contents, showing main effects (fertilization (F), bacteria (B)) and their interaction (fertilization × bacteria (F × B)). Note: significance at $p \leq 0.05$ (*), $p \leq 0.01$ (**), $p \leq 0.001$ (***), ns (not significant); Table S2: root growth parameters: root length density (RLD), root area density (RAD), root diameter, and root dry weight (mean ± S.E.; n = 3) in wheat plants under "100%N", "100%N + bact", "75%N", and "75%N + bact" treatments in various soil layers. In brackets: % variation vs. reference treatment 100%N. Letters: statistical comparison among treatments (Tukey test, $p \leq 0.05$); Table S3: rheological properties of wheat dough: strength (W), tenacity-to-extensibility ratio (P/L index), protein degradation index (PDI), and stability (n = 1, derived from 3 replicates/plots) of dough from grains of wheat plants under "100%N", "100%N + bact", "75%N", and "75%N + bact" treatments. On the right, the % variation vs. reference treatment 100%N; Table S4: macro and micro-nutrient content in grain and straw (mean ± S.E.; n = 3) of wheat plants under "100%N", "100%N + bact", "75%N", and "75%N + bact" treatments. Letters: statistical comparison among treatments (Tukey test, $p \leq 0.05$).

Author Contributions: Conceptualization, T.V. and Y.W.; methodology, F.B., A.P., G.B., F.V., S.P., V.B. and G.V.; software, F.V.; validation, A.P., T.V. and F.V.; formal analysis, A.P., G.B. and F.V.; investigation, F.V. and A.P.; resources, T.V.; data curation, A.P.; writing—original draft preparation, F.V. and P.K.B.; writing—review and editing, T.V., Y.W. and A.P.; visualization, F.V.; supervision, T.V.; project administration, T.V.; funding acquisition, T.V. and Y.W. All authors have read and agreed to the published version of the manuscript.

Funding: This research entitled "Agronomic Effects of Field Inoculation with Bacteria Consortia in Wheat Planting" was funded by the Guangzhou University, China (GISU) and the APC was funded by BIRD230559/2023 entitled "The Impact of Landscape in Shaping the Microbiota of Pollinators" from the University of Padua, Italy.

Institutional Review Board Statement: Not applicable.

Data Availability Statement: The raw data supporting the conclusions of this article will be made available by the authors on request.

Acknowledgments: We are grateful to Monica Mattarozzi from the University of Parma for supplying the ESEM images.

Conflicts of Interest: The authors declare no conflicts of interest.

References

1. Novoa, R.; Loomis, R.S. Nitrogen and plant production. *Plant Soil* **1981**, *58*, 177–204. [CrossRef]
2. Zayed, O.; Hewedy, O.A.; Abdelmoteleb, A.; Ali, M.; Youssef, M.S.; Roumia, A.F.; Yuan, Z.C. Nitrogen journey in plants: From uptake to metabolism, stress response, and microbe interaction. *Biomolecules* **2023**, *13*, 1443. [CrossRef] [PubMed]
3. Taria, S.; Arora, A.; Alam, B.; Kumar, S.; Yadav, A.; Kumar, S.; Arunachalam, A. Introduction to Plant Nitrogen Metabolism: An overview. In *Advances in Plant Nitrogen Metabolism*; CRC Press: Boca Raton, FL, USA, 2022; pp. 1–18.
4. Guo, K.; Yang, J.; Yu, N.; Luo, L.; Wang, E. Biological nitrogen fixation in cereal crops: Progress, strategies, and perspectives. *Plant Commun.* **2023**, *4*, 100499. [CrossRef] [PubMed]
5. Rosenblueth, M.; Ormeño-Orrillo, E.; López-López, A.; Rogel, M.A.; Reyes-Hernández, B.J.; Martínez-Romero, J.C.; Martínez-Romero, E. Nitrogen fixation in cereals. *Front. Microbiol.* **2018**, *9*, 1794. [CrossRef] [PubMed]
6. Paull, J.A. Century of Synthetic Fertilizer: 1909–2009. *J. Bio-Dyn. Tasmania* **2009**, *94*, 16–21.
7. David, W.I.; Agnew, G.D.; Bañares-Alcántara, R.; Barth, J.; Hansen, J.B.; Bréquigny, P.; de Joannon, M.; Stott, S.F.; Stott, C.F.; Guati-Rojo, A.; et al. 2023 roadmap on ammonia as a carbon-free fuel. *J. Phys. Energy* **2024**, *6*, 021501. [CrossRef]
8. Smith, C.; Hill, A.K.; Torrente-Murciano, L. Current and Future Role of Haber–Bosch Ammonia in a Carbon-Free Energy Landscape. *Energy Environ. Sci.* **2020**, *13*, 331–344. [CrossRef]
9. Chen, J.; Wu, H.; Qian, H. Groundwater nitrate contamination and associated health risk for the rural communities in an agricultural area of Ningxia, northwest China. *Expo. Health* **2016**, *8*, 349–359. [CrossRef]
10. Sutton, M.A.; Howard, C.M.; Erisman, J.W.; Billen, G.; Bleeker, A.; Grennfelt, P.; Grizzetti, B. (Eds.) *The European Nitrogen Assessment: Sources, Effects and Policy Perspectives*; Cambridge University Press: Cambridge, UK, 2011.
11. Lin, T.Y.; Chiu, Y.H.; Lin, Y.N.; Chang, T.H.; Lin, P.Y. Greenhouse Gas Emission Indicators, Energy Consumption Efficiency, and Optimal Carbon Emission Allowance Allocation of the EU Countries in 2030. *J. Nat. Gas Sci. Eng.* **2023**, *110*, 204902. [CrossRef]
12. Suhameena, B.; Devi, S.; Gowri, R.; Kumar, S.D. Utilization of *Azospirillum* as a Biofertilizer—An overview. *Int. J. Pharm. Sci. Rev. Res.* **2020**, *62*, 141–145.
13. Raffi, M.M.; Charyulu, P.B.B.N. *Azospirillum*-biofertilizer for sustainable cereal crop production: Current status. In *Recent Developments in Applied Microbiology and Biochemistry*; Academic Press: Cambridge, MA, USA, 2021; pp. 193–209.
14. Pereg, L.; de-Bashan, L.E.; Bashan, Y. Assessment of affinity and specificity of *Azospirillum* for plants. *Plant Soil* **2016**, *399*, 389–414. [CrossRef]
15. The, S.V.; Snyder, R.; Tegeder, M. Targeting nitrogen metabolism and transport processes to improve plant nitrogen use efficiency. *Front. Plant Sci.* **2021**, *11*, 628366. [CrossRef] [PubMed]
16. Alori, E.T.; Babalola, O.O. Microbial inoculants for improving crop quality and human health in Africa. *Front. Microbiol.* **2018**, *9*, 2213. [CrossRef] [PubMed]

17. Reed, L.; Glick, B.R. The recent use of plant-growth-promoting bacteria to promote the growth of agricultural food crops. *Agriculture* **2023**, *13*, 1089. [CrossRef]
18. Bianco, C. Plant-Growth-Promoting Bacteria. *Plants* **2024**, *13*, 1323. [CrossRef]
19. Kumari, E.; Kumari, S.; Das, S.S.; Mahapatra, M.; Sahoo, J.P. Plant Growth-Promoting Bacteria (PGPB) for sustainable agriculture: Current prospective and future challenges. *Agro. Environ. Sustain.* **2023**, *1*, 274–285. [CrossRef]
20. Castaldi, S.; Lorenz, C.; Vitale, E.; Santorufo, L.; Isticato, R.; Arena, C. Potentialities of Technosol-isolated PGPB consortium in promoting plant growth in lettuce seedlings. *Plant Soil* **2024**. [CrossRef]
21. Bhattacharyya, P.N.; Jha, D.K. Plant growth-promoting rhizobacteria (PGPR): Emergence in agriculture. *World J. Microbiol. Biotechnol.* **2012**, *28*, 1327–1350. [CrossRef]
22. Egamberdieva, D.; Lugtenberg, B. Use of plant growth-promoting rhizobacteria to alleviate salinity stress in plants. In *Use of Microbes for the Alleviation of Soil Stresses*; Miransari, M., Ed.; Springer: Berlin/Heidelberg, Germany, 2014; pp. 73–96.
23. Landeta, C.; Marchant, F. Biostimulants: Emerging Trend and Opportunities. In *Biostimulants: Exploring Sources and Applications*; Plant Life and Environment Dynamics; Ramawat, N., Bhardwaj, V., Eds.; Springer: Singapore, 2022; pp. 287–306.
24. Kumar, M.; Tomar, R.S.; Lade, H.; Paul, D. Methylotrophic bacteria in sustainable agriculture. *World J. Microbiol. Biotechnol.* **2016**, *32*, 120. [CrossRef]
25. Madhaiyan, M.; Poonguzhali, S.; Sa, T. Characterization of 1-aminocyclopropane-1-carboxylate (ACC) deaminase containing *Methylobacterium oryzae* and interactions with auxins and ACC regulation of ethylene in canola (*Brassica campestris*). *Planta* **2007**, *226*, 867–876. [CrossRef]
26. Subhaswaraj, P.; Jobina, R.; Parasuraman, P.; Siddhardha, B. Plant Growth Promoting Activity of Pink Pigmented Facultative *Methylotroph-Methylobacterium extorquens* MM2 on *Lycopersicon esculentum* L. *J. Appl. Biol. Biotechnol.* **2017**, *5*, 42–46. [CrossRef]
27. Ventorino, V.; Sannino, F.; Piccolo, A.; Cafaro, V.; Carotenuto, R.; Pepe, O. *Methylobacterium populi* VP2: Plant Growth-Promoting Bacterium Isolated from a Highly Polluted Environment for Polycyclic Aromatic Hydrocarbon (PAH) Biodegradation. *Sci. World J.* **2014**, *2014*, 931793. [CrossRef] [PubMed]
28. Van Dien, S.J.; Okubo, Y.; Hough, M.T.; Korotkova, N.; Taitano, T.; Lidstrom, M.E. Reconstruction of C3 and C4 metabolism in *Methylobacterium extorquens* AM1 using transposon mutagenesis. *Microbiology* **2003**, *149*, 601–609. [CrossRef] [PubMed]
29. Torres Vera, R.; Bernabé García, A.J.; Carmona Álvarez, F.J.; Martínez Ruiz, J.; Fernández Martín, F. Application and effectiveness of *Methylobacterium symbioticum* as a biological inoculant in maize and strawberry crops. *Folia Microbiol.* **2024**, *69*, 121–131. [CrossRef]
30. Tani, A.; Sahin, N.; Kimbara, K. *Methylobacterium oxalidis* sp. nov., isolated from leaves of *Oxalis corniculata*. *Int. J. Syst. Evol. Microbiol.* **2012**, *62*, 1647–1652. [CrossRef]
31. Omer, Z.S.; Tombolini, R.; Gerhardson, B. Plant colonization by pink-pigmented facultative methylotrophic bacteria (PPFMs). *FEMS Microbiol. Ecol.* **2004**, *47*, 319–326. [CrossRef]
32. Kutschera, U. Plant-Associated *Methylobacteria* as Co-Evolved Phytosymbionts. *Plant Signal. Behav.* **2007**, *2*, 74–78. [CrossRef]
33. Madhaiyan, M.; Alex, T.H.H.; Ngoh, S.T.; Prithiviraj, B.; Ji, L. Leaf-residing *Methylobacterium* species fix nitrogen and promote biomass and seed production in *Jatropha curcas*. *Biotechnol. Biofuels* **2015**, *8*, 222. [CrossRef]
34. Pascual, J.A.; Ros, M.; Martínez, J.; Carmona, F.; Bernabé, A.; Torres, R.; Pérez-Piqueres, A.; Sánchez-Reolid, L.; Fernández, F. *Methylobacterium symbioticum* sp. nov., a new species isolated from spores of *Glomus iranicum* var. *tenuihypharum*. *Curr. Microbiol.* **2020**, *77*, 2031–2041. [CrossRef]
35. Stabentheiner, E.; Zankel, A.; Pölt, P. Environmental scanning electron microscopy (ESEM)—A versatile tool in studying plants. *Protoplasma* **2010**, *246*, 89–99. [CrossRef]
36. Murashige, T.; Skoog, F.A. revised medium for rapid growth and bio-assays with tobacco tissue cultures. *Physiol. Plant.* **1962**, *15*, 473–497. [CrossRef]
37. Visioli, G.; D'Egidio, S.; Vamerali, T.; Mattarozzi, M.; Sanangelantoni, A.M. Culturable endophytic bacteria enhance Ni translocation in the hyperaccumulator *Noccaea caerulescens*. *Chemosphere* **2014**, *117*, 538–544. [CrossRef]
38. Murchie, E.H.; Lawson, T. Chlorophyll fluorescence analysis: A guide to good practice and understanding some new applications. *J. Exp. Bot.* **2013**, *64*, 3983–3998. [CrossRef] [PubMed]
39. Dworkin, M.; Foster, J.W. Experiments with some microorganisms which utilize ethane and hydrogen. *J. Bacteriol.* **1958**, *75*, 592–603. [CrossRef] [PubMed]
40. Bertola, M.; Mattarozzi, M.; Sanangelantoni, A.M.; Careri, M.; Visioli, G. PGPB colonizing three-year biochar-amended soil: Towards biochar-mediated biofertilization. *J. Soil Sci. Plant Nutr.* **2019**, *19*, 841–850. [CrossRef]
41. Delory, B.M.; Weidlich, E.W.; Meder, L.; Lütje, A.; van Duijnen, R.; Weidlich, R.; Temperton, V.M. Accuracy and bias of methods used for root length measurements in functional root research. *Methods Ecol. Evol.* **2017**, *8*, 1594–1606. [CrossRef]
42. Vamerali, T.; Guarise, M.; Ganis, A.; Bona, S.; Mosca, G. Analysis of root images from auger sampling with a fast procedure: A case of application to sugar beet. *Plant Soil* **2003**, *255*, 387–397. [CrossRef]
43. AACC International. *Approved Methods of Analysis*, 11th ed.; Method 46-11.02. Crude protein—Kjeldahl method, boric acid modification. Approved November 3, 1999; AACC International: St. Paul, MN, USA, 2010.
44. Visioli, G.; Bonas, U.; Dal Cortivo, C.; Pasini, G.; Marmiroli, N.; Mosca, G.; Vamerali, T. Variations in yield and gluten proteins in durum wheat varieties under late-season foliar *versus* soil application of nitrogen fertilizer in a northern Mediterranean environment. *J. Sci. Food Agric.* **2018**, *98*, 2360–2369. [CrossRef]

45. Glick, B.R.; Penrose, D.M.; Li, J. A model for the lowering of plant ethylene concentrations by plant growth-promoting bacteria. *J. Theor. Biol.* **1998**, *190*, 63–68. [CrossRef]
46. Dal Cortivo, C.; Barion, G.; Visioli, G.; Mattarozzi, M.; Mosca, G.; Vamerali, T. Increased root growth and nitrogen accumulation in common wheat following PGPR inoculation: Assessment of plant-microbe interactions by ESEM. *Agric. Ecosyst. Environ.* **2017**, *247*, 396–408. [CrossRef]
47. Yim, W.J.; Kim, K.Y.; Lee, Y.W.; Sundaram, S.P.; Lee, Y.; Sa, T.M. Real time expression of ACC oxidase and PR-protein genes mediated by *Methylobacterium* spp. in tomato plants challenged with *Xanthomonas campestris* pv. *vesicatoria*. *J. Plant Physiol.* **2014**, *171*, 1064–1075. [CrossRef] [PubMed]
48. Madhaiyan, M.; Poonguzhali, S.; Ryu, J.H.; Sa, T.M. Regulation of ethylene levels in canola (*Brassica campestris*) by 1-aminocyclopropane-1-carboxylate deaminase-containing *Methylobacterium fujisawaense*. *Planta* **2006**, *224*, 268–278. [CrossRef] [PubMed]
49. de Aquino, G.S.; Ventura, M.U.; Alexandrino, R.P.; Michelon, T.A.; Pescador, P.G.A.; Nicio, T.T.; Watanabe, V.S.; Gomes, T.D.; Oliveira, A.L.M.; Hata, F.T. Plant-promoting rhizobacteria *Methylobacterium komagatae* increases crambe yields, root system and plant height. *Ind. Crops Prod.* **2018**, *121*, 277–281. [CrossRef]
50. Xu, C.; Tao, H.; Tian, B.; Gao, Y.; Ren, J.; Wang, P. Limited irrigation improves water use efficiency and soil reservoir capacity through regulating root and canopy growth of winter wheat. *Field Crops Res.* **2016**, *196*, 268–275. [CrossRef]
51. Mardanov, A.; Samedovam, A.; Shirvany, T. Root-Shoot Relationships in Plant Adaptation to Nitrogen Deficiency. In *Root Demographics and Their Efficiencies in Sustainable Agriculture, Grasslands and Forest Ecosystems*; Box, J.E., Ed.; Developments in Plant and Soil Sciences; Springer: Dordrecht, The Netherlands, 1998; Volume 82, pp. 147–154.
52. Lv, X.; Zhang, Y.; Hu, L.; Zhang, Y.; Zhang, B.; Xia, H.; Wang, Z.; Sun, Y.; Kong, L. Low-nitrogen stress stimulates lateral root initiation and nitrogen assimilation in wheat: Roles of phytohormone signaling. *J. Plant Growth Regul.* **2021**, *40*, 436–450. [CrossRef]
53. Sivri, D.; Köksel, H.; Bushuk, W. Effects of wheat bug (*Eurygaster maura*) proteolytic enzymes on electrophoretic properties of gluten proteins. *N. Z. J. Crop Hortic. Sci.* **1998**, *26*, 117–125. [CrossRef]
54. Salis, L.; Goula, M.; Valero, J.; Gordún, E. Prolamin proteins alteration in durum wheat by species of the genus *Eurygaster* and *Aelia* (Insecta, Hemiptera). *Span. J. Agric. Res.* **2010**, *8*, 82–90. [CrossRef]
55. Wieser, H.; Kieffer, R. Correlations of the amount of gluten protein types to the technological properties of wheat flours determined on a micro-scale. *J. Cereal Sci.* **2001**, *34*, 19–27. [CrossRef]

agriculture

MDPI

Article

Effects of Poultry Manure Biochar on *Salicornia herbacea* L. Growth and Carbon Sequestration

Danbi Chun [1], Hyun Cho [2], Victor J. Hahm [2], Michelle Kim [2], Seok Won Im [3], Hong Gun Kim [1] and Young Soon Kim [1,*]

1 Institute of Carbon Technology, Jeonju University, Jeonju 55069, Republic of Korea; 202102544@jj.ac.kr (D.C.)
2 Global Prodigy Academy, Jeonju 55069, Republic of Korea; hyuncf22@gpa.ac.kr (H.C.); jisunghfa20@gpa.ac.kr (V.J.H.)
3 Koreastevia, Jeongeup-si 56216, Republic of Korea
* Correspondence: kyscjb@jj.ac.kr; Tel.: +82-63-220-3157

Abstract: In order to explore the potential of biochar produced from poultry manure for sustainable waste utilization, carbon sequestration, and agricultural development, this study examines the impact of biochar on the growth of the halophyte plant *Salicornia herbacea* L., or glasswort. Because of their properties of morphological and chemical properties, biochar has been gaining interest as a potential solution to addressing both the concerns of climate change and unsustainable agriculture. In this study, the characteristics of biochar were analyzed and its impact on plant growth by stem length was measured over 15 weeks. Poultry-based biochar was created through pyrolysis at the temperatures of 400, 500, and 700 °C. Various amounts of biochar produced from pyrolysis at 500 °C were put to soil. However, the average surface area and average pore size values of poultry manure biochar produced from temperatures 400, 500, and 700 °C were similar enough to be negligible. The biochar sample produced from the pyrolysis temperature of 500 °C had an average pore size of 17.18 nm and a surface area of 18.06 m^2/g. From weeks 4 to 15, all groups exhibited increased stem length, with the most significant differences observed between the biochar 0% (control) and biochar 10% groups, with biochar 0% and biochar 10% denoting 0% and 10% weight concentrations of biochar, respectively. While biochar 5% and biochar 7% groups showed minimal differences in stem length, biochar 10% demonstrated a significant increase, suggesting an optimal biochar percentage for enhancing plant growth. Carbon credit estimations have suggested that 1 ton of poultry manure biochar produced from pyrolysis at 500 C° equates to an estimate of 0.5248 ± 0.0580 carbon credits, the highest of all three biochar samples. All three samples (biochar produced from 400, 500, and 700 °C pyrolysis temperatures) had increased heavy metal contents and a wider range of functional groups. The findings indicate that biochar can effectively improve soil health and plant performance overall, with biochar 10% showing the most significant impact on Salicornia growth.

Keywords: biochar; glasswort; poultry manure; halophyte; growth rate; carbon credit

1. Introduction

With climate change and unsustainable agriculture, the dangers associated with soil degradation, soil contamination, livestock management, water scarcity, and food security are increasing on a global scale [1]. The Republic of Korea, in particular, aims to mitigate the concerns of global warming by reducing greenhouse gas emissions through CO_2 capture, utilization, and storage, with the end goal of becoming a carbon-free state by 2050 [2]. Among the tools available to work toward this goal is biochar. Biochar offers a potential solution to addressing the concerns of both climate change and unsustainable agriculture. Agricultural sustainability, carbon sequestration, and soil amendment have been proven to be enhanced by the use of biochar [3]. Biochar's high porosity and surface area, as well as its capacity to lessen soil acidity and nutrient leaching, are other beneficial characteristics [4].

Additionally, prior studies have demonstrated that biochar can effectively increase crop biomass and grain output regardless of precipitation conditions [5]. Overall, biochar has proven to be a versatile and highly effective resource for a variety of climates and drought conditions. With these properties, biochar has been gaining interest as a promising key to sustainability.

A variety of biomass can be pyrolyzed to produce biochar. The kind of biomass used and the production method, which can change depending on the pyrolysis temperature and kind of thermal conversion used, can affect the composition and properties of the biochar [6]. However, as biochar is mostly made up of carbon, the different types of biochar can be grouped and treated as having comparable effects at a basic level. According to Major et al., for example, different types of biochar have been predicted to have a 25% carbon recovery rate due to their high porosity [7]. In contrast, ordinary photosynthetic plants have a carbon recovery rate of 0%. Along with its ability to effectively capture CO_2, biochar has been found to positively affect soil microbial growth and enhance carbon storage, thereby promoting soil amendment [8–10]. Because of its porous structure, heavy metals may also be drawn to and retainedby biochar. This helps to clean the soil and encourage the growth of plants [11,12].

While the characteristics of different types of biochar are similar, animal manure has been identified as a particularly effective feedstock because of its rich nutrients [12]. With the continuous increase of poultry production and demand in South Korea, as well as the excessive production of poultry worldwide, poultry manure should be considered as a feedstock for biochar. Poultry overproduction combined with the improper handling of poultry manure can result in gas and odor emissions that are bad for livestock and human health, including NH_3, CH_4, N_2O, H_2S, and NO_x [13]. An estimated 220 to 280 kg of odorous ammonia are released into the atmosphere each year for every 1000 kg of chicken dung produced [14]. With China and the United States being top poultry producers and leaders in biochar application, utilizing poultry manure for biochar can help manage significant quantities of this waste [15,16]. Thus, biochar can also be used to create a recovery system to turn potentially harmful livestock matter into a valuable resource.

The diversity of halophyte species presents a wide array of potential uses for these plants. The *Salicornioideae* subfamily is one of the most diverse subfamilies of halophyte species. Within this genus is the *Salicornia*. Because these plants are high in edible oil, nutrient, and protein content, they are effective as both feedstock and food [17]. The plant *Salicornia herbacea* L. (glasswort) has been used as a model to identify the growth parameters and impact of the biochar on plants. Because of its high yields of biomass, seed, oil, and ethanol, glasswort—a crop that is widely used in the food and medicinal industries—is predicted to be extremely sustainable in the future [18]. Glasswort is a halophyte, meaning that it is resistant to extreme weather and salinities that are harmful to most other plants [19]. In a world where soil erosion, water scarcity, and global warming caused by climate change pose serious concerns for plant sustainability and biodiversity, the mechanism by which halophytes can better tolerate such conditions provides a framework for crop sustainability and resilience. The ability of halophytes to endure the increasing environmental stress and thereby maintain their production yield makes them a valuable source of food, fodder, biomass, and medicinal compounds in the foreseeable future. As halophytes, they can be grown in the absence of freshwater resources, making them an extremely versatile species. For example, *Salicornia herbacea* L. has already been implemented into the cuisine and cultures of various cultures. Moreover, there are not many studies on using livestock dung to produce biochar, since wood is typically utilized as biomass due to its high lignin content [20]. Furthermore, this study seeks to provide a clearer understanding of the differences in the characterization of cow and poultry manure biochar. We can compare the characteristics of cow and poultry manure biochar based on the results from our previous study, "Systematic Characterization of Cow Manure Biochar and Its Effect on *Salicornia herbacea* L. Growth [21]".

With the various advantages of biochar, this study aims to explore the effects of poultry-based biochar application on soil health and plant growth. Previous research done by Adekiya et al. demonstrated the benefits of hardwood-based biochar in improving the characteristics and chemical properties of soil as well as cocoyam yields [1]. Our present study follows much of the methodology presented in our previous study on cow manure biochar, providing a comparison of the thermal, elemental, and growth properties of poultry manure biochar to cow manure biochar when applied to *Salicornia herbacea* L, or glasswort. This present study seeks to expand the scope of biochar research by demonstrating the benefits of poultry manure biochar on glasswort. Furthermore, this research attempts to address the environmental concern of poultry waste management by utilizing poultry manure as a feedstock for biochar. Through this comparison and the focus on investigating poultry manure biochar in cultivating glasswort, this study aims to provide an increased understanding of the possible uses and effects of biochar.

2. Materials and Methods

2.1. Biochar Materials and Preparation

Poultry manure (PM) was acquired from a farm in Jeollabuk-do, South Korea, and stored at −20 °C to preserve its integrity. For a full day, the PM was dried in an oven set to 105 °C (VS-9500H, VISION SCIENTIFIC, Daejeon-Si, Republic of Korea). An electric furnace (OTF-1200X-S, MTI Korea, Seoul, Republic of Korea) was used to carbonize the samples for three hours. Nitrogen was injected at a rate of 10 cc min–1 to establish an atmosphere of nitrogen gas inside the furnace. A schematic diagram of the pyrolysis reactor is shown in our previous work [21]. To produce the poultry manure biochar (PMB), 400, 500, and 700 °C were used as the carbonization temperatures. Glasswort seeds were sourced from the Dasarang Co., Ltd. in Shinan-gun, Jeollanam-do, Republic of Korea, and commercial soil (55–60% coco peat, 15–20% peat moss, 5–10% perlite, 5–10% vermiculite, 5–10% zeolite G, and 5–10% zeolite P) was gathered from Nongkyung Company, Jincheon, Chungcheong-buk-do, Republic of Korea. Biochar was produced by carbonizing poultry manure, which served as the biomass raw material. This biochar was then mixed with soil to cultivate glassworts, as illustrated schematically in the workflow diagram shown in Figure 1.

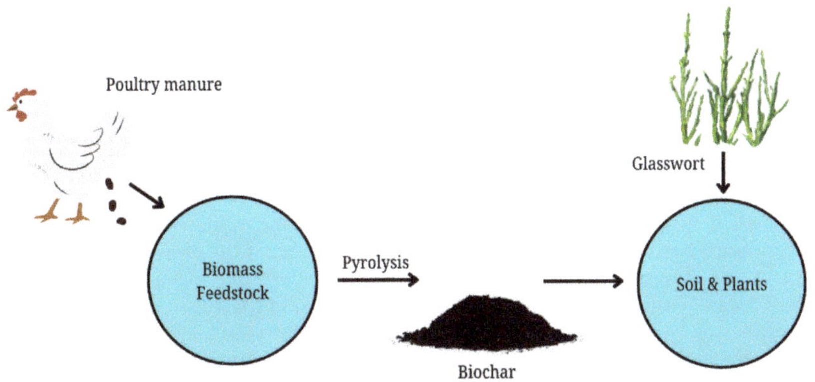

Figure 1. Workflow diagram of research methodology.

2.2. Methods for Plant Development Experiments

Plant development studies were carried out to see how biochar made from PM affected the characteristics of soil and the growth of plants. After being carbonized at 500 °C, the biochar was mixed by hand with commercial soil from Nongkyung Company at the weight concentrations of 0, 3, 5, 7, and 10% (PMB0% or control, PMB3%, PMB5%, PMB7%, and PMB10%, respectively). Biochar pyrolyzed at 500 °C was selected for use at this temperature. Our prior study [21] provides a graphic depiction of the pots and soil amalgamation.

2.3. Plant Growth Experiment Protocol

The test pots were placed in a research facility setting where the temperature was consistently maintained between 24 and 26 °C. Each pot was immersed in two centimeters of water, and regular observation ensured that the water level was continuously raised. The plants received a daily irrigation of 10 mm of water. The watering schedule was implemented with two day breaks during a four week period. To address the negative effects that the plants displayed, such as a change in color, saline water or saltwater was introduced. Glasswort measurement photos were taken with Digital Microscope using the program 'Dino Capture'. Stem length (ML0), stem thickness (DL1), and leaf thickness (DL0) values were able to be captured and are shown in Figure 2.

	ML0(mm)	DL1(mm)	DL0(mm)
1	15.699	0.684	2.148
2	11.085	0.677	1.861
3	6.634	0.692	1.168

Figure 2. Digital microscope images of glasswort growth during 4 weeks in Biochar 5% soil. The plants' ML0, DL1, and DL0 measurements are represented by the numbers 1, 2, and 3 in each image photo, respectively.

2.4. Methods for Characterization

The procedures followed in the earlier study on CMB by Shin, H. et al. [21] were referenced in defining the stages of PMB production. For a more thorough rundown of the characterization procedures, consult the mentioned study. To find the best manufacturing temperature for the biochar, thermogravimetric analysis (TGA, SDT-650, Ta Instruments, New Castle, DE, USA) was first carried out. A sample of 10 mg of the biochar was placed in an aluminum pan before the crucible was sealed to prevent the entry of external air. Samples underwent a gradual heating from 30 to 800 °C under a nitrogen gas environment. The pH of the soil was also tested, because biochar is an alkaline resource that might affect the pH of the soil (pH meter, pH8100, ETI, Worthing, West Sussex, UK). The pH of the biochar was measured by washing the biochar with deionized water, mechanically shaking the sample for an hour at 25 °C, letting it stand for 30 min, and measuring the pH value after stabilization. The oxygen, carbon, hydrogen, nitrogen, and sulfur in the biomass and biochar were measured using an elemental analyzer (Flash Smart, Thermo Fisher Scientific, Waltham, MA, USA). The heavy metal concentrations of arsenic, cadmium, copper, molybdenum, nickel, lead, and zinc were detected using another elemental spectrometer (Optima 7300 DV ICP-OES Spectrometer, Perkin Elmer, Waltham, MA, USA). Zinc levels were tested using two different instruments (MARS XPRESS 230/60, CEM, Elk River, MN, USA; Optima 8300, Perkin Elmer, Waltham, MA, USA).

All of the tested heavy metals could potentially be found in soil. Notably, zinc and copper are essential plant micronutrients. The existence of functional groups was next ascertained by Fourier transform infrared spectroscopy (FT-IR, FT-IR Spectrum Two, Perkin Elmer, Waltham, MA, USA). For FT-IR, the biochar was ground into powder form and was measured between the wavelengths of 400 to 4000 cm^{-1}. To determine the specific surface area of the biochar, a Brunauer-Emmett-Teller (BET) analysis (BELSORP-maxII, MicrotracBEL Corp., Osaka, Japan) was performed on samples pretreated at 100 °C for 24 h. The morphology of the biochar was assessed using a scanning electron microscope (SEM, CX-200TA, COXEM, Daejeon, Republic of Korea).

After the biochar was incorporated into the soil, the total organic carbon (TOC) and organic matter (OM) were measured using an analyzer (CN928, LECO Analytical Instruments, Saint Joseph, MI, USA) and an electrical conductivity meter (EC meter, ORION VERSASTAR PRO, Thermo Fisher Scientific, Waltham, MA, USA), respectively. Cation exchange capacity (CEC) was identified by the Korean Agriculture Technology Promotion Agency (KOAT) using the ammonium acetate method. Total carbon (TC) of the soil after the implementation of biochar was measured using an elemental analyzer (Flash Smart, Thermo Fisher Scientific, Waltham, MA, USA).

2.5. Statistical Analysis

The statistical analysis of the stem length (ML0), stem thickness (DL1), and leaf thickness (DL0) variables was carried out using IBM SPSS28 Statistics software. One-way ANOVA was used to detect statistical differences, and a post hoc Tukey test was performed afterward. The Tukey test was used to compare the means, and the significance threshold was set at 0.05. To check for homogeneity of variance (absolute deviation), the Levene's test was employed. Lastly, the normality of sample distributions was determined using the Shapiro-Wilk test.

2.6. Equations

For a more elaborate overview of biochar, the yield equation, Gompertz equation, and various carbon credit calculation methods were utilized to predict plant growth and carbon credits.

By determining the biomass sample's initial mass (M) and the sample mass after carbonization (m), then entering these values into the following equation, the yield of each temperature group was determined:

$$\text{Yield of biochar (\%)} = (\text{m}/\text{M}) \times 100 \quad (1)$$

The modified Gompertz equation is shown below:

$$f(t) = 10e^{-e^{((0.5\times 2.7183)\div 10)\times(5-t)+1}} \quad (2)$$

This modified Gompertz equation uses a starting value of 10 and Euler's number (approximately 2.7183) with a small value, in this case 0.5, as the power of e to create a model of continuous growth or decay [22]. The parameters used in the Gompertz value are nonspecific modifications that are meant to show the comparison between the different concentration samples.

Furthermore, Equation from Appendix 4 of the 2019 Refinement to the 2006 IPCC Guidelines for National Greenhouse Gas Inventories was used to estimate the annual change in biochar carbon stock, as indicated below [23]:

$$\Delta BC_{\text{Mineral}} = \sum(p = 1 \text{ to n}) \, (BC_{\text{TOTp}} \cdot F_{\text{Cp}} \cdot F_{\text{permp}}) \quad (3)$$

where $\Delta BC_{\text{Mineral}}$ is the total change in carbon stocks of mineral soils associated with biochar amendment, in tonnes of sequestered carbon per year; BC_{TOTp} (total production of biochar) is the mass of biochar incorporated into mineral soil during the inventory year for each biochar production type *p*, in tonnes of biochar dry matter per year; F_{Cp} (carbon content of production type p) is the organic carbon content of biochar for each production type *p*, in tonnes C tonne^{-1} of biochar dry matter; and F_{permp} (permanent carbon content of production type p) is the fraction of biochar carbon for each production type p remaining after 100 years, in tonnes sequestered C tonne^{-1} biochar C.

Carbon credits can be estimated using Puro.earth's carbon calculation method. This method follows the calculation of the sequestered carbon and multiplies the value by a factor of 44/12 to find the amount of CO_2 stored over 100 years [24]. The equation used is shown below:

$$E_{stored} = Q_{biochar} \cdot C_{org} \cdot F_p{}^{TH,Ts} \cdot 44/12 \quad (4)$$

where $Q_{biochar}$ is the amount of biochar produced over the reporting period; C_{org} is the organic C content of the biochar produced; and $F_p{}^{TH,Ts}$ is the permanence factor of biochar organic C over a given time horizon TH in a given soil at temperature T_s. While we did not have the permanence factor of the biochar, Appendix 4 of the referenced IPCC inventory provided an estimation of the 100-year permanence value based on temperature [23]. The values of $Q_{biochar}$, C_{org}, and $F_p{}^{TH,Ts}$ can be substituted for BC_{TOTp}, F_{Cp}, and $F_p{}^{TH,T}$, or $\Delta BC_{Mineral}$. In other words, the IPCC and the Puro.earth equations can be combined to create the equation used for carbon credit calculation:

$$E_{stored} = \Delta BC_{Mineral} \cdot 44/12 \quad (5)$$

3. Results and Discussion

3.1. The Properties and Characterization of Biochar

The thermal characterization of biochar is crucial for understanding how to optimize its production and application. To create biochar, samples must be carbonized. For manure, this would mean pyrolysis at a high enough temperature that the sample's moisture, which usually comes from its content of hemicellulose and cellulose, is reduced, leaving a solid product of higher energy yield per unit weight. The TGA-DTG curve allows a visual representation of the high enough pyrolysis temperature needed to create effective PMB.

Therefore, we can determine the appropriate manufacturing temperature of the biochar. Figure 3 shows the TGA-DTG curve of the PM. There is an initial weight loss of around 10% from 30 to 260 °C and another subsequent drop in weight from around 260 to 450 °C. The sample's initial weight loss was mostly caused by the loss of moisture and other light-volatile ingredients. The steepest drop in weight occurs at the subsequent drop. A large percentage of PM weight is made up of hemicellulose and cellulose, with the remaining mass being largely composed of lignin. The steep weight loss at 260 °C can be attributed to the reduced hemicellulose and cellulose at that temperature. The final weight percentage is mostly made up of the remaining lignin, which can remain until the temperature reaches 900 °C [25]. The sharp decrease in weight at 260 °C presented the onset of pyrolysis for the PM sample.

Figure S1 shows the TGA-DTG results of cow manure biochar (CMB) from the previous study. Compared to CMB, poultry manure biochar (PMB) had a significantly less steep loss of weight percentage due to moisture. The case is also the same for the reduced hemicellulose and cellulose levels that come with increasing temperature. It can also be observed that CMB maintains a higher weight (%) at 800 °C despite having steeper drops [21]. More lignin is present in biochar-maintained biomass because lignocellulosic biomass, which is made up of hemicellulose and lignin, retains its weight % even at high temperatures. The variations in lignin content between poultry and cow manure are consistent with the findings displayed in the TGA results of both investigations. PM typically has a lower lignin level of 4.2%, but CM shows a greater lignin concentration of 11.5% [26].

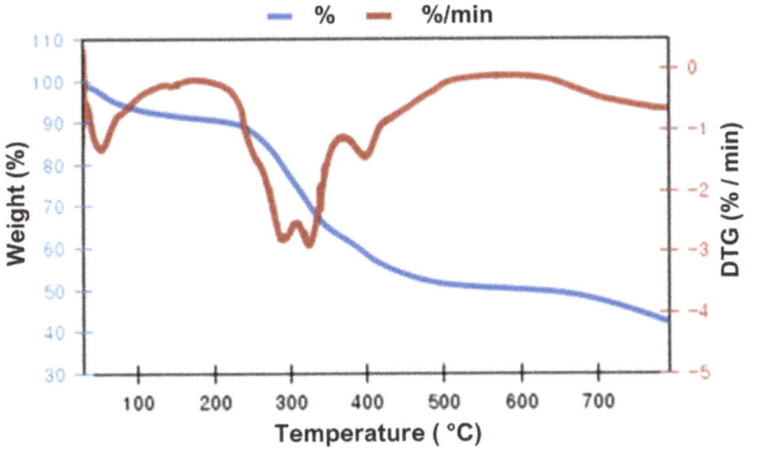

Figure 3. TGA-DTG spectra of poultry manure (PM) from 30 to 800 °C, 10 °C/min.

Based on the PM sample's TGA results, the temperatures 400, 500, and 700 °C were chosen for carbonization. Table 1 presents the yield percentage and pH of biochar. The highest yield percentage was 52.78% at 400 °C, followed by a yield of 47.20% at 500 °C, and finally 36.07% at 700 °C as the lowest yield percentage. The yield percentage at 500 °C was approximately 10.57% less than the yield percentage at 400 °C, and the yield percentage at 700 °C was approximately 23.58% less than the yield percentage at 500 °C. Table 1 also displays the pH values of PMB at 400, 500, and 700 °C, all of which are alkaline. The corresponding figures are 9.62, 12.24, and 10.30. We used a pH meter to find the biochar's pH. The pH value of PMB700 was the highest at 12.24, followed by PMB500 at 10.30 and PMB400 at 9.62. The capacity of biochar to adsorb NH_4^+ increases between pH values of 4 and 8, but declines at pH values of 9 or higher [1,3]. The impact of the different PMB concentrations on the pH of commercial garden soil was investigated by mixing it with the soil and observing the resulting changes.

Table 1. Yield and pH of poultry manure biochar (PMB) according to pyrolysis temperature.

Samples	Yield (%)	pH
PMB400	52.78	9.62
PMB500	47.20	10.30
PMB700	36.07	12.24

The elemental analysis results of the heavy metal and component content of PMB400, PMB500, and PMB700 biochar samples are displayed in Tables 1 and 2. Table 2 shows a trend of rising carbon content as the pyrolysis temperatures increased. For PMB400, PMB500, and PMB700, the carbon content was 30.51%, 34.75%, and 35.16%, respectively. The carbon content of PMB400 being lower than the content of the PM can be attributed to the loss of CO_2 and CO during carbonization. However, compared to PM, the carbon content of PMB500 and PMB700 was higher. The carbon content was shown to increase at progressively lower rates as the temperature was increased.

Table 2. Component content of poultry manure (PM) and poultry manure biochar (PMB) by %.

Samples	C	H	N	O	S
PM	33.69	4.55	5.57	30.83	0.34
PMB400	30.51	2.74	3.45	17.04	0.24
PMB500	34.75	1.80	3.19	15.89	0.13
PMB700	35.16	0.98	1.51	7.57	0.00

The metal content of PM and PMB was measured by ppm. In this analysis, Cu, Ni, and Zn were detected while As, Cd, Mo, and Pb were not detected. Both zinc and copper are essential elements for plant development and metabolism. The rest of the tested elements, conversely, can be toxic to plants when present in large amounts. The samples showed dramatic differences in the concentrations of the detected elements. In PM, PMB400, PMB500, and PMB700, respectively, the Cu and Zn values were shown to increase. The most significant increase in both Cu and Zn content was from 400 to 500 °C. The increase in Cu and Zn at different carbonization temperatures can be attributed to the continuous decrease in yields of biochar as the temperature increases in Table 1, which condenses the substance and increases the element per ppm content. The Ni content, on the other hand, exhibited irregular patterns. The highest recorded nickel content was at 400 °C with a concentration of 4.38 ppm.

3.2. *FT-IR for Functional Groups in Biochar*

The biochar's FT-IR results from this study are shown in Figure 3. The results identified the presence of various functional groups in the samples. Out of the samples (PM, PMB400, PMB500, and PMB700), PMB400 had the highest number of functional groups, followed by PMB500, PMB700, and finally PM as the sample with the fewest functional groups. These results are consistent with other research papers showing that the quantity of functional groups decreases as biochar is manufactured at higher temperatures [21,27]. The functional groups that were obtained from PMB400 were O-H, C-H_n, N=N=N, C≡N, C-H, C=O, C-H, and C-OC; those obtained from PMB500 were C-H_n, C=O, O-H, C-H, and C-OC; those obtained from PMB700 were O-H, C-O-C, and C-O; and those obtained from PM were O-H and C-O. All of the samples had a lot of bonds between carbon and other elements, with C-H being the most frequent type.

Studies have indicated that functional groups are responsible for certain biochar behaviors, such as contaminant removal and surface oxidation [28,29]. Because these functional groups are affected by and variable depending on pyrolysis temperature and the feedstock of biochar, they can be adjusted to match ideal conditions based on where the biochar is used [30,31].

Increasing lignin contents have been shown to increase the porosity, aromaticity, stability, and carbon content while decreasing the CEC, pH, and nutrient content of biochar [30]. Likewise, the results on PM biochar, Tables 1–3 and Figure 3 matched these findings. All three pH values of PMB400, PMB500, and PMB700 were higher than the pH values of CMB400, CMB550, and CMB700. Furthermore, the contents of copper and zinc were higher for PMB than CMB at all temperatures excluding 700 °C. At this temperature, CMB700 exhibited a higher Zn content (1022.702 ppm) compared to PMB700 (828.925 ppm). Finally, Figure 3 and Table 4 show a higher aromaticity for the PMB samples overall. PMB exhibited higher variance and occurrence of functional groups compared to those shown by CMB samples through FT-IR analysis.

Table 3. Content of heavy metal elements in poultry manure (PM) and poultry manure biochar (PMB) by ppm.

Samples	As	Cd	Cu	Mo	Ni	Pb	Zn
PM	n.d.	n.d.	85.03	n.d.	1.96	n.d.	522.57
PMB400	n.d.	n.d.	109.06	n.d.	4.38	n.d.	634.28
PMB500	n.d.	n.d.	145.78	n.d.	3.31	n.d.	769.20
PMB700	n.d.	n.d.	165.92	n.d.	2.54	n.d.	828.93

n.d. stands for "not detected".

Table 4. Specific surface area and average pore size of poultry manure biochar carbonized at 400 °C (PMB400), 500 °C (PMB500), and 700 °C (PMB700).

PMB	BET (m^2g^{-1})	Average Pore Size (nm)
400 °C	14.17	22.53
500 °C	18.06	17.18
700 °C	15.32	19.85

3.3. Morphological Features and Textural Properties of Biochar

Morphological investigations of the biochar samples pyrolyzed at the temperatures of 400, 500, and 700 °C were conducted using SEM, and the pore size and surface area of the samples were identified using BET. Table 4 lists the surface area and pore size measurements and Figure 4 displays the SEM results. The porosity of biochar can significantly impact the soil's capacity to absorb organic contaminants, water, and heavy metals. For example, biochar captures organic pollutants through its pores. Therefore, determining the potential of biochar for carbon sequestration requires an understanding of its porosity.

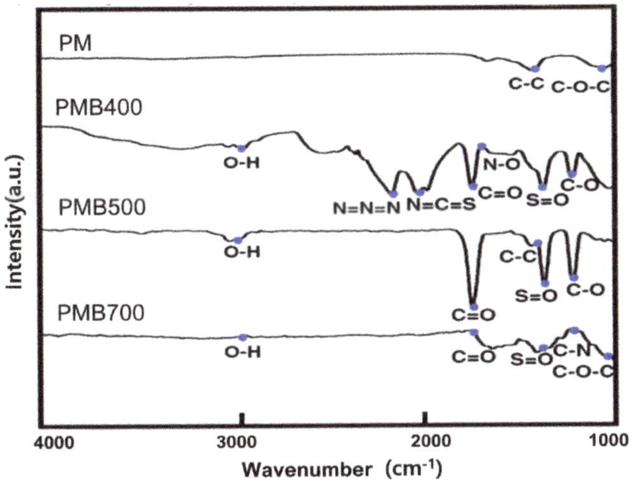

Figure 4. FT−IR of poultry manure (PM), poultry manure carbonized at 400 °C (PMB400), 500 °C (PMB500), and 700 °C (PMB700).

The pyrolysis temperature during biochar production has been shown to affect the porosity of the biochar [32]. Specifically, the porosity of biochar has been observed to be likely to decrease from 500 °C [33]. This is explained by the rise in relative ash content and the devolatilization process of organic matter that comes with increasing temperatures to above 450 °C, thus slowing down the development of pores and surface area. The results of the BET measurements loosely followed this observation. Rather than a decrease at 500 °C, the results reflected a decrease in average pore size from 400 (22.53 nm) to 500 °C (17.18 nm).

The pore sizes at these values were the largest and smallest pore sizes, respectively. In the middle sat 700 °C with an average pore size of 19.85 nm. Too high of a temperature can close smaller pores, which could account for the increase in pore size from biochar carbonized at 500 to 700 °C. These BET results confirmed the SEM images that are shown in Figure 5, in which PMB500 is shown to have smaller pores than PMB400 and PMB700, as well as PMB400 having the largest pores of all the samples.

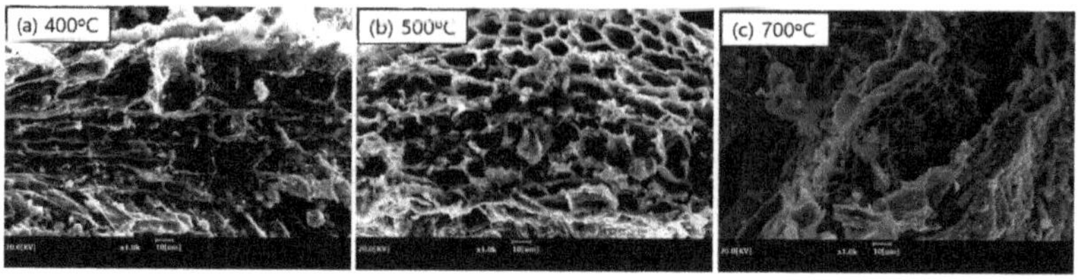

Figure 5. SEM Images of (**a**) poultry manure carbonized at 400 °C (PMB400), (**b**) 500 °C (PMB500), and (**c**) 700 °C (PMB700).

Compared to the BET results found with regard to CMB in the previous study, the BET values of PMB were significantly lower. With reference to Shin et al. [21], Table 1 shows the BET and average pore size values of CMB. For example, the highest BET value yielded in this study was 18.06 $m^2\ g^{-1}$ by PMB500, while the highest value with the cow-manure biochar was 117.57 $m^2\ g^{-1}$. PMB also had more stable average pore size values than CMB across different pyrolysis temperatures. Both followed the same pattern of the lowest average pore size appearing at the middlemost temperature (500 °C for PMB, 550 °C for CMB). These comparisons show that the BET of CMB are more affected by temperature changes than that of PMB.

3.4. Plant Growth Experiment

The plant growth experiment was run for 15 weeks, wherein the glasswort plants were grown and measured weekly. Measurements of stem length (ML0), stem thickness (DL1), and leaf thickness (DL0) were taken in millimeters. The selection of PMB500 was chosen to allow for comparisons to the effects of CMB550 from the previous study [21]. As shown in Table 5, the measurements were logged from 4 to 15 weeks of the mean and standard deviation of the growth stem length (ML0) of the different plant populations. The various populations consist of a population without biochar (control), a population containing 3 wt.% biochar (PMB3%), a population containing 5 wt.% biochar (PMB5%), a population containing 7 wt.% biochar (PMB7%), and a population containing 10 wt.% biochar (PMB10%). All groups experienced increased stem length from the 4th to the 12th week. In the control group, the mean stem length dipped from weeks 4 to 5 (11.95 to 8.22 mm) and 11 to 12 (34.93 to 33.56 mm). The mean stem length dipped from weeks 11 to 12 (66.86 to 58.42 mm) in the case of the PMB7% samples. This drop in mean stem length can be attributed to new plant growth as well as plant death. The mean stem length dipped from weeks 11 to 12 in the PMB7% group's case. Using one-way ANOVA, we found that all weeks, except weeks 4 and 10, were statistically significant. There was a p-value of under 0.001 in weeks 5, 6, 8, and 14, a p-value of under 0.01 in weeks 7, 11, and 12, and a p-value of under 0.05 in week 15. From weeks 12 to 13, the mean stem length of all plants except PMB3% increased. In this frame, the increase in plant length shows relatively low values compared to those from previous weeks. Similarly, in week 14, the mean length of all plants except PMB3% increased. Finally, from weeks 14 to 15, there is a noticeably higher increase in the mean stem length of all groups.

Table 5. Average stem length (ML0) and standard deviation (SD) in millimeters.

Week	Control	PMB3%	PMB5%	PMB7%	PMB10%	*p*-Value
4	11.95 ± 4.64 nnd (0.250) (n = 20)	10.83 ± 6.77 nnd (0.001) (n = 49)	11.29 ± 4.69 nnd (0.039) (n = 25)	11.52 ± 5.87 nnd (0.006) (n = 59)	14.74 ± 6.22 nd (0.133) (n = 22)	np
5	8.22 ± 3.75 nnd (0.001) (n = 51)	13.19 ± 8.63 nnd (0.001) (n = 50)	17.07 ± 9.15 nnd (0.005) (n = 29)	18.29 ± 10.22 nd (0.070) (n = 30)	15.88 ± 7.56 nnd (0.020) (n = 40)	***
6	12.74 ± 6.76 nnd (0.003) (n = 50)	16.93 ± 12.59 nnd (0.001) (n = 45)	24.27 ± 22.65 nnd (0.001) (n = 29)	28.35 ± 15.96 nd (0.481) (n = 23)	26.19 ± 10.14 nnd (0.058) (n = 30)	***
7	15.25 ± 9.53 nnd (0.003) (n = 31)	24.37 ± 16.87 nnd (0.001) (n = 36)	26.07 ± 22.02 nnd (0.008) (n = 20)	31.49 ± 15.45 nd (0.248) (n = 18)	28.25 ± 13.99 nd (0.065) (n = 23)	**
8	20.92 ± 11.37 nd (0.071) (n = 24)	41.15 ± 24.24 nnd (0.013) (n = 24)	39.41 ± 33.51 nnd (0.011) (n = 18)	53.21 ± 29.21 nd (0.466) (n = 18)	52.10 ± 23.91 nd (0.423) (n = 16)	***
10	32.92 ± 22.92 nnd (0.004) (n = 16)	50.76 ± 25.18 nd (0.364) (n = 18)	54.38 ± 32.53 nd (0.249) (n = 15)	61.84 ± 23.68 nd (0.504) (n = 18)	60.62 ± 23.00 nd (0.892) (n = 18)	np
11	34.93 ± 25.68 nnd (0.007) (n = 15)	60.24 ± 30.13 nnd (0.016) (n = 18)	59.41 ± 33.58 nd (0.521) (n = 14)	66.86 ± 30.85 nd (0.814) (n = 13)	79.88 ± 27.72 nd (0.999) (n = 13)	**
12	33.56 ± 20.41 nd (0.179) (n = 12)	64.52 ± 28.28 nd (0.367) (n = 17)	61.93 ± 40.91 nd (0.53) (n = 16)	58.42 ± 28.76 nd (0.410) (n = 14)	84.81 ± 33.94 nd (0.993) (n = 13)	**
13	41.37 ± 32.05 nnd (0.003) (n = 15)	52.22 ± 27.09 nd (0.073) (n = 27)	62.07 ± 31.92 nd (0.759) (n = 14)	65.29 ± 23.54 nd (0.922) (n = 11)	84.90 ± 30.18 nd (0.999) (n = 12)	**
14	43.72 ± 29.32 nd (0.249) (n = 12)	46.73 ± 22.62 nd (0.173) (n = 26)	68.46 ± 30.51 nd (0.243) (n = 13)	70.64 ± 33.49 nd (0.528) (n = 12)	87.99 ± 20.56 nd (0.154) (n = 10)	***
15	54.79 ± 30.85 nd (0.062) (n = 12)	60.59 ± 27.29 nd (0.661) (n = 16)	77.17 ± 30.16 nd (0.872) (n = 14)	77.32 ± 23.81 nd (0.223) (n = 10)	95.16 ± 22.81 nd (0.167) (n = 10)	*

p-values are denoted as * ($p < 0.05$), ** ($p < 0.01$), *** ($p < 0.001$), and np (*p* above 0.05 and therefore not meeting the probability threshold value). The samples were tested for normal distribution every week, which is shown by the nd (normal distribution) and nnd (non-normal distribution) and the specific *p*-values in the parentheses beside these notations. The replication numbers used to find the average values are denoted as (n = replication number).

When comparing the difference in stem length from weeks 4 to 15, we can see an increase in stem length in all groups. However, there is a substantial difference between the control and PMB10% groups, with mean stem length rising as the percentage of biochar increases within a group. While groups PMB5% and PMB7% did not show much subsequent difference in their mean stem length, there is a significant difference shown by PMB10%, suggesting a threshold amount for effective biochar usage. Within all weeks, only groups PMB5% and PMB10% showed a continuous increase in the mean stem length. These fluctuations in the mean within the other groups can be attributed to a decrease in the number of plants due to plant mortality.

Compared to the results shown by CMB in our previous study, there is less fluctuation in stem length. In both experiments, the plants with the 10% concentration of biochar exhibited the highest mean growth. While there was not much difference between plants with 5% biochar and 7% biochar between both studies, all plants with the biochar concentrations showed higher mean stem length than the control samples from weeks 5 to 15 [21].

Figure 6 shows the Gompertz growth models of stem length (ML0) for the control, PMB7%, and PMB10% samples from Table 5. The Gompertz models show that the proportion of biochar in the soil significantly affects the growth of glassworts. The glassworts in the control group exhibited the lowest average predicted growth, whereas those in the PMB10% group displayed the highest average predicted growth. The PMB7% group showed an intermediate average predicted growth. Compared to the control group's average stem length of 54.79 mm in week 15, the PMB10% group's average stem length in week 15 was 95.16 mm, approximately 1.74 times longer. The highest predicted growth for the PMB7% and PMB10% groups was observed between weeks 7 and 8, with another significant increase for the PMB10% group between weeks 10 and 11. The decrease in average stem length was due to the deaths of the taller sample plants in the group. However, according to predictions using the Gompertz equation, the average stem length continued to increase throughout the entire period.

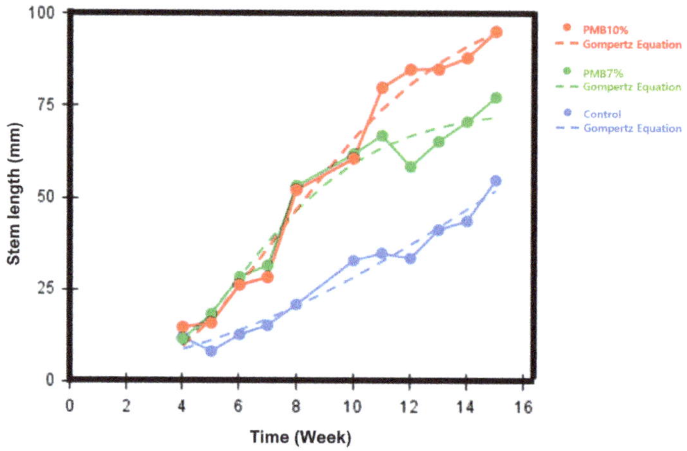

Figure 6. Growth models of the mean and standard deviations of the actual data and the fitted Gompertz curve prediction growth models of control, PMB7%, and PMB10% samples.

Additionally, the DL1 (stem thickness) and DL0 (leaf thickness) values were also recorded from weeks 4 to 16 despite not being the focus of this study. Ultimately, the ML0 (stem length) values were chosen for statistical analysis due to the clear upward trend shown in the increasing biochar concentrations. The mean and standard deviation values of DL1 and DL0 within weeks 4 to 15 are shown as (a) and (b) on Figure 7. To obtain these measurements, PMB pyrolyzed at 500 °C was incorporated into soil at different concentrations: 0% (control), 3% (PMB3%), 7% (PMB7%), and 10% (PMB10%). For both measurements, the biochar with the 10% and 7% concentrations of PMB tend to fluctuate closely with each other. The positive and negative fluctuations can be attributed to plant death. Both concentrations had the highest difference in week 12 for both DL1 and DL0. By week 15, the biochar with the 10% concentration had a slightly greater mean for DL1 while the biochar with the 7% concentration had the slightly greater mean for DL0. Furthermore, for all the weeks shown, the plants with the control sample showed lower mean DL1 and DL0 values than plants incorporated with PMB5%, PMB7%, and PMB10%.

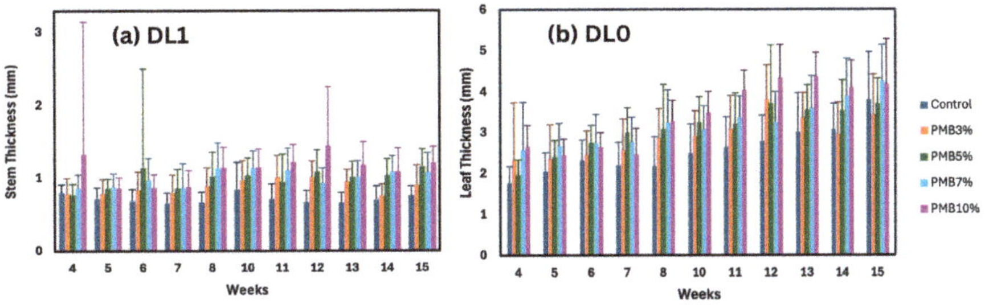

Figure 7. Mean and standard deviation values of (**a**) DL1, stem thickness; and (**b**) DL0, leaf thickness.

3.5. Soil Analysis of Biochar-Treated Soil after 15 Weeks

The soil analysis findings of the soil treated with biochar for 15 weeks are displayed in Table 6. The samples were treated with PM biochar that had been pyrolyzed at 500 °C (PMB500) to allow for comparison with CMB550. This investigation was done to look into using biochar for carbon storage. The Zn concentration of the bed soil was 8.11 mg/kg. The top soil has a cation exchange capacity (CEC) of 20.74 cmol/kg. It is evident that when the percentage of PMB grew, all analysis values other than CEC increased as well. The slow rise in total carbon (TC) indicates that using biochar may help soil store carbon for a longer period of time. Furthermore, the soil pH can be observed to increase as the cation exchange capacity (CEC) fluctuates, with the exception of PMB7%, suggesting that soil pH is not as affected by CEC as it is by biochar content. The control sample was found to have the highest CEC, yet it was also found to have the lowest soil pH. The increasing concentrations of Zn as more biochar is added shows one cumulative effect of increasing the biochar content of soil. PMB10% was shown to have the highest values of TC, TOC, OM, and Zn. Both PMB5% and PMB10% were found to have soil pH values of 7.85, which is the highest in Table 6. All samples of biochar-treated soil exhibited higher TC, TOC, OM, Zn, and soil pH values than the control pot. Furthermore, the control sample was found to have the highest CEC of all samples.

Table 6. Soil analysis of PMB500 after 15 weeks at 0%, 3%, 5%, 7%, and 10% biochar concentration.

Sample (PMB %)	TC (%)	TOC (%)	CEC (cmol Kg^{-1})	OM (g Kg^{-1})	Zn (mg Kg^{-1})	pH
Control (0%)	13.44	13.43	67.93	231.71	32.30	7.00
3%	15.21	15.20	53.35	262.22	74.03	7.55
5%	16.43	16.42	56.58	283.17	103.97	7.85
7%	18.56	18.55	61.68	319.97	150.25	7.75
10%	21.11	21.00	66.93	363.85	171.01	7.84

The TC, TOC, CEC, and OM values were similar between the CMB-treated soil from the previous study and PMB treated soil. The soil analysis results for CMB-treated soil is shown as Table S4 from our previous study [21]. For reference, the TC percentage was 15.21 and 14.04 for PMB3% and CMB3% respectively, 16.43 and 18.67 for PMB5% and CMB5%, respectively, 18.56 and 18.84 for PMB7% and CMB7%, respectively, and 21.11 and 21.40 for PMB10% and CMB10%, respectively. Between these values, there is a pattern of CMB having higher fluctuations than the PMB. However, at a biochar concentration of 10%, all values except for Zn concentration are relatively similar. CMB10% and PMB10% had TOC percentage values of 21.30 and 21.00, CEC values of 64.83 and 66.93 cmol/Kg, and OM values of 368.85 and 363.85, respectively. The highest difference is in zinc concentration. For all biochar-treated soil, zinc concentration was notably higher for CMB-treated soil than for PMB- treated soil.

It is thought that a high carbon concentration in biochar is associated with the formation and stability of soil aggregates [1]. Because the TC and TOC results between the PMB and CMB-treated soil were relatively similar, their effectiveness on stabilizing soil structure were likely similar as well. The increasing OM values for both PMB and CMB also shows that biochar reduces the dispersion ratio of soil, contributing toward soil stability. The results have also been similar for biochar created from other feedstock, such as coal [1,2]. Likewise, it is noted that biochar shares many similar effects on soil. A soil with a high dispersion ratio is prone to soil erosion and physical degradation due to its weak structures. The improvement of this ratio aligns with the increase in mean stem growth as shown in Table 5.

4. Discussion

4.1. Comparative Studies

Many comparative studies have been referred to throughout this paper. Adekiya et al. reported the soil analysis of coal biochar based on biochar concentration, similar to how our soil analysis was done for PMB [1]. The yielded OM, pH, and CEC values were similar between PMB, CMB, and coal biochar-based soil analysis [21]. From this, it can be derived that biochar causes soil to increase in alkalinity and decrease in dispersion ratio. On the other hand, the values of CEC fluctuated for all samples. This indicates that the specific biochar concentration is a crucial indicator of the biochar's ability to hold onto essential nutrients for the soil. The relative values for CEC from 0% biochar concentration to 10% biochar concentration were similar for the CMB and PMB samples. For CMB and PMB, there was a slight decrease in the CEC from these concentrations. The coal samples, however, yielded a higher value for CEC over time. Granted, the timeframe for measuring the effect of coal-based biochar was greater than the timeframe for the CMB and PMB studies. The hardwood-based biochar timeframe was greater than the timeframe for the CMB and PMB studies. Additionally, Adekiya et al. reported the effects of the hardwood-based biochar in improving soil parameters (bulk density, porosity, moisture content, mean weight diameter), as well as dispersion ratio, chemical properties, and cocoyam yield [1]. Specifically, the porosity of the biochar reduced bulk density while increasing the porosity of the soil. The improved physical and chemical characteristics of the soil were shown to have contributed significantly to the increased cocoyam yield. Likewise, the use of biochar was a significant factor in the increased yield of glasswort growth in this study.

This study also examined glasswort growth using different concentrations of PMB pyrolyzed at 400 °C. Kudrevatykh et al. reported the potential of glasswort as a sustainable resource by demonstrating its high yields of biomass, seed, oil, and ethanol. According to Calone et al., glasswort is also highly resistant to weather and salinity conditions harmful to most plants. Therefore, the cultivation of glasswort, or halophytes, could prove to be valuable in mitigating the sustainability concerns brought on by global warming and unsustainable agriculture. Poultry manure (PM) was chosen as feedstock for biochar due to its rich nutrient content and the growing production of poultry worldwide [12,15]. Additionally, this increase in production would mean an overflow of PM, which has been shown to produce an estimated 220 to 280 kg of toxic ammonia for every 1000 kg of chicken dung produced [14]. Using PM for biochar not only helps manage this waste but also utilizes its high nutrient content, making it a valuable feedstock resource for biochar. Our results showed a dramatic increase in mean plant stem growth from the control (0% biochar concentration) and PMB10% (10% biochar concentration) soil samples. Similarly, Park et al.'s study on the effect of biochar on corn productivity under drought conditions showed that all of the biomass produced was greater when biochar was used [5]. The study's findings on biochar increasing the nutrient storage of soil matches the difference in copper and zinc values based on the different temperatures used to create biochar. We utilized 400, 500, and 700 °C as the low, medium, and high temperatures in our investigation. The temperature at which the biochar was pyrolyzed had a significant influence on the Cu and Zn content levels among the numerous values examined for biochar derived from PM.

4.2. Carbon Credit Calculation

To work toward Korea's goal for a carbon neutral state by 2050, carbon credits can be considered as one approach to controlling carbon emissions. By identifying the change in carbon sequestered by the use of biochar and converting the change to the amount of CO_2 sequestered, the value of carbon credits saved can be estimated. Due to the reach of our experiment, this present study will only compare the difference in sequestered carbon values between the low, medium, and high temperature values used to pyrolyze the PMB.

Table 7 shows the change in carbon stocks by tons of sequestered carbon per year based on the temperatures in which the PMB were pyrolyzed. According to Appendix 4 of the 2019 IPCC guideline, F_{cp} was defined as 0.38 ± 49% for all samples while the F_{permp} values were defined as 0.65 ± 15% for PMB400, 0.80 ± 11% for PMB500, and 0.89 ± 13% for PMB700 [21]. The values of BC_{TOTp} were determined through the yield percentages of the different PMB samples according to Table 1, with the value being 0.5278 for PMB400, 0.4720 for PMB500, and 0.3607 for PMB700. While the calculated values on the change in sequestered carbon are not specific to the PMB source and are the result of dropping the uncertainty values of the F_{cp} value, they can be used to analyze the relative differences in sequestered carbon between PMB samples. PMB400 was found to yield an estimated change in sequestered carbon of 0.1305 ± 0.0194 tons per year with a lower and upper bound of 0.111 and 0.1499. PMB500 yielded an estimated change of 0.1432 ± 0.0158 tons of sequestered carbon per year. The lower and upper bound was 0.1274 and 0.159. Finally, PMB700 showed an estimated change of 0.1217 ± 0.0159 tons of sequestered carbon per year. The lower and upper bound of this change was 0.1376 and 0.1058. In terms of sequestered carbon, PMB500 demonstrated the largest relative change at the upper bound, whereas PMB700 demonstrated the largest relative change at the lower bound.

Table 7. Estimated Change in carbon stocks by tonne of sequestered Carbon per year for 400, 500, and 700 °C.

PMB	BC_{TOTp}	F_{Cp}	F_{permp}	Change in Sequestered Carbon ($\Delta BC_{Mineral}$)	Change in Sequestered CO_2 (E_{stored})
400 °C	0.53	0.38	0.65 ± 15%	0.13 ± 0.02	0.49 ± 0.0711
500 °C	0.47	0.38	0.80 ± 11%	0.14 ± 0.02	0.52 ± 0.06
700 °C	0.36	0.38	0.89 ± 13%	0.12 ± 0.02	0.45 ± 0.06

Using these values, carbon credits can be estimated using the derived equation from Section 2.6, Equations. Based on the results yielded by this equation, PMB400, PMB500, and PMB700 yielded a change in sequestered CO_2 of 0.4780 ± 0.0711, 0.5248 ± 0.0580, and 0.4462 ± 0.0583; respectively. Out of the samples, PMB500 had the highest lower and upper bound values with the values of 0.4671 and 0.5830 tonnes of change in sequestered CO_2, followed by PMB400, which had the lower and upper bound values of 0.4074 and 0.5496 tonnes of change. With the lowest lower and upper bounds was the PMB700 sample, with bounds of 0.3879 and 0.5045 tonnes of change. With the CO_2 equivalent of the change in sequestered carbon, we can identify the carbon credit value of the poultry manure-based biochar pyrolyzed at different temperatures. Carbon credits usually represent one metric ton (tonne) of CO_2. Therefore, since the changes in sequestered CO_2 were measured by tonne per year, these figures correspond exactly to the yearly carbon credits produced by pyrolyzing biochar made from chicken manure at 400, 500, and 700 °C.

5. Conclusions

Through statistical analysis and comparison, this study seeks to shed more light on how PMB affects soil. Biochar was produced from the pyrolysis of poultry manure samples at 400, 500, and 700 °C.The yield, BET, average pore size, and pH level components of each sample were examined. Cow and poultry manure-based biochar differed in their lignocellulosic and elemental compositions, as revealed by the TGA-DTG and elemental analysis of the PMB. While stem thickness, stem length, and leaf length values were measured during the experiment period, this study only focused on statistically analyzing stem length for its constant upward growth within the higher concentrations of biochar. Nevertheless, this study showcased the successful incorporation of PMB500 within soil through this constant upward growth.

Additionally, the carbon credits of the different PMB samples were estimated from the values and Equation 4Ap.1 from Appendix 4 of the 2019 Refinement to the 2006 IPCC Guidelines for National Greenhouse Gas Inventories and the formula from Puro.earth. The combined equations yielded the change in sequestered CO_2 per tonne of PMB incorporated into soil, thereby identifying the carbon credits earned using poultry manure pyrolyzed at 400, 500, and 700 °C. The results estimated that PMB pyrolyzed at 500 °C yielded the highest change in sequestered CO_2 per tonne. Therefore, PMB500 was found to have the highest carbon credit values of the samples used. Because carbon credits could prove to be a helpful step toward reducing the emission of CO_2, even an estimation of the carbon credits of biochar samples can be a useful tool for directing further analysis on the basis of incorporating biochar into agricultural practices. While the application, morphological features, and heavy mental content of the PMB samples proved promising, further research into biochar is essential to understand its characteristics based on the feedstock used. Although the BET and average pore size values of PMB400, PMB500, and PMB700 were similar, PMB500 has the potential to be the optimal biochar out of these samples for its relatively high BET value and change in sequestered CO_2 per tonne. Further insight into the chemical reactions and optimal uses of biochar is necessary to realize its potential and to be able to maximize its benefits for sustainability and soil types. Biochar has the potential to become a key element in sustainable practices through continued research endeavors within the field.

Supplementary Materials: The following supporting information can be downloaded at: https://www.mdpi.com/article/10.3390/agriculture14091590/s1, Figure S1: TGA-DTG Spectra of cow manure (CM) from 30 to 800 °C, 10 °C/min.

Author Contributions: Conceptualization, D.C. and Y.S.K.; methodology, D.C., S.W.I. and Y.S.K.; validation, Y.S.K. and H.G.K.; formal analysis, H.C., V.J.H., M.K. and H.G.K.; investigation, H.C., V.J.H., M.K. and H.G.K.; resources, S.W.I. and Y.S.K.; writing—original draft preparation, Y.S.K.; writing—review and editing, H.C., V.J.H., M.K. and Y.S.K.; visualization, H.C., V.J.H. and M.K.; supervision, Y.S.K.; All authors have read and agreed to the published version of the manuscript.

Funding: This research was supported by "Regional Innovation Strategy" through Research Foundation of Korea (NRF) funded by the ministry of Education (MOE) (2023RIS-008), as well as the National Research Foundation of Korea (NRF) grant funded by the Korean government (MSIT) (No. 2016R1A6A1A03012069 and No. 2020R1A2C1102174).

Institutional Review Board Statement: Not applicable.

Data Availability Statement: The data presented in this study are available on request from the corresponding author.

Conflicts of Interest: The authors declare no conflicts of interest.

References

1. Adekiya, A.O.; Agbede, T.M.; Olayanju, A.; Ejue, W.S.; Adekanye, T.A.; Adenusi, T.T.; Ayeni, J.F. Effect of biochar on soil properties, soil loss, and Cocoyam yield on a tropical sandy loam alfisol. *Sci. World J.* **2020**, *2020*, 9391630. [CrossRef] [PubMed]
2. Kim, H.; McJeon, H.; Jung, D.; Lee, H.; Bergero, C.; Eom, J. Integrated Assessment Modeling of Korea's 2050 Carbon Neutrality Technology Pathways. *Energy Clim. Chang.* **2022**, *3*, 100075. [CrossRef]
3. Alling, V.; Hale, S.E.; Martinsen, V.; Mulder, J.; Smebye, A.; Breedveld, G.D.; Cornelissen, G. The role of biochar in retaining nutrients in amended tropical soils. *J. Plant Nutr. Soil Sci.* **2014**, *177*, 671–680. [CrossRef]
4. Brtnicky, M.; Datta, R.; Holatko, J.; Bielska, L.; Gusiatin, Z.M.; Kucerik, J.; Hammerschmiedt, T.; Danish, S.; Radziemska, M.; Mravcova, L.; et al. A critical review of the possible adverse effects of biochar in the soil environment. *Sci. Total Environ.* **2021**, *796*, 148756. [CrossRef] [PubMed]

5. Park, J.H.; Yun, J.J.; Kim, S.H.; Park, J.H.; Acharya, B.S.; Cho, J.S.; Kang, S.W. Biochar improves soil properties and corn productivity under drought conditions in South Korea. *Biochar* **2023**, *5*, 66. [CrossRef]
6. Weng, Z.H.; Liu, X.; Eldridge, S.; Wang, H.; Rose, T.; Rose, M.; Rust, J.; Singh, B.P.; Tavakkoli, E.; Tang, C.; et al. Priming of soil organic carbon induced by sugarcane residues and its biochar control the source of nitrogen for plant uptake: A dual 13C and 15N isotope three-source-partitioning study. *Soil Biol. Biochem.* **2020**, *146*, 107792. [CrossRef]
7. Major, J.; Lehmann, J.; Rondon, M.; Goodale, C. Fate of soil-applied black carbon: Downward migration, leaching, and soil respiration. *Glob. Change Bio.* **2010**, *16*, 1366–1379. [CrossRef]
8. Wu, M.; Han, X.; Zhong, T.; Yuan, M.; Wu, W. Soil organic carbon content affects the stability of biochar in paddy soil. *Agric. Ecosyst. Environ.* **2016**, *23*, 59–66. [CrossRef]
9. Woo, S.H. Biochar for carbon sequestration. *Clean Technol.* **2013**, *19*, 201–211. [CrossRef]
10. Schulz, H.; Dunst, G.; Glase, B. Positive effects of composted biochar on Plant growth and soil fertility. *Agron. Sustain. Dev.* **2013**, *33*, 817–827. [CrossRef]
11. Oo, A.Z.; Sudo, S.; Akiyama, H.; Win, K.T.; Shibata, A.; Yamamoto, A. Effect of dolomite and biochar addition on N_2O and CO_2 emissions from acidic tea field soil. *PLoS ONE* **2018**, *13*, 0192235. [CrossRef] [PubMed]
12. Cao, X.; Ma, L.; Gao, B.; Harris, W. Dairy-Manure derived biochar effectively sorbs lead and atrazine. *Environ. Sci. Technol.* **2009**, *43*, 3285–3291. [CrossRef] [PubMed]
13. Dróżdż, D.; Wystalska, K.; Malińska, K.; Grosser, A.; Grobelak, A.; Kacprzak, M. Management of poultry manure in Poland—Current state and future perspectives. *J. Environ. Manag.* **2020**, *264*, 110327. [CrossRef]
14. Tańczuk, M.; Junga, R.; Kolasa-Więcek, A.; Niemiec, P. Assessment of the energy potential of chicken manure in Poland. *Energies* **2019**, *12*, 1244. [CrossRef]
15. Production—Chicken Meat. U.S. Department of Agriculture. Available online: https://fas.usda.gov/data/production/commodity/0115000 (accessed on 7 May 2024).
16. Cha, J.S.; Park, S.H.; Jung, S.-C.; Ryu, C.; Jeon, J.-K.; Shin, M.-C.; Park, Y.-K. Production and utilization of biochar: A review. *J. Ind. Eng. Chem.* **2016**, *40*, 1–15. [CrossRef]
17. Cárdenas-Pérez, S.; Piernik, A.; Chanona-Pérez, J.J.; Girgore, M.N.; Perea-Flores, M.J. An overview of the emerging trends of the *Salicornia* L. genus as a sustainable crop. *Environ. Exp. Bot.* **2021**, *191*, 104606. [CrossRef]
18. Kudrevatykh, I.Y.; Kalinin, P.I.; Mitenko, G.V.; Alekseev, A.O. The role of plant in the formation of the topsoil chemical composition in different climatic conditions of steppe landscape. *Plant Soil* **2021**, *465*, 453–472. [CrossRef]
19. Calone, R.; Mircea, D.-M.; González-Orenga, S.; Boscaiu, M.; Lambertini, C.; Barbanti, L.; Vicente, O. Recovery from Salinity and Drought Stress in the Perennial *Sarcocornia fruticosa* vs. the Annual *Salicornia europaea* and *S. veneta*. *Plants* **2022**, *11*, 1058. [CrossRef]
20. Majda, M. Engineering lignin: Getting more from woody biomass. *Plant Physiol.* **2022**, *188*, 926–927. [CrossRef]
21. Shin, H.; Chun, D.; Cho, I.-R.; Hanif, M.A.; Kang, S.-S.; Kwac, L.K.; Kim, H.G.; Kim, Y.S. Systematic Characterization of Cow Manure Biochar and Its Effect on *Salicornia herbacea* L. Growth. *Sustainability* **2024**, *16*, 3396. [CrossRef]
22. Jia, X.; Xi, B.D.; Li, M.L.; Yang, Y.; Wang, Y. Meraptoreomic analysis of the functional insights into microbial communities of combined hydrogen and methane production by anaerobic fermentation from reed straw. *PLoS ONE* **2017**, *12*, e0183158. [CrossRef] [PubMed]
23. Intergovernmental Panel on Climate Change (IPCC). Biochar Appendix. In *2019 Refinement to the 2006 IPCC Guidelines for National Greenhouse Gas Inventories: Volume 4 Agriculture, Forestry and Other Land Use*; Intergovernmental Panel on Climate Change: Geneva, Switzerland, 2019; Available online: https://www.ipcc-nggip.iges.or.jp/public/2019rf/pdf/4_Volume4/19R_V4_Ch02_Ap4_Biochar.pdf (accessed on 18 May 2019.).
24. Han, K.H.; Yun, S.I.; Kwak, J.H.; Lee, S.I. A review on international carbon credit certification methodologies for biochar as a soil amendment. *Korean J. Soil Sci. Fertil.* **2023**, *56*, 572–594. [CrossRef]
25. Yang, H.; Yan, R.; Chen, H.; Lee, D.H.; Zheng, C. Characteristic of hemisellulose, cellulose and lignin pyrolysis. *Fuel* **2007**, *86*, 1781–1788. [CrossRef]
26. Orlando, M.-Q.; Borja, V.-M. Pretreatment of animal manure biomass to improve biogas production: A review. *Energies* **2020**, *13*, 3573. [CrossRef]
27. Beusch, C. Biochar as a soil ameliorant: How biochar properties benefit soil fertility—A review. *J. Environ. Prot. Sci.* **2021**, *9*, 28–46. [CrossRef]
28. Lehmann, J.; Joseph, S. *Biochar for Environmental Management: Science, Technology and Implementation*; Routledge: Abdington, UK, 2015; pp. 38–62.
29. Godwin, P.M.; Pan, Y.; Xiao, H.; Afzal, M.T. Progress in Preparation and Application of Modified Biochar for Improving Heavy Metal Ion Removal From Wastewater. *J. Bioresour. Bioprod.* **2019**, *4*, 31–42. [CrossRef]
30. Janu, R.; Mrlik, V.; Ribitsch, D.; Hofman, J.; Sedláček, P.; Bielská, L.; Soja, G. Biochar surface functional groups as affected by biomass feedstock, biochar composition, and pyrolysis temperature. *Carbon Resour. Convers.* **2021**, *4*, 36–46. [CrossRef]
31. Yang, C.; Liu, J.; Lu, S. Pyrolysis temperature affects pore characteristics of rice straw and canola silk biochars and biochar-amended soils. *Geoderma* **2021**, *397*, 0016–7061. [CrossRef]

32. Singh, Y.S.P.; Bhandari, S.; Bhatta, D.; Poudel, A.; Bhattarai, S.; Yadav, P.; Ghimire, N.; Paudel, P.; Paudel, P.; Shrestha, J.; et al. Biochar Application: A Sustainable Approach to Improve Soil Health. *J. Agric. Food Res.* **2023**, *11*, 100498. [CrossRef]
33. Edeh, I.G.; Masek, O.; Fusseis, F. 4D structural changes and pore network model of biomass during pyrolysis. *Sci. Rep.* **2023**, *13*, 22863. [CrossRef]

Disclaimer/Publisher's Note: The statements, opinions and data contained in all publications are solely those of the individual author(s) and contributor(s) and not of MDPI and/or the editor(s). MDPI and/or the editor(s) disclaim responsibility for any injury to people or property resulting from any ideas, methods, instructions or products referred to in the content.

MDPI AG
Grosspeteranlage 5
4052 Basel
Switzerland
Tel.: +41 61 683 77 34

Agriculture Editorial Office
E-mail: agriculture@mdpi.com
www.mdpi.com/journal/agriculture

www.ingramcontent.com/pod-product-compliance
Lightning Source LLC
Chambersburg PA
CBHW061833210926
51850CB00006B/881